Network Design

Management and Technical Perspectives

Second Edition

OTHER AUERBACH PUBLICATIONS

Agent-Based Manufacturing and Control Systems: New Agile Manufacturing Solutions for Achieving Peak Performance
Massimo Paolucci and Roberto Sacile
ISBN: 1574443364

Curing the Patch Management Headache
Felicia M. Nicastro
ISBN: 0849328543

Cyber Crime Investigator's Field Guide, Second Edition
Bruce Middleton
ISBN: 0849327687

Disassembly Modeling for Assembly, Maintenance, Reuse and Recycling
A. J. D. Lambert and Surendra M. Gupta
ISBN: 1574443348

The Ethical Hack: A Framework for Business Value Penetration Testing
James S. Tiller
ISBN: 084931609X

Fundamentals of DSL Technology
Philip Golden, Herve Dedieu,
and Krista Jacobsen
ISBN: 0849319137

The HIPAA Program Reference Handbook
Ross Leo
ISBN: 0849322111

Implementing the IT Balanced Scorecard: Aligning IT with Corporate Strategy
Jessica Keyes
ISBN: 0849326214

Information Security Fundamentals
Thomas R. Peltier, Justin Peltier,
and John A. Blackley
ISBN: 0849319579

Information Security Management Handbook, Fifth Edition, Volume 2
Harold F. Tipton and Micki Krause
ISBN: 0849332109

Introduction to Management of Reverse Logistics and Closed Loop Supply Chain Processes
Donald F. Blumberg
ISBN: 1574443607

Maximizing ROI on Software Development
Vijay Sikka
ISBN: 0849323126

Mobile Computing Handbook
Imad Mahgoub and Mohammad Ilyas
ISBN: 0849319714

MPLS for Metropolitan Area Networks
Nam-Kee Tan
ISBN: 084932212X

Multimedia Security Handbook
Borko Furht and Darko Kirovski
ISBN: 0849327733

Network Design: Management and Technical Perspectives, Second Edition
Teresa C. Piliouras
ISBN: 0849316081

Network Security Technologies, Second Edition
Kwok T. Fung
ISBN: 0849330270

Outsourcing Software Development Offshore: Making It Work
Tandy Gold
ISBN: 0849319439

Quality Management Systems: A Handbook for Product Development Organizations
Vivek Nanda
ISBN: 1574443526

A Practical Guide to Security Assessments
Sudhanshu Kairab
ISBN: 0849317061

The Real-Time Enterprise
Dimitris N. Chorafas
ISBN: 0849327776

Software Testing and Continuous Quality Improvement, Second Edition
William E. Lewis
ISBN: 0849325242

Supply Chain Architecture: A Blueprint for Networking the Flow of Material, Information, and Cash
William T. Walker
ISBN: 1574443577

The Windows Serial Port Programming Handbook
Ying Bai
ISBN: 0849322138

AUERBACH PUBLICATIONS

www.auerbach-publications.com
To Order Call: 1-800-272-7737 • Fax: 1-800-374-3401
E-mail: orders@crcpress.com

Network Design

*Management and
Technical Perspectives*

Second Edition

Teresa C. Piliouras

AUERBACH PUBLICATIONS

A CRC Press Company
Boca Raton   London   New York   Washington, D.C.

Library of Congress Cataloging-in-Publication Data

Piliouras, Teresa C. Mann.
   Network design : management and technical perspectives / Teresa C. Mann Piliouras.--
2nd ed.
      p. cm.
   Includes bibliographical references and index.
   ISBN 0-8493-1608-1
      1. Telecommunication. 2. Computer networks. I. Title.

TK5101.P55 2004
004.6--dc22

2004046435

Visit the Auerbach Web site at www.auerbach-publications.com

© 2005 by CRC Press LLC
Auerbach is an imprint of CRC Press LLC

No claim to original U.S. Government works
International Standard Book Number 0-8493-1608-1
Library of Congress Card Number 2004046435
Printed in the United States of America 1 2 3 4 5 6 7 8 9 0

Contents

About the Author

Teresa C. Piliouras is Vice President of TCR. She has extensive experience managing and implementing large-scale IT and data mining projects for such companies as Accenture (formerly Andersen Consulting), Epsilon, PepsiCo, SIAC, Boehringer Ingelheim, and many others. She is active in the field of bioinformatics, which is amenable to treatment by many network design algorithms. She is Adjunct Professor of Computer Science at Polytechnic University and Iona College, where she teaches courses on data mining, database design, and Internet technologies. She holds a Ph.D. in operations research from Polytechnic University, an MBA from Iona College, and a B.S. from the University of Illinois. She has lectured and published extensively. She can be reached at tcp@tcrinc.com.

Preface

There are hundreds of protocols and technologies used in telecommunications. Protocols run the full gamut from application level to physical level. It is overwhelming to try to keep track of them. This book is a broad survey of the major networking protocols and technologies and how they interrelate, integrate, migrate, and segregate functionality. It attempts to put into perspective the fundamental issues that managers and engineers should focus upon when designing a telecommunications strategy and selecting technologies.

For managers, this book provides technology overviews, case studies, and tools for decision making and technology evaluation. The book discusses major trends and influences in society, government, legislation, industry, and technology. By providing an historical context, managers can better understand the evolution of IT systems and how they can be migrated to capitalize on new technologies as they become available. Guidelines and recommendations are given for technology selection, outsourcing (with templates, checklists, and recommendations), disaster recovery, business continuity, and security. Numerous sources of free information are cited so the reader can keep abreast of important new developments that may impact network design and implementation.

For engineers, this book provides a review of the major protocols up and down the OSI protocol stack and how they relate to network design strategies. Some of the topics discussed in this book include Internet standards, protocols, and implementation (e.g., VoIP, TCP/IP, etc.); client/server and distributed networking (e.g., COM+, CORBA, and XML/Web services); value added networking services (e.g., ATM, packet switching, Frame Relay, VPNs, ISDN, Gigabit Ethernet, DSL, cable, and dial-up access); disaster recovery and business continuity technologies (e.g., tape, SAN, NAS, etc.); legacy mainframe technologies and migration to TCP/IP,

MAN technologies (e.g., SONET, DWDM, RPR, and MPLS); and WAN (wireless, public, and private) and LAN networking.

For engineers wanting to peek under the technology covers even more, this book provides insights into the mathematical underpinnings and theoretical basis for routing, network design, reliability, and performance analysis.

By providing coverage of all these areas in one book, the author hopes that this will help managers and technical professionals better understand and appreciate each other's perspectives in the network design process. Thus, this book intends to help bridge the communication gap that often exists between managers and technical staff involved in the design and implementation of networks.

Network managers need a sound grounding in basic design principles to effectively manage, plan, and assess the plethora of new technologies and equipment available for designing networks. They must also understand how requirements should be formulated and specified for design engineers. In turn, network designers and engineers need a sound grounding in basic management principles to fully understand how organizational requirements are best reflected in design recommendations.

Specific implementations of technology are presented through the use of case studies. In the case studies, both management and technical considerations are discussed. This approach is used to demystify the design process, which traditionally — on anything other than an ad-hoc basis — has been limited to the purview of highly trained and specialized professionals. By describing the *lingua franca* of both managers and design engineers in common terms, it is hoped that each will gain a better understanding of the total network design process.

The material for this book is based on the author's experiences as an industry consultant and professor teaching undergraduate- and graduate-level network design and management courses. In each of these roles, the author has dealt with network managers and with hands-on network designers. Frequently, there is sub-optimal communication between these two groups of professionals. For example, network managers frequently express a lack of confidence in evaluating whether or not organizational objectives will be satisfied by the recommendations proposed by design engineers. Design engineers, on the other hand, voice concerns that "management does not have any idea of what I am doing and what is involved," and thus lack confidence that the goals established by management are realistic. Network managers say they need to understand the technical jargon and basic approaches that network designers use so they can evaluate vendor offerings and staff recommendations. Managers want this perspective *without* getting too immersed in technical details. Network implementers say that they need and want practical advice on how to apply sound design principles in the context of a *realistic* design scenario, where organizational, budgetary, political, and other considerations must enter into the design process. Thus, this book seeks to help management and design professionals work *together* toward achieving their respective goals in the network design process.

Acknowledgments

Any undertaking such as this gives one pause to reflect on how much one owes to others. In particular, I would like to acknowledge the contributions of the following people:

I owe a substantial debt of gratitude to Byron Piliouras for his ongoing feedback and technical assistance throughout the preparation of this manuscript, and for his constant support and encouragement. He made numerous contributions throughout the book in the form of case studies, diagrams, and interviews, which reflect his vast industry experience with all aspects of network design. He had extraordinary patience while I picked his brain as frequently as possible.

I give my sincerest appreciation to Rich O'Hanley, CRC Press, for his timely support, excellent advice, and kind words of encouragement. The staff at CRC Press are consummate professionals, and I have the highest regard for them. Without their support, this book certainly would not have been possible.

I wish to thank Joseph Akalski for his contributions and the case study he provided for Appendix A.

I thank Dr. Maurice Cohen for the wonderful cover art. He is an award-winning scientist and artist whose research and art reflect his contributions to the field of chaos theory.

I want to thank Matt Durgin, of Sane Solutions, for his quick responses and help in providing graphical screen displays and information on the NetTracker products.

I also want to acknowledge the many useful comments, suggestions, and proofreading help provided by the following individuals: Ross Bettinger, John Braun, and Anudhra Padhi. Ross Bettinger spent many intense nights proofing the manuscript as we came close to press time, which never seemed to dim his ability to catch lots of errata and typos. John

Braun helped update the sections on the Internet, SOHOs, and VPNs, and I appreciate his many thoughtful remarks.

I also wish to thank Drs. Kornel Terplan, Aaron Kershenbaum, and Robert Cahn. Many of the techniques used to design networks were pioneered and developed by these individuals, and the field as a whole has been profoundly impacted by their ongoing accomplishments and contributions. This edition continues to reflect the imprint and enduring contributions of Dr. Terplan as co-author of the first edition.

Finally, I thank my family for their love and support.

Chapter 1

Making the Business Case for the Network

1.1 Management Overview of Network Design

The basic goal of network design is to interconnect various software and hardware devices so that resources can be shared and distributed. Despite the apparent simplicity of this goal, network design is a very complex task that involves balancing a multitude of managerial and technical considerations. This chapter focuses on the managerial decisions involved in planning and designing a network.

Business concerns and philosophy have a profound impact on the network planning process. In some organizations, expenses associated with the network are viewed as "overhead." Taken to an extreme, this view can lead to the perspective that anything but the most basic expenditures on the network are superfluous to the primary business. In this type of environment, it is not uncommon to observe a lack of formal commitment and managerial sponsorship of the network. End users, acting on their own initiative, may buy and install network components and software without working with any central budgeting and planning authority. Although formal budget allocations for the network may not exist, this does not mean that the network is "free." An informal, reactive network planning process is often associated with frequent downtime because the network can be dismantled or changed on a whim, without regard to how the changes might impact the people using it. Thus, there is an opportunity cost associated with the lost staff productivity resulting from the network downtime. There is also an opportunity cost associated with lost productivity resulting from the diversion of staff from their primary job functions to support the network. Furthermore, when changes to the network are not carefully planned and implemented, it compounds the difficulty of maintaining the network in a cost-effective way. For organizations of any substantial size lacking a formal network planning process, it is not uncommon for audits to reveal millions invested in technology that is not effectively utilized and that does not support ongoing organizational needs. Although this is network design at its worst, it is not that uncommon. The moral of this is that ignoring or avoiding direct consideration of the true network costs does not make them go away. A see-no-evil/hear-no-evil/speak-no-evil strategy only "works" when decision makers are not accountable for their actions. In a cost-conscious, competitive business climate that focuses increasing scrutiny on inefficient processes ripe for reengineering, this is a risky approach.

In contrast, there are organizations that view the network as the corporate lifeblood. In this environment there is considerable management accountability for and scrutiny of the network planning and implementation. Increased recognition of the network's importance to the bottom line improves the chances that the network(s) will be well planned and executed. For example, it is vital to the New York Stock Exchange that its networks perform reliably even under conditions of extreme stress.[1] High-profile, high-performance networks require thorough planning to ensure that they can meet the demands placed upon them. A systems approach is essential to ensuring a comprehensive assessment of critical network requirements.

A *systems approach* means that the requirements are considered from a global perspective that encompasses both top-down and bottom-up views. The discussion that follows outlines a general methodology for performing a systems analysis of the network requirements, from a business perspective. The business perspective is a top-down, big-picture view of how the design will impact the organization. This discussion continues, from a technical perspective, in the chapters that follow. The technical perspective is a bottom-up, narrowly focused view concentrating on essential design details.

1.1.1 Define the Business Objectives

Defining the business objectives is a vital first step in the network planning process. A logical start to the top-down analysis is to define the business objectives served by the network. The business objectives should relate to the strategic focus of the organization. There may be many motivations for building and implementing a network. After the business objectives have been made explicit, they can be prioritized, and objective criteria can be developed for measuring the success of the network implementation. The objectives will also help determine the type of network needed, and the level of expenditure and support that is appropriate. The business objectives have many impacts on technical decisions regarding the selection of technology, the performance requirements, and required resource commitments. When the objectives are poorly defined, there are often many complications down the road.

Many business objectives relate to gaining and maintaining competitive advantage. According to [CARR03], four basic strategies to sustain competitive advantage are:

1. Low-cost leadership
2. Focus on market niche
3. Product and service differentiation
4. Strategic alliances and linkages with partners

Telecommunications technology has transformed business models by supporting all of the above strategies. For example, many small businesses demonstrate low-cost leadership using Internet-based storefronts to offer products and services with less overhead than their larger competitors. If a business has an effective Web site and well-organized internal processes, it is not obvious whether that business has one or a thousand service representatives to support its customers' needs. Thus, even a small business can have a strong presence and international reach.

Interland,[2] a full-service Web hosting provider for small businesses, believes there is vast market potential for software tools to create an online presence for the 20 million businesses in the United States with fewer than ten employees. Interland provides a basic service — for as little as $23 a month — that includes a Web site, a dot.com domain name, and 30 e-mail-message accounts. It also offers hundreds of design templates and "the ability to add and edit pictures and publish and update text without having to program in HTML, or Hypertext Markup Language" so that small companies can easily develop, customize, and maintain their own Web sites with very little cost or training. [LOHR03]

Amazon.com, established in 1994, is another example of how telecommunications technology can be used to capitalize on these strategies for competitive advantage. Amazon.com minimizes its capital outlay by being an exclusively online retailer that holds very little inventory. It outsources almost all of its operations, except for information technology (IT). Integral to Amazon.com's success is its use of sophisticated Customer Relationship Management (CRM) technology to collect and analyze customer buying habits and preferences. Amazon.com leverages customer preference information it collects to effectively target market niches and to make tailored product recommendations to customers when they return to the site. This is called *personalization*. Amazon.com is continuously expanding the breadth and depth of its product offerings through strategic alliances and partnerships with book publishers, retailers (of clothing, toys, electronics, etc.), and technology service providers.

Other business objectives that telecommunications can support include:

- *Compliance with legislation and regulatory requirements that may mandate fundamental business process changes.* An example of this is the Health Portability and Accountability Act of 1996 (HIPAA), which is discussed in more detail in Section 1.5.1. HIPAA has profound and far-reaching impacts on the processing and handling of information and patient records in the healthcare industry.
- *Improved outreach and accessibility.* Many nonprofit and government agencies provide information and services through the World Wide Web because it is a cost-effective means to reach their intended audience.
- *Enhanced marketing efforts to reach new customers and to reduce attrition and churn of existing customers.* This might include customer satisfaction, loyalty, and targeted marketing programs.

New business objectives can have a significant impact on processes, procedures, and systems. Once the business objectives are fully understood, tactical and operational strategies can be developed for the network implementation. To fully understand these impacts, the organization should evaluate:

- Current processes and business practices
- Changes required in current processes and business practices to achieve the desired business objective(s)
- Process, resource, technology, staffing, and organizational requirements for successful implementation

Meetings and idea-generating sessions with decision makers, planners, and other key players can generate a lot of potential business objectives. If the list is long, the "80/20" or Pareto rule can be used to whittle it down to a manageable size.[3] This rule is used to focus on important concerns and to avoid distraction by trivial or overly difficult ones.

To apply the 80/20 rule in the context of management planning, first start by identifying business objectives that are redundant or similar, so they can be aggregated. Objectives that, upon further reflection, appear unimportant should be dropped. Each business objective should be evaluated with

Table 1.1   Identifying High Importance Business Objectives

Pareto Analysis Matrix	High Importance Business Objectives	Low Importance Business Objectives
High difficulty	Business objective A Business objective B Business objective C ... (Note: these objectives need careful selection and further evaluation)	Business objective X Business objective Y Business objective Z ... (Note: these objectives are definitely not worth pursuing)
Low difficulty	Business objective P Business objective D Business objective Q ... (Note: these objectives should be pursued)	Business objective Ψ Business objective Ω Business objective Φ ... (Note: these objectives are not worth pursuing)

respect to its potential value and difficulty. Table 1.1 suggests a format for collecting and presenting the results of this evaluation. Business objectives with "High Importance/Low Difficulty" ratings (i.e., the "80 percent" solution group) are the most desirable, followed by those having "High Importance/High Difficulty" (i.e., the "20 percent" solution group). High importance objectives are selected for further review and scrutiny. Other objectives on the evaluation list are rejected from further consideration because the effort they require is not warranted by their potential return and risk to the organization.

The surviving business objectives should be carefully evaluated with respect to risk factors, required effort, and the availability of time and resources. Other metrics may be appropriate, depending on the organization and the nature of the project. Each organization should develop its own evaluation metrics based on the input of planners and decision makers and the resulting organizational consensus. Table 1.2 presents a method for scoring each business goal. To use this table, each decision factor should be entered in a separate row under the column labeled "Decision Factor." The example shows risks the organization wants to avoid, with each decision factor measured on a scale from 0- to 100. Decision factors with high scores (i.e., 100) reflect low-risk, desirable outcomes; and those with low scores (i.e., 0) reflect high-risk, undesirable outcomes. Each decision factor, in turn, should be weighted by its relative importance. *The total of all the weights for all the decision factors must sum to one (1).* Note that the final score obtained for the example shown in Table 1.2 should be interpreted as a rank order score. Thus, this score

Table 1.2 Scoring Business Objectives: An Example

Business Objective:	Decision Factor	Decision Factor Score (Max. Score = 100; Min. Score = 0) (a)	Decision Factor Weighting (%) (b)	Decision Factor Total Score (c) = (a) * (b)
Provide online shopping capability to DoD for radioactive materials	(Substitute appropriate metrics as required)	(Note: This score is typically determined by organizational consensus)	(Note: The relative weights of the factors with respect to each other are determined by Saaty or other method)	
	Legal liability if executed poorly	$(a_1) = 0$	$(b_1) = .50$	$(c_1) = 0.0$
	Required man-hours	$(a_2) = 10$	$(b_2) = .10$	$(c_2) = 1.0$
	Time required	$(a_3) = 50$	$(b_3) = .05$	$(c_3) = 2.5$
	Capital costs	$(a_4) = 2$	$(b_4) = .30$	$(c_4) = .60$
	Other factors	$(a_5) = 90$	$(b_5) = .05$	$(c_5) = 4.5$
Total		(Note: "100" = Low Risk, "0" = High Risk)	**100.00%** (Note: the total of *all* decision weights must exactly equal 100%)	$\Sigma (c_i) =$ (0 +1 + 2.5 + .6 + 4.5) = 8.6

has meaning only in relationship to the scores assigned to other alternatives under consideration. The scoring is not like high-school grading where a score of "90 to 100" is an A, a score of "80 to 79" is a B, etc. If an alternative A_1 receives a score of "8.6" and this is the highest computed score for all alternatives, then this means that A_1 is the best alternative, based on the decision factors and the weighting used. The score does not mean that that A_1 meets a minimum threshold of acceptance; this must be determined independently. If the next highest ranked alternative,

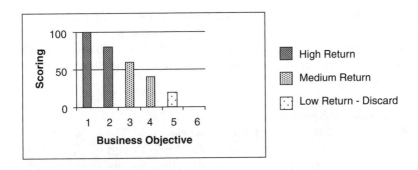

Figure 1.1   Scoring business objectives using Pareto analysis.

B_1, receives a score of "4," this cannot be interpreted that A_1 is twice as good as B_1. It only means that A_1 is to be preferred over B_1. Human judgment is needed to determine the number of final alternatives that will be accepted for consideration, after all the alternatives have been ranked. In some cases, after all the alternatives have been scored and ranked, the decision makers may decide that none of them adequately addresses the requirements at hand. In this case, all the alternatives may be rejected and new alternatives may need to be devised.

A number of methodologies are recommended in the literature to develop relative weights for decision factors.[4] One of the more well known is Saaty's Hierarchical Pairwise comparison method. [SAAT80] Other methods based on fuzzy logic, which incorporate the notion of subjective evaluation factors, are presented in [RUBI92] and [RUBI99]. Once the decision factors and their corresponding weights have been agreed upon within the organization, they should be applied consistently to each and every business objective under consideration. After a total score is calculated for each business objective, Pareto analysis can be repeated to further refine the list of business objectives upon which the organization will concentrate. As presented in this example, high scoring business objectives are preferred over lower scoring ones. This scoring process is illustrated in Figure 1.1.

Scoring each business goal provides a basis for selecting which projects should be undertaken based on the underlying value to the organization. Mathematical scoring helps ensure consistent evaluation across a set of choices. Scoring is a useful tool, but the process should not be applied too stringently because the scoring process may obscure some of the subtleties in the selection process. Scoring should be used as a gross filter, not a fine one. The human element and common sense should prevail during the decision-making process. The intent is to engage the appropriate stakeholders in discussion and to achieve consensus on the organizational focus. This process helps ensure that technology is not being adopted for the sake of technology, but for what it can do for the organization. This is the foundation for an effective technology selection process.

In summary, the business objectives and related findings should be formalized in a written document. We recommend that this report include the following sections:

- Executive Summary
- Business Objectives of the Network
  - *Strategic Objectives and Goals:* this is driven by upper management and their vision and direction for the company or institution.
  - *Tactical Objectives and Goals:* this concerns meeting budget objectives and planning directives.
  - *Operational Objectives and Goals:* this focuses on short-term, day-to-day operational concerns.
  - Desired Competitive Advantage(s)
  - Evaluation Metrics and Their Relative Importance
- Major Functional Units and Staff Affecting and Affected by Proposed Network
- Project Infrastructure
  - Senior Executive Sponsor(s)
  - Project Sponsor
  - Project Manager
  - Team Members
  - Key End Users
  - Support Services and Facilities

1.1.2 Determine Potential Risks, Dependencies, Costs, and Benefits

After the business objectives are well understood, their feasibility should be evaluated in light of the proposed network project. This involves a thorough and systematic evaluation of the resources, people, procedures, and environmental aspects of the network project to determine potential vulnerabilities and interdependencies.

The feasibility of the project should be considered relative to a number of risk factors, including:

- *Technology-related risk factors,* which may include:
  - The use of new and emerging technologies, which are inherently more risky than the use of proven, well-established technologies, and may cause unforeseen delays or expense.

- Limitations in the technology relative to its intended use. For example, the use of public network infrastructures, such as the Internet, poses different risks and trade-offs than the use of private network infrastructures.
■ *Personnel or labor risk factors,* which may include:
  - The organization's experience with similar projects. Often, the network implementation will run into difficulty because the organization is not well positioned to support network mainte-nance. The ability to manage and maintain the network once it is in place is an important determinant of whether or not the project will succeed.
  - The degree of organizational support and commitment to the network project. Lack of executive follow-through is a major reason large IT projects fail. In part, this is due to the fact that "major IT projects require radical change in an organization... including changes in both job functions and time-honored ways of doing business. Radical change can't occur if people are thinking, 'Does the boss really care about this?'" [WAHL01]
■ *Security risk factors,* which may include:
  - Vulnerability to unauthorized outside intrusion via computer viruses, Trojan horses, denial-of-service attacks, and hacker invasions.
  - Vulnerability to inappropriate use of corporate resources by "authorized" system users. An example of this risk is an employee with access to sensitive corporate documents who might illegally copy and make them available to competitors.
■ *Financial risk factors,* which may include:
  - Tangible financial impacts resulting from lost profits, budget changes, etc., due to project or network failure. Research has shown that companies with an average market capitalization of $27 billion per year lose "$670 million in market value over a two-day period after announcing an IT related problem, [such] as a hardware or software glitch." [WAHL01]
  - Intangible impacts resulting from loss of customer goodwill, industry perceptions, etc., due to project or network failure. Although these can be very difficult to estimate, they can be very significant. When Citibank publicly admitted in 1994 that hackers had broken into its network, making $10 million in illegal transfers, its banking rivals were quick to court top customers with claims that its computer systems were more secure. [BEHA97] Now, having learned from Citibank's experi-ence, the financial industry keeps security breaches unpubli-cized. "...companies are increasingly adept at covering up

breaches because disclosing vulnerabilities can leave them open to more attacks and more bad publicity. The dozen FTC investigators working on Internet security cases rely mainly on reports from the news media and individual users." [TEDE03]

■ *Disaster risk factors,* resulting from occurrences such as:
  - Natural catastrophes, such as floods, earthquakes, hurricanes, etc.
  - Acts of war
  - Fire
  - Power outages

■ *Legislation, regulatory, and liability risk factors,* which may include:
  - Legislation that may either create or eliminate markets and avenues for competition. For example, the United States House and Senate are considering Internet gaming laws that would have far-reaching implications for various local, state, and federal agencies. These laws, if passed, would also "have the potential to impact multiple markets (banking, ISPs, and the U.S.$26 billion gaming industry, for example)." [DAVI03]
  - New regulatory requirements that may force major business process and infrastructure changes to ensure the compliance required for continued operation. For example, The Financial Modernization Act of 1999, also known as the "Gramm-Leach-Bliley Act" or GLB Act, has provisions to protect the personal financial information of consumers held by various types of institutions, including banks, securities firms, insurance companies, and institutions offering the following services: lending, brokering, consumer loans, transferring or safeguarding money, individual tax return preparation, financial advice or credit counseling, residential real estate settlement services, and consumer debt collection. [FTC03] It has a major impact on the networking infrastructure, and processes and procedures of these institutions.

After identifying potential risks, it is important to calculate their probability of occurrence and quantify their impacts so the organization can develop mitigating strategies. These strategies — which should be developed as early in the planning process as appropriate — involve risk reduction, control, and transfer. Risk reduction and control might involve, for example, the adoption of a disaster recovery and data continuance strategy. Risk transfer involves shifting the burden of exposure to a third party, often through insurance or outsourcing.

Risks can be evaluated using the techniques outlined in Section 1.1.1. A decision matrix, listing each risk, associated decision factors, and the relative weight of each decision factor, can be summarized in a table similar to Table

Formula for Success

| Save
Time | Save
Money | Make
Money | Good
Business |

Figure 1.2 Making the Business Case for the Network

1.2. The results can be used to determine which risks require special attention and further action (i.e., reduction, control, or transfer).

A cost-benefit analysis should also be performed to determine expenditure levels and payback time periods required for a viable project. Companies are under considerable pressure to do more with less, and IT budgets have suffered significant cutbacks. Many companies are maintaining profitability, not by growth but through attrition and savings. The business model has become: if last year you spent a dollar, this year spend 80 cents to do the same or more.

According to the Gartner Group [BABA02], the IT services industry is over $520 billion and steadily increasing. Technology spending is too large a percentage of most corporate budgets for CEOs and upper management to ignore. Increasingly, this is reflected by the fact that more CIOs (Chief Information Officers) have reporting relationships directly under CFOs (Chief Financial Officers). This, in turn, changes the dynamics of how IT projects are approved and increases the emphasis on more tangible and shorter ROI[5] than ever before. Required payback periods of 12 to 14 months are not uncommon.

The perceived value of an IT project's success largely depends on the financial returns realized by the organization. Careful financial analysis of proposed costs and benefits is essential to obtaining senior management buy-in throughout the network implementation life cycle. Return on investment (ROI) is a powerful, nearly universal, and well-understood metric to help justify new initiatives to demonstrate how they will make money, save money, or save time. Figure 1.2 summarizes the goal of ROI analysis. Appendix A presents a detailed case study and ROI analysis for a proposed network solution to demonstrate how ROI is used to determine payback periods. The ROI analysis of a major network infrastructure project usually includes a comparison of the available options, such as:

■ Do nothing.
■ Postpone project initiative.
■ Outsource.
■ Build a custom solution.
■ Buy off-the-shelf products and integrate components.
■ Combine some of the above options.

Typically, large corporate clients have an internal, core IT group to consolidate vendor contracts and to get high-volume discounts. To get instant ROI, big companies are doing more project management and are consolidating from perhaps as many as 25 to 30 down to 3 to 5 IT vendors to leverage their ability to get volume discounts. As companies experience reductions in IT staffing, they have even more incentive to consolidate vendors and to sign multi-year contracts because it is easier to manage and minimizes risk.

Although it is generally cheaper to use fewer vendors, this may also mean a loss of flexibility or opportunities to embrace new technology. Oftentimes, company headquarters will tell business units to work internally with centralized corporate buying groups. If a business unit can demonstrate a better business case with a seller or provider outside the approved vendor list, they usually will be allowed to use them. However, when it comes to problem resolution and project management, the business unit must sink or swim on its own because, in general, it will not get the same level of corporate backing as if it had used internally sanctioned vendors. The centralized IT function should retain a firm hand in overseeing vendor arrangements outside approved lists. If too many business units negotiate vendor contracts separately, the company will not be able to achieve the same high-volume discounts they could otherwise.

IT staffs are cautious about giving up the rights of ownership to too much of their organization because they do not want to outsource their own jobs. However, companies — both large and small — are looking for providers to perform services for them that are not a part of their core competencies. Web hosting, data co-location, and disaster recovery services are some of the areas companies are increasingly looking to outsource in an effort to achieve cost reductions in IT.

It can be extremely difficult to properly estimate costs and ROI, particularly when the benefits are intangible, new technologies are being used, or a new system is being implemented. For example, quality is difficult to justify in quantifiable, tangible terms, and yet it is an especially important aspect of computer and network security. For example, many companies currently using ATM[6] and Frame technology[7] are considering using the Internet for their network infrastructure, because IP/VPNs[8] can cut corporate network costs as much as 30 to 40 percent. Older Frame Relay and ATM networks are based on private, dedicated lines over which the company has complete end-to-end control. However, if the public Internet is used to transport confidential and sensitive company data, additional measures are usually needed to provide adequate protection against security breaches and loss or corruption of critical data. If companies are going to *save* 30 percent by using a VPN, they should also bear in mind that they will likely need to *spend* at least another 5 percent for additional security, such as firewalls, to ensure the quality and integrity of the network.

Depending on how company budgets are structured, security might be funded by the CSO,[9] as opposed to the CIO,[10] creating organizational difficulties when trying to get additional security expenditures approved for a new information technology initiative. To deal with these issues, it is important to demonstrate how scrimping on security to save a few thousand dollars might lead to security breaches with sizable, tangible costs. For example, if it can be shown that one security breach might cause a $50 million loss that could be prevented with a $150,000 firewall, and you are saving $2 million by switching from Frame Relay to a VPN, it should be much easier to justify the funds needed for the firewall.

Outside organizations are often used to demonstrate potential risks and vulnerabilities from hackers and insiders. Once the need for security is evident and quantifiable, it is easier to sell incremental increases in the security budget to upper management. Sometimes the perceived risk of using Internet-based networking leads the organization to reject it as an option and to keep its network infrastructure private. This might occur if business units do not want to fund additional costs to secure the network or to risk possible compromise of sensitive, confidential information. This is particularly true in risk-averse companies, such as banks and other financial institutions.

Merrill Lynch is an example of a company that uses ROI as the cornerstone of its technology planning. Merrill Lynch has over 2000 active IT projects at any one time. According to Marvin Balliet, Merrill Lynch's Chief Financial Officer for the Global Technology and Services Group, ROI is used as part of an "overall governance model" to help select and manage technology projects. Any technology initiative over $2 million requires a business case or appropriation request that has been completed according to a standard format. After a budget request is approved, a quarterly report must be prepared on the project status. This report must include information on the project budget, scope, user involvement, and an updated ROI estimate of the project value. [MCEA02]

In summary, the findings of the risk, dependency, and ROI analysis should be formalized in a written document. We recommend that the following sections be considered for inclusion in this report:

- Executive Summary
- Project Scope
- Business Issues Requiring Further Study
- Major Dependencies
  - Other Corporate Initiatives That May Affect Project
  - Skill Requirements
  - Technology Issues
  - Timeframe Requirements
- Assessment of Project Risks
  - Potential Impacts on the Business as a Whole
  - Types of Potential Risks
  - Recommended Strategies for Risk Reduction, Control, and Transfer
- Cost-Benefit Analysis
  - Quantified Tangible and Intangible Benefits
  - Impacts on Profitability
  - Impacts of Hidden Costs
  - Impacts of Tangible and Intangible Risks
  - Analysis of Procurement Options (e.g., lease or purchase)
  - ROI Analysis of Networking Alternatives
- Comparison of Alternative Solutions
  - Key Rating Factors and Relative Importance Evaluation
  - Ranked Alternatives by Weighted Factor Analysis
  - Recommended Solution(s)

1.1.3 Identify Project Requirements

At this stage in the network planning process, information is collected on what the proposed network is supposed to do. Because many technical decisions will be based on this information, it is important that the requirements analysis be as accurate as possible. This step will likely involve interviewing decision makers and staff who have a significant role in planning or using the network. It will also involve collecting information from various sources on the following:

- Estimated traffic patterns and flow
- Application programs and services to be supported by the network
- Destination and number of proposed system users
- Estimated equipment and line costs
- Network reliability requirements
- Network security requirements
- Network delay requirements

In general, it is not easy to collect the information needed to perform a complete requirements analysis. Chapter 8 discusses at length why this is so, and how the requirements affect the network design. However, the main reason the data collection is time consuming and complex relates to the fact that the data needed is often not available or is not in the form needed for analysis. Considerable effort is usually required to estimate and derive the real parameters needed to design the network.

The extent to which a company relies on outside vendors to document the project requirements largely depends on the available internal resources, staffing, and skills sets. Small and medium-sized companies usually do not have the resources to plan extensive network projects and rely more heavily upon vendors for assistance in preparing this documentation. Larger companies are better staffed and wish to maintain more control over the requirements definition process. They typically take a more proactive role in defining the project requirements before vendors are asked to present proposals. In large companies, project requirements are assessed on an ongoing basis and as part of budgetary planning.

Engaging end users in the discussion of project requirements can provide important insight into the types of applications that must be supported, as well as problems that need to be addressed. End-user discussions can also be helpful in clarifying requirement priorities. However, end-user involvement must be managed carefully so as not to encourage unrealistic expectations about the network implementation.

Requirements specification usually evolves over time with a Request for Information (RFI), Request for Quote (RFQ), and finally a Request for Proposal (RFP), from qualifying vendors.[11] Chapter 6 discusses this process in more detail.

Ultimately, the project requirements should be documented in a written report that consolidates the findings. We recommend the following sections for inclusion in this report:

- Management Summary
- Project Requirements
  - Functional Requirements
- Performance Requirements
  - Low, Normal, and Peak Usage and Capacity Requirements
  - Response Time, Delay, and Queuing Constraints
  - Availability and Uptime Requirements
  - Reliability and Transmission Accuracy Requirements
- Specialized Application and Software Requirements
- Potential Growth, Expansion, and Scalability Requirements

- Distance and Geographic Requirements
- Backup and Redundancy Requirements
  - Budgetary Constraints and Restrictions
  - Quality Assurance and Control Requirements
  - Security and Control Requirements
  - Test Plan Requirements
  - Resource Requirements
- Partial List of Supporting Documentation
  - Estimated Traffic Patterns and Flow
  - Current and Future Application Programs, and Services Supported
  - Current and Future Network Destinations and Sites
  - Current and Future Network Node Types (i.e., external, internal, suppliers, etc.)
  - Protocols Required to Support Applications and Network Services
  - Current and Future Number of Proposed System Users
  - Existing Legacy Equipment (servers, routers, etc.)
  - Switching and Gateway Requirements
  - Estimated Equipment and Line Costs
  - Number of Existing and Planned Lines and Capacity
  - Mix and Types of Data Transmitted (i.e., percent data traffic, percent voice traffic, percent video traffic, etc.)

1.1.4 Develop Project Implementation Approach

Once the business objectives, project feasibility, and major requirements have been determined, a strategy for moving forward on the network design and implementation should be developed. A number of factors must be considered when formulating the project approach. Among the most important factors are the business and technical constraints and risks. Constraints and risks heavily influence decisions on whether or not to proceed with the design using outside consulting help, or with in-house staff, or with some combination of the two.

Tactics to manage the potential project risks should be developed early on in the project, particularly if the risks are great. One method of dealing with risk is to create an implementation plan that is evolutionary, as opposed to revolutionary. This suggests a phased implementation approach that might, for example, specify how existing systems are to coexist or be transformed to operate in the new network environment. Pilot projects may be helpful in evaluating various network options and can be planned at this stage.

Business processes affected by the network implementation should also be examined and those needing refinement should be identified. Strategies for making the workflow adjustments should be addressed in the project plan. To promote awareness and support of the network, it may be necessary to incorporate educational and internal communications programs into the project plan.

Given the pace at which technology is evolving, it is important to accept change as a fact of life and to plan accordingly. This involves planning reviews at critical points throughout the network planning and implementation process to identify unforeseen problems or changes that must be accommodated by mid-course adjustments. The urge to resist "scope creep" must remain strong while making these mid-course adjustments, to prevent the project requirements from changing and burgeoning out of control.

The implementation strategy should be formalized in a written document. We recommend the following sections for this report:

■ Management Summary
■ Assumptions and Major Constraints
■ Workflow and Organizational Impacts
■ Project Approach
  – Risk Management Strategies
  – Project Staffing
  – Project Schedule
  – Transition and Deployment Plan
  – Change Control Procedures

After completing a top-down business analysis of the network plan, the organization is well positioned to begin a detailed analysis of the technical network requirements and to develop network alternatives for consideration. The chapters that follow discuss this process in detail.

1.2 Strategic Positioning Using Networks

The planning process described in the preceding sections provides the foundation to understand how relevant technologies are aligned with the business objectives. It also ensures that the organization has acceptable trade-offs between risks, costs, and benefits. This is needed to decide whether and how new technology makes sense in the organizational context.

This section presents a number of case studies to demonstrate how effective use of technology can transform markets, foster demand, and create strategic advantage. It also shows how important it is in today's environment to keep pace with technological developments, even if it does not make sense to be an industry innovator. Finally, this section presents a method for calculating potential rewards to be derived from using or developing a new technology.

1.2.1 Case Studies

Some companies are very successful being innovators and early adopters of technology. Other companies, in the same market niche, are not. Sometimes, the more successful long-term strategy is to wait until a technology matures before adopting it. It is not always easy to decide when to adopt a new technology.

Early adopters derive the benefits of new technology and sometimes the ability to capture new market share. However, they must also be ready to contend with problems of interoperability and less than perfect solutions and performance, because there are usually a lot of bugs and other issues to work out in the early releases of any new technology service or product.

Conservative adopters of technology seek to minimize risk. Conservative adopters wait until the prices and the risks of new technology go down. They watch to see what works and what the dominant trends are in the marketplace before making a commitment to the technology. Large companies have used this strategy very effectively by copying a small innovator and bringing to bear large resources to dominate and capture a market once it has started to germinate.

Late adopters of technology operate in a reactionary mode. Even when faced with massive customer defections due to their lack of innovation, they may be unwilling or unable to change. These organizations do not employ technology as a strategic weapon.

Small companies can be nimble and quick to develop new business models using innovative technology and networking infrastructures. When successful, they reap the benefits of being first to market with the new capabilities. However, if they cannot scale to meet demand, they will be unable to capitalize on the market they helped create. Large, well-capitalized organizations that know how to manage technology effectively are the best positioned to transform markets and establish industry leadership. However, as the case studies below illustrate, both large and small companies have successes and failures.

The United States military has always been a major source of technological invention. The development of the Internet is a direct result of DARPA's efforts to develop a robust computing network to ensure national security. However, the military is increasingly using commercially available network technology when it is easy to deploy and cost effective. "During the [Desert Storm] Gulf War, for example, ground commanders lacked timely satellite photos to prepare for combat because the four computer systems handling the pictures couldn't talk to one another..." [WALL95] When the United States invaded Iraq in 2003, American solders used GPS (Global Positioning System) devices with wireless hands-free radios and notebook computers to analyze digital maps made from satellite photographs. [SCHN03] The government has commissioned MCI to build a $45 million wireless phone network in Iraq based on the European GSM wireless standard. The network will have 19 cell towers servicing up to 10,000 mobile phones and will be used by reconstruction officials in Baghdad. [ASSO03] Loren Thompson, military analyst at Lexington Institute in Virginia, says that "The single most important advance that the U.S. military has made since Desert Storm has been to hugely improve the coordination of its forces... all [has] changed courtesy of the information revolution." [PUZZ03]

Federal Express (FedEx) is an industry example of a technology pioneer. FedEx's information technology infrastructure is based on a long-term strategy fostered by the company founder, Fred Smith. The core of FedEx's service is designed around sophisticated network and tracking systems. FedEx's success in implementing fast, effective tracking has been a major factor in the erosion of the United States Postal Service's (USPS) share of the very profitable overnight mail delivery market. FedEx was the first commercial express shipping and mail service to offer its customers the ability to track letters and packages using the Internet, free of charge. The USPS, UPS (United Parcel Service), and other competitors have had to make considerable investments in technology to remain competitive. FedEx's leadership role in the shipping and mailing industry is a direct reflection of its investment in and use of technology.

Wal-Mart is another renowned case study. Wal-Mart is the nation's largest private employer, with more than one million employees. [KERS03] The company started as a small discount store in the 1930s. From the beginning, the company founder, Sam Walton, realized the importance of tracking inventory, sales, and trends and required management reporting on a weekly and monthly basis. Although he was "notoriously cheap, Walton could be convinced to spend money on things that would save the company money in the long run and to allow it to grow." [BASE00] In 1977, Wal-Mart implemented its first computer network, which it used to automate the collection of sales data and to link stores to distribution centers and to headquarters. By the 1980s, it had implemented EDI[12] and the largest privately owned satellite network in the country. The satellite network was used for internal communications and for credit card authorizations, which drastically reduced customer transaction times. By the 1990s, Wal-Mart had sophisticated systems to track customer shopping habits and preferences. [ORTE00] In contrast, Kmart's first store opened in 1962, the same year as Wal-Mart. In 1963, Kmart had 53 stores, while Wal-Mart had one. However, Kmart did not gather data as quickly or as efficiently as Wal-Mart, nor did it invest as heavily in an information infrastructure. Even in the 1990s, Kmart analysts could not automatically generate reports providing a corporatewide view of supply and demand. Instead, they would feed data from multiple reports into a spreadsheet to aggregate figures. Although Kmart has made significant investments in technology over the past decade, it has not been enough to keep the company from Chapter 11 bankruptcy, which was filed in 2002 and from which it only recently emerged.

Yahoo! and Google are examples of companies that started as small, informal ventures. Today, they are both recognized as industry leaders for their technological capabilities and networking infrastructure. In 1994, two Stanford Ph.D. candidates, David Filo and Jerry Yang, devised a tool to keep track of their favorite Internet links. The popularity of their idea eventually led to Yahoo!'s incorporation and an initial public offering in 1996. Yahoo! is now the largest portal on the Web. Yahoo! also offers a variety of online enterprise services, including Corporate Yahoo!, a customized enterprise portal toolkit for "audio and video streaming, store hosting and management, and Web site services." [YAHO03] Google was incorporated in 1998 by Sergey Brin and Larry Page, also students at Stanford University. Google is now the world's largest search engine, with an index of more than three billion Web pages, and handles over 200 million search queries a day. Yahoo! directs search requests from its site to Google. Google has based its business model on two revenue streams: search services and advertising programs. [GOOG03] Various analysts estimate that Google had revenues in excess of $1 billion in 2003 and in 2004, it became a publicly held company.

Microsoft is a premier example of a company that is very strategic in choosing when to be first to market and when to be second. Microsoft created its monopoly position by a strategic legal and business alliance with IBM to sell personal computers with its operating system. This allowed the Windows operating system to become a *de facto* industry standard, even if IBM personal computers did not. When UUNet approached Bill Gates in 1995 for funding, he was quick to recognize the huge potential of the Internet as a vehicle to promote worldwide demand for personal computers. Microsoft acquired a 15 percent equity position in the then-private UUNet, and paid "for the deployment of 40,000 modems on the network, the largest such project on the Internet at the time. The software giant also became a banker for UUNet, lending the company almost $40 million at favorable interest rates." [CHAN97] Microsoft also funded UUNet's expansion into Europe. Today, UUNet is the largest Internet provider in the world, and the global market for personal computers running Windows operating systems is legendary.

Microsoft has solidified its market position by steadily augmenting the functionality of its operating system. In 1995, during the early days of the World Wide Web, Netscape was the leading Internet browser, with over an 80 percent market share. The same year, Microsoft introduced Internet Explorer as part of Windows 95. Microsoft made slow gains by steadily surpassing Netscape's functionality and eroding its market share. Things reached a head in 2002 when Netscape announced major layoffs and restructuring and its parent company, America Online (AOL), initiated an antitrust lawsuit against Microsoft. Finally, in May 2003, Microsoft and AOL announced a $750 million settlement, in which AOL will continue distributing Microsoft's Internet Explorer with its online service but will not have to pay royalties to Microsoft for a period of seven years. There is widespread industry speculation that this agreement will seal Netscape's fate and reflects a lack of commitment to Netscape on the part of its parent company. At the time of this writing, Netscape has about 5 percent market share, while Internet Explorer has almost 95 percent.

Microsoft and AOL announced as part of their settlement that they will work together to improve the interoperability of their instant messaging (IM) products. It will be interesting to watch this scenario unfold. AOL has about 100 million registered home users. [SAUN03] AOL's Instant Messenger™ provides a real-time, text-based chat capability that only AOL subscribers can use. It is in AOL's best interests to keep its instant messaging system proprietary as long as possible, because it is a leading reason why customers prefer AOL's paid service. According to Genelle Hung, analyst with Radicati Group, "IM is the killer application. There is no other feature that keeps people coming back [for AOL]." [NEWY03]

Because AOL's legacy offering is based on a dial-up Internet connection, it is not suited to bandwidth-intensive applications such as video. In contrast, Microsoft's MSN Windows Messenger is based on the IETF standard, Session Initiation Protocol (SIP),[13] and comes bundled as part of Windows XP. This protocol provides the capability of supporting APIs and real-time multimedia applications. It, too, is a free service to consumers and is supported by Microsoft's MSN network. Microsoft claims to have over 100 million users of its IM service. [MICR03a]

AOL is faced with a serious threat to its market as consumers migrate to other, lower cost narrowband options and to broadband DSL and cable modems, which are up to 100 times faster than dial-up modem connections. DSL dominates the global broadband market and represents about 62 percent of the broadband modems sold in 2002. Currently, about 27 percent of all U.S. home Internet connections are broadband; and by 2008, this figure is projected to grow to about 60 percent. In Europe, about 7.5 percent of all household Internet connections are broadband. [GREEN03] AOL is offering the RoadRunner service — based on an all-optical broadband IP data network — to allow it to compete in the broadband market. [AOL03]

In the corporate arena, there are over 30 players in the instant messaging (IM) market. [GREY02] Enterprise IM (EIM) products provide security and network management capabilities (with message logging, archiving, and monitoring, etc.) not found in consumer IM products. However, they are not free[14] and provide a significant revenue opportunity for vendors. Key players to watch include:

- AOL's Enterprise AIM services
- IBM's Sametime
- Yahoo's Messenger Enterprise Edition
- Microsoft's MSN Windows Messenger

Currently, IBM's Sametime is the dominant player in the enterprise IM market, controlling over 70 percent of the market. [KEIZ03] AOL's AIM can be accessed by Lotus Notes' Sametime through a contact list that is kept and accessed separately. AOL and IBM have announced plans to improve the interoperability of their respective proprietary products. According to Gartner Group, IBM has a limited window of opportunity to grow this market before Microsoft captures market share with its enterprise IM product — Microsoft Real-Time Communications Server 2003 Standard Edition (a.k.a. "Greenwich"). [KEIZ03]

As a condition of approving AOL's merger with Time Warner, the FCC requires AOL to open its IM network to competitors if it launches advanced high-speed IM services. [HU03] This was a condition for which Microsoft lobbied vigorously. AOL is currently petitioning the FCC to try to remove this restriction and to allow it to pursue its proprietary, closed approach to high bandwidth IM.

Microsoft views both consumer and enterprise instant messaging as an extension of its overall product architecture. Like EIM and Sametime, Microsoft's RTC IM strategy is based on IETF (SIP) standards and offers such state-of-the-art capabilities as photographic file sharing, audio conversations, Voice-over-IP telephone calls, e-mail, application sharing, and co-browsing. *Co-browsing* allows customer service personnel, for example, to interact directly with a customer's Web browser. Using this capability, a customer service representative might demonstrate order entry functions or answer questions directly on the customer's screen. *Collaborative browsing* integrates IM chats, e-mail, fax, telephone, and Internet phone contact in a single interaction. This capability is fully integrated and plug-and-play with the Windows XP operating system and Microsoft applications (e.g., Outlook, Word, PowerPoint, Excel, Project, etc.). Summing up, Microsoft has devised a multi-pronged strategy to make its Windows platform ever more compelling, fully integrated, and functional in supporting end-user voice, video, chat, e-mail, and software applications, thereby ensuring that it will remain a dominant player in the market. In the view expressed by Microsoft's Chief Executive, Steven Ballmer, hardware is becoming a commodity but software is not. Microsoft's strategy is to deliver powerful "Integrated Innovation" to simplify the automation of business tasks. [LOHR03B]

Lands' End (landsend.com) is another company that has made strategic use of technology and, in so doing, has transformed the business model for the clothing industry. From the beginning, Lands' End has been a direct marketing catalog company and, until its recent acquisition by Sears, did not sell clothing in stores. Lands' End works directly with mills, manufacturers, and customers. Customers now have the option of shopping by phone, mail, fax, the Web, or in Sears stores. Since Lands' End launched its Web site in 1995, it has introduced a number of industry innovations, including: [LANDS03b]

- *My Virtual Model*™. This is a tool that allows customers to create a three-dimensional model of their bodies. This model can be used to "try on" clothes that are offered on the Web site.
- *My Personal Shopper.* Customers are asked to enter data about their clothing preferences. This data is used, in turn, to allow a "personal shopper" to suggest items and outfits.
- *Lands' End Custom*™. After customers supply their measurements and information on their body type, computer-operated cutting machines are used to make custom tailored and fitted clothes.

Today, landsend.com sells more clothing than any other Web site in the world. Until Lands' End proved the contrary, the "common" wisdom was that the clothing industry was not likely to achieve success on the Internet because shoppers would want to touch, feel, and try on clothes before making purchases. Lands' End's well-designed Web site makes it easy to display merchandise, receive suggestions, and request fabric samples. Strategic use of technology enabled Lands' End to effectively counter barriers to entry in the E-commerce market.

Looking at Web-based retailing on the whole, the year 2002 was the first time Web-based retailers made a profit, on $76 billion in sales. Among the most profitable were Web merchants, like Lands' End, with preexisting catalog operations, with profits of about 22 percent. As a group, online-only retailers had losses averaging about 16 percent. The key to profitability for many online companies is to control shipping and customer service costs. The former can be addressed by contract negotiations and better shipping strategies, while the latter can be addressed using online tracking and e-mail notification to reduce customer calls. "Whereas the heralded 'first movers' of consumer E-commerce captured Wall Street's fleeting attention, the E-tailers still in business have captured something of more lasting value: the expertise that accompanies experience and the time to take advantage of it." [TEDE03] According to Shop.org, an online retailers' association, the biggest online sales growth in 2002 was in health and beauty (93 percent), apparel (54 percent), and flowers, cards, and gifts (>50 percent).

The effects of the Internet have become so pervasive that even if companies or organizations choose not to have a presence on the Web, they can be profoundly impacted by it, for good or bad. For example, in 2002, Intuit released TurboTax with anti-piracy controls, which greatly upset many customers who expressed their displeasure on Internet forums. Intuit responded to the vehement outcries by agreeing to discuss potential product changes online before implementing them in the future. The movie "My Big Fat Greek Wedding" benefited from early favorable reviews posted on Web sites, which helped compensate for its small advertising budget. [THOM03]

New Internet shopping paradigms are also shaping marketing, customer loyalty, and advertising programs. Reflecting the rising influence of informed online shoppers, Amazon.com has eliminated its entire advertising budget for television and print. Instead, it is offering customers free shipping on qualifying orders. To make their Web sites less intrusive and annoying, some "E-tailers" are converting to behavior-based advertising. Instead of asking personal questions, they monitor where site visitors click. Based on inferred visitor preferences, screen content and ads are modified accordingly. The *Wall Street Journal Online* and the *New York Times* are examples of companies currently using behavior-based advertising on their Web sites. [IVES03]

1.2.2 Calculation of Technology's Strategic Value

Calculating the possible gains from technology in tangible financial terms is an imprecise and challenging task. It involves careful consideration of the following:

- How does your product or service relate to the target markets you wish to serve?
- Among the buyers of your product or service, what is the mix that will embrace innovation and change? What percentage of each are early adopters, conservative adopters, and late adopters?

- What types of innovation are you proposing — product or process innovation?
- What is the time horizon for implementing the new technology — near term, a year, or longer?
- What barriers, if any, exist to keep your rivals from adopting a similar strategy to compete?
  - Pricing
  - Cost to convert from existing to new technology
  - Similarity of features or functions between existing and new technology
  - Regulatory requirements
  - Availability of technology
- What is the size of your company and experience with technology?

The answers to the above questions can be used as input to a discounted cash flow model to determine the potential value of a new technology to the enterprise. Discounted cash flow can be computed in a number of ways. The Adoption-Diffusion Model, developed by the Department of Agriculture, is a well-documented method, with the following steps:

1. Estimate profits to be gained by using the new technology in the initial stages. This may come from incremental revenue, return on capital, or cost reductions.
2. Forecast the time period over which the product or service will be commoditized as competitors and conservative and late adopters enter the market to drive down prices. The Adoption-Diffusion Model is based on the premise that the mix of early, conservative, and late adopters will affect the timing and mix of customer buying. The proportion of each buyer type and the time before they will enter the market as buyers should be estimated.
3. Compute expected decreases in profit flows as the product or service becomes a commodity.
4. Calculate the total discounted cash flow based on the forecasted life cycle of the product, service, or project. This calculation incorporates the projected revenue streams for each buyer type (i.e., early, conservative, and late adopters).

In short, there are many ways a network can enhance an organization's competitiveness. Networks can be used to offer new services and capabilities, streamline operations, enhance image and visibility, increase productivity and reduce costs, and improve customer service. Organizations with creativity and insight have produced networks that offer dramatic advantages over competitors that are not as effective in their use of technology. The race to adopt new network technology continues to quicken as more and more organizations extend their reach and influence through the use of global networks. Organizations, large or small, cannot remain complacent in their use of technology if they are to remain competitive in today's environment.

1.3 Dealing with Major Design Challenges

Increasingly, a company's ability to adapt to changing technological, market, legislative, economic, social, and homeland security conditions rests squarely on its IT and networking infrastructure. Network design, by its very nature, has always been complex and multifaceted, posing many organizational and technical challenges. However, these challenges continue to escalate as companies are forced to do more with less and to react in shorter and shorter timeframes. IT budgets for staffing and capital expenditures have been hit hard in recent years. To keep costs down and maintain corporate profitability, many companies are postponing routine network maintenance and upgrades. It is a fact of life that companies no longer invest in technology for the sake of technology — immediate business returns and tangible results are needed to induce corporate IT investment. The proliferation of new technologies — especially public Internet technologies — and the impetus to change must be carefully weighed against the benefits and risks.

The following sections discuss the major issues facing corporations as they plan their technology infrastructure. Common sense and good management principles should dictate the strategies for meeting these challenges.

1.3.1 Organizational Concerns and Recommendations

1.3.1.1 Senior Management Involvement and Support

A recent survey by CIO Insight of 500 chief executive officers revealed the following: [ZIFF03]

- 73 percent say their companies are highly innovative
- 65 percent agree the economy is slowing technology adoption
- 41 percent of technologies under evaluation are adopted
- 53 percent involve business executives in technology adoption
- 40 percent say the primary goal in adopting technologies is cost cutting

These findings reflect the importance of having senior management that is in tune with the overall business environment and capable of thinking strategically about where the company needs to be in the marketplace. CIOs are faced with significant challenges because there are substantial risks (both personally and to the corporation) in leading technology initiatives to transform old ways of doing business. The ability to successfully lead these initiatives requires an entrepreneurial spirit and a thorough understanding of the market and the company's operation. Senior management must also be able to clearly communicate business objectives so that bottom-line requirements can be factored into the planning and execution of the network strategy. Senior management plays a vital role in ensuring that a process is in place to make this happen.

CIOs and senior management help translate organizational goals into an actionable IT strategy and provide a focal point for responsibility and accountability within the organization. This encompasses a number of dimensions, including:

- Implementing a corporatewide planning process that engages senior management in strategic decision making
- Aligning IT plans with organizational goals
- Instituting capital planning and investment analysis of IT initiatives
- Formulating and implementing a strategic IT architecture
- Ensuring effective IT infrastructure and staffing is in place to meet organizational objectives
- Ongoing measurement and oversight of IT ROI and contribution to overall business performance

In summary, it is increasingly vital to overall corporate success to have senior management with the vision, technical competence, and skill sets to effectively manage and deploy complex network technologies in a dynamic and quickly changing environment. Senior management has a pervasive effect on organizational culture, and serves as an invaluable role model to encourage integrity, ethics, leadership, and innovation.

1.3.1.2 Recruiting, Training, and Maintaining Skilled Staff

The organization should strive to hire, train, and retain skilled managers and staff who understand technology and how it can be used to satisfy organizational objectives. This is not easy, given the highly competitive job market for network specialists and the rapid proliferation of new networking technologies.

Technology for technology's sake is no longer an appropriate organizational goal. However, some employees may push to implement a new technology because it is perceived as a better path for personal growth than other, perhaps more appropriate but older or mundane options. Conversely, there may be a need to adopt new technologies that current staff members are reluctant to embrace because they are resistant to change. Dealing with these matters presents a challenge to technical managers, particularly at a time when companies are cutting training budgets and holding the line on salary increases.

This type of dilemma is fairly common. For example, Voice-over-IP (VoIP) is emerging as a new way to consolidate both voice and data traffic over the same network, because the cost of doing so is very competitive. Traditionally, corporate PBX networks have been operated by "voice" staff who are very familiar with trouble tickets and resolution of QoS (Quality of Service) issues. With the migration to VoIP, MIS organizations must train "voice" staff on IP fundamentals. "Data" staff must also be trained to take voice trouble tickets on data networks, which is something they are not accustomed to doing. Packet delivery with Frame Relay, ATM, or TCP/IP protocols is judged on such metrics as packet loss and MTTR. With voice transmission, the ear picks up jitter — much like wow and flutter on a tape deck — if there are problems with packet delay. However, these problems are very difficult to trace in a traditional data network. It is likely that voice MIS staff will start reporting to the data MIS staff once voice traffic starts to go over the Internet. The end result is that initially this will result in a mismatch of skill sets, which will require retraining and organizational realignment. The rapid pace of new technology is likely to have significant, ongoing impacts on the organization and its structure.

Some strategies for dealing with training and staff retention include:

- Work with Human Resources to develop methods for hiring and retaining good staff.
- Where necessary, augment existing staff with consultants and vendor support.
- Encourage cross-disciplinary training and project efforts to foster depth of knowledge and application within the organization. This will avoid islands of staff expertise in specialized areas that are not easily generalized. Cross-fertilization also helps employees see the big picture and how technology relates to the business objectives.
- Use training and internal communication to reduce the fears of those affected by new technology projects.
- Encourage and offer ongoing education to help staff remain current with new trends in technology.
- A voluminous amount of technical information is available from a variety of sources such as vendor/telco/consultant presentations,

conferences, technical books (such as this text!), industry magazines, and the Internet. Turn to these sources on a regular basis
to help keep up with new developments in the industry.

1.3.1.3 Effective Program and Project Management

One of the more common reasons technology initiatives fail is poor project
management. Therefore, it is imperative that the organization develop
internal standards, methods, and procedures to promote effective planning
and project management. There should also be a well-publicized organizational commitment to do things the "right" way, which means adhering
to standardized processes and procedures even when there are substantial
pressures to take risky shortcuts.

A well-defined management review process for allocating staff and
capital resources to technology innovation projects is an essential aspect
of project management. This will likely involve regular project and capital
review meetings with senior management to obtain their buy-in and
support of IT initiatives.

Once a project has started, reporting mechanisms should be in place to
track progress against milestones, budget allocations, and projected ROI.
Projects that achieve targeted outcomes should receive continued funding
and management sponsorship. Projects that are failing should be cut short.

During the planning process, potentially serious political and organizational issues should also be identified. For example, some people may
feel threatened if they believe a proposed network or technology initiative
will compromise their power, influence, or the nature of their job. Consequently, they may attempt to hinder the project's progress. The organization
must confront these fears and develop strategies for dealing with them.

1.3.2 Technology Concerns and Recommendations

1.3.2.1 Keeping Abreast of New Developments

The organization does not operate in a vacuum and must be attuned to
changes going on around it. For example, there may be legislative,
economic, or social pressures that mandate changes in the way the
company does business. This, in turn, may have a profound impact on
company strategy and technology adoption. Keeping abreast of new
developments and relating them to organizational requirements is a formidable task, and it is rare that an organization will have all the expertise
in-house it needs to do this well.

Often, consultants and outside vendors are needed to help plan and
implement the network. It is much easier to manage the activities of the

consultants if the organization has a firm grip on the business objectives and requirements. However, sometimes consultants are needed to help develop and specify the business objectives and requirements. Although outside consultants offer benefits such as expertise and objectivity, they also present their own set of challenges. For example, it is important to develop a "technology transfer" plan when working with outside consultants, to make sure that in-house staff can carry on as needed after the consultant leaves.

Throughout the 1970s and 1980s, if an organization wanted a network, it could call IBM and IBM would design that network. It was a common adage that "the only risk is not buying IBM." However, for the foreseeable future, there will be increasing numbers of network vendors in the marketplace, and a decreasing likelihood that any one vendor will satisfy all the organization's network requirements. Although often unavoidable, using multiple vendors can pose problems, particularly when there are problems with the network implementation and each vendor is pointing a finger at the other. Because it is increasingly likely that a particular network vendor will provide only part of the network solution, it is incumbent upon the network design team to make sure that the *overall* network requirements are addressed.

It is important to maintain ongoing awareness of new product offerings from service providers that provide differentiated, value-added services. This can be accomplished by implementing an ongoing RFI, RFQ, and RFP process. This involves regular interaction with key service providers and vendors. It is also a good way to guarantee that the organization is getting the best deals possible on products and services.

1.3.2.2 Maintaining a Secure Networking Environment

With the growing reliance on public infrastructures and Internet access, and the threat of disasters and terrorist activities, there is a heightened awareness for the need to implement security safeguards to protect sensitive company data. Many companies have responded by creating Chief Security Officer (CSO) positions to ensure corporatewide oversight of all aspects of security. Government has taken a stronger stance with the enactment of the Federal Information Security Management Act (FISMA). FISMA mandates the creation of CSO positions within U.S. federal agencies to oversee all security policies and practices. [USDE03]

Typically, CSO responsibilities include:

- Identifying security goals and objectives consistent with corporate strategy
- Overseeing the protection of tangible assets, intellectual property, computer systems, networking infrastructure, and people

- Managing security policy, standards, guidelines, and procedures to ensure ongoing security maintenance
- Establishing relationships with local, state, government, and federal law enforcement agencies, as appropriate
- Overseeing investigation of security breaches and organizational response to ensure appropriate disciplinary, legal, and corrective actions

Some organizations require or prefer security managers who are CISSP certified. CISSP certification involves passing an examination that covers the following areas: [(ISC)²]

- Access Control Systems and Methodology
- Applications and Systems Development
- Business Continuity Planning
- Cryptography
- Law, Investigation and Ethics
- Operations Security
- Physical Security
- Security Architecture and Models
- Security Management Practices
- Telecommunications, Network and Internet Security

CISSP recertification is required every three years, along with ongoing continuing professional education (CPE) credits. Certification is limited to individuals who have a minimum of three (with a college degree) or four (without a college degree) years of active experience in one of the ten security domains listed above.

There are always new ways to compromise network security. It is an ongoing process to stay on top of the potential vulnerabilities to which the company is exposed and to maintain proper vigilance. A good security plan is multifaceted and includes both low- and high-tech defenses. Processes and procedures must also be in place to ensure compliance with security requirements. For example, Microsoft NT and XP come with many built-in security features; however, they are not activated when the operating systems are installed. It is incumbent upon the security administrator to enforce the enabling and use of security capabilities provided by the network.

Employees must do their part to safeguard and protect all company resources. For example, some people who use IM and wireless networking technologies are not aware of the potential security risk they pose. Employees have been known to use an AOL IM chat to communicate sensitive data to other employees without realizing that the data is being sent in pure, unencrypted ASCII text — a format that can fairly easily be read by unauthorized hackers. For example, Stifel Nicolaus, a brokerage firm in St. Louis, made headlines when it was discovered that its institutional brokers had been using AOL Instant Messenger clients, which were deployed without permission. The Director of IT encountered stiff opposition to an outright ban on the practice, so he was forced to quickly deploy a gateway from IMLogic to manage, log, and archive IM messages. [FONT02]

Hackers have also been known to eavesdrop on unsecured, unencrypted Internet communications over a wireless network. This is fairly easy to do using hot spot sniffers[15] — such as KISMET, NetStumbler, DStumbler, and Wellenreiter. Employees must take network security threats very seriously and should be provided with the tools to secure their computing environment. Many companies have adopted strict policies and have instituted employee awareness programs to guard against these types of potential security leaks.

Many corporations rely on some form of outside verification and validation of their IT and networking infrastructure. This helps to side-step political and internal pressures to circumvent security when there is also considerable pressure to implement new systems and networking quickly and cheaply.

The goal of network designers should be to seamlessly integrate security services between different network environments and technologies. To support employee demand for IM chats, for example, might mean implementing enterprise IM software. In the case of wireless networking, employees should be given PKI encryption or products such as Citrix Solutions for Workforce Mobility. If this is not feasible, the organization might need to provide (albeit slow) WAN connections and dial-up links for "road-warrior" employees to ensure the necessary degree of security in their communications. As suggested by these examples, the network environment and types of required access have significant impacts on the security approach the organization should employ.

A number of standards organizations are developing recommendations to help industry and government implement best practices with respect to security. The National Institute of Standards and Technology (NIST) and the National Security Agency (NSA) have established a program under the National Information Assurance Partnership (NIAP) known as the NIAP Common Criteria Evaluation and Validation Scheme for IT Security (CCEVS). It "is a partnership between the public and private sectors. This program is being implemented to help consumers select commercial off-the-shelf information technology (IT) products that meet their security requirements and to help manufacturers of those products gain acceptance in the global marketplace." [CCEV04] The CCEVS defines open, publicly available standards for *"Protection Profiles"* and *"Security Targets."* Protection Profiles specify consumer security requirements for a wide range of applications (i.e., E-commerce, etc.) and products (i.e., firewalls, etc.). *"Security Targets* are the security objectives of a specific product or system, known as the Target of Evaluation (TOE). The Target can conform to one or more Protection Profiles as part of its evaluation. A TOE is assessed using the Common Evaluation Methodology and assigned an Evaluation Assurance Level (EAL)." [NORT02] Seven EAL levels have been defined:

- *EAL-1:* relates a product's performance and conformance with published documentation and specifications.
- *EAL-2:* relates to an evaluation of a product's design, history, and testing.
- *EAL-3:* relates to an independent evaluation and verification of a product that is still in design phase to assess its potential vulnerabilities, environment controls, and configuration management.
- *EAL-4:* relates to a very in-depth evaluation of the target design and implementation.
- *EALs 5-7:* relate to even stricter levels of evaluation to assess the Target's ability to withstand attacks and misuse. They are used to critique products used in high-risk environments. In the United States, these evaluation levels can only be performed by the National Security Agency (NSA) for the U.S. Government.

The ISSA (Information Systems Security Association) is a nonprofit group of more than 100 volunteers and vertical industry representatives, including EDS, Dell Computer Corporation, Forrester Research Inc., Symantec and Washington Mutual, and others. ISSA's main goals are to "ensure the confidentiality, integrity and availability of information resources." [ISSA03] ISSA organizes international conferences, local chapter meetings, and seminars to promote awareness of security issues and to provide a worldwide forum for discussion. It supports a Web page for information dissemination to ISSA members.

ISSA has also announced that it is working on the completion of Generally Accepted Information Security Principles (GAISP). GAISP focuses on operational recommendations based on three sets of principles:

1. *Pervasive Principles,* which are designed for governance and executive-level management to help them develop an effective information security strategy.
2. *Broad Functional Principles,* which define "recommended tactics from a management perspective" based on the ISO 17799 security architecture.

3. *Detailed Principles,* which are still in progress. They are intended to provide very detailed recommendations on the "way to do everything from securing a firewall to physical security. For the first time, instead of having to meet an abstract goal, professionals will get rationales, examples, cross-references and detailed how-to instructions." [SCHW03]

The ISO 17799 standard has also been developed to help define an information security program. It provides a comprehensive checklist of security requirements and precautions, from which organizations can pick and choose what makes the most sense given their needs and particular situations. The ISO standard addresses the following: [WALS03]

- *Security policy.* This spells out an organization's expectations and commitment to security.
- *Security organization.* This defines the security management structure, responsibilities, and security breach response process.
- *Asset classification and control.* This provides a detailed accounting and inventory of the IT infrastructure and corresponding security requirements.
- *Personnel security.* This includes written security requirements and guidelines, employee background checks, and signed agreements to protect intellectual property and other company assets.
- *Physical and environmental security.* This deals with backup, recovery, and security as it relates to the IT networking infrastructure, physical facilities, and people.
- *Communications and operations management.* This addresses prevention and detection of security breaches.
- *Access control.* This involves establishing processes and procedures to properly restrict user access according to their security clearance.
- *Systems development and maintenance.* This establishes a development process that incorporates security considerations throughout the system life cycle.
- *Business continuity management.* This creates contingency plans to recover from catastrophic failures and disasters.
- *Compliance.* This ensures that all regulatory and legal requirements are met.

CASPR (Commonly Accepted Security Practices and Regulations) is another organization that offers advice and recommendations on security processes and procedures through the free dissemination of white papers and other security-related documents on its Web site. [CASP03]

The IT Governance Institute and the Information Systems Audit and Control Foundation sponsors the Control Objectives for Information and Related Technology (COBIT) project and is another source of advice for implementing security policies, processes, and procedures. The COBIT project is supervised by a Project Steering Committee comprised of "international representatives from industry, academia, government and the security and control profession." [CoBIT03] This organization offers free information downloads or, for a modest fee, one can purchase print and CD-ROM versions of its standards. These standards and recommendations are intended to be broadly applicable worlwide to "management, users, and IS audit, control and security practitioners." [CoBIT03] For example, the COBIT *Management Guidelines* discusses Maturity Models ("to help determine the stages and expectation levels of control and compare them against industry norms"), Critical Success Factors, Key Goal Indicators, and Key Performance Indicators. The complete set of standards also includes: [CoBIT03]

- *Executive Summary.* This summarizes key security concepts for senior management.
- *Framework.* This explains how to control delivery of IT information through 34 high-level control objectives. It also relates key evaluation criteria (effectiveness, efficiency, confidentiality, integrity, availability, compliance, and reliability) to the IT process.
- *Control Objectives.* This provides a basis for policy development by implementing 318 specific, detailed control objectives throughout 34 defined IT processes.
- *Audit Guidelines (ISACA members only).* This suggests activities to support 34 high-level IT control objectives and elucidates risks if the control objectives are not met.
- *Implementation Tool Set.* This provides case studies and presentations to help implement COBIT standards.

1.3.2.3 Managing Complexity

Perhaps the foremost challenge is the sheer multiplicity of options that must be considered when designing a network. Added to this is the fact that networks of today continue to grow in size, scope, and complexity. On top of this, the networking options available are in a constant state of flux. Change is the only constant in the world of high technology.

The key to managing complexity is to simplify wherever possible. Some of the major tactics that companies are using to reduce complexity in their networking infrastructure include:

- *Select protocols and equipment based on ease of use and manageability.* Many network and system products are not designed for easy management and control, and are difficult to install, upgrade, configure, or monitor. The difficulty only increases as equipment, protocol, and software incompatibilities are encountered. Unless there is a compelling reason to do otherwise, industry standard protocols and uniform implementations should be considered. Otherwise, it can be very difficult to manage the inevitable evolution that the network infrastructure will undergo to support new requirements and technologies.

- *Use plug-and-play components wherever possible.* The intent is to minimize the time and money spent customizing hardware and software to support the organizational requirements. Commodity components may reduce risk and improve performance and reliability over a custom solution.

- *Transition from proprietary to open standards.* This provides the organization with more options in future technology migrations. An example of where this is happening on a significant scale is the migration from legacy SNA networks to TCP/IP (this is discussed in more detail later in Chapter 2). It is important to pick the right open standard(s) because there is no guarantee that an open standard will catch on in the market and be supported in the future by a majority of vendors/providers, particularly if a new, improved open standard supplants it. It is also important to make sure that the quality, functionality, and robustness needed by the organization are supported by the standard. Open tools — such as GNU Emacs, a text and code editor — are very widely used. Others — such as IBM's Eclipse Platform for building integrated development environments (IDEs) — are supported by hundreds of tool vendors and provide robust development capabilities.

- *Use client/server and object-oriented application development methodologies to promote code reusability and modularity.* This allows thorough testing and customization, as appropriate, of subcomponents.

- *Integrate voice, data, and video on single IP platform.* An example of this is migration from separate PBX and data networks to a single VoIP data network to handle both voice and data traffic. Companies are using this approach to cut costs and to reduce the number of protocols and equipment that must be supported.

- *Automate data center and network management tasks.* Fortune 200 companies are increasingly looking at data center automation tools to reduce IT staffing costs and to improve efficiencies. An example of this is Opsware, Inc., a leading provider of data automation software. Opsware is designed to support cross-platform data center environments and a variety of management functions on servers and applications. This includes system provisioning, application provisioning, change and configuration management, asset and license tracking, patch installation, disaster recovery, and security updates. It consists of two major components: (1) Automation Platform, which provides a centralized mechanism to encapsulate business logic for managing a distributed infrastructure and its associated applications; and (2) Intelligent Software Modules (ISMs), which are specific products to manage provisioning, deployment, configuration, and modification of servers and software designed to work with standard equipment, and database and application software. [OPSW03]

■ *Utilize middleware to provide network and application control based on predefined business rules.* Middleware supports a broad range of application-specific, information-exchange, and network management functions. Middleware can be used to automate business operations by integrating front-end and back-end applications. For example, middleware can be used to allow the coexistence of new Web-based applications and legacy SNA. It is also used in E-commerce to link payment, accounting, and shipping systems with Internet-based, customer front-end applications. Middleware can also be used for server load balancing, network performance monitoring, and network recovery actions.

1.4 Similarities and Differences between Voice and Data Network Design and Planning

Traditionally, there have been clear distinctions and choices between voice and data networks. Selection of one or the other was determined by the type of traffic carried, cost and performance characteristics, and the types of equipment and services that could be supported by each type of network. However, since the 1990s and beyond, the distinctions between voice and data networks are increasingly clouded, as a variety of integrated voice and data alternatives have become available. New carrier, service, and technology options continue to proliferate. These options support not only voice and data, but also an increasingly diverse array of such applications as color facsimile, video conferencing and surveillance, high-definition television distribution, and LAN-to-LAN and MAN-to-MAN connectivity.

It is easy for an organization to install a "plain vanilla" voice network. To hook up to the public switched telephone network (PSTN), one has simply to call the telephone company (a.k.a. telco). This type of voice network provides interconnectivity between PSTN subscribers. In the early days prior to the advent of the computer, the PSTN could only support voice traffic. The introduction of computers and the ever-growing need for network connectivity led to the development of the modem (MOdulator/DEModulator). The modem is used to convert digital data signals from the computer into analog signals that can be carried on the telephone network, and vice versa. Thus, using a modem, two or more computers can send and receive data to and from each other over the PSTN. In addition to being easy to implement, this type of connectivity offers: [MULL97]

- High reliability
- Shared cost structure with large customer base
- Price stability, if calls are limited to the same local exchange
- Limited capital investment
- Few management and control responsibilities

Yet another option for implementing a voice network is to utilize a private branch exchange (PBX). There are two major types of PBXs: analog and digital. An analog PBX handles telephones directly and uses modems for data transmission. A digital PBX handles data transmission directly, and uses codecs[16] to connect telephones. Thus, both voice and data can be carried on a PBX with the right equipment. When compared to plain old telephone service (POTS) or Centrex service,[17] a PBX may be cost effective and may offer useful features not otherwise available. For example, some PBXs offer the capability to connect a direct high-speed line to a local area network, dynamic bandwidth allocation to support virtually any desired data rate, and connectivity to other networks, including packet switched networks. PBXs are usually private, meaning that the network is used exclusively by the organization paying for it. This is in contrast to the public telephone network, which at any one time can be shared by many.

A key feature of the PSTN and PBX systems just discussed is that they employ circuit switching. Circuit switching establishes a temporary, exclusive connection between channels. Because the communications channels are dedicated for the duration of the call, if all the channels or lines are busy, any additional incoming calls will be blocked (and one will hear a busy tone). When circuit switched networks are used to transmit data *exclusively,* they become increasingly less cost effective. One reason for this is that during a typical terminal-to-host data connection, the line is idle much of the time. A second reason is that circuit switching transmits data at a constant rate. Thus, both the sending and receiving equipment must be synchronized to transmit at the same rate. This is not always possible in a diverse computing environment with many types of computers, networks, and protocols.

Other emerging network technologies include digital subscriber lines, computer telephony, and integrated transmission of data and voice over TV cables.

Digital Subscriber Lines (xDSL) provide a means of mixing data, voice, and video transmissions over phone lines. DSL now has over 30 million subscribers worldwide. DSL technology is experiencing a slowdown in the United States; however, it is a very popular service overseas. For example, France Telecom has announced that it will provide the service to any town with a population of 5000 or more.

There are several different types of DSL from which to choose, each suited for different applications. DSL technologies run on existing copper phone lines and use special, sophisticated modulation techniques to increase transmission rates.

Asymmetric Digital Subscriber Line (ADSL) is the most publicized DSL scheme. It is commonly used to link branch offices and telecommuters in need of high-speed intranet and Internet access. The word "asymmetric" refers to the fact that ADSL allows more bandwidth downstream (to the consumer) than upstream (from the consumer). Downstream, ADSL supports speeds of 1.5 to 8 Mbps, depending on the line quality, distance, and wire gauge. Upstream rates range between 16 and 640 Kbps, depending on line quality, distance, and wire gauge. ADSL can move data up to 18,000 feet at T1 rates using standard 24-gauge wire. At distances of 12,000 feet or less, the maximum transmission speed is 8 Mbps.

ADSL offers other benefits. First, ADSL equipment is being installed at the carrier's central office to offload overburdened voice switches by moving data traffic from the PSTN onto data networks. This is increasingly important as Internet usage continues to rise. Second, the ADSL power supply is carried over the copper wire, so ADSL works even when the local power fails. This is an advantage over ISDN, which requires a local power supply and a separate phone line for comparable service guarantees. A third benefit is that ADSL furnishes three information channels — two for data and one for voice. Thus, data performance is not impacted by voice calls. The carrier rollout plans for this service are very aggressive, and it is expected to be widely available by the end of this decade.

Rate-Adaptive Digital Subscriber Line (RADSL) provides the same transmission rates as ADSL. However, as its name suggests, RADSL adjusts its transmission speed according to the length and quality of the local line. The connection speed is established when the line synchs up or is set by a signal from the central office. RADSL applications are similar to those using ADSL and include Internet, intranets, video-on-demand, database access, remote LAN access, and lifeline phone services.

High-bit-rate Digital Subscriber Line (HDSL) technology is symmetric, meaning that it furnishes the same amount of bandwidth both upstream and downstream. The most mature of the xDSL approaches, HDSL has already been implemented in the telco feeder plants — which provide the lines that extend from the central office to remote nodes — and in campus environments. Because of its speed — T1 over two twisted pairs of wiring and E1 over three twisted pairs of wiring — telcos commonly deploy HDSL as an alternative to T1/E1 with repeaters. At 15,000 feet, the operating distance of HDSL is shorter than that of ADSL. However, carriers can install signal repeaters to extend HDSL's useful range (typically by 3000 to 4000 feet). HDSL's reliance on two or three wire-pairs makes it ideal for connecting PBXs, inter-exchange carrier POPs (point of presence), Internet servers, and campus networks. In addition, carriers are beginning to offer HDSL as a way to carry digital traffic within the local loop, between two telco central offices and customer premise equipment. HDSL's symmetry makes this an attractive option for high-bandwidth services like multimedia, but its availability is still very limited.

Single-line Digital Subscriber Line (SDSL) is essentially the same as HDSL, with two exceptions. SDSL uses a single wire-pair and has a maximum operating range of 10,000 feet. Because it is symmetric and needs only one twisted pair, SDSL is suitable for applications such as video teleconferencing or collaborative computing, which require identical downstream and upstream speeds. The standards for SDSL are still under development.

Very high-bit rate Digital Subscriber Line (VDSL) is the fastest DSL technology. It delivers downstream rates between 13 and 52 Mbps and upstream rates between 1.5 and 2.3 Mbps over a single wire-pair. The maximum operating distance, however, is only 000 to 4000 feet. In addition to supporting the same applications as ADSL, VDSL, with its additional bandwidth, can potentially enable carriers to deliver high-definition television (HDTV).

A number of critical issues must be resolved before DSL technologies achieve widespread deployment. First, the standards are still under development and must be firmly established. Issues that must be resolved relate to interoperability, security, eliminating interference with ham radio signals, and lowering the power systems requirements from the present 8 to 12 Watts down to 2 to 3 Watts. A nontechnical but nonetheless important factor in the success of DSL technology is how well the carriers can translate the successes they have realized in their technology trials to success in market trials and commercial deployment.

Packet switching was introduced to overcome the shortcomings of circuit switching when transmitting data exclusively. In packet switching, the data is broken into small bits or "packets" before being transmitted. In addition to containing a small portion of the data, each packet contains a source and destination address for routing purposes. The first routing protocols developed to support packet switching are defined as X.25. Although X.25 packet switching protocols support reliable data delivery on both analog and digital lines, the introduction of packet switching led to the development of digital networks dedicated exclusively to data — that is, packet switched data networks (PSDNs).

There are many choices and options for designing a data network. One choice is whether the data network is to be public or private. In a private network, the organization owns or leases the network facilities (or communications links) from a communications carrier.[18] In a public network, the organization shares facilities provided by a telco or communications carrier with other users, through a switched service that becomes active at the time it is used. As is the case when deciding between a PBX and a PSTN, cost, security, and performance may be factors in swaying the decision to one type of network or the other.

Telephone companies throughout the world are continually adding and upgrading their services to drive demand. An almost overwhelming array of choices for networking is available. Standard telephone service — known as narrowband communication services (i.e., sub-voice-grade pathways with a speed range of 100 to 200 bps) — is still available. However, an alternative to standard telephone service offered by the telcos is Integrated Services Digital Network (ISDN).

ISDN uses digital technology to transmit data (including packet switched data), voice, video, facsimile, and image transmissions through standard twisted-pair telephone wire. ISDN has become widely available throughout the United States; however, there are still impediments to its adoption that have not been totally resolved. One obstacle is that there are often delays of several weeks or months before ISDN lines can be installed. Another potential obstacle is that some organizations have bandwidth requirements that exceed the capacity of ISDN (ISDN link speed can vary from 64 Kbps[19] to 1.544 Mbps[20]). Cost may be another factor to discourage use of ISDN, which is more expensive than regular phone service. In addition, the costs to upgrade to ISDN may be prohibitive. For example, ISDN does not work with standard modems operating at speeds of 28.8 Kbps. Therefore, 28.8-Kbps modem users are forced to upgrade their equipment to take advantage of ISDN. When considering ISDN, note that complex equipment set-up procedures are sometimes needed to accurately determine when to place an ISDN call. For example, when using ISDN to back up router traffic, proper triggers must be programmed into the router to ensure data restoration is initiated upon link failure rather than a routing protocol update to avoid unnecessary and expensive ISDN calls. These router and management updates can significantly add to the cost and difficulty of using ISDN.

Despite these obstacles, ISDN is increasingly being used to support telecommuting, remote access, and Internet access. As ISDN becomes more affordable, it will become especially attractive as a means to access the Internet. Banks are among the major users of ISDN, as well as small businesses installing LAN-to-LAN connections. ISDN lines can be more economical than WAN leased lines — which are typically T1 lines operating at 1.544 Mbps — when data transmission is intermittent, because with ISDN you only pay for the time you use. Dedicated lines are also not needed because ISDN allows one to make on-demand network connections. ISDN connections offer higher speed than modem connections (which are too slow for LAN-to-LAN communications) at a lower cost than that of private leased lines.

Broadband ISDN (BISDN) is another networking option. BISDN is an extension of ISDN that was developed by the telcos to allow very high rates of data transmission. With BISDN, channel speeds of 155 Mbps and 622 Mbps are available. BISDN is a technology designed to support switched, semi-permanent, and permanent broadband connectivity for point-to-point and point-to-multipoint applications.

Current packet switching technology is simply not adequate to support such applications as broadband video telephony, corporate videoconferencing, video surveillance, high-speed digital transmission of images, and TV distribution. In contrast, the deployment of BISDN relies on Synchronous Optical NETwork (SONET), a standard for extremely high data transmission rates using optical fiber. SONET supports transmission rates between OC-1 (51.8 Mbps) and OC-192 (9.6 Gbps[21]), and a variety of transmission signals, including voice data, video, and metropolitan area networking (MAN).

SONET also supports Asynchronous Transfer Mode (ATM) standards. ATM is yet another standard developed to provide high-speed (1.544 Mbps [T1] and higher) data transport for WANs. ATM is based on a cell switching technique that uses fixed-size 53-byte cells analogous to packets. The BISDN, SONET, and ATM standards were developed to work in conjunction with each other to support demanding data transport requirements.

Virtual private networks (VPNs) offer yet another option available through the telcos. A VPN is used to simulate a private network. Originally, VPNs were offered for voice traffic. However, such major players as AT&T, MCI, and others are offering VPNs with switched data services that can be packaged with voice services. VPN services often provide an economical alternative to private line services, particularly for those organizations operating international networks, because charges are generally based on usage. VPN can be an attractive choice for organizations that anticipate adding new sites to the network because it may be easier to modify the VPN than to modify a private line network. Features supported by VPNs include:

- Account codes, which can be used for internal accounting and billing purposes
- Authorization codes, which can be used to restrict access and calling privileges

- Call screening based on a number of criteria, such as geographic location, trunk group, etc.
- Network overflow control
- Alternative routing capabilities when the network becomes congested

Multinational corporations are the primary users of VPN services because VPNs offer the flexibility needed to expand and modify global networks. VPNs are particularly cost effective in providing connectivity to low-volume remote sites and to traveling employees. They also offer an effective alternative to a private meshed network, which for multinational companies can be extremely expensive.

In summary, there are a variety of choices for data and voice networks. Increasingly, hybrid solutions are available, including voice and data networks, private and public networks, central office telco services, and private PBX services. The line between voice and data networking is becoming increasingly blurred. Traditionally, voice networks supported voice and data (with the aid of a modem). However, most WAN implementations have been optimized for data transport. Over time, the capabilities of voice networks have increased, so that they now support a wide range of services and a wide variety of transmissions, including voice, data, video, and imaging. Switched public networks have become an attractive option to conventional private networks. However, there may be reasons — special requirements, security and control, reliability, backup needs — that a private line network solution is best.

There are many networking options available in the marketplace. However, the pros and cons of these options are not always readily apparent without careful analysis of the network requirements. Complicating the decision is that the costs of the network options are also difficult to compute. Service providers may charge based on usage, distance, type of service, bandwidth requirements, special deals, or some combination of these. Ultimately, the decision to use a public voice/data network or a private voice/data network must be based on the goals of the network and the business.

1.5 Major Trends and Technology Enablers

This section discusses major trends in the telecommunications industry that are shaping how networks are designed and implemented. A common theme is the extraordinary pace at which society, industry, and technology are changing and how difficult it is to maintain the appropriate trade-offs and boundaries between the respective special interests of each.

1.5.1 Societal Trends

There are several major trends in society that are influencing the adoption and design of networking infrastructures and technologies.

1.5.1.1 Computer Literacy

> "Literacy and learning are the foundation of democracy and development."

[WHIT02]

Digital computing has become integral to the fabric of society. This is increasingly reflected in home, educational, legislative, governmental, and other initiatives all over the globe. More people than ever are using computers in every aspect of their lives. This trend is chronicled in "A Nation Online: How Americans Are Expanding Their Use of the Internet," a comprehensive study of Internet usage based on a survey of over 137,000 people and published by the United States Department of Commerce National Telecommunications and Information Administration (NTIA) and the Economics and Statistics Administration. This study identified major ways in which the Internet is used and reflects the pervasive influence of computers on American life: [NTIA02]

- Internet use in the United States is growing by over two million new users per month, and is increasing across all income, education, age, race, ethnic, and gender classifications. However, families with annual incomes over $75,000 were much more likely to be online than the very poorest, by a margin of about four to one.
- More than half of the U.S. population uses the Internet, with children and teenagers using it more than any other group.
- Approximately 80 percent of Americans access the Internet through dial-up connections. However, broadband usage is growing rapidly.
- Using a computer and the Internet at work is associated with a much higher rate of home computer ownership and Internet use.

Consistent with these findings, a recent Harris Interactive poll of more than 2000 computer users aged 13 and above discovered that computers have become the most important electronic medium in American homes, particularly for entertainment. Among their findings: "Almost half of those surveyed said their computer is more important than their television (43 percent), while nearly two thirds said their computer outranks their CD player (63 percent), stereo (61 percent) or DVD player (59 percent)." [SELL03]

Many people are concerned about the "digital divide" between those who have access to computing resources and those who do not. Technical literacy is deemed an essential skill, much like reading, writing, and arithmetic, to function productively in society and to compete in the job market. The former Speaker of the House of Representatives, Newt Gingrich, has even suggested that every child in America should be given a laptop computer. [WARS03] To mitigate against socio-economic imbalances caused by computer illiteracy or lack of access to computer resources, Melinda and Bill Gates have established a foundation that partners with public libraries. Through this foundation, they have pledged $200 million to provide access to computers, software, and the Internet to poor and rural communities in the United States, with Microsoft pledging a matching $200 million in software grants. In other countries, similar programs have been implemented, although not on the same scale. For example, the municipal government of New Delhi, in collaboration with the (Indian) National Institute of Information Technology, has embarked on an experimental program to install outdoor kiosks with computer stations and Internet access in rural areas. The service is free and available to all; and children, in particular, are encouraged to use it. [WARS03] Although the outcome of these programs is yet to be determined, it is clear that technology is reshaping not only how information is disseminated, but also the very dynamics of society.

1.5.1.2 Security

This subsection discusses the major trends in security and risks to businesses and individuals arising from an inadequately secured network, as well as the strategies, resources, and free sources of information to help safeguard against these risks. There is also discussion of the challenges in balancing free speech, the need for privacy, and individual rights against enterprise and governmental efforts to protect against criminals, terrorists, and miscreants.

Enterprises — particularly in highly regulated industries — have long been aware of the potential for security breaches in their networking infrastructure and the impossibility of perfect protection against all perils, even if they do not admit it publicly. However, the general public is only recently waking up to the pervasiveness of security threats and the need to adequately secure their computer environment. The terrorist attacks of September 11, 2001, against the United States were a catalyst for focusing society's consciousness on this matter and have spurred massive government intervention to address the need for better national, enterprise, and personal security.

In its eighth annual "2003 Computer Crime and Security Survey," the Computer Security Institute (CSI) and the San Francisco Federal Bureau of Investigation's (FBI) Computer Intrusion Squad investigated security break-ins within corporations, government agencies, financial institutions, medical institutions, and universities. Key findings reported in their survey include: [CSIF03]

- 75 percent of the respondents reported financial losses due to computer breaches. Of these, 47 percent quantified their losses according to category, reporting that theft of proprietary information resulted in the highest losses, followed by denial of service.
- 78 percent reported that their Internet connection was the most frequent point of attack, by a margin of more than two-to-one over those reporting their internal systems as the most frequent point of computer security attack. This trend has held steady over the past half-dozen years.
- 30 percent reported security break-ins to law enforcement. This number is up since 1996, when only 16 percent acknowledged reporting security incidents to law enforcement.

The most prevalent attacks and abuses reported in the survey include: [CSIF03]

- 82 percent detected computer viruses.
- 80 percent detected employee abuse of network access privileges (e.g., downloading pornography or pirated software, or inappropriate use of e-mail systems).
- 59 percent reported laptop thefts.
- 42 percent detected denial-of-service (DoS) attacks.

The trend toward rising and increasingly sophisticated unauthorized intruder attacks on computer networks, particularly those exposed to Internet traffic, is further validated by CERT®. The CERT Coordination Center Web site (www.cert.org) reports a number of disturbing trends on the rise, including:

- Use of automated tools to scan and propagate attacks on a mass scale
- More evasive attack strategies to avoid detection, which might involve:
  - "Anti-forensics" techniques to hide that an attack has occurred
  - "Dynamic behavior" to intelligently modify the sequence of the attack steps to foil detection

- – "Modularity" to allow rapid change and modification of the attack components to make it more difficult to disable the intruder
- ■ Increased exploitation of firewall vulnerabilities (using ActiveX controls, IPP, etc.)
- ■ Increased threat of DoS, worms,[22] and Internet Domain Name System (DNS) attacks

As the results of these surveys illustrate, there are numerous threats to the security of a network, along with significant financial ramifications for the enterprise. It takes skill, imagination, common sense, and constant vigilance to ward off potentially serious network intrusions. One of the first steps in developing a plan to secure the network is to recognize the types of security breaches that might occur. For example, this includes unauthorized transmission, interception, or alteration of data on the network.

It is not difficult to see that there are many vulnerable points in any network. Just a few potential vulnerable points would include:

- ■ *Sniffers.* These devices can be attached to a network to intercept transmissions and to monitor and analyze network traffic. The operation of a sniffer is transparent to network users, who are unaware that the data they send and receive is being intercepted. Network managers have long used sniffers to detect congestion and manage traffic flow on a network. However, in the wrong hands, sniffers can be used to read sensitive data, including the source and destination addresses of network traffic, which can be used to compromise network security. There are analogous devices to snoop on wireless networks. For example, Kismet is an 802.11 wireless network sniffer that can be used to identify wireless networks within a specific locale and to collect raw packets of data in transmission. Kismet is available as a free software download, and is one of many such wireless sniffers. Encryption and wireless firewalls are two ways to thwart this type of security invasion.
- ■ *LAN network interface cards (NICs) and workstations.* These can be modified to intercept transmissions that are not intended for them.
- ■ *Unattended workstations that are connected to the network.* These can be accessed by unauthorized people. This includes people who walk up to a computer that has been left turned on and start to use it without permission. It also includes "remote" access by hackers using a broadband or wireless connection to use someone else's computer or network device.

■ *Network management and monitoring equipment.* These can be used to access sensitive information on the network.
■ *Leased lines.* These can be tapped ("Find a circuit number beginning with FDDC or FDDN, and all you need is a modem and a recording device to steal information." [DATA8_96]

One strategy to deal with network security is to physically secure sensitive access points to the network and to limit access to authorized users. This limits the access that outsiders have to the network; however, it does not eliminate the threat from insiders within the organization. The ways of compromising a network's security are too numerous to fully enumerate, and new techniques to thwart network security are continually being invented. A variety of strategies must be used in even a basic network security plan to assure that an adequate level of security is in place to ward off the most likely forms of attack.

This is not easy, and sometimes it is made more difficult by the network itself. For example, LANs have certain features that compound the difficulty of securing them. First, by its very design, a LAN can transmit to any device on the network, and thus any device on the network can potentially be used to intercept information. Second, the addressing method used by LANs is not secure, and it is fairly easy for a knowledgeable culprit to modify a LAN transmission so it appears to be coming from someplace else. Thus, the opportunities for unauthorized interception, modification of data, and masquerading are compounded in a LAN environment.

In LAN and WAN environments, one way security can be improved is to determine from where the potential threats are most likely to come. If internal traffic is perceived as trustworthy, while the outside traffic is not, then secure routers or bridges can be used as firewalls between the two. It should be noted that an IP-based network firewall does not lock out intruding computer viruses, because it does not actually inspect the contents of the data. This deficiency has led to the development and employment of application-layer firewalls that inspect data streams for malicious content. If internal traffic is perceived to be potentially untrustworthy, then other security measures must be implemented at each end terminal. This can be accomplished using data encryption schemes, which encode the data in such a way that it can only be decoded by an authorized person (or a very clever, unauthorized person using lots of computing power to decrypt the message). As companies migrate from private to public infrastructures, including the Internet, more sophisticated security measures are often needed.

It should be emphasized that many security violations are not the result of high-tech break-ins. More likely, they are the result of "low-tech" opportunistic events, such as finding a system password taped to a monitor; or they are the result of unintentional error(s). Common security threats often rely on the naiveté of computer users. For example, Richard Smith, a computer sleuth, uncovered a scheme involving about a thousand hijacked personal computers that were being used to host pornographic Web sites. He discovered the scheme after receiving an e-mail purporting to be from PayPal.com. The e-mail asked him to verify his password and other personal data on a Web site that was set up to look like PayPal.com, but it was not. The ruse worked on so many that the hackers were able to continually move their Web sites to different machines they had taken over, thereby avoiding detection and capture. [SULL03b] Firewalls should be installed to protect against this type of intrusion, and people should be educated not to respond to unsolicited e-mails of this sort.

Sometimes, computer users have a false sense of security when dealing with an "authorized" online third party, with unfortunate consequences. We present an example, which is by no means an isolated case, to illustrate the point. With over 15 million members, PayPal.com is the *de facto* Internet electronic payment system, and is owned by eBay.com. In the first quarter of 2003, eBay posted a record $5.32 billion in goods sold, and is widely regarded as the industry standard for online auctions. [KONR03] Buyers and sellers who have been cheated on eBay.com can be surprised to learn they do not have the same legal rights with PayPal.com that they have with a bank or credit card company, especially when it comes to disputing fraudulent transactions. The law firm of Jacoby and Myers brought a class-action lawsuit against PayPal.com for not adequately protecting and informing consumers of their rights, and for unilaterally freezing their accounts after they have reported being a victim of fraud. Consumers who make PayPal.com payments using Visa or MasterCard can directly dispute fraudulent claims with their credit card companies. Consumers who make payments with other credit cards may not be so lucky — particularly if these payments are treated as cash advances; and the credit card company is out of the picture if there are problems with the transaction. Buyers who are defrauded must adhere to PayPal.com's resolution process, which can be lengthy and does not guarantee full restitution. [SULL03] This story is intended to show that an online transaction does not *necessarily* afford the same protections as a "bricks and mortar" transaction, and this should not be taken for granted. *Caveat emptor!*

Because there are many potential ways and places where a network can be compromised, a security plan must be *comprehensive* if it is to be effective. Ideally, proactive measures should be used to prevent problems before they occur. These should be supplemented with reactive measures that involve system monitoring and automatic alert generation when security breaches are detected. Finally, passive measures can be used after a security breach to try to track and identify the problem source. An

Table 1.3   Security Threats and Recommended Precautions

Security Threat	Recommended Security Precaution
Computer virus	Install anti-virus protection and keep it up-to-date. Automatic update services are the most convenient and reliable way to ensure protection is current.
Trojan Horse	Do not open or execute unsolicited e-mail attachments. Install software only from trusted vendors and suppliers. Use security tools, such as TripWire®, to detect changes in system files. Do not grant administrator privileges to applications that do not require it.
Mobile code, which can be used to spy on the Web sites you have visited and to gather information or run malicious code on your computer	Disable Java, JavaScript, and ActiveX in your Web browser and e-mail unless you know and trust the source.
E-mail spoofing	Do not respond to unsolicited e-mail requesting sensitive information, even if it appears to be coming from a trusted source.

example of a passive measure is to collect data on network usage that can be reviewed as needed.

Table 1.3 presents some of the more common types of security risks and recommended precautions to guard against them.

Briefly summarizing, other basic WAN security measures include:

- Digital IDs, biometrics, and passwords
- Access control, at the file, program, and database level
- Encryption, for system log-in and file handling
- E-mail scanners
- Intrusion detection systems
- Tunneling using IP/Sec
- Network layer security for IP VPNs
- Sophisticated firewalls that operate at the application level
- Dummy and misleading points of entry to create "vulnerable" areas to trap hackers and to mislead them from breaching production systems

These security measures raise network costs and the complexity of the network maintenance and administration. Companies must carefully weigh these costs against potential risks to determine the most appropriate level of protection to take.

There are numerous Web sites providing free information on the latest security threats and ways to avoid them. The following Web sites are especially recommended:

- *http://www.fdic.gov/regulations/information/index.html*. This site is maintained by the Federal Deposit Insurance Corporation (FDIC) and provides information on federal regulatory examination procedures and requirements that specifically relate to data processing in the banking and financial industry.
- *http://www.ftc.gov/infosecurity*. This site is maintained by the Federal Trade Commission.
- *http://www.nipc.gov/warnings/computertips.htm*. This site is maintained by the National Infrastructure Protection Center.
- *http://www.sans.org/*. This site is maintained by the SANS (SysAdmin, Audit, Network, Security) Institute, a cooperative research and education organization. According to the Web site, "more than 156,000 security professionals, auditors, system administrators, and network administrators... share the lessons they are learning...."
- *http://www.cert.org/tech_tips/home_networks.html*. This site is maintained by The CERT® Coordination Center (CERT/CC) as part of a federally funded research and development center operated by Carnegie Mellon University.
- *http://www.csrc.nist.gov*. This site is maintained by the Computer Security Division (CSD) within the NIST Information Technology Laboratory.
- *http://w3.org/security*. This Web site is supported by the World Wide Web Consortium.
- *http://microsoft.com/security*. This Web site is operated by Microsoft and provides patches to fix security holes, etc.
- *http://www.staysafeonline.info/*. This site is maintained by the National Cyber Security Alliance.
- *http://www.cio.com/research/security*. This site is maintained by CIO.com's Security and Privacy Research Center.
- *https://grc.com/x/ne.dll?bh0bkyd2*. This site is maintained by Shields UP!, and provides tools to check for open computer port sites that may be a source of vulnerability when using the Internet.

The rise in security break-ins coincides with the rise in collection of vast storehouses of distributed databases. Enterprises, the federal government, and companies selling data about companies/individuals are all actively engaged in collecting data on just about anything and everything. Yet, the Federal Trade Commission — in its report entitled *Privacy Online: Fair Information Practices in the Electronic Marketplace* — has found that only 20 percent of the Web sites it randomly sampled have "implemented four fair information practices: notice, choice, access, and security." [FTC03] This has created an unprecedented level of risk for information misuse and invasion of privacy. This is particularly so in the United States, where there is no explicit constitutional protection for individual privacy. Stories of identify theft, for example, are increasingly common as hackers use their resources to plunder databases and to capitalize on network security leaks. However, "Individuals rarely win lawsuits against private parties for privacy violations. In general, privacy claims against nongovernmental entities must be worded in terms of property loss... the Supreme Court has stated that any expectations of privacy must derive their legitimacy from laws governing real or personal property." [CALO03]

After the September 11, 2001, terrorist attacks against the United States, the U.S. Government passed new legislation and instituted a number of wide-ranging initiatives to guard against future assaults. The passage of the Patriot Act in 2001 shifted the balance of privacy rights toward the federal government, and granted law enforcement agencies the right to collect information from Internet service providers on the Internet activities of subscribers merely by asking for it. According to the American Civil Liberties Union (ACLU), this legislation gives the FBI the right to "access your most private medical records, your library records, and your student records... and can prevent anyone from telling you it was done." [AMER03]

The Terrorism Information Awareness (TIA) is another program with far-reaching implications on the privacy rights of Americans. The mission of TIA is to detect, classify, and identify foreign terrorists through the formation of an extensive counter-terrorism information system. This system will have *extensive* data-gathering and analytical capabilities, for the purpose of generating warnings in response to triggering events or thresholds. [DARP03] This system potentially could have access to the details of every transaction a person makes. The ACLU is challenging this program on the basis that the program does not adequately protect the privacy rights of the individual, nor does it provide sufficient safeguards against abuse. According to Jay Stanley (ACLU) in a statement to the Technology And Privacy Advisory Committee of the Department of Defense, "Like the atomic bomb, that tool will change the world just by being brought into existence, because it is so powerful that even if it is not used (or not used fully), the mere prospect of such use will change everyone's behavior. Given that reality, numerous questions arise." [STAN03] These questions relate to what data will be collected, and how it will be analyzed, distributed, used, and accessed. As pointed out by the DoD (Department of Defense) in its own analysis of TIA, there is "potential harm that could result from misuse" of the data it would aggregate. [DoD, TIA Report, p. 33] "But what it does not do is acknowledge the likelihood that ever-expanding *authorized* uses will destroy our privacy." [STAN03]

Other countries have differing views on what is acceptable with respect to privacy and free speech, thus complicating international trade and practice. Most of Europe expressly forbids the collection of telemarketing data and other personal information without the express permission of consumers. In another example, French courts ruled against Yahoo! several years ago, after Nazi memorabilia was offered for sale on the company's auction site. Yahoo! complied with the ruling and has stopped all online auctions of Nazi artifacts and hate-related materials. The Far East and Middle East also have views on acceptable behavior that differ from Western practices. As international E-commerce and Internet communication continues to grow, these issues will continue to be the subject of much debate and legislation. This is especially true because the developments in technology continue to outpace our ability to adapt safeguards that have served in the past.

1.5.1.3 Legislation and Regulation

This subsection presents a broad survey of recent legislation and regulations that impact the telecommunications sector. For comprehensive, ongoing information on legislation and regulations that are under consideration, are pending, or have been approved and enacted, the reader is referred to the following Web sites:

- Federal resources:
  - *http://www.whitehouse.gov/omb/egov/about_glance.htm.* This is the link to the President's official E-government initiatives Web site.
  - *http://www.regulations.gov.* This is a link to a federally sponsored portal designed to promote public access, review, and comment on federal documents pertaining to regulatory information published in the *Federal Register,* the U.S. Government's legal newspaper. The portal is a result of the President's E-Rulemaking Initiative.
  - *http://www.uschamber.com/capwiz/load.asp?p=/chamber/issues/ bills.* This is the link to the United States Chamber of Commerce Web site. This organization's membership consists of "more than

3 million businesses, nearly 3000 state and local chambers, 830 associations, and over 90 American Chambers of Commerce abroad."

– *http://www.naruc.org/AboutNaruc/whatis.shtml*. This is the link to the Web site for the National Association of Regulatory Utility Commissioners (NARUC), a nonprofit organization of governmental agencies involved in the regulation of telecommunications, energy, and water utilities and carriers in all 50 states, the District of Columbia, Puerto Rico, and the Virgin Islands. NARUC's mission, as mandated by federal law, is to serve the public interest by ensuring that services provided in these industries are fair, competitively priced, and nondiscriminatory.

■ State resources:

– *http://www.ncsl.org/programs/lis/cip/cipcomm/news0902.htm*. This is the Web site for the online newsletter of the National Conference of State Legislatures (NCSL). It provides electronic tools for research on state issues, and access to publications and other relevant information.

■ Industry and other resources:

– *http://www.tiaonline.org/policy/whatsnew.cfm*. This is the link to the Web site for the Telecommunications Industry Association (TIA), a major U.S. nonprofit trade association. The TIA works to foster the development of standards, and to promote a competitive market for manufacturers and suppliers of global communications and IT products and services. The TIA submits comments to the FCC and other federal agencies on telecommunications standards, laws, regulations, and policies.

– *http://www.usta.org/ustainfo.html*. This is the link to the Web site to the United States Telecom Association (USTA), a major trade association representing 1200 service providers and suppliers for the telecom industry. The USTA lobbies and supports the interests of the telecommunications industry in Washington, D.C.

– *http://www.congress.org/congressorg/issuesaction/billlist/?issue= 39*. This is the link to the Web site of Congress.org, and is a service of Capitol Advantage. Capitol Advantage's Internet advocacy tool — Capwiz™ — can be used to identify and communicate with elected representatives at all levels of government. It also provides electronic access to information on "issues and actions" in government.

The complexity and diversity of technology and its impacts are similarly mirrored in the laws and regulations governing the industry. The effects of the 1996 Telecommunications Act on telcos and service providers have been profound. Government continues to exert its influence in regulating new developments and in shaping the marketplace, in ways that are neither entirely predictable nor consistent. The September 11, 2001, terrorist attacks on the United States have led to a rash of new legislation, at all levels of government, focused on security. To encourage accountability and provide added incentive to corporations to adhere to good network security practices, a new concept of "downstream liability" is emerging in computer law. If a company fails to properly secure its network, and hackers exploit this weakness, using it as a springboard to hack into yet other companies, the company may be held liable, particularly if the resulting damages are substantial. [BEHA97]

Other state and federal legislation and regulations have focused on consumer and child protection. The states, as expected, have not been uniform in their response in protecting individual rights to privacy and security. For example, New Hampshire could become the first state to provide legal protection for people who tap into and eavesdrop on wireless networks. House Bill 495 has been introduced; it would effectively legalize "war driving — motoring through inhabited areas while scanning for open wireless access points." [MCWI03] In contrast, New York, Idaho, Pennsylvania, Colorado, and other states have made it a crime to knowingly access any computer network without authorization and have stiffened the penalties for doing so. In other developments, states have been setting the stage for increased sales and use tax collection on Internet sales. There have also been a number of state initiatives to support cyber-schools. Cyber-schools provide Internet services to augment home-based schooling. There is a lot of debate on how to ensure funding and proper oversight of cyber-schools, and legislation in this area is contentious. Other legislative efforts have been directed toward expanding "911" services in rural areas and in cellular phones. Legislation also continues on behalf of people with disabilities to ensure services are offered to support their special needs. Other legislative and regulatory highlights include:

- ■ *Terrorism.* Key legislation and regulations in this area include:
  - – *USA Patriot Act:* the Uniting and Strengthening America by Providing Appropriate Tools Required to Intercept and Obstruct Terrorism Act of 2001. This legislation is wide sweeping and greatly extends the legal use of electronic surveillance. It authorizes, for example, the use of "roving wiretaps" that follow a person under surveillance wherever he or she might go. It also extends allowable surveillance to include e-mail and URLs visited on the Internet.

– *Sound Practices to Strengthen the Resilience of the U.S. Financial System publication by the Board of Governors of the Federal Reserve System (Docket No. R-1128), the Office of the Comptroller of the Currency (Docket No. 02-13), and the Securities and Exchange Commission (Release No. 34-46432; File No. S7-32-02).* The U.S. financial market and its participants are highly interdependent, and thus the failure of one can have far-ranging impacts on the market as a whole. The purpose of this white paper is to advise financial institutions on sound business practices they should employ, in light of the events of September 11, 2001. Four critical business practices are defined: *(1) Identify clearing and settlement activities supporting critical financial markets. (2) Ascertain appropriate recovery objectives for processes and applications identified in the previous step (1). (3) Maintain geographically dispersed resources so as to meet recovery objectives.* This is not an easy business practice to implement because financial institutions are based on synchronous operations. For example, as a trade is processed (e.g., for stocks, bonds, etc.), it updates production database(s), while other transactions may be held in queue as needed to preserve database integrity and concurrency. Existing technology only allows synchronous remote copy operations (which can effectively shadow current operations to support near instantaneous backup) within a distance of about 200 kilometers, or about 125 miles. The original recommendations proposed moving remote backup facilities as much as 300 miles from New York City. Because most planned backup facilities were only within about a 30-mile radius of New York City, this proposal generated considerable controversy within the financial community. It has since been relaxed to tolerate asynchronous remote copy operations as an alternative, because this can occur at virtually any distance. If communications are asynchronous, the production and the backup databases may not match, due to time delays in processing. The potential loss of financial transactions worth millions of dollars, due to a major network failure, may well exceed manyfold the cost of the network. Even if a rebate can be obtained from carrier(s) for network outages, this amount will likely be dwarfed by the losses sustained by a brokerage or other trading institution. Asynchronous backups do not entirely protect against loss of data in high-speed, high-volume financial applications. Therefore, they are not a complete solution to the problem of ensuring continuity of operation in the financial markets in the event of a major catastrophe. Some institutions are planning or implementing a bunker data center within about 100 miles of their main data center based on synchronous communications, with asynchronous connectivity to a secondary data center at the proposed 250- to 300-mile limit. For obvious reasons, the location of bunker data centers is a closely guarded secret and their operation is unlikely to be outsourced. The Sound Practices paper recommends that backup sites be served by an entirely different infrastructure (i.e., transportation, telecommunications, water supply and electric power facilities) from the primary site(s). It further recommends that recovery operations be independent of and unaffected by the evacuation or inaccessibility of staff at the primary site. For practical reasons, at this distance, separate staffing and infrastructure are likely a necessity. This is potentially an area that institutions may wish to outsource to service providers. Service providers can accommodate a range of solutions — from full to partial support of the backup requirements. Satisfaction of this third requirement has raised vocal protests from the financial community, which has suffered serious financial pressures and market downturns and is reluctant to take on significant added costs to their operations. *(4) Test backup and recovery processes and procedures routinely and thoroughly.* This should encompass inclusive tests of the connectivity between primary, backup, and other core clearing and settlement sites. The goal is to recover from and resume clearing and settlement activities on the same business day and within two hours after a failure or disruption. This latter requirement represents a significant challenge to the financial industry. The full text of this inter-agency white paper can be found at http://www.sec.gov/news/studies/34-47638.htm.

- *FISMA:* the Federal Information Security Management Act. This act mandates that federal agencies secure their respective IT infrastructures by identifying potential risks and implementing appropriate precautions, processes, and procedures. It requires adherence to standards developed by the National Institute of Standards and Technology (NIST). These standards set guidelines and thresholds for IT security relating to security controls, data integrity, access control, and auditing. The act requires the Office of Management and Budget to enforce mandatory compliance with these standards.

■ *Privacy:* Key legislation and regulations in this area include:
- *Health Insurance Portability and Accountability Act (HIPAA) of 1996.* This act mandates that the Department of Health and

Human Services (HHS) develop a unique health identifier for each individual, employer, health plan, and provider and establish and put forth guidelines for protecting the confidentiality of personal medical information. Covered entities (with the exception of small health plans) must comply by April 20, 2005, and small health plans by April 20, 2006. The final HHS ruling specifically states that technology is moving faster than the regulators and therefore they have eschewed technology-specific directives in their recommendations. The text of the *Implementation Specifications* from the HHS final ruling are available in a 289-page PDF document, which can be downloaded at: http://www.cms.hhs.gov/regulations/hipaa/cms0003-5/0049f-econ-ofr-2-12-03.pdf. In the past, the healthcare industry has been very secretive about releasing information to the patient, although it is their own personal data. The government is forcing the industry to change and to make this information accessible online. Many physicians still write medical notes using paper and pencil and the records are not transcribed, so this necessitates a major change in the way many practitioners handle record-keeping. This legislation considers IM chats, collaborative processing, and other correspondence part of a person's medical records and requires that it also be made available online for the patient review. This requirement is generating a huge demand for SANs and NAS devices in the healthcare industry. In general, insurance and healthcare companies have made conservative investments in their IT infrastructures. Most of their innovative use of technology has been in medical diagnostics and testing (i.e., simulation and modeling, high-tech lasers for precise surgery, CAT scans, MRIs, etc.). The industry must scale a sizable learning curve in using technology to support business processes. The financial industry is ahead of the healthcare industry in this area, having used E-commerce, EDI, and other advanced communications to support demanding applications for many years.

– *The Financial Modernization Act of 1999.* This is also known as the Gramm-Leach-Bliley Act or GLB Act, and requires financial institutions to protect and secure the personal information of their customers. As part of its implementation of the GLB Act, the Federal Trade Commission has issued the Safeguards Rule, which is available online at www.ftc.gov/privacy/glbact. Institutions that must adhere to the regulations are listed in Section 313.3(k) of the FTC's Privacy Rule Page, and can also be found at this Web site. The regulations require financial institutions to develop a written security plan, and when implementing the plan, to consider employee management and training, information systems, and management of system failures. Additional information on these regulations can be found at http://www.ftc.gov/infosecurity.

– *Privacy infringement laws.* Tools for electronic surveillance and snooping are readily available in a variety of consumer devices. For example, new camera phones have hit the market; these are extremely small and inconspicuous, and can easily be used for voyeuristic pictures. Strategy Analytics predicts that 42 million camera cell phones will be sold worldwide in 2003, and the number will reach 218 million by 2008. Some health clubs in the United States have banned the use of these phones to prevent locker room photography. The courts are developing laws that will give individuals the right to sue for privacy infringement if these phones are misused. [HANL03]

■ *Telemarketing.* Key legislation and regulations in this area include:
– *"Do-Not-Call Implementation Act."* Signed into law on March 11, 2003, this act requires telemarketers to keep lists of customers requesting that they not receive telephone solicitations. Under penalty of law and heavy fines, telemarketers are prohibited from contacting consumers on the do-not-call list. Direct marketing and telemarketing associations have vehemently opposed this legislation, and a number of legal challenges are still pending. However, consumers have generally welcomed this legislation, and a majority of states have enacted similar do-not-call legislation. This type of legislation is expected to significantly decrease the call volume of telemarketers by about 30 to 50 percent. This, in turn, has potentially negative revenue impacts on telcos and service providers as call minutes and volumes are reduced. It also may also affect products and bundling of services telcos plan to offer to telemarketers, because service providers are already suffering from serious market downturns and increased competition. Consumers can add their names and telephone numbers to the do-not-call list by going to this Web site: http://www.donotcall.gov.

■ *Pornography.* Key legislation, regulations, and developments in this area include:
– *The U.S. Supreme Court has upheld Megan's Law.* This law allows states to publish the names and photographs of convicted sex offenders on the Internet. A majority of states already post sex offender registry information on the Internet.

- *Children's Online Privacy Protection Act (COPPA)*. This law requires commercial Web sites to obtain parental permission before collecting information from a site that a child might use.
- *Children's Internet Protection Act*. This law was upheld in June 2003 by the U.S. Supreme Court. It requires libraries receiving federal assistance for Internet access to use software that filters or blocks obscene and pornographic material.

■ *Identity theft*. Key developments in this area include:
- Stronger legislation is being passed, particularly at the state level, to protect against identity theft. This includes efforts to stop companies from using employee or customer Social Security numbers on identification cards, bank statements, and mailed documents. A number of states (including Illinois and Pennsylvania) are passing legislation to allow restitution, recovery of court costs, attorney fees, and damages to victims in cases of identity theft.

■ *Sales tax*. Key legislation in this area includes:
- *Streamlined sales and use tax agreement*. As of July 2003, 20 states have enacted legislation to conform to this agreement, which is aimed at simplifying and standardizing state sales and use tax collection. The agreement goes into effect when at least ten states have been certified that they are in compliance with its terms. This is a likely prelude to the enactment of legislation to collect sales tax on goods and services sold over the Internet. Currently, consumers in one state can purchase goods or services on the Internet in another state without paying local sales tax. According to a University of Tennessee study, $440 billion worth of revenue will be lost to states in Internet sales between 2001 and 2011 because of the moratorium on taxing of E-commerce purchases. [LEVI03] This represents a substantial revenue drain on states, and they are lobbying to change the situation. Target, Toys "R"Us, Wal-Mart, and other E-commerce merchants have recently announced that they will charge sales tax for online purchases at their Web sites, but only on items purchased in states that have a physical presence (such as a retail store, distribution center, or call center) in that state. This practice is consistent with a 1992 U.S. Supreme Court decision on states' rights to charge sales tax. [LEVI03]
- *Tax on digital goods and services in the European Union*. The European Union has begun to levy a new tax on digital goods and services imported by consumers. For the past few years, physical goods sold online in European Union nations have been subject to import taxes. Typically, these taxes are collected

by shippers at the time of delivery. But until recently, taxes have not been collected on software downloads, auction listings, or other Internet services. This is beginning to impact E-commerce sellers in the United States. For example, in response to new international requirements, Ebay.com will now calculate and remit taxes to the appropriate government each time an Ebay.com seller pays the auction site for a listing. Other sellers are following suit. [TEDE03b]

■ *Accounting industry reform.* Key legislation in this area includes:
  - *Sarbanes-Oxley Act.* This legislation requires the SEC to enjoin public companies to establish codes of ethics for senior management or, if not, to explain why. All annual reports for fiscal years ending on or after July 15, 2003, must include ethics codes. MCI, in an effort to reestablish credibility in light of massive accounting irregularities at WorldCom (as it was formerly known), has gone further by establishing an ethics office. The legislation has a number of other provisions, including prohibition against accounting firms providing non-audit services to clients they are auditing, corporate punishment for retaliation against Wall Street analysts who write negative market reports, whistleblower protection for employees reporting violations of securities laws, etc.

1.5.2 Service Provider Trends

This subsection discusses major impacts on the dynamics of service provider competition as well as trends in the products and services offered in the marketplace. Key highlights of this discussion include:

■ Industry consolidation and market collapse as unintended side effects of the 1996 Telecommunications Act.
■ Migration of enterprise demand from traditional, circuit-based, high-margin voice services toward low-margin, data and wireless services. This trend is fueled by VoIP and the convergence of telecommunications technologies.
■ Efforts by service providers to develop name-brand, bundled offerings and value-added services with improved margins.
■ Emergence of cable companies as a serious competitor to traditional voice and data telcos and other service providers.

In its early days, the telephone system operated as a government-sponsored monopoly. In many countries, this describes the current state of the telecommunications industry. In the United States, until fairly recently, voice calls comprised the predominant form of traffic, for both consumers and enterprises. Local calls were generally handled through a single RBOC (Regional Bell Operating Company), while long-distance transmission was generally handled through an interexchange carrier (IXC) or a local access and transport area (LATA).

In February 1996, the U.S. Congress passed the Telecommunications Act of 1996. This act was one of the most important pieces of legislation since the Communications Act of 1994. The primary intent of the legislation was to promote competition among various providers of communications services and to break up the AT&T monopoly of the phone system. Key provisions in this legislation included: [LEVI96]

- New obligations were mandated for local exchange carriers (LECs), requiring them to interconnect to competitors, unbundle their services, and establish wholesale rates so that competitors could resell LEC services to fill gaps in their own service offerings.
- States and municipalities were barred from restricting any provider offering interstate or intrastate communications services. It encouraged the removal of barriers that would discourage small businesses and entrepreneurs from offering services in the telecommunications market.
- All remaining restrictions in the 1982 Consent Decree that broke up the Bell System were lifted. This allowed the Bell Operating Companies (BOCs) to provide telephone service (e.g., private lines, local calling, etc.) outside their own regions, and within their own regions if certain conditions were met. It also allowed the BOCs to manufacture terminal and other networking equipment and to enter new markets that were previously forbidden (e.g., electronic publishing).
- Barriers separating the cable and the phone industries were removed, and provisions were made to encourage cable companies to enter the local exchange and local access markets.
- Section 502 of the act made it illegal to send indecent communications to minors, to send obscene material through communication devices, or to use a communications device to harass.

Although the legislation did promote competition, it was generally not in the manner intended. In 2002, the American telecommunications market collapsed,[23] with rippling effects throughout the economy. In part, this collapse was due to over-capacity in provider networks, lack of customer demand, and accounting scandals (particularly in the case of WorldCom, which admitted to over $9 billion in fraudulent accounting practices).

However, the effects of the 1996 Telecommunications Act also played a significant role.

The act included provisions that required local telephone companies to share portions of their networks — that is, to provide "open-access" — with competitors at regulated rates. This was designed to encourage smaller carriers to enter the marketplace. Many market entrants bought capacity from major providers (such as WorldCom) and resold it. This did not encourage market entrants to invest in new infrastructure based on sound business models, because capacity could be leased so cheaply from competitors. It did spur unsustainable competition, leading to market implosion in 2002.

As a result of this legislation, CLECs (competitive local exchange carriers)[24] were created to compete directly with existing local telephone companies (ILECs; incumbent local exchange carriers). MCI,[25] North American Telecom, and Level 3 Communications are examples of CLECs. The term ILEC (incumbent local exchange carrier)[26] is used to describe a telephone company that provided local service at the time the Telecommunications Act of 1996 was enacted, and includes the regional Bell Operating Companies (RBOCs).[27]

In theory, as a result of the 1996 Telecommunications Act, IXCs could also compete in LEC markets as long as they opened up their markets to ILECs, and vice versa. However, ILECs are heavily unionized and are not easy markets to enter. In a local exchange, one has to wire each and every house. A LEC area in a metropolitan city with a population of two million has more nodes than an entire ICX global network — more, in fact, than any IXC in existence. In contrast, IXCs can create worldwide networks with only 100 to 500 nodes. It is also fairly easy to obtain long-haul rights-of-way from power companies, railroads, energy companies, and gas lines to install fiber optic cable, because fiber optic cable is nonvolatile and photonic (and can safely be run near volatile gases, liquids, and electrical lines without fear of interference). This is how an IXC can quickly create markets in 20 to 30 cities just by buying bandwidth on preexisting network highways. To break this log jam and encourage competition in the LEC market, the government forced LECs to sell bandwidth to the house and rights-of-way to competitors, giving consumers the option of selecting their local, long distance, and cell phone service with a provider of their choice. The FCC ruling requiring LECs to open their infrastructure (including loops, switches, lines, etc.) to competitors at regulated, discount rates is referred to as UNE-P (Unbundled Network Element-Platform). In a debate that is likely to continue, LECs argue that UNE-P represents a subsidy to their competitors, while competitors argue that UNE-P was the LEC's price of admission into the long-distance market.

In actuality, the competition spawned by the 1996 Telecommunications Act was only in the cloud (i.e., the IXC market) and not in the local exchange market. LECs (local exchange carriers) effectively hold a monopoly position, with a huge installed customer base and a heavily depreciated, subsidized infrastructure. Because of the considerable investment and labor required to run facilities to each residence, LECs maintained their incumbent markets.

When IXCs, which are not unionized, were opened to competition, the resulting market turmoil put intense pressure on IXC wholesalers. They responded with price reductions. Many secondary IXC market players bought cheap bandwidth and sold it under their own name. However, in fact, it was provided by one of a few industry wholesalers. Wholesale providers sold bulk bandwidth to resellers and, competing against them, they also sold to retail markets. This resulted in a downward spiral that drove retail prices very close to wholesale. Wholesalers offer buffet-style service — providing lots of food with few cooks — to resellers. Retailers and resellers, who by now had razor-thin margins, had to provide full service to their big customers, at higher cost. Ultimately, large customers were risk averse and reluctant to go with small providers that did not have a sound track record. Instead, they would negotiate a very low price with a start-up, and would use that as a basis for negotiating terms with a large, full-service provider. This was not a sustainable situation for either wholesalers or retailers. It lasted as long as venture capitalists continued to infuse cash into IXCs as an investment vehicle, and was not based on goods sold. Eventually, many small start-ups that had bought up stores of wholesale capacity folded because of insufficient customer demand.

Putting the situation another way, pursuant to the 1996 Telecommunications Act, the IXCs followed the strategy: "First secure the land. Then cultivate the deals." They "bought" business without requiring healthy profit margins on goods and services sold. Much like the times of the Gold Rush, the emphasis was not on digging the gold, but on staking as many claims as possible, and even paying the person with a claim next to yours twice what they paid to buy even more. The thinking was, "Why worry? There is plenty of gold!" But after buying the hills of gold, IXCs discovered they were sitting with huge assets, heavy debt loads, and too few customers to buy the gold. Some IXCs wrote off their assets, while still others filed for bankruptcy. The tangible, real assets were bought back and reabsorbed by remaining players as the IXC market regained equilibrium. For the most part, in the aftermath of this market upheaval, the original, large service providers have reasserted their leadership positions, albeit in depressed market conditions.

As a result of this consolidation, there is no longer as much competition in the IXC market. There are still a few niche providers selling services on very, very thin margins to maintain market share; but on the whole, providers are in a somewhat better position to tell customers, "You are going to have to pay a fair price for a fair service." This promises to be a difficult adjustment period for customers, who are used to free services, signing bonuses, and waivers from contractual "Terms and Conditions" in exchange for their business. Some providers are demanding signed contracts or they will charge tariff rates. In this environment, it is likely that rates will go up (or at least not go down) as service providers attempt to regain healthier profit margins. This consolidation is analogous to the history of the railroad, which overbuilt in the early days of expansion to the West. Eventually, the market crashed, a few big players picked up the pieces, and only a few dominant railroad companies remained.

The incentive for local telcos (i.e., Verizon, SBC Communications, Qwest, and BellSouth)[28] to move into long distance has happened only recently, as they are seeing flat or negative growth in their own markets. Large companies are looking to reduce phone costs in a variety of ways. For example, to reduce costly 800 service and its roaming charges, companies are using local calling based RAS (Remote Access Service)[29]. By installing a RAS phonebook with local calling numbers on their computers, remote employees can dial in to a RAS server using a standard telephone line and a modem (or modem pool, ISDN card, X.25, PPTP, etc.). A RAS server performs authentication and maintains persistent connections until the session is terminated. This is illustrated in Figure 1.3 and Figure 1.4. Companies are also looking at VoIP, IP/PBXs,[30] and other technologies to reduce long-distance and local phone charges.

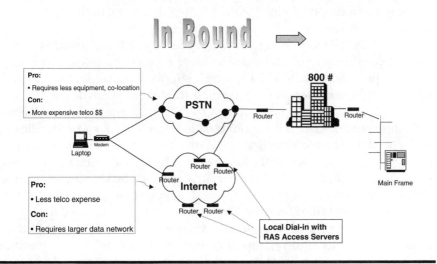

Figure 1.3   800 Service versus Dial-up Service and RAS

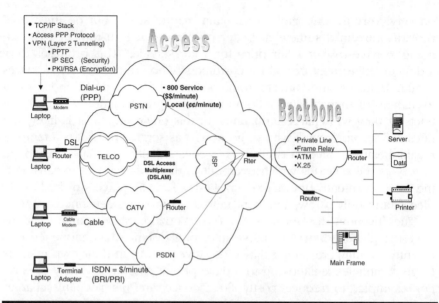

Figure 1.4   RAS Dial-up from Remote Site to Corporate WAN

Cellular phone use in residential markets, along with e-mail and cable, have reduced the demand for second phone lines. Using free minutes, one can make cellular phone calls all weekend without using a regular phone. According to Yankee Group, the average American wireless subscriber logged 490 minutes a month in the first quarter of 2002, as opposed to 480 minutes a month per person for wired voice calls. With the advent of local number portability (LNP),[31] this is likely to further encourage adoption of wireless cellular service. [ROSE03]

The RBOC's DSL offering for high-speed, broadband Internet access has lost considerable market share to cable companies and to other long-distance providers (such as MCI, which supports almost half the world's Internet and e-mail services). According to Strategy Analytics, at the end of 2002, only a third of the country's 18 million broadband users utilized DSL, and predict DSL will continue to lose ground as cable bundles its high-speed Internet connection with attractive consumer-oriented packages. Making matters worse for the RBOCs, while cable company operating margins on high-speed data connections are estimated to be as high as 60 percent (before interest, taxes, and amortization), the RBOCs lose money on every DSL subscriber. They will not break even until they reach a critical mass of subscribers so that economies of scale apply. Verizon estimates it will have to sign up three million subscribers before reaching the break-even point. [BLAC03] In the event that the RBOCs do capture revenue for a second phone line dedicated to Internet dial-up, it is a mixed blessing. The PSTN was sized and engineered to provide the capacity needed to support calls averaging five minutes on the busiest day of the year. When using the Internet, people can stay online for hours, tying up network capacity beyond its original design. These factors, and market declines in the telecom sector, have cut into RBOC revenues.

The RBOCs have been lobbying to expand and compete in long-distance markets, with some success. The Verizon merger with Bell Atlantic, NYNEX, and GTE was approved, which in one fell swoop consolidated markets from Washington, D.C., up to Maine. This represents about half of the United States' spending on telecommunications. Through a series of mergers, there are now four super Bells or RBOCs, down from seven in 1984, tightening their hold on local markets. The RBOCs are getting approved for long-distance rights at the same time companies like MCI and other big CLECs are moving into local service. Each is going into the other's markets. However, because overall IT growth is down, the market as a whole is stalled. Growth in the IT industry is likely to be in the single digits (approximately 4 percent or less) over the next year, with growth of no more than 6 to 7 percent over the next couple of years (through 2005). [LOHR03C]

Although local Bell companies are starting to get approval to provide long-distance service in multiple states, they do not yet have a national infrastructure to tie it together for data and Internet services. Typically, they buy these services from IXCs (such as MCI and AT&T). National IXCs — such as AT&T, MCI, and Sprint — are trying to gain ground in regional markets. After losing its merger deal with Comcast, AT&T remains an IXC and has not been as aggressive in pursuing regional markets. Sprint still has a heavy debt load, which hampers its ability for market expansion. MCI, once it emerges from bankruptcy, appears well positioned to support demand for integrated voice, data, and Internet services.

In summary, both local and long-distance service providers are entering each other's markets; but until user demand increases, this will not happen with the frenzy of the 1990s. A measure has been introduced in Congress — The Broadband Regulatory Parity Act of 2002 — to require regulatory fairness in the treatment of broadband service providers. The intent is to establish a more level playing field for providers and to redress past market imbalances caused by the Telecommunications Act of 1996. With the FCC ruling in 2002 that the Bells will not have to share installed fiber with competitors, as they are compelled to do with copper phone lines, the definition of fair competition continues to evolve along an erratic path.

Most of the legislative attention has focused on the LECs and the IXCs, and how to foster competition within and between their respective markets. Cable companies have not received nearly as much attention. However, this is likely to change as cable companies are gaining market share and introducing lucrative and attractive new services. Like CLECs, cable operators service the home market, with a large base of over 73 million homes. [MOWR02] Their fiber optic infrastructure has enormous bandwidth capacity. This fact has not been lost on consumers who are using cable and cable modems to access the Internet. Some analysts estimate that this market will grow to 40 million subscribers by 2007. [BATH03] The Yankee Research Group is also predicting that the corresponding growth in DSL subscribers — the RBOC's equivalent broadband service — will be less than half as much. [MOWR02] In addition to Internet access, cable technology is also well suited to support other popular value-added services, such as video-on-demand and personal video recorders (PVRs). According to a survey by the Leichtman Research Group, 30 percent of all Americans spend $15 or more on video or DVD rentals each month; and of these, approximately 66 percent would use video-on-demand more frequently if it were available. The survey concluded by saying that "As more cable companies offer PVR functions in their set-top boxes, satellite companies will lose an important competitive differentiation." [MOWR02]

Cable's role has largely been limited to one-way data transmission. Cable has been successful in providing broadband Internet access; but in doing so, it only provides physical access. It is the ISP (Internet service provider) that provides the intelligence for managing the communications. If cable television service fails or is spotty, it has not been perceived with the same urgency as a failure with the telephone service. As evidence, consider the fact that the PSTN was designed so that even in a power failure the telephone still works. Given its history, infrastructure, and technology, the cable industry has only very recently been in a position to offer QoS (Quality of Service), SLAs (service level agreements), and other service guarantees needed to support two-way, demanding voice and data requirements. Even so, cable operators are neophytes in providing the same level of customer service that the public has come to expect from RBOCs. With less than two million telephony customers, cable has not been a significant player in the voice market. [BATH03]

However, some cable companies have ambitious plans to move into voice communications. For example, both Cox Communications and AT&T Broadband are now offering voice-over-cable service. They are "planning to expand their phone operations in 2004 using Internet technology that's cheaper and packed with features like inexpensive second and third phone lines. At the current pace, the cable companies will probably have 30 percent of the phone market over the next decade…" [ROSE03]

In March 1998, a new standard — DOCSIS (Data Over Cable Service Interface Specifications) 1.0 — was ratified by the International Telecommunications Union (ITU-TS). This standard is very important to the cable industry as it provides a basis for global deployment of advanced multimedia services. This protocol provides bi-directional, interactive transmission between a cable company and a personal computer or television set. It supports downstream user transmission rates up to 27 Mbps (however, because some cable companies use T1 connections to the Internet or share bandwidth with other users, this lowers the effective transmission rate accordingly) and upstream transmission rates in the range of 500 Kbps to 2.5 Mbps. A number of extensions to the standard have been announced, including DOCSIS 1.1 and DOCSIS 2.0. Improvements provided by DOCSIS 1.1 include: (1) doubling the upstream transmission speed of DOCSIS 1.0; (2) improved security; (3) new services (such as tiered data services, low latency services for voice, and multimedia service for streaming video, interactive games, etc.); and (4) enhanced QoS features. DOCSIS 2.0 provides additional QoS capabilities needed to support IP telephony and multimedia applications. It also provides a six-fold improvement in the upstream transmission speed over DOCSIS 1.0. DOCSIS versions are backward compatible. A cable modem can be upgraded from older to newer versions of DOCSIS with a program modification to its EEPROM[32] memory. A new version of the standard — eDOCSIS — has also been developed to allow embedding DOCSIS in other devices, such as personal computers, set-top boxes, television receivers, and PacketCable MTAs (multimedia adapters).

PacketCable is an IP service platform initiative led by CableLabs that builds upon DOCSIS 1.1. It is designed to support interoperable device interfaces for real-time multimedia services (such as interactive game, streaming video, video telephony, voice-mail, e-mail, etc.) over cable utilizing a managed IP backbone. It provides interconnection between a cable network and the PSTN with QoS guarantees. A number of versions of this standard have been released. The PacketCable 1.1 specification provides lifeline service functionality. PacketCable 1.2, completed in 2000, uses the Session Initiation Protocol (SIP) to interconnect local cable zones. It is designed to allow cable Multiple System Operators (MSOs) to exchange voice traffic on their respective IP networks without handing calls off to the PSTN, thereby avoiding ILEC access charges. PacketCable 1.3 supports stand-alone client devices (i.e., MTAs). MTAs provide voice compression, packetization, security, and call signaling on telephone handsets and fax machines through an RJ-11 connector. The first MTA products were certified for PacketCable 1.0 compliance in December 2002. [CABL03] CableLabs is also developing a standard — CableHome — as a further extension to DOCSIS and PacketCable. The purpose of this standard is to enable remote configuration and management of CableHome-compliant devices, such as residential gateways, bridges, and client adapters. It is designed for both residential and business use. [KIST02]

Until recently, telcos and cable operators serviced different market sectors with distinct products and services. However, as can be seen from

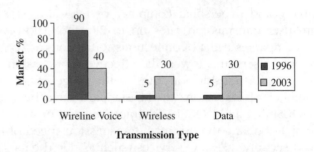

Figure 1.5   Changes in Market Composition since 1996 [BATH03]

the previous discussion, this is no longer the case. Cable companies, particularly large ones like Cox Communications, are positioning themselves to capture both residential and (particularly small and medium sized) business markets. Not only are LECs and IXCs moving into each other's territory but they are now faced with serious competition from cable operators.

The positioning of the cable operators to capture small and medium-sized business customers is a portent of the industry direction as a whole. In the past, service providers typically favored big customers with large IT budgets. These companies buy a lot of private line, Frame Relay, ATM, and voice technology from service providers. For these customers, this technology entails minimal risk. It is low cost and can easily be maintained and controlled in-house. However, for service providers, this is not a growth market, as typically these customers want the same or similar services for 10 percent less each year.

To reduce costs and simplify their networks, large customers are migrating their voice traffic to data networks, and their data networks to TCP/IP. This is shown in Figure 1.5. According to Frank Iaana, president of AT&T's Network Services Group, complete migration to TCP/IP will take time because he believes that "...it will take a long time for the end points — the phones — to be converted to IP. In fact, there are still local loops that are analog. The edge of the network moves the slowest, so convergence will take some time." [REND03]

Despite declining profit margins, service providers want to keep their large customers because they require minimal investment to maintain. From a provider's perspective, this customer base also represents steady, substantial revenue flow. To retain this market segment, service providers must continue to make large capital outlays to upgrade their networks to support faster data transmission (which now contains a larger mix of voice, multimedia, etc.). Fiber optics, DWDM (Dense Wavelength Division Multiplexing), and MPLS (Multi-Protocol Layer Switching) are some of the technologies carriers are incorporating into their network infrastructures to make this happen. In the long run, this will make it easier and cheaper to offer customers new services, thus providing new revenue opportunities.

VoIP (Voice-over-IP) — which allows convergence of voice, data, and video on a single data network — is a very significant market threat to the existing telco business model because it eases market entry to competitors such as Vonage and Net-to-Phone. Vonage subscribers use a Cisco ATA-186[33] phone adapter to convert analog phone signals to digital signals, so they can be transmitted over a broadband connection and onto the Vonage phone network. Set-up requires plugging a Cisco ATA-186 into a phone, a power supply, and a broadband connection. The Cisco ATA-186 device is similar in size to a personal CD player, and has been designed for portability. Given a high-speed Internet connection, it can be used anywhere in the world to avoid paying long-distance or international dialing rates. The Vonage service is independent of phone and cable companies and is separate from the means of transmission. Therein lies the threat to cable and phone companies.

Currently, the perceived growth opportunity in telecom is in the small and medium-sized business segment, and service providers have shifted their focus accordingly. This is due, in large part, to the sheer number of potential customers in this market. In contrast, there is a relatively small pool of very large customers; so if a major client stops buying, it is difficult to make up lost revenue with another client. Typically, this loss must be recouped by snatching business from a competitor. IXCs have not quite positioned themselves to capture a mass market but they are seeking a broader and more varied clientele than in the past. As a stock analyst would say, they are pursuing a diversified portfolio to minimize risk. Although small and medium-sized business markets do not have the IT budget of a big company, they also do not get volume discounts. This improves service provider profit margins.

Small companies are looking to service providers to help them offer features and functions to customers so they can effectively compete against larger companies. Their motto is: "Our service is as good as the big company's and is a whole lot cheaper!" To this end, small companies want sophisticated products and services tailored to their needs. Service providers are responding with bundled offerings to support a gamut of requirements. For example, it is possible to buy cell phone, DSL, three-way calling, conference calling, and Internet service as a packaged service. In the past, these services would be purchased separately. One vendor might be used for voice, another for cell phone service, and yet another for Internet service — all selected based on the cheapest price. Service providers are trying to capture as much of this business as possible by providing attractive package deals. On some services they are making a profit, and on some they are not. The end user cannot buy the services separately, so the net result is that the profit margins can be controlled. End users, particularly in the consumer and small and medium-sized business markets, find packages attractive because this gives them a single point of contact, a single bill, and an easier implementation and restoration process. This is particularly appealing to small businesses that do not have the IT staff to support sophisticated technology infrastructures. Marketing studies have shown that customers who buy bundled product offerings are more loyal and easier to retain than customers who do not buy bundled offerings. This provides added incentive to service providers to lock in small and medium-sized customers with attractive products and services (such as voice-mail, 800 services, calls centers, etc.).

In conclusion, RBOCs own the wire infrastructure into residences and businesses, thereby holding a monopoly on local phone services. Their control of access to the "last mile" severely limits the entry of other market participants, who must pay access fees to connect to their networks to provide local phone services. The Telecommunications Act of 1996 attempted to open network competition, but with limited success. CLECs have not been successful in establishing themselves in local markets, and the RBOCs continue to exert dominance in their markets. New players entering the IXC market after 1996 overextended themselves; and when the capital market collapsed in 2001, many of them went belly-up. Where legislation has failed to open markets, technology may succeed. Technological innovation — particularly transmission of voice, data, and video on one pipe — has led to major industry changes. Many companies are (or are considering) migrating their voice traffic onto private IP data networks to eliminate local access calls and to enable new functionality. Although this mode of transport is not as reliable as older technologies, it is cheaper and more flexible. It also separates services from transport mode, thus opening the door to new service providers. Although it is a significant potential challenge to ICXs, VoIP is a relatively minor threat to the "last mile" access position occupied by RBOCs because their focus is on local voice services. Cable companies are entering the market with bundled services they can provide on their own infrastructure, posing a more direct threat to the RBOCs. "Without question, the cable companies would look more favorably on investing in telecom voice service if regulation favored facilities-based competitors." [BATH03] Government policy and legislation has a huge influence on the telecom market and in shaping the direction of industry innovation, and will continue to do so in the future.

1.5.3 Technology Enablers

Historically, telecommunications vendors have sought to capture and protect market niches through strategic innovation implemented in proprietary technology. This strategy made it more difficult for customers to migrate to other solutions offered by competitors. As companies come under increased pressure to reduce the cost, complexity, and deployment time of their networks, they need more flexibility in adopting cost-effective new technologies to achieve these goals. Some customer requirements are sufficiently unique and demanding as to require a proprietary approach. However, where more generic approaches suffice, customers are putting increased pressure on vendors to employ open standards. They are also demanding more choice and plug-and-play functionality between and across vendor offerings. Customers are also placing greater emphasis on component reusability and scalability when making buying decisions. This has had major impacts on the business models of industry suppliers of hardware, software, and other services. It has made it more challenging for vendors to provide high-margin, value-added products and services, and to control commoditization of their offerings.

Moore's Law[34] encapsulates the belief that technological innovation will lead to continuous improvements in computing and communications bandwidth, keeping the pressure on vendors to innovate. Technology's inexorable progress continues to manifest itself in major improvements in both the raw power and built-in intelligence of new telecommunications software and hardware products. Rapid convergence in the application of technologies across a spectrum of systems, devices, and network implementations is a reflection of these capabilities. Convergence also reflects market demands to which vendors are responding by bundling and marketing attractive products and services. Recognizing that few will have the clout or market dominance of Microsoft or IBM, many vendors seek to lock in customers and maintain competitive advantage by providing compelling product offerings on open platforms, with value-added proprietary extensions.

The discussion of these themes continues in the subsections that follow. In particular, the focus is on the following enablers that are having a profound impact on the future use and development of technology:

- Technology convergence
- High bandwidth technologies
- Wireless computing
- Open standards
- Machine intelligence

1.5.3.1 Technology Convergence

Technology convergence refers to the consolidation of disparate implementation approaches onto a single platform or approach (which may be provided by software, hardware, or service structure). The ultimate promise of convergence is a seamless, natural man–machine interface with the end user. Convergence is also being used to increase the efficiency of computer and networking resources. Convergence is increasingly reflected at all levels in the telecommunications industry, including:

- *Telco infrastructure.* An example of this, as discussed in previous subsections, is the use of DWDM, RPR, and other technologies to extend the capacity of existing telco SONET infrastructure. This allows service providers to transport a diverse traffic mix — which may include voice, data, and multimedia — over ATM, IP, or Ethernet on the same fiber. This type of convergence reduces the telco's infrastructure costs and allows new QoS and product offerings to be configured more easily, if not dynamically.
- *IP networks.* As an example of this, enterprises are migrating legacy SNA and LAN data to TCP/IP. This reduces the enterprise's costs and the difficulty of maintaining the IT infrastructure by reducing the number of protocols that must be supported. VoIP is another related example of this type of convergence that allows voice and data traffic to be combined on the same data circuits. This allows enterprises to save on long-distance and access charges for phone service.
- *Integrated devices.* This trend is apparent in the number of devices available in the marketplace that support multiple functions. For example: (1) major printer manufacturers now offer devices that combine printing, fax, scanning, and copying functions in one device. These multi-function printers are priced very competitively to be attractive to the home and SOHO user. (2) Softphones demonstrate computer–telephone integration (CTI). They run on a personal computer or laptop, and are being used to replace standard office phones (which may cost hundreds of dollars). They support unified messaging and provide a single PC interface to voice, e-mail, and fax messages. Softphone technology is not widely deployed but it is especially attractive for call centers. Call center operators can use CTI to receive incoming calls, identify the caller through caller ID, and interact with the caller, while performing database look-ups and other processing functions on their computers. (3) PDAs, phones, and pagers are also available that support wireless Internet connections, video, and other capabilities. Siemens has announced a new product — OpenScape — that works with Microsoft's Real Time Server. It is designed to enable an array of end-user devices (including a pager, PDA, PC, Blackberry, etc.) to be selectively controlled and accessed using predefined rules. For example, if multiple devices are ringing at once, OpenScape

can assist end users in deciding which application to open. It can also be used to filter incoming content and to determine the availability of contacts prior to sending an e-mail or making a phone call. [SIEM03] These are just a few of many product offerings that serve a multiplicity of functions.

■ *Integrated applications.* This type of convergence leverages software to support and span multiple applications. As described previously, Microsoft has SIP-enabled Windows XP to integrate telephony, instant messaging, chat, e-mail, and other applications on a desktop or laptop computer. As another example, Canon Inc. has recently announced its MEAP (Multifunctional Embedded Application Platform) product architecture. MEAP is an open Java development platform that uses the Open Services Framework (OSGi). It is designed to allow "creation of multiple, customized software applications that run inside the device without the need for additional external hardware. With Canon MEAP, customers will be empowered to tailor their multifunction system to their specific workflow processes, and will have considerable flexibility and choice with respect to how they integrate products into their enterprise." [CANO03] This is only a small sample of the many products reflecting software convergence.

■ *Service provider convergence.* Increasingly, providers are entering new markets that were previously — due to regulatory, technological, historical, and other factors — off-limits to them. Service providers now have the opportunity to participate in expanded markets and to diversify product offerings for competitive advantage. LECs and IXCs are now moving into both local and long-distance markets. As another example, cable operators are offering combined cable TV, phone, and broadband services to home and business subscribers. Local phone companies are no longer the sole providers of "last mile" service to the home or business, as wireless services, cable operators, and others are being allowed into this market by new legislation and regulations.

The evaluation matrix presented previously in this chapter can be used to assess and quantify the usefulness of convergence from either a vendor or consumer perspective. Some of the factors driving the decision to employ a "converged" solution are listed below. These factors should be weighted according to the enterprise priorities when making a final determination on the value of the converged technology:

■ *ROI.* Convergence should provide positive ROI benefits to the enterprise. This might include the promise of lowered costs and increased revenues. To aid the ROI analysis, the factors listed below should be quantified in dollar terms, to the extent possible. If this is not possible, a subjective assessment should be made of the benefits that can be achieved by a simplified, converged platform or service.

■ *Complexity.* Reduced complexity might translate into such benefits as reduced training time, simplified provisioning of services, reduced maintenance, and improved network management and control.

■ *Improved or new services.* This relates to how convergence supports improvements or new features and functionality in a product or service offering. This advantage is substantive if it provides new revenue opportunities, helps to retain market share, or improves user productivity.

■ *Human factors.* This relates to the ease and appeal of using the converged service or product. For example, although PDAs offer much the same type of functionality as a paper calendar or personal computer software, their small size and combination of functions make them convenient to use, and thus desirable. Human factor improvements in the user experience may include both objective and subjective aspects. For example, some people want a PDA because it can confer status within their peer group. On a less subjective basis, small compact PDAs, cell phones, and pagers are often preferred over bulkier models because they are easier to carry. Consumers are often willing to pay more or more readily adopt a product that is ergonomic or aesthetically pleasing.

■ *Reliability.* This factor may also encompass such dimensions as quality, security, and uptime. In assessing this factor, the organization should determine whether convergence will lead to increased or decreased reliability.

■ *Productivity.* This factor assesses the value of the technology in improving productivity. As with other evaluation factors, this factor may be positive or negative, depending on the quality of the solution and its implementation in practice.

■ *Ease of adoption.* This subjective evaluation factor can be quantified, at least in part, by calculating the costs associated with transitioning to the new technology. The ease of adoption may be influenced by the degree to which the new solution must coexist with current technology and infrastructure.

■ *Open or proprietary standards.* When assessing this factor, bear in mind that an open or proprietary standard is not good or bad, per se, but only in the relative context of the organization and how it hinders or promotes its goals.

- *Legal and regulatory requirements.* It can be challenging to assess the dimensions of this factor because context is important in determining the rules that apply. For example, government regulation has limited the scope of LEC, IXC, and cable services, and has mandated some forms of convergence legal, while others illegal. Intellectual property rights may also be a consideration in deciding whether or not to adopt a converged technology. For example, Napster and Web sites for sharing music files are increasingly targets of lawsuits by the recording industry. The president of the Recording Industry Association of America (RIAA) has announced that the RIAA will go after anyone who downloads music illegally, "regardless of personal circumstances."[35] [ASSO03B]

1.5.3.2 High-Bandwidth Technologies

High-bandwidth technologies both feed and are fueled by increasingly sophisticated new applications requiring multi-tasking, large file sharing, and collaborative processing. The following high-bandwidth technologies occupy center stage in the unfolding drama of telecommunication's evolution:

- *Gigabit Ethernet and 10 Gigabit Ethernet.* This technology has brought broadband speed to the desktop. Within the enterprise, Gigabit Ethernet is rapidly becoming a LAN standard, and Dell and Compaq have announced business desktops and workstations with integrated Gigabit Ethernet. 10 Gigabit Ethernet is being used to extend the reach of the LAN into the MAN, supporting connectivity over longer distances. The IEEE 802.3ae Task Force was created to extend 10 Gigabit Ethernet to include WAN connectivity to provider carrier class switching and routing of Ethernet traffic. "Almost every service provider now offers Ethernet service in some form, efforts to allow Ethernet to work in concert with legacy public network technologies are moving forward on a number of fronts, and those carriers offering best-effort Ethernet are starting to make plans to deliver quality of service guarantees." [XCHAN03] Most of these offerings take the form of point-to-point and switched multi-point services. Point-to-Point and Gigabit Ethernet are particularly attractive to carriers because this service costs them up to 40 percent less in operational and capital spending, as compared to legacy TDM private line services. [XCHA03] Demand for this technology is strong, particularly in the financial and healthcare markets.
- *Fiber optics.* This technology supports the infrastructure that service providers use to offer WAN services. DWDM is being used to enhance the capabilities of SONET transport at gigabit speeds,

along with MPLS and RPR. Nippon Telegraph and Telephone Corporation (NTT), Nortel Networks, and Lucent currently offer terabit Ethernet networking equipment, and efforts are underway in the industry to develop 10 Terabit Ethernet, which is projected to become a commercial reality by 2010. [GILH03] It has been a major challenge in the past to interface WAN and LAN traffic. However, this is being facilitated by the dominance of Ethernet in the market and by standards that are being developed to interface the two. Carriers are also upgrading their core routers, high-capacity switches, programmable network processors,[36] and other network equipment to allow them to offer products and services that historically have been performed in-house on CPE (customer premise equipment). This sets the stage for outsourcing intelligence to the cloud (i.e., the service provider's network).

■ *Broadband access.* This technology provides high-speed, low-cost access to the Internet or corporate network, and includes cable, DSL, Ka band satellite, Fiber-to-the-Home (FTTH), broadband wireless, and wireless local area networks. According to Morgan Stanley, broadband usage had reached a critical mass of over 44 million subscribers at the end of 2002. [MEEK03] Given the demand and the state of the technology, broadband access is eventually expected to become ubiquitous worldwide. In the past, RBOCs were slow to offer broadband service other than DSL (which can be provided cheaply over existing wire phone lines). Because the RBOCs control access to the "last mile"[37] in the past, this has limited consumer connectivity options. The emergence of cable operators, IXCs, and other providers with competitive offerings has given consumers more choice and wider availability of broadband services. Because cable service was originally designed to carry unidirectional entertainment content, it is not widely deployed in commercial settings. Cable operators must address concerns with the reliability of their service, power supply, and infrastructure management before they become a major player in supporting corporate network requirements.

1.5.3.3 Wireless Technologies

Wireless communications unhooks the user from the network, providing the ultimate in convenience for a mobile society. Wireless communication is available in many types of popular devices, including:

■ Pagers
■ Cellular phones (including smart phones, PCS, and Web-enabled i-mode phones supporting color video, etc.)

- Personal data assistants (PDAs)
- Blackberry devices
- GPS devices
- Cordless computer peripherals (e.g., mouse, keyboard, printers, etc.)
- Satellite television
- Wireless LANs

Wireless communications have been developed to support LAN, WAN, and personal area network (PAN) applications.

The IEEE is very active in promoting industry acceptance and open standards for wireless local area networks (Wireless LANs), wireless personal area networks (Wireless PANs), wireless metropolitan area networks (Wireless MANs), broadband wireless access standards, and wireless sensor standards. The major IEEE standards for wireless communications include:[38]

- *802.11*™. These specifications deal with the physical and media access control layers and provide a basis for device compatibility across multiple vendors. They define the air-wave interface between and among wireless clients and a base access station.
- *802.15*™. These standards are designed to address Wireless PANs requirements for low-complexity and low-power consumption. The working group overseeing these standards also collaborates with the Bluetooth Special Interest Group. Four IEEE 802.15 standards projects are currently in development: (1) IEEE Std 802.15.1™-2002 — 1 Mbps WPAN/Bluetooth v1.x derivative work; (2) P802.15.2™ — Recommended Practice for Coexistence in Unlicensed Bands; (3) P802.15.3™ — 20+ Mbps High Rate WPAN for Multimedia and Digital Imaging and P802.15.3a™ — 110+ Mbps Higher Rate Alternative PHY for 802.15.3; and (4) P802.15.4™ — 200 Kbps max for interactive toys, sensor, and automation needs. [WIRE03] The Bluetooth specification can be downloaded at http://www.bluetooth.com/dev/specifications.asp.
- *802.16*™. These specifications "support the development of fixed broadband wireless access systems to enable rapid worldwide deployment of innovative, cost-effective and interoperable multivendor broadband wireless access products." [WIRE03]
- *P1451.5*™. These wireless sensor standards were designed to promote device compatibility across multiple vendors, by defining a wireless transducer communication interface. [WIRE03]

One should take into account the following considerations when evaluating wireless technology for enterprise use:

■ *Security.* Wireless networks present unique security challenges. WLANs can be accessed with omni-directional antennas up to a quarter of a mile. It is also fairly easy to install an unauthorized wireless access point on a hardwired LAN because of the nature of LAN transmissions. Complete transmission security requires an end-to-end solution (such as encryption).

■ *Interference.* Wireless networks are susceptible to interference and are therefore less reliable than a hardwired network connection.

■ *Limited power, processing, and display capabilities.* The battery life, screen size, and processing capabilities (typically a 16-bit processor) of wireless devices are not as robust as similar hardwired devices. Complex tasks and large data transfers may not be suitable for these types of devices.

■ *Standards.* Major standards for wireless technology are fairly new and are still evolving. This makes it more difficult to develop a strategy with respect to technology adoption, particularly because this also implies rapid product obsolescence.

1.5.3.4 Open Internet Standards

Widespread adoption of the Internet has profoundly impacted almost every aspect of the telecommunications industry. Ubiquitous Internet technologies have made network computing available to almost every person on the planet. Within the enterprise, Internet technologies provide a means to integrate legacy infrastructures, while deploying new ones that take advantage of ongoing technological improvements. The result has been the creation of a huge user base desirous of more and more Internet capabilities to support mission-critical, business-to-business (B2B) and business-to-consumer (B2C) network applications. The demand for and availability of powerful enterprise network solutions has converged at a very rapid rate, encouraging yet more innovation and leverage of Internet technologies. This is fostered by numerous standards bodies and organizations that exist to promote the use of Internet technology and its continued innovation and development. In particular, Web services have emerged as an important new set of open standards for deploying high-volume, distributed networking across a disparate range of applications and networks. This momentum has forced major vendors in a monopoly or near-monopoly position (such as Microsoft) to adapt their product strategies to incorporate open Internet protocols.

1.5.3.5 Intelligent Technologies

Machine intelligence is found in numerous network devices and software, such as routers, network management devices, CPU chips (Intel derived its name from "intelligent electronics"), operating system software, and intelligent agents. Machine intelligence refers to the automation of decision making.[39] It has become so important and pervasive that a new measure — Machine Intelligence Quotient (MIQ) — has been defined to rate the comparative intelligence of software/hardware implementations that perform control, reliability assessment, and fault diagnosis tasks. Machine intelligence offers enhanced functionality and finely tuned behaviors in response to dynamic conditions. This type of automation provides an unprecedented opportunity for machines to carry out more and more tasks that were once the sole domain of humans. Early adopters of smart technology are already reaping benefits.

The applications of smart network technology are almost limitless. For example, some companies, such as Ford Motor Co. and Prada "are using the Web to allow machines to monitor each other, track products as they move through warehouses, and even make decisions without human intervention." [MULL03] As another example, Beckman Coulter Inc., a manufacturer of blood analyzers and medical equipment, "links the machines it sells to a computer back in its factory. The computer, unlike humans, works every minute of the week to monitor that everything is running smoothly. When a problem crops up, the computer alerts a Beckman technician, who can often make repairs before the machine breaks down. Beckman expects this system to save it as much as $1 million annually. But the far larger benefit is customer satisfaction for a company that fixes the machines it sells before they show signs of malfunctioning." [MULL03] Wal-Mart and the DoD are other examples on the forefront of technology, requiring that their suppliers utilize RFID[40] technology, in various forms, at the case and the palette level by 2005. Many agree that the advantages of RFID technology, when compared to barcodes, are worth the investment. If you are big enough, you can even get others to help you pay for the investment. Although Wal-Mart and the DoD must set up an infrastructure to deal with reading and processing the RFID data, a large part of the cost is being passed on to their suppliers. Another exciting smart technology is "smart dust." Researchers at the University of California at Berkeley and Palo Alto Research Center Inc. have developed microscopic chips equipped with antennas that can be used to detect toxic chemicals, biological agents, or other airborne conditions. According to Professor Michael Sailor, who headed up the research effort, the chips were designed to be as inconspicuous and as easily embedded or dispersed as dust. The chips are capable of sending information that "can be read like a series of barcodes by a laser ... to tell us if the cloud that's coming toward us is filled with anthrax bacteria...." [REGE01]

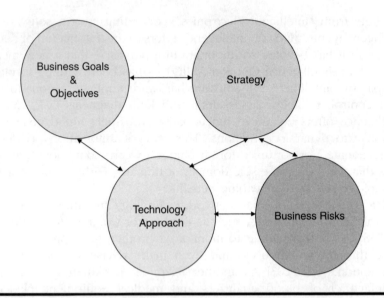

Figure 1.6   Interrelationships in Technology Planning

1.6 Summary

This chapter introduced a systems methodology for network planning and implementation. This methodology involves:

- Defining the business needs and objectives
- Assessing the major risks and dependencies
- Defining the project requirements
- Designing an implementation strategy

The planning methodology is not strictly linear, as each aspect is interrelated. This is illustrated in Figure 1.6. It is most effective when it is standardized and enforced across the organization.

This chapter surveyed major network technologies and architectures, and identified similarities and differences in the planning and implementation of these networks.

An examination of the trends with major impacts on the design and planning of networks revealed:

- Increasing demand for fast, high-capacity networks to support larger and more complex networks handling larger and more complex applications
- Increasing adoption of open-based standards, such as the Internet- and TCP/IP-based protocols

- Emerging concerns with security and control over network activity
- Evolving regulatory environment to promote growth and security in the telecommunications industry

Notes

1. As a result of the September 11, 2001, terrorist attack, the physical building, telecommunication, and access facilities of the World Trade Center (WTC) were destroyed. The main Verizon switching station, located near the base of the WTC, was obliterated. Within four and a half days, the switching hub was rebuilt; and six days after the attack, on Monday, September 17, the NYSE reopened.
2. Dan Bricklin is the chief technology officer at Interland. He is also famous as the co-developer of VisiCalc, the first electronic spreadsheet program for the personal computer. The company Web site is http://www.inter-land.net
3. Pareto's Rule: In the late 18th century, Vilfredo Pareto discovered that most of the European wealth was held by a fraction of the population. The actual formula Pareto developed is: $\log N = \log A + m \log x$, where N is the number of wage earners who receive incomes higher than x, and A and m are constants. In short, he discovered that 80 percent of the wealth was owned by 20 percent of the population. A modern equivalent of the 80/20 rule is: 80 percent of the work in data mining is preparation and 20 percent is the actual modeling.
4. These methods are discussed in more detail in Appendix C: Subjective Evaluation and Ranking Procedures. Saaty's hierarchical comparison method and other fuzzy logic weighting approaches are also reviewed in Appendix C.
5. ROI: Return On Investment. This is a widely used financial metric for evaluating and comparing alternative spending proposals. For more detail on its meaning and calculation, see Appendix A.
6. ATM: Asynchronous Transfer Mode. This is a switching technique for transmitting cells (or packets) or data. ATM was developed to support high-speed data transport. See Chapter 2 for a detailed discussion of this technology.
7. Frame Relay: a packet-based transmission protocol operating at layer 2 of the OSI Reference Model. See Chapter 2 for a detailed discussion of this technology.
8. IP/VPNs: IP-based virtual private networks provide secure tunneling for data transmission over an unsecured network (i.e., the Internet). They are based on an IP network for communication. They are private because transmissions are encrypted for security. They are virtual because they do not require dedicated lines. Several protocols support IP/VPNs, including PPTP, L2TP, L2F, and IPSec. See Chapter 2 for a detailed discussion of IP/VPN technology.

9. CSO: Chief Security Officer (a.k.a. Corporate Security Officer), an emerging watchdog position to ensure corporatewide security planning and oversight.

10. CIO: Corporate Information Officer, who typically reports to a CEO (Chief Executive Officer) or a CFO (Chief Financial Officer), and sometimes to the newly created position of CSO (Corporate Security Officer).

11. RFIs, RFQs, and RFPs: *Requests for Information* (RFIs) are typically informal requests to vendors for information on basic product or service features and functions. *Requests for Quotes* (RFQs) are at the next level of detail and provide a preliminary cost estimate. The *Request for Proposal* (RFP) is a formal, legal document with highly detailed specifications and a proposal for work. These documents are discussed in detail in Chapter 6: Outsourcing.

12. EDI: Electronic Data Interchange. The American National Standards Institute (ANSI) has approved the X12 Standards to define mechanisms for data exchange between companies.

13. SIP: Session Initiation Protocol. This is a signaling protocol designed to support Internet-based telephony, videoconferencing, presence, events notification, and instant messaging. A SIP-enabled device can initiate call set-up, routing, and authentication between peer endpoints in an IP domain. In contrast, H.323 is an older telephony protocol that provides similar functionality but depends on a centralized phone switch or service to perform call processing and control functions.

14. Jabber is a free, open source instant messaging system based on XML protocols that use SSL for security. Some companies may wish to consider this as an IM alternative. More information on Jabber can be obtained at http://www.jabber.org/.

15. Hot spot sniffers: These devices provide wireless network discovery and auditing. Some of these devices enable one to analyze the consistency and signal strength of a wireless network, discover access points and network interface cards, record the network location with GPS, recover encryption keys, browse applications, and perform other monitoring functions. A "hot spot" refers to a wireless network.

16. Codec: a device analogous to a modem that allows a digital line to process analog signals.

17. Centrex service: This service allows lines to be rented at the local phone company's central office. It provides the same types of functions and features as a private automatic branch exchange (PABX) or private branch exchange (PBX). Note that a PABX is an *automatic* telephone switching system within a private enterprise. In contrast, a PBX requires a live operator to switch calls. Most private branch exchanges are automatic, so the abbreviation "PBX" is usually synonymous with "PABX."

18. Communications carrier: refers to a communications company providing circuits or lines to carry network traffic; also referred to as a "common carrier."

19. Kbps = 1024 bits per second.

20. Mbps = 1,000,000 bits per second.

21. Gbps = One billion bits (1,000,000,000) per second.

22. Worm: a computer worm is self-replicating malicious code. A well-known example is the Code Red worm.
23. Since October 2001, 17 companies, including Global Crossing and World-Com, with a combined market capitalization of $96 billion, filed for bank-ruptcy protection. [MALI02]
24. CLECs: competitive local exchange carriers; sometimes also referred to as CAPs (competitive access providers).
25. Note that MCI is also an interexchange carrier (IXC).
26. The term ILEC is frequently used synonymously with LEC.
27. The RBOCs were created from the former Bell Operating Companies when the Bell System was dismantled by a 1983 consent decree.
28. Note that SBC Communications owns Southwestern Bell, Pacific Bell, Ameritech, and SNET. Verizon owns Bell Atlantic, NYNEX, and GTE. Qwest owns US West.
29. Windows NT RAS (Remote Access Service) is used to connect remote workers to corporate WANs. It appears as a Dial-up Networking icon on the desktop.
30. VoIP-PBX: an IP-based PBX system supporting Ethernet telephones on a LAN and external V-o-IP gateway servers. It supports a variety of telephone devices, including SIP and H.323 terminals, PC-based "softphones," and proprietary IP telephones.
31. Local network portability (LNP) allows a subscriber to transfer a hardwired phone number to a wireless cell phone.
32. EEPROM: electronically erasable programmable read-only memory. It can be erased and reprogrammed repeatedly through the application of electric voltage. It does not have to be removed from the device (in this case, the cable modem) that houses it when it is modified.
33. ATA: Analog Telephone Adapter.
34. Moore's Law. In 1965, Intel's founder, Gordon Moore, noted that the number of transistors per square inch on an integrated circuit had doubled every year and he predicted the trend would continue for the foreseeable future.
35. In a crackdown by the recording industry, subpoenas and threats of expensive lawsuits are being served to parents, grandparents, and room-mates of people suspected of illegally sharing music. Some music sharers are unaware that what they are doing is illegal or puts them at risk for legal entanglements, and have been shocked when they receive notices from the RIAA. The RIAA has issued numerous subpoenas, and expects to file lawsuits seeking damages ranging from $750 to $150,000 for each pirated song. The RIAA prepared its subpoena lists by matching the Internet address of music downloaders with names and addresses of ISP subscriber records. [ASSO03b]
36. Programmable network processor: a device that can perform packet pro-cessing, filtering, switching, and other functions. It operates at "hardware" speed although it is software based, and it can be reprogrammed. Net-working equipment that uses a programmable network processor can be easily upgraded. This allows new capabilities to be added after the equip-ment is installed.

37. The "last mile" refers to the network connection between the home or business office and the telco.
38. Detailed documentation on these specifications can be found at the IEEE Web site: http://standards.ieee.org/wireless/overview.html#802.11.
39. In some disciplines of artificial intelligence, machine intelligence refers specifically to pattern recognition and data mining. Here, we use the term "machine intelligence" in a broad context.
40. RFID: the acronym for radio frequency identification. This technology uses electromagnetic or electrostatic frequencies for identification purposes, typically as a replacement for barcodes. "The advantage of RFID is that it does not require direct contact or line-of-sight scanning. An RFID system consists of three components: an antenna and transceiver (often combined into one reader) and a transponder (the tag). The antenna uses radio frequency waves to transmit a signal that activates the transponder. When activated, the tag transmits data back to the antenna. The data is used to notify a programmable logic controller that an action should occur." [SEAR04]

References

[AMER03]   American Civil Liberties Union, 2003, http://www.aclu.org/SafeandFree/SafeandFree.cfm?ID=12126&c=207.

[AOL03]   America Online, www.aol.com.

[ASSO03]   Associated Press, Iraq contract fuels MCI critics, *Connecticut Post*, May 22, 2003, p. A20.

[AXNE03]   Axner, D., Gigabit Ethernet Goes on MAN Hunt, IT World.com.

[BABA02]   Babaie, E., De Souza, R., Clark, L., Hale, K., Heng, J., and Narisawa, R., Worldwide IT Services Market Forecast, 2001–2005, Market Statistics, Gartner Group Report, July 17, 2002.

[BASE03]   Baseline Magazine.com, How Kmart Fell Behind, *Baseline Magazine*, December 10, 01. www.baselinemag.com/article2/0,3959,657355,00.asp.

[BEHA97]   Behar, R., Who's Reading Your E-Mail?, *Fortune Magazine*, February 3, 1997.

[CALO03]   Caloyannides, M., Privacy vs. Information Technology, IEEE Security and Privacy, 2003, http://computer.org/security/v1n1/j1prv.htm.

[CARR03]   Carr, H. and Snyder, C., *The Management of Telecommunications: Business Solutions to Business Problems Enabled by Voice and Data Communications, Second Edition,* McGraw-Hill, New York, 2003, pp. 13–14.

[CASP03]   CASPR, The CASPR Project, copyright 2003, http://www.caspr.org/.

[CCEV04]   CCEVS, *About Us (NIAP CCEVS),* copyright 2004, http://niap.nist.gov/cc-scheme/aboutus.html.

[CHAN97]   Chandrasekaran, R., Making UUNet into a Very Big Deal, *Washington Post*, September 29, 1997, F12, http://www.washingtonpost.com/wp-srv/business/longterm/post200/stories98/uunet.htm.

[CIAM02]   Ciamp, M., *Guide to Wireless Communications*, Course Technology, 2002, p. 181.

[CISC00] Cisco Systems, CiscoWorks Blue SNA View Version 3.0 Data Sheet, http://www.cisco.com/warp/public/cc/pd/wr2k/wrib/wribsa/prodlit/snav3_qp.pdf.

[CISC01] Cisco Systems, Positioning SNA Switching Services and Data-Link Switching Plus, 2001, pp. 1–8. http://www.cisco.com/warp/public/cc/pd/ibsw/ibdlsw/tech/dlsw_wp.pdf.

[CISCO01B] Cisco Data Systems, Schroder, Salomon Smith Barney, Citigroup Global IPMulticasting Network uses Cisco IOS® Multicast Technology to Deliver Real-Time Trading Data, Cisco Systems, 2001, http://www.cisco.com/warp/public/732/press/docs/citigroup_cp_final1.pdf.

[CISC02] Cisco Systems, IBM Communications Controller Replacement Design Guide, 1999–2002, pp. 1–20, http://www.cisco.com/warp/public/cc/so/neso/ibso/data/datacent/ibmcm_wp.pdf.

[CISC03] Cisco Systems, Cisco-IBM Internetworking, January 6, 2003, pp. 1–20, http://www.cisco.com/warp/public/534/18.pdf.

[CLAV03] Clavenna, S. and Heywood, P., 2002 Top Ten: Technology Trends, *Light Reading*, January 11, 2003, www.lightreading.com/document.asp?doc_id=26508&site=lightreading.

[CoBIT03] COBIT, 2003, http://www.isaca.org/cobit.htm.

[COLE93] Cole, G., IEEE 802.6 metropolitan area network standard, *Datapro*, McGraw-Hill, New York, 7, May 1993.

[COOP97] Ramo, J.C., Welcome to the wired world as the globe's political and business elites meet in Davos, Switzerland, to ponder the implications of the information revolution, the future is glowing — and growing — all around them, *Time*, Vol. 149, No. 5, February 3, 1997.

[DARP03] DARPA, Information Awareness Office, Terrorism Information Awareness (TIA) System, 2003, http://www.darpa.mil/iao/TIASystems.htm.

[DATA95] Datapro Information Services Group, Financial planning for network managers, *Datapro*, McGraw-Hill Companies, Inc., New York, February 1997.

[DATA8_96] Datapro Information Services Group, Sources of leaks in LANs and WANs, *Datapro*, McGraw-Hill Companies, Inc., New York, August 1996.

[DATA96] Datapro Information Services Group, High speed local area networks, *Datapro*, McGraw-Hill Companies, Inc., New York, November 1996.

[DAVE01] Davenport, T., The New, New IT Strategy, *CIO Magazine*, May 1, 2001, www.cio.com/archive/050101/davenport_content.html.

[DAVI03] Davids, A., Tech Bills Updates January–March 2003, IEEE Distributed Systems Online, http://dsonline.computer.org/techbills/janmar03.htm.

[DYME94] Dymek, W., IBM systems network architecture (SNA), *Datapro*, McGraw-Hill Companies, Inc., New York, April 1994.

[FEDE03] Feder, B., Information On-Ramp Crosses a Digital Divide, *New York Times*, July 8, 2003, p. C1, C6.

[FONT02] Fontana, J., Biz users prep for IM onslaught, *Network World Fusion*, November 18, 2002, http://www.nwfusion.com/news/2002/1118microsoftim.html.

[FOUN02] Foundry Networks, Telecom Ottawa selected Foundry to power its Metropolitan Area Network, one of the world's first standards-based 10-Gigabit Ethernet MANs, July 24, 2002, http://www.foundrynet.com/about/newsevents/releases/pr7_24_02b.html.

[FTC03]   Federal Trade Commission, 2003, http://www.ftc.gov/privacy/glbact/.

[GOOG03]   Google Press Center, *Google Inc. Company Overview*, 2003. http://www.google.com/press/overview.html.

[GREEN03]   Greenspan, R., *Broadband's Reach Gets Broader, atnewyork.com*, February 7, 2003, http://www.atnewyork.com/news/article.php/1580601.

[GREY02]   Grey, M., AOL Enters Enterprise IM Market, Dominance Not Assured, Gartner Group, November 12, 2002, http://techupdate.zdnet.com/techupdate/stories/main/0,14179,2897233,00.html.

[HAFN03]   Hafner, K., Internet Access for the Cost of a Cup of Coffee, *New York Times*, May 8, 2003, p. G5.

[HAFN03b]   Hafner, K. and Preysman, D., Special Visa's Use for Tech Workers is Challenged, *New York Times*, Business Day, May 30, 2003, p. C1.

[HANL03]   Hanluain, D., Forget F-stops: These Cameras Have Area Codes, *New York Times*, July 3, 2003, p. G1.

[HARRI03]   Harris, R., Telecommunications Industry Dynamics: Implications for Policy Making, http://www.connectillinois.org/issues/harris.htm, 2000–2003.

[HEIN03]   Heingartner, D., Roving the Globe, Laptops Alight on Wireless Hot Spots, *New York Times*, June 5, 2003, p. G4.

[HU03]   Hu, J., "AOL asks FCC to lift IM restriction," CNET News.com, April 4, 2003, http://news.com.com/2102-1032_3-995595.html?tag=ni_print.

[IBM94]   *IBM Networking Solutions*, Volume 3, Edition 1a, Chapter 15, Multiprotocol Transport Networking Solutions, 1994.

[IBM95]   IBM International Technical Support Organziation, Multiprotocol Transport Networking (MPTN) Architecture Tutorial and Product Implementations, November 1995, *IBM Redbook*, http://publib-b.boulder.ibm.com/Redbooks.nsf/RedbookAbstracts/sg244170.html?Open.

[IBM02]   IBM, Hardware Withdrawal: All Models of IBM 3745 Communication Controllers and IBM 3746 Nways Multiprotocol Controllers Plus Selected Features and MESs, *Product Announcement*, February 26, 2002, http://www.networking.ibm.com/announce/022602.html.

[IBM03]   IBM, IBM System Network Architecture Protocols, *Internetworking Technologies Handbook*, 2003, Chapter 37, http://www.networking.ibm.com/app.

[IEC03]   International Engineering Consortium, MultiProtocol Label Switching (MPLS), 2003, http://www.iec.org/cgi-bin/acrobat.pl?filecode=94.

[IEEE02]   IEEE 802.16 Working Group, *Broadband Wireless Access: An Introduction to the Technology Behind the IEEE 802.16 WirelessMAN™ Standard*, May 24, 2002, http://grouper.ieee.org/groups/802/16/pub/backgrounder.html.

[INTE03]   About Internet2®, 2003, www.internet2.edu.

[ISSA03]   About ISSA, *ISSA*, 2003, http://www.issa.org/aboutissa.html.

[(ISC)²]   CISSP Certification Examination — Applicant Requirements, (ISC)², http://www.isc2.org/cgi-bin/content.cgi?page=43.

[IVES03]   Ives, N., Advertising Online Profiling, separating the car buff from the travel seeker, is a new tool to lure advertisers, *New York Times*, June 16, 2003, p. C10.

[KEIZ03]   Keizer, G., IBM, AOL Forge Deal to Integrate IM Platforms, *TechWeb News*, February 3, 2002, www.techweb.com/wire/story/TWB20030203S0005.

[KERS03] Kershaw, S., Wal-Mart Sets a New Policy That Protects Gay Workers, *New York Times*, July 2, 2003, p. A1.

[KONR03] Konrad, R., eBay posts first-quarter profit, edges past expectations, *Associated Press*, April 22, 2003, http://www.charlotte.com/mld/charlotte/business/special_packages/despair_in_mill_town/5691269.htm.

[LAND03] Landler, M., With Wireless, English City Reaches Across the Digital Divide, *New York Times*, Business Day, May, 31, 2003, p. C1, C3.

[LANDS03b] Landsend.com, About Lands' End, 2003, http://www.landsend.com/cd/fp/help/0,,1_36877_36883_37027___,00.html#landsenddotcom.

[LEVI96] Levine, H., Blaszak, J., Block, E., and Boothby, C., The Telecommunications Act of 1996, *Datapro*, McGraw-Hill, Inc., New York, June 1996.

[LEVI03] Levinson, M., Coming to a State Near You: Internet Sales Tax, *CIO Magazine*, April 15, 2003.

[LEWI98] Lewis, S., IBM eNetwork Communications Server for OS/2 Warp Quick Beginnings, Version 6, GC31-8189-03, December 1998. http://www-3.ibm.com/software/network/commserver/library/publications/csos2_60/HTML/qbhtml.htm.

[LOHR03] Lohr, S., A Once and Present Innovator, Still Pushing Buttons, *New York Times*, May 6, 2003, pp. G1, G11.

[LOHR03C] Lohr, S., Has Technology Lost Its 'Special' Status?, *New York Times*, May 16, 2003, p. C1.

[LOHR03B] Lohr, S., Microsoft Leader Tells Workers of IBM–LINEX Threat, *New York Times*, June 15, 2003, p. C6.

[KEIS89] Keiser, G., *Local Area Networks*, McGraw-Hill, New York, 1989.

[MART94] Martin, J., with Chapman, K. and Leben, J., *Local Area Networks, second edition*, Prentice Hall, Englewood, NJ, 1994.

[MCEA02] McEachern, C., No Proof, No Project, *Wall Street & Technology Online*, November 11, 2002, http://www.wallstreetandtech.com/story/inDepth/WST20021111S0002.

[MCWI03] McWilliams, B., Licensed to War Drive in NH, *Wired News*, April 29, 2003, http://www.wired.com/news/print/0,1294,586651,00.htm.

[METC00] Metcalfe, B., From the Ether: Quality multimedia in real time is the impetus for Enkido's 768 service, *InfoWorld*, April 28, 2000, http://archive.infoworld.com/articles/op/xml/00/05/01/000501opmetcalfe.xml.

[MFN99] MFN, Metromedia Fiber Network Leases Dark Fiber Infrastructure to America Online, MetroMedia Fiber Press Release, New York, February 2, 1999, http://www.mfn.com/news/pr/19990202_Lease_AOL.shtm.

[MICR03a] Microsoft Press Release, 100 Million Customers and Counting: MSN Messenger Extends Worldwide Lead Among Instant Messaging Providers, May 12, 2003, http://www.microsoft.com/presspass/press/2003/may03/05-12100MillionPR.asp.

[MICR03] Microsoft SNA Server Product Unit, SNA Open Gateway Architecture, A Technology White Paper, 2003, http://www.microsoft.com/technet/archive/default.asp?url=/technet/archive/sna/plan/soga.asp.

[MINO93] Minoli, D., *Broadband Network Analysis and Design*, Artech House, Boston, 1993.

[MPLS02]   MPLS Forum, *GMPLS Interoperability Event Test Plan and Results*, 2002, http://www.mplsforum.org/interop/GMPLSwhitepaper_Final1009021.pdf.

[MULL03]   Mullaney, T., Green, H., Arndt, M., Hoft, R., and Himelstein, L., The E-biz Surprise, *Business Week Online*, May 12, 2003, http://www.business-week.com/magazine/content/03_19/b3832601.htm.

[MULL96]   Muller, N., Federal and state telecom reform initiatives, *DataPro*, McGraw-Hill, New York, April 1996.

[MULL97]   Muller, N. and Costello, R., Planning voice networks, *DataPro*, McGraw-Hill, New York, 3, 1997.

[MULL03]   Mullaney, J., Green, H., Arndt. M, Hof, R., and Himelstein, L., The E-Biz Surprise, *Business Week*, May 12, 2003, http://www.businessweek.com/magazine/content/03_19/b3832601.htm.

[NEWY03]   *New York Times*, AOL Seems to Get Edge in the Truce with Microsoft, *New York Times*, May 31, 2003, p. C14.

[NORT02]   Nortel Networks, *An Overview of the International Common Criteria for Information Technology Security Evaluation,* 2002, http://www.nortelnetworks.com/solutions/security/collateral/nn101441-0802.pdf.

[NORT03]   Nortel Networks, Enabling Technologies, 2003, http://www.nortelnetworks.com/corporate/technology/oe/protocols.html.

[NTIA02]   United States Commerce Department's National Telecommunications and Information Administration (NTIA) and the Economics and Statistics Administration, *A Nation Online: How Americans Are Expanding Their Use of the Internet*, posted February 5, 2002, http://www.ntia.doc.gov/reports.html.

[OPEN02]   OpenConnect®, *What is Web-to-Host?*, 2002, http://www.openconnect.com/products/mandate.html.

[OPPE01]   Oppenheimer, Priscilla, *Top-Down Network Design*, Cisco Systems, 2001.

[OPSW03]   www.opsware.com, 2003.

[ORTE00]   Ortega, B., In Sam We Trust, The Untold Story of Sam Walton and Wal-Mart, The World's Most Powerful Retailer, Three Rivers Press, (CA), 2000.

[OWEN91]   Owen, J., Data networking concepts, *Datapro*, McGraw-Hill Companies, New York, 14, 1997.

[POLA03]   Polaris Communications and Microsoft Corporation, TCP-IP-to-SNA Connectivity with Channel Attached Gateways, A White Paper, 2003, http://www.microsoft.com/technet/archive/default.asp?url=/technet/archive/sna/evaluate/polaris.asp.

[PUZZ03]   Puzzanghera, J., Tech tools alter battle tactics, SiliconValley.com, February 19, 2003.

[REES99]   Reeser, T., RDP or ICA – Which Should You Choose?, *Windows&.Net Magazine*, 1999, http://www.winnetmag.com/Articles/Index.cfm?ArticleID=5151&pg=1&show=526.

[REGE01]   Regents of the University of California, *UCSD Researchers Fabricate Tiny 'Smart Dust' Particles Capable of Detecting Bioterrorist and Chemical Agents*, 2001, http://ucsdnews.ucsd.edu/newsrel/science/mcsmartdust.htm.

[RESI01]   Resilient Packet Ring Alliance, An Introduction to Resilient Packet Ring Technology, October 2001, www.rpralliance.org.

[RESI03]   Resilient Packet Ring Alliance, RPR Business Case Study: A Comparison of RPR and SONET, April 2003, www.rpralliance.org.

[ROGE95]   Rogers, E., *Diffusion of Innovations*, 4th edition, Free Press, London, 1995.

[ROZH03]   Rozhon, T., Latest SARS Victim is Clothing Industry, *New York Times*, May 17, 2003, p. A6.

[ROUT00]   Routt, T., SNA and TCP/IP Convergence — A S/390 Perspective, *BCR ACCESS,* a supplement to *Business Communications Review,* August 2000, pp. 18–24, http://www-1.ibm.com/servers/eserver/zseries/networking/pdf/Vedacom.pdf.

[RUBI92]   Rubinson, T., A Fuzzy Multiple Attribute Design and Decision Procedure for Long Term Network Planning, Ph.D. dissertation, 1992.

[RUBI99]   Rubinson (Piliouras), T. and Geotsi, G., Estimation of Subjective Preferences Using Fuzzy Logic and Genetic Algorithms, *Information Processing and Management of Uncertainty in Knowledge-Based Systems,* Granada, Spain, July 1999.

[RUMB99]   Optimizing Host Access in Server Based Environments, Rumba White Paper, December 6, 1999, www.rumba.com.

[RYAN97]   Ryan, G., Editor, Dense Wavelength Division Multiplexing, Ciena Corporation, 1997.

[SAAT80]   Saaty, T., *The Analytic Hierarchy Process*, McGraw-Hill, New York, 1980.

[SAUN97]   Stephen S., The Brightest Ideas in Networking, *Data Communications,* McGraw-Hill, New York, 1997, p. 1.

[SAUN03]   Saunders, C., Lotus Sametime Takes Notes, *Instant Messaging Planet.com,* April 29, 2003, http://www.instantmessagingplanet.com/enterprise/print.php/2198801.

[SCHA95]   Schatt, S., Local routers and brouters, *Linking LANs,* McGraw-Hill, New York, 1995, pp. 91–101.

[SCHN03]   Schneiderman, R., High Flying Defense Systems Poised to Take Off, *Electronic Design,* June 16, 2003, ED Online ID #5073, www.elecdesign.com/articles/Print.cfm?ArticleID=5073.

[SCHO02]   Schoolar, D., Plenty of Light Left in Optical MAN Services, Cahners In-Stat Press Release, June 12, 2002, http://global.mci.com/news/marketdata/network/.

[SCHW03]   Schwartz, M., ISSA tries to breathe new life into old idea for an overarching security standard, *Information Security® Magazine,* May 2003, http://www.infosecuritymag.com/2003/may/gaisp.shtml.

[SEAR00]   SearchNetworking.com, Glossary: OC-768, August 9, 2000, http://searchnetworking.techtarget.com/gDefinition/0,294236,sid7_gci294572,00.html.

[SEAR04]   searchNetworking.com, RFID Definitions, January 16, 2004, http://searchnetworking.techtarget.com/sDefinition/0,,sid7_gci805987,00.html.

[SELL03]   Sellers, D., Poll: Americans look to computers for home entertainment MacCentral, May 20, 2003, http://maccentral.macworld.com/news/2003/05/20/poll/.

[SMI97]   Smith, M., Virtual private network (VPN) services: overview, *Datapro,* Mc-Graw-Hill, New York, 1997.

[SPIR02]   Spirent Communications, BGP/MPLS Virtual Private Networks Performance and Security over the Internet, 2002, http://www.spirentcom.com/analysis/index.cfm?WS=27&D=2&wt=2.

[SPOH93]   Spohn, D., *Data Network Design*, McGraw-Hill, New York, 1993.

[STALL94]   Stallings, W., *Data and Computer Communications, Fourth Edition*, Macmillan Publishing, New York, 1994.

[STAN03]   Stanley, J., ACLU Statement on Terrorist Information Awareness Before the Department of Defense Technology and Privacy Advisory Committee, June 19, 2003, http://www.aclu.org/SafeandFree/SafeandFree.cfm?ID=12945&c=206.

[STAR96]   Stark, M., Encryption for a Small Planet, *BYTE*, McGraw-Hill, New York, March 1996.

[STEI96]   Steinke, S., The Internet as Your WAN, *LAN Magazine*, Miller Freeman, Inc., Vol. 11, No. 11, pp. 47–52, October 1996.

[SKVA97]   Skvarla, C., ISDN Services in the U.S: Overview, *Datapro*, Mc-Graw-Hill, New York, April 1996.

[SULL03]   Sullivan, B., Credit Card Fraud Help Policies Vary, *MSNBC.com*, January 14, 2003, http://www.msnbc.com/news/858786.asp.

[SULL03b]   Sullivan, B., Could Your Computer Be a Criminal?, *MSNBC.com*, July 15, 2003, http://www.msnbc.com/news/939227.asp.

[TEDE03]   Tedeschi, B., E-Commerce Report Filling Cracks in E-Tailing Promise to Protect Privacy, *New York Times*, Business Day, June 30, 2003, p. G5.

[TEDE03a]   Tedeschi, B., E-Commerce Report Web Merchants Broke Even in 2002 after Streamlining Their Operations to Help Staunch Losses, *New York Times*, Business Day, May 19, 2003, p. C5.

[TEDE03b]   Tedeschi, B., Europe Set to Begin Digital Tax, *New York Times*, Business Day, June 23, 2003.

[THOM03]   Thompson, N., More Companies Pay Heed to Their "Word of Mouse" Reputation, *New York Times*, June 23, 2003, p. C4.

[TITTE01]   Tittel, E. and Johnson, D., *Guide to Networking Essentials*, Course Technology, 2001.

[USDE03]   United States Department of Homeland Security, Federal Information Security Management Act (FISMA), www.feddcirc.gov/library/legislation/FISMA.html.

[UTTE96]   Utterback, J., *Mastering the Dynamics of Innovation*, Harvard Business School Press, September 1996.

[UUNE03]   UUNET, History of UUNET, 2003, http://www.uunet.co.za/_aboutuunet/SAhistory/SAhistory.asp.

[WAHL01]   Wahlgren, E., The Digital Age Storms the Corner Office, *Business Week Online*, September 6, 2001.

[WALL95]   Waller, D., Spies in Cyberspace, *Time*, 145, 11, March 20, 1995.

[WALS03]   Walsh, L., Security Standards Standard Practice ISO 17799 Aims to Provide Best Practices for Security, But Leaves Many Yearning for More, *Information Security*, March 2002, http://www.infosecuritymag.com/2002/mar/iso17799.shtml.

[WARS03]   Warschauer, M., Demystifying the Digital Divide, *Scientific American*, August 2003, pp. 42–47.

[WELS03]   Welsh, W., Deficits Don't Deter New CIOs, *Washington Technology*, May 12, 2003.

[WHIT02]   White House, *National Security Strategy*, September 2002.

[WU01]   Wu, T., *How Ethernet, RPR, and MPLS Work Together: The Unified Future of Metro Area Networking*, Riverstone Networks, 2001.

[YAHO03]   Yahoo.com, The History of Yahoo! — How It All Started..., 2003, http://docs.yahoo.com/info/misc/history.html.

[ZIFF03]   Ziff Davis Media Inc., Are New Technologies Adding Business Value?, *CIO Insight,* June 16, 2003.

Chapter 2

Wide Area Network Design and Planning

2.1 Management Overview of WAN Network Design

This chapter focuses on the topological design of WAN data networks. This includes a presentation of the techniques used to place nodes and links in a network to minimize cost and achieve performance objectives.[1] The ultimate result of this process is a set of recommendations for placing specific nodes and links in a network, and the selection of specific technologies to implement the network.

The selection of line speeds and technologies is largely predicated on the type of network infrastructure (i.e., whether it is centralized or decentralized) and the performance requirements of the end-users and applications which it must support. Until fairly recently, only a few line choices were available. One could choose between private lines (operating at DDS,[2] DS1,[3] or DS3[4] speeds) or public lines (operating X.25 at 56 Kbps or 64 Kbps). Now there are many high-bandwidth options available, including Frame Relay, Asynchronous Transfer Mode (ATM), ISDN, SONET, and DWDM. The sections that follow discuss the pros and cons of these choices for various network scenarios, and ways that they can be used in combination to provide a total network solution.

All these choices compound the difficulty of designing the network. The network designer must generate several alternative designs for *each* link type and technology considered to achieve a sense of the price/performance trade-offs associated with each option. The decisions and comparisons involved in planning all but the smallest networks are sufficiently complex to warrant the aid of automated network design tools. Chapter 7 surveys network design tools that offer a variety of design techniques for producing networks with different link types and technologies.

When implementing the network, the organization must decide whether or not to use in-house or outside staff and resources, or an appropriate mix. After identifying several promising designs, an organization may wish to solicit bids from various service providers. The major carriers will develop bids according to an organization's needs, which may include design, implementation, and even operation of the network. These bids provide very useful information on the expected network performance and cost. The organizational budget, in-house skills, security requirements, and role of the telecommunication and MIS functions to the survival of the company are major decision factors in deciding to use outside services to implement all or part of the network.

The first-year benefits of outsourcing network implementation and management are usually positive because the difficult task of coordinating these functions is shifted to another party. Upper management sees a drop in payroll and related employee benefits, and the telecom manager does not need to tie up all of his or her capital on one-time purchases. Because technology is always changing, contracts with outsourcers are likely to be subject to ongoing negotiation and mutual misunderstandings. However, outsourcing may help position the company to adopt new technologies as they become available and to grow to meet expanding network demands.

When a company elects to buy network equipment, the one-time purchase costs are typically amortized over a three- or five-year period. However, network technology is often outdated within 12 to 18 months. Thus, many telecom managers find they need updated customer premise equipment (CPE) before they have fully depreciated their equipment. One way to avoid this situation is to lease the network equipment with the stipulation that upgrades will be made available on a continual basis. This way, the network manager no longer needs to maintain an obsolete depreciable asset on the network accounting budget. At the end of the lease, the equipment can usually be bought outright if the network manager so desires. However, leasing can be expensive. In addition, if the equipment is leased, there may be a cyclic need to replace the equipment as the lease period expires. If the organization owns the equipment, the communications/IS department can continue to function while plans are made to update it.

From the outsourcer's point of view, the first year of network support requires a high outlay of capital to cover its start-up costs. Typically, the outsourcer does not realize a profit until several years into the contract. Coincidentally, this is about the same time that the customer is growing tired of paying monthly costs for outdated equipment and services. At this time, the customer may want to renegotiate the contract or determine its liability for breaking or breaching the contract. Conversely, the outsourcer wants to keep the contract active because this is when the majority of the return on investment (ROI) is realized.

Outsourcing offers many potential benefits, including:

- Reduced communication overhead costs
- Reduced labor and training costs
- Improved availability of skilled network management personnel
- Economies of scale in buying services and equipment
- Competitive edge due to the outsourcer's reliance on customer satisfaction and bottom-line performance
- Reduced demands on general management for day-to-day operational decision making
- Better control of large, fixed-cost expenditures through variable cost contracts tailored to the organization's yearly activities

However, outsourcing may also be associated with the following problems:

- It may be difficult to accurately determine the outsourcer's qualifications and track record.
- There may be limited options for rectifying a non-working alliance.
- There may be judgment differences between the outsourcer and the organization due to rapid technology changes and corporate culture issues.
- The outsourcer may lack the critical mass and capital market assets needed to properly service large contracts.

In summary, one must weigh several factors when choosing a network design. The final network selection will reflect decisions on:

- *Organizational goals.* The role of the network in the organization must be clearly defined. If the applications supported by the network dominate the organizational requirements, then it may be necessary to develop specialized networks, or sub-networks, to provide optimal support for each type of application. This approach may involve trading simplicity and cost for increased network performance. In contrast, when simplicity and operational control are paramount, then price and performance may matter less. This change in emphasis can have significant impacts on the type of design favored. Therefore, the organization must choose the design strategy carefully to ensure that it is consistent with the business goals. Enterprise networking can be strongly influenced by the need or desire for cost reductions, improved utilization of existing network facilities, implementation of new applications and services, simplified network infrastructure, support of diverse network or internetworking requirements, and network reliability.

■ *Use of public versus private leased lines.* The organization's need for control, specialized services, and cost economies may dictate the choice between public and private lines. For example, private lines are noted for their high degree of reliability and predictable performance. However, Frame Relay public lines (which use statistical multiplexing to increase the efficiency of the line utilization, in contrast to private lines that use the less-efficient, static time-division form of multiplexing) may offer an attractive cost-effective solution for certain transmission requirements.

■ *Type of service.* The mix of traffic carried by the network helps determine the type of service (e.g., DDS lines, Frame Relay, etc.) needed. For example, Frame Relay only supports data transmissions, and thus it may not be suitable for an organization transmitting voice, data, and possibly video traffic. The respective strengths and limitations of each type of service must be weighed relative to the organization's needs.

■ *Carrier selection.* If most of the network locations are in the same city, then it may be advisable to use a local carrier. However, if most of the network locations are geographically dispersed over state or country boundaries, then an IEC[5] will be needed. Some carriers only offer a limited number of services. Depending on the requirements, this can impact the selection of the network carrier.

■ *Use of in-house versus outside staff and resources.* The organization's skill sets, budget, future growth, and specialized needs must be considered when determining the appropriate mix of internal and outside staffing and resources. Suggestions for evaluating outsourcing options are provided in Chapter 6.

2.2 Technical Overview of WAN Network Design

The secret to managing the complexity of network design is: divide and conquer! Most networks can conceptually be divided into hierarchical levels. Recognizing the number of operational levels needed in the network is a critical first step in the design process. For example, a backbone network is composed of one or more links that interconnect clusters of lower-speed lines. The clusters of lower-speed lines are called access networks. The backbone can be thought of as one level of the network topology. The access networks can be thought of as a second level of the network topology.

After decomposing the network design problem into the appropriate number of levels, the next challenge is to decide what type of topology is appropriate to support the traffic that must be carried on that portion of the network. Sometimes, more than one topology is possible and should be explored. Selecting a topology for each network segment is important because the design techniques are classified according to the type of network they produce.

For example, a backbone network is designed to carry substantially greater traffic than the access network(s), and therefore it is not surprising that the respective topologies of each may be very different. Backbone networks are usually designed with a mesh[6] topology, while local access networks are often designed as trees. A mesh topology provides greater routing diversity and reliability than a tree. However, for low-volume traffic, the cost savings offered by a tree topology can more than compensate for the loss of reliability and performance. When designing a network comprised of a backbone and a number of access networks, the standard approach is to design the backbone separately from each of the access networks. After optimizing the design of each respective component, the total network solution can be put together.

It is possible that several networks will need to be interconnected. The strategy for designing this type of inter-network is very similar to that just described. Each sub-network should be designed and optimized with respect to its particular requirements. Once the sub-networks have been designed, the interconnection between the networks can be treated as a separate design problem. When all these problems have been solved, the design solution is complete.

The subsections that follow describe methods for decomposing and solving centralized and distributed network design problems.

2.2.1 Centralized Network Design

In a centralized network, a single site or processing unit controls all communications. All communications from one device to another go through the central site using a static, predetermined routing path. An example of a centralized network is given in Figure 2.1. Typically devices, or nodes, in this type of network include the central processing unit, concentrators, and computer terminal devices.

Mainframe systems, dating back to the 1970s, typify centralized networks. In this environment, end users — that is, persons or application programs — communicate with each other through communication links between devices to and from a central mainframe computer. This pattern

Layer 7	Application	Operating System & Application Software (e.g., E-Mail, Word Processing, Database Software)
Layer 6	Presentation	Network Services: Translation, Compression, Encryption, etc. (e.g., AFP)
Layer 5	Session	Establish, Maintain, Terminate Sessions Between User Apps (e.g., FTP, SNMP)
Layer 4	Transport	Flow Control, Error Recovery, & Reliable End-to-End Network Communication (e.g., TCP, UDP)
Layer 3	Network	Data Transmission & Switching (e.g., IP, IPX)
Layer 2	Data Link	Data Encoding & Framing, Error Detection & Control (MAC, LAN Drivers)
Layer 1	Physical	Physical Interface (e.g., FDDI, Token Ring, Ethernet)

Figure 2.1   Example of centralized legacy SNA network.

of connectivity forms a star-like topology. With this type of configuration, most of the application processing and associated data maintenance is performed at a central computer facility. From a network management perspective, this configuration is attractive because it is conducive to centralized control and oversight of the network by the Management Information Systems (MIS) or Information Technology (IT) department. This, in turn, makes it easier to manage daily computer operations and to control security. If a link fails, only that part of the network is affected and fault is usually fairly easy to isolate. The star configuration may also offer economies of scale, because support for one processing site is usually cheaper than support for multiple processing sites. A star topology is also cheaper than a mesh, or fully interconnected topology. A major drawback of centralization is that it creates a single point of failure. A disaster at the main computer site can disable the entire network.

IBM's legacy SNA networks dominate this market. The next two subsections review the SNA architecture, how it is configured, and how it is being migrated and integrated with new technologies.

Subsequent subsections review the mathematical solution techniques for each of the three major sub-problems associated with centralized network design. From a technical perspective, centralized network design decomposes into three general sub-problems:

1. Concentrator location
2. Terminal assignment
3. Multi-point line layout

Each of these sub-problems is solved in turn, using any of a number of heuristic design algorithms. This chapter presents some of the better-known and most popular heuristic design algorithms. The results of each sub-problem are then combined to form an overall design solution. The heuristic design algorithms provide an initial low-cost solution, which can be fine-tuned as necessary. In practice, it is advisable to use several different algorithms to solve the same problem(s) to get as much insight as possible into what the "best" solution looks like.

The concentrator location problem, as its name implies, involves placing concentrators at key points in the network. Likewise, the terminal assignment problem involves connecting terminals to the closest or most central concentrator. The multi-point line layout problem involves clustering terminal devices on a shared line to minimize line costs. After the multi-point lines are designed, they are treated as a single terminal, and terminal assignment algorithms can be used to connect them to the closest concentrator. Thus, the interconnection of the devices to a concentrator is solved separately from the interconnection of the concentrators to each other or to the central site. The order in which these problems are solved is likely to have a significant bearing on the final outcome. It may be useful to design the network assuming in a first pass that the concentrator locations are taken as given, for the purposes of terminal assignment and multi-point line layout. Conversely, the problem can be re-solved, assuming that the terminal locations and multi-point lines are taken as given, in order to determine optimal concentrator placements.

After putting together all the pieces, the resulting network can be analyzed. Node and line utilization can be checked. There may be nodes or links with too little traffic going through them to be cost effective and thus it may make sense to remove them from the final design. It may also be advisable to place links in strategic places to provide additional capacity or redundancy in the network.

As stated previously, a centralized network uses a static path for routing traffic between devices. As defined in Chapter 8, a minimal spanning tree is the cheapest way to connect all the nodes in a network. Therefore, the lowest-cost routing path in a centralized network corresponds to a minimal spanning tree. Chapter 8 introduces Kruskal's algorithm as a means to construct a minimal spanning tree. The only problem with Kruskal's algorithm as presented there is that it does not take constraints on line capacity into account.

Obviously, the multi-point line layout should reflect the capacities of the lines to be used (which typically correspond to DDS lines operating at speeds up to 56 Kbps). When capacity constraints must be observed, the multi-point layout problem becomes a Capacitated Minimal Spanning Tree (CMST) problem. The following subsections introduce four CMST algorithms: (1) Prim's algorithm, (2) a revised CMST version of Kruskal's algorithm, (3) Esau-Williams (E-W) CMST algorithm, and (4) the Unified algorithm. Later, we demonstrate how these techniques can be used to compute a CMST for multi-point line layout. Briefly summarizing, these algorithms take the terminal and concentrator locations as given, as well as the line speed being considered for the multi-point line. From this information, the algorithms construct a multi-point line layout that minimizes costs and avoids violating capacity and cycling constraints (a minimal spanning tree, by definition, does not allow cycling because cycling would introduce redundant links).

CMST algorithms assume that there is no constraint on the number of lines coming into the terminal or concentrator. In actual practice, there may be restrictions on the number of incoming lines that can be handled. In addition, CMST algorithms rely on differences in link costs when designing the multi-point line. However, there may be situations where large numbers of terminals in the same building or location have similar link costs. In these situations, CMST algorithms are not suitable, and bin-packing algorithms should be used instead to design the multi-point line. Bin-packing algorithms are used to "pack" as many terminals as possible on a single line to minimize the number of multi-point lines created. Later, we demonstrate three bin-packing algorithms for designing multi-point lines:

1. First Fit Decreasing
2. Best Fit Decreasing
3. Worst Fit Decreasing

The terminal assignment problem, which involves associating a multi-point line or a terminal to a concentrator, may be easy to solve if there is only one natural center to which to connect. If more than one concentrator or center location is possible, a simplistic approach is to assign each terminal to the nearest *feasible* center. This is a type of greedy algorithm that may, unfortunately, strand the last terminals connected far from the concentrator or center. One way to compensate for this problem is to use a modified form of the Esau-William's algorithm. In this approach, the value of a trade-off function is computed for each terminal. This trade-off calculates the difference in cost between connecting a terminal to the first choice concentrator and the second nearest choice concentrator. When this difference is large, preference should be given to the first choice connection. This form of the terminal assignment algorithm is described in more detail later in this chapter. When a design contains stranded terminals, far from a natural center, it may also indicate that there is insufficient concentrator coverage in the network. It may be possible to better solve this problem by adding new or more strategically placed concentrators in the network.

The concentrator location sub-problem, as its name implies, involves determining where concentrators should be placed and which terminals should be associated with each concentrator. There are several types of concentrator location algorithms. The Center of Mass (COM) algorithm is used when there are no candidate sites for concentrators. The COM algorithm proposes concentrator sites based on natural traffic centers. The results of the COM algorithm must be checked for feasibility, because the algorithm may recommend a location that is not practical. Section 2.2.5.2 reviews the COM algorithm in detail. In addition to the COM algorithm, Section 2.2.5.2 introduces two other concentrator location procedures: the ADD and the DROP algorithms. Both the ADD and the DROP algorithms assume that the following information is available:

- Set of terminal locations (i)
- Set of potential concentrator locations (j)
- Cost matrix specifying the cost, c_{ij}, to connect terminal i to concentrator j for all terminals i and concentrators j
- Cost matrix of d_j, the cost of placing a concentrator at location j

From this information, these algorithms construct a set of concentrator locations and a set of terminal assignments for each concentrator location.

In addition to selecting concentrator sites in a centralized network, the ADD and DROP algorithms can also be used to identify where backbone nodes should be placed in a distributed network. Although the ADD and DROP algorithms may yield similar results, there is no guarantee or expectation that they will.

2.2.1.1 Legacy System Network Architecture (SNA)

Large IBM mainframe systems have dominated the centralized corporate network environment for more than 30 years. Although new technologies are supplanting and transforming this environment, it is still true that "much of today's business-critical applications and data remain on IBM mainframes." [CISC02] IBM's proprietary framework for supporting communication between devices (such as terminals,[7] mainframe/host processors, routers,[8] cluster controllers,[9] and front-end processors[10]) is called Systems Network Architecture (SNA). Figure 2.1 illustrates a prototypical SNA mainframe network.

The development of SNA predated the advent of open networking standards, personal computers, and distributed computing. SNA was designed to allow thousands of inexpensive end-user terminals to be shared and connected to host computers that were far too expensive to be reserved for individual use. Key benefits of SNA include:

- Efficient network and resource utilization
- Highly scalable network capabilities for supporting tens of thousands of users and OLTP (online transaction processing) applications
- Reliable uptime performance with guaranteed Quality of Service (QoS)
- Comprehensive built-in network management capabilities

Despite the lure of new technologies, SNA is still used by many banks, financial institutions, and other corporations because it provides security, manageability, and ongoing operation of entrenched legacy hardware and software with minimal new investment. For example, batch payroll processing applications — which place a premium on security over speed — are commonly run in SNA environments.

SNA was the first end-to-end communications protocol between end-user terminals and mainframes using private, leased lines. In the early days of SNA, a typical wide area network used private lines with speeds of 4800 to 9600 bps. SNA was designed to conserve bandwidth while guaranteeing response time for mission-critical applications. It did so by prespecifying all devices and traffic routing paths. In a manner analogous to making a phone call, SNA uses a connection-oriented transport technology that establishes a secure transmission path between source and destination nodes. Data streams are sent along predefined routes and arrive at the destination node in sequence. This scheme guarantees data integrity and delivery while minimizing overhead and eliminating broadcast traffic.[11]

To better understand its strengths and weakness, it is helpful to understand the structure of the SNA. SNA is based on the following major elements:

- *End users,* who generate source and destination information requests.
- *Logical Units (LUs), Physical Units (PUs), and System Services Control Points (SSCPs).* LUs are ports for end-user access and represent an application to the network. They also manage data transmission between end users. PUs are responsible for managing and controlling device resources within a node and are implemented on the host by SNA access methods and network control programs. The SSCP is a network addressable unit residing in the host processor, and is implemented by the access method. Among other things, SSCPs manage the network configuration, information flow and routing, and session activation. In SNA, end users are considered external to the network and must be represented by an LU. A session is a logical connection between two LUs that is required before communication can begin. Different session types are defined between different types of LUs (i.e., between application programs and a printer, between an application program and a

single display terminal, etc.). A session profile is characterized by the session type and features (e.g., priority, class of service, etc.) of its virtual route.

■ *Path control network,* which controls the dataflow through LUs, PUs, and SSCPs.

Figure 2.2 provides a high-level summary of these relationships. The LU (which resides in a display terminal) is used to sign on to the network and to establish a connection to an application program through the PCN (Patch Control Network). A session is established, which is a logical connection between two network addressable units. The PCN routes the traffic between the end users. The application program, in turn, has an LU. When the application program LU receives a request from the PCN, it processes the request and returns a reply back through the PCN. This processing is transparent to the end users. Sessions are continuously monitored to ensure consistent and low response times. If a node fails to reply to acknowledge control messages (for example, a node fails to respond to a poll) within a predetermined time interval, a session "time-out" occurs and the connection is broken. This scheme ensures consistent response times; however, it also has major impacts on how and whether SNA can be integrated with other non-SNA protocols. These implications are discussed further in this subsection and the next.

Access method software resides on the host processor and provides an interface between application programs and the network. Access method software is invoked by applications software (and the application programmer) to perform various network tasks using a single interface to communications facilities running on the operating system. The access method is responsible for connecting and disconnecting sessions, and maintaining network configuration and management information. ACF/VTAM

OSI Layer 7	Application	Includes network applications (e.g., FTP, SMTP, & http)
OSI Layer 4	Transport	Host to host data transmission (e.g., TCP,UDP)
OSI Layer 3	Network	Data Transmission & Switching (e.g., IP, IPX)
OSI Layer 2	Data Link	Data Transmission Between Network Node (e.g., ppp and Ethernet)
OSI Layer 1	Physical	Physical Interface (e.g., FDDI, Token Ring, Ethernet)

Figure 2.2   SNA logical units and path control network.

(Advanced Communications Facility/Virtual Telecommunications Access Method) is an example of a legacy SNA access method. VTAM tracks control messages between all NCPs (Network Control Program)[12] and determines session routes and data flows between each terminal and application. VTAM controls access to applications and prevents unauthorized program use. NCCF (Network Communications Control Facility) is an example of a subsystem under ACF/VTAM that provides an interface to network management programs to the network operators.

CICS, TSO, and IMS are other commonly used SNA application subsystems that may also reside on the host processor. They support functions such as interactive program development, database retrieval and updates, remote batch processing, and graphical display on terminals and printers.

SNA was originally designed to support three types of network connections:

1. Data channels connecting the host computer and other devices — such as printers, disk drives, etc. — within a data center
2. Token Ring LANs connecting devices in a campus within a one- to three-mile radius
3. Private or public switched dial lines operating in full- or half-duplex mode in point-to-point or multi-point configurations using Synchronous Data Link Control (SDLC) or High Level Data Link Control (HDLC) protocols.

SDLC provides serial, bit-by-bit synchronous transmission between buffered nodes operating on a common data link under centralized control. A single node — designated the primary node — controls the data link by issuing commands to a secondary node. The secondary node must respond to these commands within a predetermined time window. If it does not, the primary node interprets this as an error condition and initiates recovery actions.

SNA WANs were traditionally based on synchronous private lines. However, X.25 and Frame Relay have also been used for many years as alternative transport media. The CCITT X.25 protocol defines an interface for transmission between data terminal equipment (DTE), data circuit terminating equipment (DCE), and a public packet switched network. The X.25 protocol operates at the bottom three OSI layers. The IBM X.25 NCP Packet Switching Interface (NPSI) is needed to interface SNA with a private or public (PSDN) X.25 packet network. NPSI is a program residing in IBM communications controllers that acts as a gateway between an SNA host and the PSDN. It allows both SNA and non-SNA users to access an SNA host via a PSDN. NPSI encapsulates SNA data in X.25 packets and passes them to the X.25 network for transmission. On the destination end, the NPSI removes the X.25 packet information so the transmission can be handed back to and handled by SNA.

Conceptually, Frame Relay is similar to an X.25 packet-switched data network (A more detailed discussion of X.25 and Frame Relay is presented later in this chapter). Compared to a private line WAN, Frame Relay can offer significant savings and may make sense if the business requirements allow use of a public network for data transport. Because Frame Relay is based on a statistical multiplexing strategy, over-subscription and network congestion are possible. FRADs (Frame Relay Access Devices) and IBM BNNs (Boundary Network Nodes) are used to compensate for these incompatibilities and to adapt Frame Relay for use with SNA. BNNs are designed to allow a cluster controller residing on a Frame Relay connection to access the mainframe, making remote access to Frame Relay transparent to VTAM. FRADs are used to do protocol conversion to BNN formats on Frame Relay for legacy SNA devices. Thus, SNA devices appear to be cluster controllers or some other SNA switched resource to the NCP (Network Control Program). A process of encapsulation/de-encapsulation, as described above for X.25 networks, is followed according to Frame Relay and SNA protocol standards.

Desktop computing proliferated in the 1980s and 1990s, pushing the development of local area networks (LANs) to support high-speed connectivity between terminals, printers, and other devices. *Token Ring*[13] was developed by IBM, so it is not surprising that this was the first LAN protocol supported by SNA. IBM enhanced SNA to support Ethernet and FDDI as these two protocols gained dominance in the marketplace.

In contrast with SNA, LAN traffic requires higher bandwidth to support broadcast messages used by the protocols to exchange routing information, to locate devices, and to perform other coordination and network functions. In addition, LANs are designed to support large file transfers that can easily dominate a low-speed line. SNA traffic is typically interactive with message sizes under 1200 bytes. Although SNA is designed to prioritize traffic, it was not designed to do so *between* IBM and non-IBM protocols. Bridges can be used to extend and interconnect Token Ring or Ethernet network segments, but this is not a good solution for connecting remote locations that use low-speed serial lines. This is because SNA sessions require an acknowledgment within a prescribed interval or else they will time out, breaking the network connection. Routers are designed to deal with this SNA requirement and are typically used to interconnect LANs.

IBM has continually refined SNA over the years to support evolving market demands and networking trends. SNA evolved to allow terminal access to multiple mainframes, interconnection of multiple independent networks, and thousands of connections. Other network paradigms that SNA supports include multi-protocol networking (such as AppleTalk,[14] TCP/IP,[15] etc.), very high bandwidth[16] networking (using ATM, etc.), network management functions, and open standards (to encourage other vendors to integrate with IBM products and services).

SNA also supports IBM's distributed Advanced Peer-to-Peer Networking (APPN) and Advanced Program-to-Program Computing (APPC). APPN allows devices to communicate directly with each other without going through a central host facility, as well as internetworking of independent local area networks. These protocols provide peer-to-peer networking. Like SNA, APPN was originally connection oriented to ensure a high level of network performance. IBM introduced High-Performance Routing (HPR) to augment APPN with connection-less type service. HPR provides a number of benefits, including automatic end-to-end rerouting around network failures and congestion control. [IBM03]

Legacy SNA is distinguished from more recent versions of SNA, including APPN and APPC, by its strict and complete control over every network resource. It is based on a master–slave relationship between the host computer and terminal devices. It is also distinguished by its hierarchical as opposed to peer-to-peer architecture. Traditional SNA defined two categories of nodes: sub-area and peripheral nodes. Sub-area nodes provide intermediate node routing, address mapping and other network services. Peripheral nodes communicate with other nodes via sub-area nodes. Legacy SNA environments may employ equipment, such as FEPs and AS400s,[17] that IBM has discontinued because, nonetheless, it still works well for long-lived, mission-critical software applications. Section 2.2.1.2 discusses strategies for migrating SNA networks to more open networking environments.

2.2.1.2 Migration of SNA Networks to Open Networking Standards

In a recent product announcement, IBM acknowledged it will no longer market communications controllers that were the heart of many SNA networks: [IBM02]

> "...in many companies [the] focus has shifted from System Networked Architecture (SNA) networking to TCP/IP and other Internet technologies. In some cases, much of the SNA application traffic now runs over IP-based Wide Area Networks using routers and technologies such as TN3270 and data link switching (DLSw). The explosive growth of the Internet and TCP/IP traffic have resulted in a severe decline in the demand for new 3745 and 3746 Communication Controllers. Thus, the products are being withdrawn from marketing...."

The purpose of this section is to discuss ways in which companies are responding to the obsolescence of legacy IBM product offerings and the push toward new networking standards, such as TCP/IP.

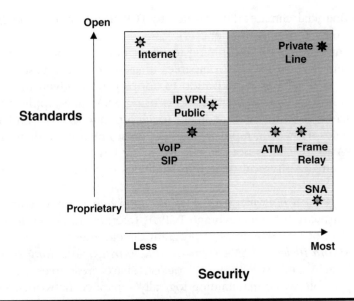

Figure 2.3   Examples of open standards versus security.

The key factor in determining a company's strategy for migrating legacy SNA network and applications is the bottom line. What are the costs and the offsetting benefits to compel a major change in infrastructure? To evaluate these trade-offs, it is helpful to analyze the relative merits of various networking alternatives. Figure 2.3 illustrates one such possible trade-off: the relative security between various proprietary and open standards (which provide flexibility and compatibility with emerging new Internet standards). Other considerations that may drive or retard movement towards TCP/IP include:

- What are the costs to convert or upgrade to TCP/IP?
- What functional requirements driving the business will be supported by the new technology? Is there a need for more speed, interconnectivity, and integration?
- What opportunities will be lost by forgoing new technology?

Many legacy SNA systems are maintained by institutions — such as banks, financial institutions, trading firms, etc. — which are, by nature, very risk averse and do not embrace radical change easily. There are substantial costs to convert from SNA because it involves a major overhaul of the hardware and software. Companies will continue to run legacy SNA applications as long as they can do so cheaply. When does a TCP/IP solution clearly dominate SNA? Although hardware and software vendors are promoting open protocols as a platform for innovative solutions, in today's financial climate, the real key to TCP/IP adoption is proving that it opens new business frontiers.

The migration path and timing to adopt new technology depends on the current infrastructure. Companies that have not invested in SNA certainly do not have a motivation to adopt it now, given the direction of the industry. Even in organizations using SNA, new applications and infrastructure investments are likely to incorporate TCP/IP and Internet technologies. However, rapid, wholesale transformation of all applications from SNA to TCP/IP is not realistic for many large SNA shops.

To summarize, typical migration strategies of SNA to TCP/IP include:

■ *Develop a platform and strategic approach* for new software and hardware that will leverage TCP/IP, Internet, and open technology. Do not use SNA in new program development.

■ *Maintain legacy SNA software and hardware as long as possible.* If SNA and TCP/IP must coexist, share "resources as much as possible while maintaining logically separated networks. This may include shared LAN segments, WAN links, bridges, routers, switches, servers, and storage devices. However, each network's workstations and applications remain transparent and unavailable to others...." [ROUT00] In this environment, there might be desktop personal computers — running TN3270 Server SNA client emulation software — connected to Token Ring or Ethernet switches, while routers connect SNA terminals and controllers to the mainframe. This allows existing SNA applications to operate alongside non-SNA applications with minimal investment.

■ *Interface SNA and TCP/IP using encapsulation, conversion, or thin-client solutions.*

 – Encapsulation. This approach involves wrapping header information around an SNA packet so it can routed through a TCP/IP backbone. Routers are used to encapsulate SNA and APPN traffic within TCP/IP packets according to one of three main protocols — remote source-route bridging (RSRB), data-link switching (DLSW), or data-link switching plus (DLSW+). Routers, acting as access points between the SNA and TCP/IP networks, implement DLSw (or one of the other protocols) to collapse the appearance of the TCP/IP network to SNA so that it looks like a single link. The router encapsulates the SNA data into TCP/IP packets, sends it through the network, "spoofing" each end of an active SNA session while transmission occurs across the network. Spoofing means that although the data transmission has not actually been received at the destination end and is still in transit, the router sends acknowledgments back to the sender that all is well (although at some later point the transmission may fail). This is required to prevent SNA session time-outs. Once data is received by a router on the destination end, it is de-encapsulated back into SNA format.

– Protocol conversion. This approach is supported provided by gateways and involves changing the data transmission of one protocol into another. A gateway must be able to handle differences that may exist in the function and mapping between all protocol levels, while remaining transparent at each end of the connection. For example, an SNA gateway node might be used to connect a TCP/IP LAN to an IBM mainframe. The gateway converts packet headers traveling from TCP/IP into SNA into the equivalent SNA protocol header format. This conversion is very CPU intensive. Gateways are often used to relieve the load on routers, so that routers can perform other network functions more effectively. IBM's Sockets over SNA MPTN is an example of a gateway that connects an SNA network and a TCP/IP network. When it forwards data from TCP/IP, it uses MPTN protocols; and when it forwards data from SNA to an IP host in a TCP/IP, it uses TCP/IP protocols. [IBM94], [LEWI98]

– Screen scraping/thin-client solutions. In this approach, a screen scraping API is used to display only keystrokes and screen updates on the IBM 3270 terminal, separating the display of the program results from the actual application data and its processing. The API emulates and creates a virtual IBM 3270, creating a bridge to execute user transactions and to intercept commands and data. This approach moves the management and execution of the application *from* the host or workstation *to* a server, allowing IBM 3270 terminals to connect to SNA, Windows, and other non-SNA applications. This method is widely applicable and compatible with many applications and network protocols, and is supported by a number of commercial products. For example, Microsoft's Windows NT® Terminal Server Edition allows multiple users to run applications separately on a server. Citrix's Independent Computing Architecture (ICA®) is another example of a protocol to separate application logic from the user interface. From an IBM 3270 emulation session, using Rumba 2000 Citrix and Terminal server edition, a user can execute a command to a host computer to establish an APPC connection. After establishing the connection, the host activates a transaction program (TP) residing on the workstation or terminal. "This program displays the image sent from the host. The host knows which PC to send the document to by using the PC's LU address." [RUMB99] Rumba's Communications Monitor can be used to provide real-time network management of PU and LU sessions for large numbers of users. Rumba's TP Director associates incoming requests to a mainframe or AS/400 host with the user making the request, mapping usernames with LU addresses. "TP is started only on that user's desktop and from there on out, the image viewer works normally. For SNA environments, RUMBA Office for Citrix and Windows Terminal Server includes the RUMBA router which has been optimized for these operating systems." [RUMB99].

■ *Keep IP networks private, except for selected Web-to-host applications.*
  – In general, the industry migration has been from private SNA to private IP network protocols. This shift allows the organization to capitalize on the benefits of open technology, without the security risk associated with using a public infrastructure.
  – Web-to-host applications. In some cases, corporations want to permit IP clients to exchange data over intranets or the Internet. Applications that might require this type of access "range from pre-qualifying loan applications, offering advertising subscriptions, providing tax account status, online library catalogs to business partners, order tracking/inventory information and intranet applications for mobile users." [OPEN02] Web-to-host applications allow a browser to access information on the mainframe. The browser communicates via the Internet using the Hypertext Transfer Protocol (HTTP). Because this approach opens the corporation to more potential unauthorized intrusion and security risk, it also implies that adequate precautions (such as firewalls, etc.) must be in place. There are a number of commercial products to support this capability. IBM's Enterprise Extender software for routers allows IP clients to access SNA host computers and servers, and is one possible option. Open-Connect® has a number of product offerings — including Web-Connect Pro — to implement Web-to-host connectivity.

Figure 2.4 illustrates several SNA and TCP/IP internetworking strategies described above.

In addition to internetworking compatibility, security is a major concern in migrating from SNA to TCP/IP. Security is enhanced by minimizing the number of external and multi-protocol access points allowed into the network and by implementing multiple safeguards designed to work in concert. Security precautions can be implemented on several levels. Suggestions for securing transmission of TCP/IP and SNA traffic include:

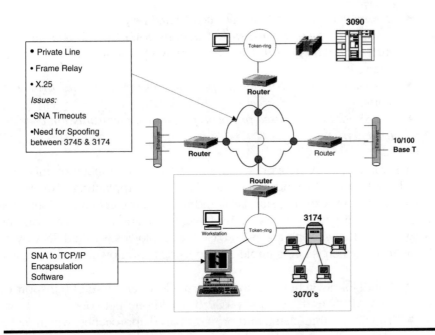

Figure 2.4   Example of SNA and TCP/IP internetworking.

■ *Require user IDs and passwords.* These should be used for each application. Security tools such as IBM's RACF (Resource Access Control Facility) should be used. RACF is used to identify, verify, authorize, and control system users. It also provides auditing and other security management capabilities.

■ *Use host access control products, such as IBM's ACF/2 (Advanced Communications Function) and Communications Server products.* The Advanced Communications Function/Network Control Program (ACF/NCP) controls network attachments, performs error recovery, and routes data. IBM is replacing ACF with Communications Server products, which are designed to work in conjunction with TCP/IP to provide a full range of security and network management capabilities.

■ *Implement firewall and gateway security.* Routers can provide packet-level filtering at the OSI network level. Circuit-level gateways operate at the OSI session layer and can be used to verify the legitimacy of a requested session. They also can be used to hide, and thereby help secure, information about the private network because data flowing through the gateway appears to have originated there. Application-level gateways (or proxies) can also be used to filter packets at the application level and to log user activity.

> ■ *If remote dial-in or Internet access must be provided to users, use additional security measures such as RADIUS (Nortel's Remote Authentication Dial-in User Services). A VPN should also be implemented.*
>
> ■ *Use encryption techniques to provide secure end-to-end transmissions.*
>
> ■ *Use SSL or SHTTP based protocols for securing transmission over the Internet.*

In summary, SNA networks still support a vast number of mission-critical applications around the world. TCP/IP is supplanting SNA as the new corporate networking standard, and use of intranets and the Internet is becoming ubiquitous. Access to IBM mainframe and AS/400 applications using a TCP/IP backbone and Web-based technologies is called SNA-over-IP (SNAoIP), which is being implemented using the following technologies:

■ Data Link Switching (DLSw), which allows end-to-end transport of SNA/APPN traffic across a TCP/IP backbone network
■ Tn3270(E), thin-client, and browser-based approaches, which make SNA applications accessible from TCP/IP clients running on desktops and workstations (as opposed to SNA/3270 dumb terminals or emulators)
■ Routers and middleware gateways, which perform messaging and other conversion and security functions to mediate communication between SNA and TCP/IP applications

Migration and integration of these technologies can generate tremendous savings for the corporation. However, this conversion must preserve essential elements of the SNA environment: dependability, manageability, and security.

2.2.2 Distributed Network Design

A distributed network disperses the computing and communication capabilities throughout the network, allowing devices to communicate directly with each other. The decision to operate a distributed network versus a centralized network is largely driven by geographic, operational, economic, and performance considerations. Distributed networks can be used to:

■ Reduce the impact of a single point failure on the function of the network as a whole, thereby providing diversity, backup, and disaster recovery.

- Facilitate phased, incremental system upgrades without fear that the entire network will be disrupted.
- Reduce reliance on any one vendor, encouraging competitive buying opportunities.
- Reduce bottlenecks at high-volume usage centers by distributing workloads.
- Provide improved functionality and service.

The subsections that follow discuss these types of distributed networks: Metropolitan Area Networks (MANs), Wide Area Networks (WANs), and client/server networks.

Also reviewed are the mathematical solution techniques associated with distributed network design. In a distributed network, traffic is no longer routed in a static pattern through a central host processor (as it is in centralized networks). Instead, traffic may flow between nodes, over various possible routes. This communication pattern gives rise to a mesh topology. In general, distributed network design is harder analytically and more computationally complex than centralized design.

Distributed network design can be decomposed into three major sub-problems:

1. Develop network topology (which specifies node and link connections).
2. Assign traffic flows over the network links.
3. Size the line capacities, based on the topology and estimated traffic flows.

Solving these problems — especially the latter two — is difficult because the link flows are inextricably tied to the line capacities. In addition, in a mesh topology, there may be many possible routing paths for the traffic flows. Without explicit knowledge of the traffic flows, it is difficult to estimate the required line capacities, and vice versa. Mathematically, the simultaneous solution of both (2) and (3) above is very complex. According to [KERS93], the expected computational complexity of any mesh design algorithm is at best "$O(N^5)$ because there are $O(N^2)$ candidate links to consider, and consideration of a link would involve doing a routing, which requires finding paths between all pairs of nodes, itself an $O(N^3)$ procedure." As discussed in Chapter 8, a procedure with a computational complexity of $O(N^5)$ is considered very complex indeed, and is impractical for solving problems with a large number N of nodes.

Integer programming techniques can be used to find exact, optimal distributed design solutions in cases where the network does not exceed more than a few dozen nodes. [KERS93, p. 305] In this context, an optimal solution is one that can be mathematically proven to be the lowest-cost design that does not violate cycling and capacity constraints. For a more comprehensive survey of distributed network design methods, the reader is referred to [KERS93].

Within the literature, one mesh topology algorithm stands out for its low computational complexity and high-quality results — the Mentor algorithm. Remarkably, the computational complexity of the Mentor algorithm is $O(N^2)$. Its low computational complexity is achieved by *implicitly* routing traffic as the algorithm proceeds. Considerable attention is given to the Mentor algorithm because it is so powerful and because it provides consistently good results (see Section 2.2.6.2). Consistent with standard methodology, the Mentor design algorithm decomposes the design problem into several smaller problems. The first problem it tackles is selecting locations for the network backbone. The algorithm then proceeds to assign links, based on certain traffic routing assumptions. The simplifying assumptions that Mentor uses are based on the following insights about the characteristics of "good" distributed network designs: [KERS93], [CAHN98]

■ Traffic flows follow short, direct paths as opposed to long, circuitous ones.

■ Links should be highly utilized, as long as performance is not impacted adversely.

■ Long, high-capacity links are used in preference to short or low-capacity links to achieve cost economies of scale.

Although these characteristics are not wholly compatible with each other, Mentor attempts to achieve a reasonable balance between them. Briefly summarizing, the Mentor algorithm takes as given the terminal and concentrator locations, a maximum limit on the amount of usable line capacity that can be used, and a cost matrix specifying the cost of links between each of the terminal and concentrator locations. The output of the algorithm is a design layout that minimizes cost without violating the prespecified line capacity restrictions. Several variations of the Mentor algorithm used to reflect different uses of technology (e.g., multiplexer-based networks, router-based networks, etc.) are discussed in Section 2.2.6.2.

2.2.2.1 Metropolitan Area Networks

The protocols used in a LAN do not support network coverage over a large, dispersed area. The need for high rates of data transport over longer distances led to the development of metropolitan area networks (MANs). MANs are designed to support high data transmission rates (in the same range as LANs or higher) and interconnectivity across an entire city or campus. The demand for MANs is increasing commensurately with the growing popularity of complex and high bandwidth applications, such as:

- High-definition television
- High-quality radiology and medical imaging
- Internet streaming
- Multimedia applications
- Videoconferencing
- Voice-over-IP (VoIP)

MANs can be connected to other MANs, WANs, or LANs using routers and bridges.[18] Generally, bridges are the preferred medium for interconnecting MANs, except in cases where multiple network types and protocols will be supported. In this case, routers are generally used. MANs can be operated as public or private networks. A private network, with dedicated lines,[19] can be more cost effective than a public network with switched lines[20] when the transmission volume is sufficiently large. A public MAN is comprised of a transport and an access network component. The transport network component is maintained by a telco or communication provider, while the access network component is usually maintained by the subscriber organization. Figure 2.5 presents a conceptual overview of the service provider infrastructure. Service providers offer a broad range of transport options to satisfy a diverse set of customers and requirements.

2.2.2.1.1 IEEE MAN Standards

The first IEEE 802.6™ standard for MAN specification was withdrawn and superceded by other IEEE standards. This standard was based on Distributed Queue Dual Bus (DQDB) technology and was designed to support data packet transmission at rates of 1.5 to 155 Mbps. DQDB uses two optical fiber buses to connect similar types of network components, as does FDDI, but at a higher transmission rate. In this scheme, the buses are separate and uni-directional, and use the same signaling frequency. Network nodes are tied to each other as a logical bus. Thus, each node is located both upstream and downstream from other bus nodes. DQDB technology is based on the fact that each station knows about frames queued at all other stations. Under DQDB, end nodes continuously transmit empty data frames around each ring. Whenever a station on one of the network nodes has something to send, it generates a frame request on the bus that is carrying traffic away from it. If redundancy is required in the network, a DQDB can be configured as a looped bus in which the two ends are co-located. If a fault occurs in one of the buses, the nodes

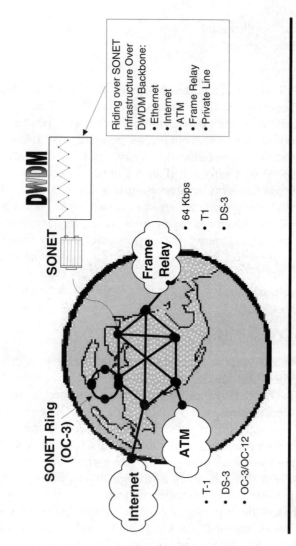

Figure 2.5   Conceptual overview of service provider infrastructure.

at either end of the fault can act as both beginning and end points. DQDB is obsolete and slow by today's standards.

The most recent MAN IEEE standards — 802.16™ — are designed to support broadband wireless MAN communications. In particular, the IEEE 802.16™ (Air Interface for Fixed Broadband Wireless Access Systems) and IEEE 802.16.2™ (IEEE Recommended Practice for Local and Metropolitan Area Networks Coexistence of Fixed Broadband Wireless Access Systems) standards define protocols for wireless transmission between a subscriber transceiver and (generally, a service provider's) base transceiver station. These standards support QoS mechanisms for voice, data, and video traffic. They are designed with developing countries in mind because they allow creation of high-capacity wireless access networks in just weeks even if a wired infrastructure is lacking. [IEEE02] IEEE 802.16™ specifically deals with the first-mile/last-mile connection between the service provider's point-of-presence (POP) and the customer's premise equipment (CPE). Until recently, access to the MAN was based on copper wire, using time division multiplexing (TDM)[21] at fairly low speeds. Thus, the availability of high-speed wireless and fiber access is an important development for high bandwidth communication.

2.2.2.1.2 Ethernet-to-MAN Connectivity

Because Ethernet traffic dominates the LAN scene, Gigabit Ethernet and 10 Gigabit Ethernet backbones are often natural choices to transmit traffic from high-speed Ethernet LANs onto a fiber-based MAN. 10 Gigabit Ethernet offers transport over longer distances on dark fiber than Gigabit Ethernet — 40 kilometers versus 5 kilometers. 10 Gigabit Ethernet is compatible with standard SONET transmission techniques and equipment in a MAN environment. Because it is designed to operate at a rate equivalent to SONET's OC-192, it is possible to transmit Ethernet data over SONET using optical add/drop multiplexers (OADMs) and repeaters. Although 10 Gigabit Ethernet is significantly faster and more scalable than Gigabit Ethernet, it only runs on optical fiber in full-duplex mode. Gigabit Ethernet works on both copper and fiber facilities. The network designer must decide which is more appropriate, given the network environment, requirements, and plans for the future.

If Ethernet data dominates the traffic to be sent over the MAN, there is little or no incentive to embrace transport on T1, Frame Relay, or ATM as this will entail additional network complexity (in the way of protocols, equipment, and staffing skills needed to maintain the network). However, Ethernet was not designed to provide the same levels of QoS and reliability as legacy WAN protocols. It also does not have a weighted, fair allocation scheme to prevent an application from sending very large amounts of data and monopolizing the network resources. Although Ethernet has cost and simplicity advantages over T1, Frame Relay, and ATM technologies in transporting large, bursty (or sporadic and changing) volumes of data traffic, it also has disadvantages that require compensating, enabling new technologies. These technologies include Resilient Packet Ring (RPR), Multi-Protocol Label Switching (MPLS), and Dense Wavelength Division Multiplexing (DWDM).

2.2.2.1.3 SONET, DWDM, RPR, and MPLS Technologies in MANs

Whether from (10) Gigabit Ethernet, wireless, or some other type of access link, LAN traffic is handed to the MAN for transport and eventual hand-off to the customer premise equipment (CPE) at the destination site. For over a decade, service providers have based their MAN transport networks on SONET/SDH technology. Figure 2.6 illustrates a traditional SONET MAN architecture connecting two data centers. Figure 2.7 illustrates a more state-of-the-art MAN architecture connecting two data centers using DWDM on a SONET ring.

The SONET/SDH standard is based on optical fiber technology. The term "SONET" is used in North America, while the term "SDH" is used in Europe and Japan. SONET/SDH circuits offer far superior performance and reliability than microwave or cable circuits. In essence, SONET/SDH is an integrated network standard that allows all types of traffic to be transported on a single fiber optic cable. Because SONET/SDH is a worldwide standard, it is possible for different vendors to interface their equipment without conversion. SONET/SDH efficiently combines, consolidates, and segregates traffic from different locations through one facility. This ability — known as grooming — eliminates backhauling[22] and other inefficient techniques used in some carrier networks. SONET/SDH also eliminates back-to-back multiplexing overhead by using new techniques in the grooming process. These techniques are implemented in equipment known as add-drop multiplexers (ADM). SONET/SDH employs digital transmission schemes and supports efficient Time Division Multiplexing (TDM) operations.

SONET was specifically designed to provide carrier-grade service and offers these benefits:

- QoS mechanisms
- Automatic restoration and recovery
- Full network management capabilities
- Compatibility with a diverse and broad array of vendor equipment and offerings
- Support for a full range of transport protocols (e.g., T1, Frame Relay, ATM, etc.)

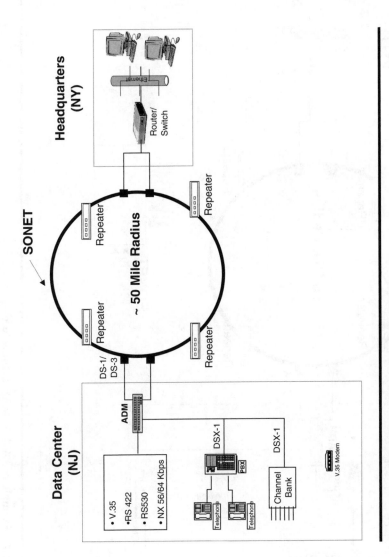

Figure 2.6   Traditional SONET MAN architecture.

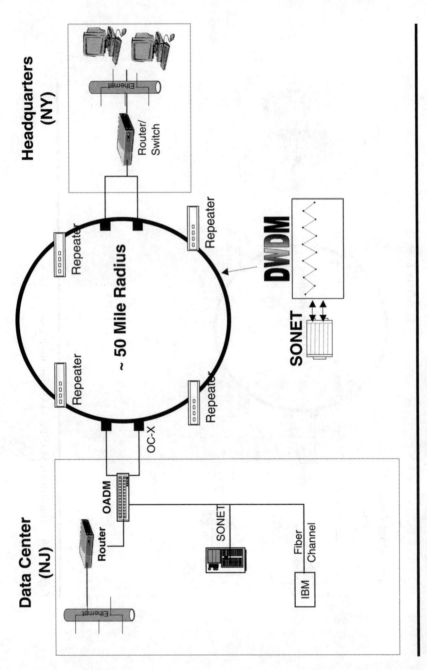

Figure 2.7 Example of DWDM MAN architecture.

These capabilities come at a cost, because a substantial amount of overhead is built into SONET to perform these functions. SONET was designed for highly reliable, circuit-switched, point-to-point transport of voice traffic and low-bandwidth data transmissions. However, SONET is not designed for high-volume, bursty data traffic. Because SONET is circuit based, reserved bandwidth cannot be used when it is idle to support other services. This is because SONET does not allow statistical multiplexing between circuits. In addition, the payload or data portion of a SONET frame has a fixed format and size (from 87 bytes for OC-1 to 16,704 bytes for OC-192); so if the data transmission does not fit exactly into the standard SONET frame size, this will lead to the creation of frames that are not fully loaded and, thus, wasted capacity. [AXNE03] SONET does not handle multi-cast traffic efficiently either. It sends separate, multiple copies of multicast packets on separate circuits allocated to each destination. LAN protocols send a lot of broadcast traffic, so sending LAN traffic over SONET is a mismatch with SONET's design. SONET's very high reliability also comes with very high overhead because it typically reserves half the fiber ring bandwidth for restoration and recovery. Effectively, this amounts to approximately 50 percent of the ring capacity. SONET overhead also comes in the form of equipment — such as FRADs, ATM Integrated Access Devices (IADs), or routers — needed to provide LAN to MAN/WAN interfaces to SONET.

In the past, when service providers needed to add capacity to their backbone, they would lay down new fiber rings. However, this is no longer cost effective, given the economic pressures on service providers and customers' insatiable demands for bandwidth. In response, an alternative technology — Dense Wavelength Division Multiplexing (DWDM) — has been developed to stretch the bandwidth capabilities of SONET. DWDM transmits data as pulses of light at a specified wavelength (or lambda). Each wavelength or lambda represents a separate data channel, much like a radio channel that operates at an assigned frequency. DWDM is a method of densely packing wavelength frequencies, thus increasing the bandwidth capacity of existing fiber. This is illustrated in Figure 2.8.

Adding new DWDM network terminal equipment to increase the number of channels on a SONET infrastructure can be expensive. However, it is much less expensive than installing new fiber. Using traditional SONET, transmission of 40 Gbps over 1000 miles would require 16 fiber pairs and approximately 272 regenerators. In contrast, a 16-channel DWDM system would require a single fiber pair, and 4 amplifiers every 193 miles. [RYAN97] "DWDM systems can support more than 150 wavelengths, each carrying up to 10 Gbps. Such systems provide more than a terabit per second of data transmission on one optical strand.... " [NORT03] Although DWDM technology is not yet widely deployed, it is being used by an

Years of Usage	Optical Carrier Level	Data Rate
1985-2000	OC-1	51.84 Mbps
	OC-3	155 Mbps
	OC-12	622 Mbps
	OC-48	2.1 Gbps
	OC-192	9.6 Gbps
> 2000	OC-256	13.3 Gbps
	OC-768	39.8 Gbps
In Progress	OC-768	39.8 Gbps multiples (w/ DWDM)

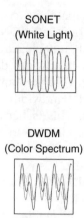

SONET
(White Light)

DWDM
(Color Spectrum)

Figure 2.8   SONET and DWDM data transmission rates.

increasing number of large companies. Enkido — the first company to offer OC-768 — provides 40-Gbps service on one lambda and one optical channel of a DWDM fiber. Enkido's customers include Deutsche Telecom, NBC, Disney, Department of Defense Advanced Research Projects Agency, NASA, and Nippon TV. Enkido is targeting customers running high-quality, real-time multimedia applications. [METC00] New DWDM systems are in development to support transmission speeds in excess of 10 trillion bits per second (10 Tbps) per fiber over OC-768. "This translates into the theoretical capability of one fiber to support, simultaneously, an active Internet connection to every household in the U.S." [SEAR00]

Because DWDM supports simultaneous transmission of many data streams, it allows carriers to provide different levels of service to different customers on a shared fiber ring. DWDM is also designed to work with SONET/SDH, ATM, Frame Relay, and other protocols, so that existing SONET infrastructure can be preserved.

It is important to note that DWDM is a layer 1 photonic multiplexing scheme. It does not provide layer 2 media access control (MAC) for transporting packets across a shared medium. Therefore, it lacks error checking and control over packet transmission on the fiber ring. To address this deficiency, Resilient Packet Ring (RPR) technology was developed to augment SONET in concert with DWDM. RPR is based on the IEEE 802.17™ standard, which provides a media access control (MAC) layer for data transport on a MAN. RPR is independent of the physical layer and can be implemented over SONET, dark fiber, and DWDM. It integrates transport, routing, and switching functions in a single platform, providing significant savings to both carriers and customers. Traditional SONET networks require costly point-of-presence interfaces on the core router.

Using RPR, port connections are consolidated, resulting in an approximately 50 percent capital expenditure reduction and lowered operating expenses. [RESI03] Cisco offers a similar scheme using SRP (Spatial Reuse Protocol) and DPT (Dynamic Packet Transport), which it supports on most of its Gigabit Ethernet and SONET products.

RPR is based on the premise that each packet on the ring will eventually reach its destination, irrespective of the path it takes. Because all nodes are aware of the status of all other nodes on the ring, packet handling on the ring can be reduced to three basic actions: (1) insertion of a packet on the ring, (2) forwarding a packet through the ring, and (3) removing a packet from the ring. This simplification greatly reduces the amount of processing needed at each node to determine packet routing.

A key feature of RPR is its SONET-like 50-millisecond restoration and recovery capabilities. RPR automatically detects fiber cuts, node failures, and other outages on the ring and reroutes traffic around them. SONET reserves half the ring topology for backup purposes to guard against these types of failures. In contrast, RPR uses both fiber rings to carry traffic during normal operation and reverts to using half the ring only when a failure occurs. RPR's recovery scheme is based on wrap and steer protection. If a failure is detected on one ring, traffic is "wrapped" around adjacent nodes and rerouted in the opposite direction onto a second ring. Figure 2.9 portrays this protection mechanism. If a ring segment fails,

Figure 2.9   RPR wrap recovery mechanism.

"steer" protection notifies all nodes in the ring. Each node updates its network topology references and, if necessary, steers traffic around the failure by transmitting on a different ring.

According to the Resilient Packet Ring Alliance, [RESI03] the real benefit of RPR for carriers transmitting voice and data traffic is the scalability and cost advantage it affords relative to SONET/SDH as the number of users and the bandwidth requirements increase. This benefit is realized through:

- Support of broadcast and multicast traffic
- RPR implementations over Ethernet to provide variable-length frames to minimize overhead as compared to SONET
- Implementation of dynamic mechanisms for restoration and recovery on a MAN ring
- Unification of voice and data transport over the same infrastructure
- Lowered carrier equipment costs and ability to leverage existing fiber infrastructure
- Efficient bandwidth management based on statistical multiplexing
- Simplified service provisioning and ability for carrier to offer customers to purchase bandwidth as they need it, in small increments
- Spatial reuse, which allows simultaneous transmission between nodes on the network as long as the transmission paths do not overlap
- Packet classification and QoS guarantees to guard against delay and jitter to support time-sensitive and high-availability applications

MPLS is another protocol designed to work in tandem with RPR and DWDM to support carrier grades of service. MPLS (Multi-Protocol Label Switching) is an IETF specification for enhancing the efficiency and resiliency of packet routing across different types of topologies, applications, hardware, and transport protocols (including IP, ATM, and Frame Relay — hence the derivation of the name "Multi-Protocol"). Most telecom service providers are implementing MPLS on packet-based networks to perform traffic engineering and value-added services. In essence, MPLS is used to convert a connection-less IP path to a connection-oriented virtual circuit path. Some of the major advantages of MPLS include: [IEC03]

- Integration of IP and ATM in the network
- Support for virtual private networks (VPNs)
- Flexibility and robust recovery procedures to reroute traffic around network failures and congestion
- Dynamic creation of traffic engineered QoS and class-of-service (CoS) paths through an IP network, which allows service providers to manage traffic flows based on priority and service level agreements, so that premium service subscribers can be given preference on the network

MPLS distributes the complexity of processing packets to optimize the function of edge and core routers in the service provider's network. Services such as QoS, VPNs, encryption, and authentication are applied to packets upon entry to the IP network, so that "individual routers are not required to support all these services for all of the service provider's customers." [SPIR02] MPLS assigns labels that are appended to IP packets by a label edge router (LER) configured on the network edge. These labels contain a wealth of information, including the packet destination and path, bandwidth, delay, source IP address, layer 4 socket number, and the required level of service. Routers in the network core — called label switch routers (LSRs) — process the label information and pass along the packets. Having layer 2 information about network links available at layer 3 (IP) allows the LSRs to process and forward packets much faster than if they had to look up the packet destination in a routing table. Because MPLS is connection oriented, packets going to the same destination follow the same path through the network. Thus, routers only need to switch (as opposed to route) packets along a predefined LSP (label switch path).

Extensions to MPLS, as reflected in GMPLS (Generalized Multiple Protocol Label Switching), are under development to support more capabilities and multiple switching types (TDM, lambda, and fiber). GMPLS is also being developed to make it easier for service providers to do provisioning and network management. [MPLS02]

2.2.2.1.4 Key Users and Basic MAN Configurations

Having reviewed some of the major MAN technologies, the discussion now focuses on key users of the technology — who are in every market sector! "Overall U.S. optical MAN services revenues will grow from $1.9 billion in 2001 to $6.9 billion by 2006, as both consumers and business users move to higher speed Internet connections, businesses outsource corporate IT functions, and the localization of data traffic continues." [SCHO02] The main types of MAN users are:

- *Academic, government, and research institutions.* There are numerous examples of MANs supporting these types of organizations around the world. To name a few, these include:
  - *WinstonNet.* This fiber optic MAN connects numerous sites around Winston-Salem, North Carolina, to Internet2.[23] WinstonNet interconnects libraries, universities, hospitals, research institutions, schools, recreation centers, churches, and local government facilities. It is being used in many ways. For example, it is used to (1) provide training and high-speed Internet access to people without home computers, (2) link engineering and biotechnology

businesses that need to send and receive huge amounts of data, and (3) allow people to pay parking tickets, etc. [FEDE03]

- *PREN.* The Portland Research and Education Network (PREN) connects OGI, OHSU, and PSU — the three principal research and education institutions in the Portland metropolitan area — to Internet2.

- *Harrisburg, Pennsylvania.* The city has implemented a MAN to interconnect state agencies and over 25,000 users. [WELS03]

- *TANet.* Taiwan Academic Network (TANet) connects 12 regional networks and over six million users in the largest Asian MAN.

- *AbMAN.* Aberdeen Metropolitan Network in Scotland.

- *CAMAN.* Cluj Academic Metropolitan Area Network in Romania.

■ *Companies with large bandwidth requirements.* Corporate users are utilizing MAN technology to interconnect computing sites and data centers within a limited campus, particularly for disaster recovery and business continuance. A few examples of large corporations using MANs include:

- *Citigroup and Salomon Smith Barney* are implementing a MAN to interconnect all their London sites using Cisco routers and software. [CISCO01B]

- *America Online* has leased a dark fiber infrastructure from Metromedia Fiber for a corporate MAN. [MFN99]

- *Telecom Ottawa* is using a 10-Gigabit Ethernet MAN installed by Foundry Networks. [FOUN02]

■ *Small to medium-sized companies.* These businesses are using MANs for many of the same reasons as large companies. However, small companies have correspondingly smaller budgets, staffing, and resources. One of the main differences between large and small companies is how they deploy and implement MAN technology, as discussed in more detail below.

When deploying a MAN, several key decisions must be made. The first is to decide how to access the MAN from the end-user sites. The next decision is to determine the underlying technology and network configuration to carry traffic across the MAN. Most MAN configurations consist of fiber rings or point-to-point connections. Ring connections usually carry traffic between multiple sites. Point-to-point connections are generally used when there is a large amount of traffic and sufficient justification to install dedicated facilities between just two sites. Finally, a decision must be made as to how much of the deployment should be outsourced. Every aspect of the MAN installation and maintenance can be outsourced if necessary. Service providers offer a full suite of services (including QoS, security, network management, VPN, resiliency and restoration capabilities)

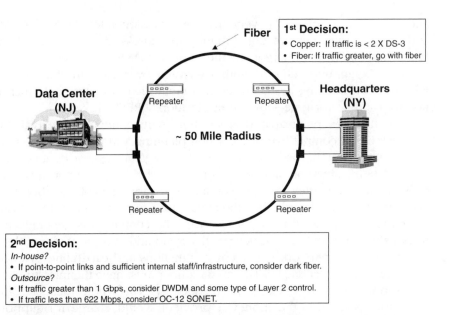

Figure 2.10   MAN Configuration and Decision Points

to suit every budget and requirement. The appropriate course to take in making these decisions ultimately relates to the size and the sophistication of the organization and the type of network configuration needed to connect the end sites. Figure 2.10 provides a synopsis of these decision points.

Telcos and service providers work with large companies on an individual case basis (ICB) to install the level of MAN infrastructure needed. If a company is already using lit buildings,[24] this facilitates connection to the MAN. In addition, if the end sites are near existing fiber rings or other facilities that the service provider has already installed, this will be less costly than installing an entirely new infrastructure tailored to support the company's individual requirements. However, if the customer's bandwidth requirements would consume a significant portion of the provider's available capacity, a dedicated network solution may be necessary. This is because providers typically do not want to sell their entire network capacity to one client. There are several possible choices for high-capacity access to a MAN infrastructure: OC-12 to OC-192, dark fiber,[25] DWDM, Gigabit Ethernet, and 10 Gigabit Ethernet. The latter two are suited for LAN protocol handoffs, while the former choices are suited for WAN protocol hand-offs.

SONET remains a low-cost solution for MAN transport of voice and private line traffic. However, it is not as cost effective for bursty data traffic and does not scale well at very high transmission rates. SONET ADMs and Ethernet switches are not designed to address the need for a MAC layer in a MAN environment. In the case of SONET, this results in the network being underutilized with excessive overhead. In the case of Ethernet switches, this results in nondeterministic traffic control, which is not appropriate for time-sensitive applications. [RESI01]

Dark fiber is very cheap and a lot of third-party contractors are available to install it at a competitive price. When planning a MAN across an entire metropolitan region, it is imperative to have legal access, or right of way, to the data transmission facilities. Therefore, before a contractor or service provider lays fiber, they must also obtain necessary rights of way, bonding, licenses, etc. When laying fiber, the contractor's responsibility is to provide an infrastructure, not a managed service. The contractor is generally not responsible for installing the devices that send and control photonic flow, or for providing network management functions and capabilities. Hence, large companies that use DWDM on dark fiber must have the infrastructure and sophistication to manage the technology. The whole idea of using dark fiber is to save the money a provider would charge for support, maintenance, troubleshooting, and repair service by performing these functions in-house. It should be noted that not all types of older fiber are compatible with DWDM. If a fiber upgrade is needed, this should be factored into the cost analysis of this approach.

Typically, dark fiber installations are used for simple, point-to-point or ring topologies that do not require fancy routing. For example, dark fiber might be used to provide a simple link to carry bulk data traffic between two sites exchanging interactive or large file transfers. Increasingly, Gigabit Ethernet is being used on dark fiber MANs. 10 Gigabit Ethernet is also being used because it is capable of supporting network attached storage (NAS)[26] and storage area networks (SANs).[27]

RPR over fiber rings is also used for short- and long-distance MAN connectivity. In this scenario, RPR is used as a substitution for a layer 2/3 protocol to provide switching/routing, restoration, and network services capabilities over DWDM. For LAN interconnectivity, Ethernet runs over a DWDM infrastructure, which routes via RPR or similar protocol. In this configuration, optical add/drop multiplexers (OADMs) are used to replace SONET ADM equipment at end-user sites. This is shown in Figure 2.11. This reduces the cost and complexity of the network as capacity is added. Currently, fiber rings can support upwards of 80 lambda wavelength channels each operating at OC-192, with trials being performed utilizing OC-768 technology. If there is sufficient demand for point-to-point links over DWDM, the only equipment needed is a customer premise device to convert the traffic to specific wavelengths and to perform multiplexing. Point-to-point networks can be implemented with (for additional network

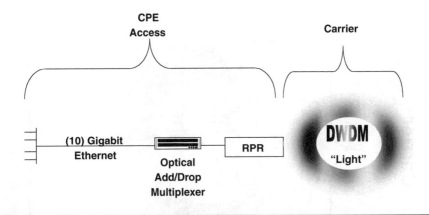

Figure 2.11   Customer premise equipment to carrier: Gigabit Ethernet over RPR and DWDM.

management capabilities) or without OADMs. This is a typical approach offered by telcos and carriers.

Service providers also offer DWDM transport with MPLS. This approach allows QoS and value-added services for time-sensitive and demanding applications, and builds on the provider's existing SONET infrastructure. In Figure 2.12, customer premise equipment running Gigabit Ethernet traffic is transported over DWDM using MPLS and RPR.

Companies using TCP/IP must decide what type of IP packet format should carry traffic in the fiber ring: IP over ATM over SONET, or IP over SONET, or IP over (10) Gigabit Ethernet. If other transport protocols are used, a similar decision must be made. This determination has significant implications on the network protocols and equipment that must be supported. Conversely, the network protocols and equipment have significant impacts on the range of appropriate packet formats. As traditional circuit-switched networks are being transitioned to IP networks, DWDM is becoming increasingly important in handling IP and ATM traffic.

Smaller businesses and organizations do not have the bandwidth requirements or resources to justify a custom MAN solution; however, they still need capacity in excess of T1 speeds (1.544 Mbps). As Fortune 1000 companies load their networks with 10 to 100 Mbps and higher of traffic, ATM and Frame Relay are no longer cost effective or fast enough for MAN transport. This market segment typically needs high-speed Ethernet connectivity to transmit voice, data, and video applications. In the past, these companies based their network infrastructure on copper because they did not have the money for fiber (to support DS-3–OC-12 speeds or more). Now they are asking service providers to build the MAN infrastructure for them and to provide access with Gigabit Ethernet hand-offs. Service providers

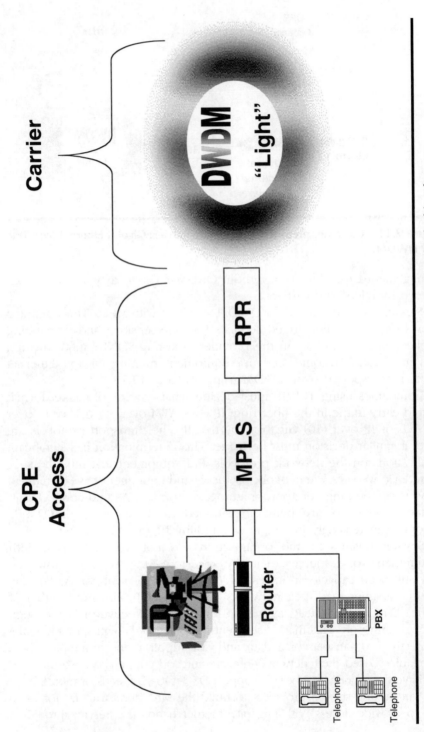

Figure 2.12 Customer premise equipment to carrier: DWDM, MPLS, and RPR over Gigabit Ethernet.

are encouraging this by offering high-bandwidth Ethernet switching capabilities that are very competitively priced relative to SONET. DWDM over dark fiber does not make sense for small companies because they typically cannot troubleshoot and fix it if anything goes wrong. Outsourcing the MAN deployment can be very attractive to smaller businesses because they lack the economies of scale, resources, and skill sets to do this in a cost-effective way.

In summary, MANs are being transformed by an array of new technologies, which include Gigabit Ethernet, DWDM, RPR, MPLS, and wireless communications standards. These technologies offer a simplified network based on fewer network elements, protocols, and fiber. They provide greater speed, reliability, flexibility, and class-of-service advantages, allowing service providers to deliver new MAN product offerings to a variety of consumers at lower cost. The demand for these services is being fueled by increasingly sophisticated and demanding applications.

2.2.2.2 Wide Area Networks

Wide area networks (WANs) connect computers and other devices separated by very large distances. WANs support data transmission between cities, states, and countries. The most commonly used devices for interconnecting WANs and LANs include bridges, routers, extenders for remote access, brouters, and gateways. Although WANs support transmission of many of the same types of applications as MANs, they typically do so at a lower speed. For example, some of the more common line speeds used in WANs are 1.544 Mbps (T1),[28] 44.746 Mbps (T3),[29] and 51.84 Mbps (OC-1).[30]

In today's WANs, traditional analog lines are being phased out in favor of all-digital networking facilities. The adoption of digital technology — based on Frame Relay, ATM, and SONET — is fueled by rapidly dropping costs and its high reliability, speed, and versatility. In addition to supporting a variety of data transmissions — voice, video, imaging, etc. — digital technologies provide a means to interconnect a multitude of LANs, MANs, and WANs in a variety of ways to create networks that are truly global in reach.

Several decades ago, transmission circuits were far more error prone than they are today, thus necessitating the use of various protocols to compensate for the transmission problems. However, as optical fiber circuits have gained in popularity, the need for older, slower, more resource-intensive protocols has become much less important. Frame Relay takes advantage of improved circuit reliability by eliminating many unnecessary error checking, correction, editing, and retransmission features that have been part of many data networks for the past two decades. Frame Relay has been available for many years. However, the flexibility that

Frame Relay offers in allocating bandwidth on demand is somewhat new. Frame Relay is one of several fast packet-switching options available today.

Asynchronous Transfer Mode (ATM) provides high-speed, low-delay multiplexing and switching. In addition, ATM supports simultaneous transmission of any type of user traffic (such as voice, data, or video). ATM segments and multiplexes user traffic into small, fixed-length cells. Each cell contains a header, which contains virtual circuit identifiers. An ATM network uses these identifiers to relay the traffic through high-speed switches from the sending customer premise equipment (CPE) to the receiving CPE. ATM is able to achieve high transport speeds because it performs very little error detection and correction, and because queuing delays are minimized by the small cell size used to transmit data. ATM is most effective, in terms of speed and reliability, when used in conjunction with SONET-based fiber optic circuits.

In the new millennium, many new applications are emerging that substantially increase the traffic and demand on the network. Some of these applications include teleradiology,[31] CAD/CAM drawings, video, VoIP, and high-definition television. In addition, networks are larger and more complex as corporations have grown, merged, and increased their outreach. This has led to the need for WAN-to-WAN, LAN-to-WAN, and LAN-to-LAN interconnectivity. Some large companies with high-bandwidth applications and demanding network requirements now need connection speeds of OC-192 (10 Gbps) and higher. This is forcing service providers to migrate their infrastructure to fiber optics and new technologies — such as DWDM and MPLS — to meet customer demands in a cost-effective manner.

2.2.2.2.1 Private WANs

Private WANs are based on communications facilities used exclusively by the organization that owns or leases them from a service provider. Private WANs offer flexibility, security, and control over internal data communications. A private WAN is a good solution when these considerations are paramount.

During the late 1960s and throughout the 1970s, WANs based on IBM's SNA, and TCP/IP or X.25 packet switching dominated the corporate scene. WANs built at this time consisted of such components as terminals, processors, modems, multiplexers,[32] concentrators,[33] and communications links. Efficiency and speed were less important in these early WANs than cost-effective connectivity. The most common applications involved connecting users to remote resources to perform tasks such as database retrieval and maintenance, batch file processing, online data entry, etc. These applications are much less time sensitive than the demanding applications of today (e.g., videoconferencing). In the 1980s and 1990s, the most commonly used communication link was the T1, which was used to build large corporate WAN backbone networks. A "backbone" typically consists of a collection of high-speed lines that transmit and consolidate traffic from a number of other lines operating at lower speeds. The lower-speed, dedicated lines were typically used to transmit data to and from mainframes and dumb terminals, at a rate of 9.6 Kbps or 19.2 Kbps. The consolidation of traffic onto lines with a larger capacity offered many throughput and cost advantages over many lines with smaller capacity. WAN networks such as these flourished as T1 costs steadily declined over the years. T1-based networks remain the most representative type of private WAN. Only a handful of companies are using very-high-speed fiber optic and DWDM technologies to support a private WAN infrastructure.

The most widely used WAN protocols — X.25, Frame Relay, ATM — are based on statistical, packet-based service. The selection of a particular protocol is influenced by how much "time" you need to buy. If you cannot tolerate delay on the network, you can pay for a guaranteed level of service by using a TDM solution. ATM is analogous to buying a box seat at a ball game; it provides QoS guarantees and provides higher bandwidth, but at a higher cost than X.25 or Frame Relay. If you are not worried about maximizing your seating capacity and you need QoS, then you will have to pay a higher price for this than for a ticket on top of the bleachers.

The most typical user of X.25 is a large, risk-averse organization with widely dispersed locations. These types of users are commonly found in government, finance, insurance, transportation, utilities, and retail industries. For example, X.25 is often used by banks that must ensure timely and reliable large-scale fund transfers. Although X.25 is an old protocol, it is also an extremely reliable service, which is why it remains in use today for these types of applications.

As long as the transmission lines are clean (e.g., they are fiber based), one does not need X.25 with its extensive error checking capabilities, assuming one can also tolerate some time delay. If the network in use is "cleaner," this also implies that one does not need OSI layer 3 and 4 error detection and correction capabilities. Layer 2, on which Frame Relay is based, pushed the burden of error correction and flow control to the CPE. Frame Relay uses a simple rule: if there is any problem (due to transmission errors or network congestion) with a frame, discard it. Intelligence built into end devices — such as PCs, host computers, etc. — detects the data loss and performs the necessary recovery actions. The major difference between Frame Relay and X.25 is that X.25 guarantees data integrity and network managed flow control; however, in doing so, X.25 introduces some delay. Frame Relay offers faster transmission time but does not provide a guarantee of data integrity. Thus, one of the issues that must be considered when evaluating Frame Relay is the need for data integrity. Frame Relay can be a cost-effective single solution if all the locations to be interconnected are widely geographically dispersed. Frame Relay is also being used to save on long-distance costs by integrating the voice and data traffic over the network.

2.2.2.2.2 Public WANs

A public WAN is based on an infrastructure that is supported by a service provider and which is shared by many users (i.e., the "public"). The major legacy public, switched facilities available today are Frame Relay and Asynchronous Transfer Mode (ATM). An appealing aspect of switched-based networking is that it essentially involves outsourcing the network maintenance to the communications carrier. These services operate at speeds ranging from 10 Mbps to 2.5 Gbps.

The demand for bandwidth and WAN connectivity to LANs (which operate at much higher speeds than traditional WANs), combined with the growth in IP data traffic, is having profound impacts on service provider infrastructures. Legacy WAN protocols are designed to ensure network performance and QoS; however, these time division multiplexing (TDM) ATM-based networks add unnecessary overhead and do not scale well. In response, telcos and service providers are converting to MPLS and DWDM technology as a way to offer performance, manageability, and capacity on existing fiber infrastructure. These technologies allow an organization to interconnect LANs, and to consolidate all the data traffic (which might, for example, include a mix of TCP/IP, SNA, and Appletalk traffic) on a single backbone network. This improves the network scalability, tolerance for growth, and the availability of new services.

Frame Relay provides T1/E1 access rates, typically at a substantially lower cost than comparable T1/E1 private leased line solutions. Public Frame Relay networks are widely used. Frame Relay networks are also being used to implement high-speed virtual private networks (VPNs) capable of supporting high bit-rate transmission.

Switched Multi-Megabit Data Service (SMDS) is yet another option for high-speed connectivity, although it is very rarely used and is not widely available. SBC/Pacific Bell is one of a few providers still offering the service. SMDS is designed to connect LANs, MANs, and WANs. It is designed to ease the geographic limitations that currently exist with low-speed WANs. SMDS is a connectionless packet switching service that provides LAN-like performance beyond a subscriber's location. SMDS can also be used to interconnect low-speed WANs that are geographically dispersed. SMDS provides a convenient, high-speed interface to an existing customer network. SMDS is positioned as a *service*, and not as a method for designing or building networks. SMDS is targeted for large customers and sophisticated applications that need a lot of bandwidth, but not all the time. SMDS is good at handling data applications that transfer information in a bursty manner. SMDS can also be used to support many interactive applications (with the exception of real-time, full-motion video applications, for which SMDS is too slow). For example, it takes only one to two seconds to send a high-quality color graphic image over an SMDS network. For many applications, this speed is more than adequate. One of the major reasons SMDS never really took off in the marketplace is that it requires CPE that is expensive in relation to that required by Frame Relay (e.g., FRADs or router upgrades).

2.2.2.2.3 Wireless WANs

Wireless WANs (WWANs) cover a *much* larger area than wireless local area networks (WLANs), and may span the better part of the globe. This technology is not new; however, its adoption has been somewhat slow. In part, this is due to the fact that mobile communications systems are undergoing rapid technological changes, and many different protocols and standards are being used. Wireless WANs are well suited for critical, low-volume transmissions; however, they are not generally suitable or cost effective for large file transfers or demanding multimedia applications. They are also prone to interference, particularly in inclement weather and mountainous terrain. Despite these limitations, corporations are using wireless WANs as a convenient way to give employees access to systems, applications, databases, and the like while they are away from the office. Although WWANs are not widely employed, the demand for them is growing as more and more people want and expect network connectivity at any time, from any place.

Wireless WANs support both voice and data transmissions in a variety of ways using digital cellular (or PCS), radio signals over analog, ham radio, satellite, microwave, infrared, and laser communication. An end user connects to a WWAN using a wireless modem embedded in a laptop, PDA,[34] mobile phone, or similar device. Service providers offer access to their WWAN for a fee. Most typically, the wireless modem transmits and receives signals from a radio tower, which in turn sends the signal to a switching center. From there, the signal is transferred to a public or private network link for eventual delivery to the end-site within a corporation's private network.

Although it is not yet a formal standard, the Wireless Application Protocol (WAP) has recently been developed to allow small devices — such as PDAs, mobile phones, pagers, etc. — to access Internet e-mail, newsgroups, and Internet Relay Chats (IRC).[35] The protocol supports HTML and XML through the Wireless Markup Language (WML). WML is designed to optimize presentation of Web content on very small screens. The request for Web content is sent from a wireless device through the wireless network to a WAP gateway. At the WAP gateway, the request is processed and the required information retrieved and returned back to the mobile device. Cellular communication will continue to be the preferred medium for the wireless consumer and business market. A cellular system usually works within a completely defined network (which operates using protocols for setting up and clearing calls and tracking mobile units) through a wide geographical area. Cellular systems communicate within large cells — with a radius in the kilometer range — with seamless hand-offs as the mobile user moves from cell to cell. A cellular system incorporates the PSTN (public switched telephone network), the MTSO (mobile telephone switch office), the MSU (mobile subscriber unit), and the cell site and antenna that handle radio transmissions.

There are many types of cellular systems. The two major North American analog voice systems include:

1. *AMPS* (Advanced Mobile Phone Service), which was first offered in the early 1980s and is used worldwide for analog cellular transmissions, is used for voice traffic.
2. *NAMPS* (Narrowband Analog Mobile Phone Service) is a second-generation phone system that combines voice processing with digital signaling. This triples the carrying capacity relative to AMPS. This service is available in the United States and overseas.

TDMA — which is similar to TDM in that it is a time division multiple access scheme — is the digital standard in cellular communications. TDMA has the same 30-MHz frequency band and number of channel allocations as NAMP but is more efficient and uses digital voice encoders for signal compression. Compared to analog cellular service, TDM offers increased capacity and superior transmission security. TDMA is used in three main digital wireless technologies; however, each TDMA implementation is different and they are not interoperable:

1. *D-AMPS*. This is the North American digital cellular system.
2. *GSM*. The Global System for Mobile communication is the most widely used of the three TDMA digital technologies. It is available in 120 countries and is the *de facto* wireless telephone standard in Europe. Many GSM service providers have made roaming agreements with other foreign operators so GSM users can often use their mobile phones when they travel to other countries. GSM supports both telephony and data transmission. It also supports a

number of other subscriber services, including dual-tone multi-frequency (DTMF) for tone signaling, voice mail, fax mail, short message services, CCITT Group 3 facsimile, and others. It also provides supplementary services such as call forwarding, caller ID, multi-party service, call waiting, and others. GPRS (General Packet Radio Services) is one of several promising enhancements to GSM that provides high-speed data transmission. GPRS provides wireless, packet-based service at rates between 56 and 114 Kbps. It can be used for continuous Internet connectivity, videoconferencing, and other multimedia applications. GPRS will allow mobile users to access a virtual private network instead of using a dial-up connection. It is also designed to work with Bluetooth, IP, and X.25.

3. *PCS.* Personal Communications Systems (PCS) are being used by telecommunications providers to help reduce cellular churn, joining customers who are in close vendor proximity by providing end-to-end service. PCS systems use the same technology as digital cellular but occupy higher frequency bands. American Personal Communications (APC), a subsidiary of Sprint, uses GSM as the basis for its personal communications service (PCS).

CDMA (Code Division Multiple Access) is a competing digital cellular technology based on spread-spectrum frequency assignments. In contrast to TDMA, CDMA does not assign a specific frequency to each mobile user. Audio signals are digitized and then transmitted on a range of frequencies that are varied according to a defined code. The digitized signals can only be interpreted by a receiver programmed to use the same code. Because the number of possible frequency-sequencing codes is in the trillions, for practical purposes, this ensures the privacy of the transmission. CDMA networks employ a scheme called "soft hand-off" to minimize signal degradation as the mobile user travels between cells in the network. WCDMA (Wideband Code-Division Multiple Access) is a wideband ITU standard based on CDMA. It is also known as IMT-2000 Direct Spread. WCDMA is a third-generation (3G) mobile wireless technology designed to carry voice, images, data, and video communications at speeds up to 384 Kbps.

In comparing TDMA and CDMA technologies, each has its relative merits. CDMA is more efficient and allows the cellular network to support more users. It also provides more secure, higher-quality transmission. TDMA has less chance for user interference, and, being an older and more established technology, is cheaper. Major wireless vendors support both digital technologies. Some of these vendors include Qualcomm, Ericsson, Motorola, AT&T, and Nortel.

In summary, WANs can be built from a mix of public and private line facilities, and from components such as routers, bridges, terminals, hosts, etc. In the past, private WAN networks were dominated by T1 facilities and by legacy protocols such as Frame Relay, SMDS, and ATM services. With the introduction of MPLS and DWDM operating on optical fiber, packet switching services support the transport of many data types at very high speeds, from OC-192 (10 Gbps) up to OC-768 (40 Gbps). Wireless WANs are also being used to provide voice and data transmission to the Internet, VPNs, and other corporate networks. As the workforce becomes more mobile, and the technology becomes cheaper and more readily available, wireless WANs will have increasing importance in the planning of enterprise networking infrastructures.

2.2.3 Star Topology Networks

2.2.3.1 Overview

The star topology is the simplest and easiest type of network to design. It was one of the first network topologies to emerge in early centralized networks.

A star topology features a single internetworking central hub/switch that provides access from leaf internetworks into the backbone with access to each other only through the core networking device. Each station in a Wide Area Network (WAN) star topology is directly connected to a common central hub or switch that utilizes circuit-switched technology to communicate with each end-station or device. Star networks commonly employ digital Private Branch Exchange (PBX) and digital data switches as the central networking device. Often, a time division multiplexer (TDM) is used to aggregate the PBX and data switch traffic onto larger bandwidth facilities to reduce network costs.

Other examples of a star topology include local area networks (LANs) and metropolitan area networks (MANs) that employ packet broadcasting through a central node that connects to each end-station by means of two unidirectional point-to-point links. In this case, all transmissions to the central switch are transmitted to all of the other end-stations. Although this LAN/MAN topology physically resembles a star, logically it performs like a bus.

Star topologies are recommended for:

- Small, centralized networks
- Voice networks
- Message switching
- Switched networks accessing backbone networks

The advantages of a star approach include:

- Straightforward topology for centrally controlled networks
- Simplified network management
- Isolates line failures to a single site
- Predictable network performance
- Easy to expand

Some of the disadvantages associated with star topologies include:

- High circuit cost
- Requires a central switching point
- Central switch presents a single point of failure
- Topology is not scalable

2.2.3.2 Design Considerations and Techniques

Star topologies help ensure that network delay is kept to a minimum, and are good for voice and fax applications that are sensitive to time delays. Delay is also a critical design factor when using satellite facilities that can themselves introduce signal delays of up to 250 milliseconds, resulting in annoying voice echoes. To combat this phenomenon, many popular multiplexer manufacturers provide built-in echo cancellation circuitry. Digital signal processors (DSPs) are often used to perform voice compression, to reduce the bandwidth needed between sites. However, DSP algorithms can add up to 100 to 150 milliseconds of delay (50 to 75 milliseconds for each end) and can produce a "pseudo-echo." Over the years, this delay has been reduced by more efficient DSP design and sampling techniques, and digitized bit reassembly. This has reduced the need, in some cases, for a star topology in voice-sensitive applications.

In a star topology, the main design decision is where to place the center node. In many cases, this decision is determined by pragmatic considerations, such as the location of existing data processing and telecommunication facilities. In the event that there are no guidelines for selecting the center location, clustering and center of mass algorithms (see Section 2.2.5.2 for examples) can be used to find the most central location for the node placement.

One such algorithm is Prim's algorithm. The first step of Prim's algorithm involves computing a network center, based on estimated traffic flows. The second step involves assigning links to the center node. Depending on a prespecified design parameter, Prim's algorithm produces a star-like or tree-like topology (thus, Prim's algorithm is also useful for designing shared, multi-point lines). The difference between a star-like and a tree-like topology is that in the former, all the links converge toward a central location, while in the latter, they do not. Prim's algorithm takes as given the locations of the sources and destinations of traffic, estimated traffic flows between these locations, and line costs to connect the sources and destinations.

Prim's algorithm:

Step 1 Find the network median:

(1a) Calculate a weight for each node i, as the sum of all traffic leaving and entering the node:

$$Node\_Weight_i \;=\; \sum_{j=1}^{n} \left(T_{j,i} + T_{i,j} \right) \qquad (2.1)$$

where
i is a selected node from a total set of nodes n
$T_{i,j}$ is the traffic flowing from node i to node j
$T_{j,i}$ is the traffic flowing from node j into node i.

(1b) Calculate a figure of merit for each node i as:

$$Figure\ of\ Merit_i = \sum_{for\ j \neq i, j=1}^{n} \left(Node\_Weight_j \times Cost_{i,j} \right) \qquad (2.2)$$

where
i is a selected node from a total set of nodes n
Node_Weight$_j$ is as defined above
Cost$_{ij}$ is the cost of a link between node i and j.

Note that in Equation 2.2, i *cannot* equal j.

(1c) Sort the figures of merit from low to high.

(1d) Select the node with the *smallest* number as the network center, designated c.

Step 2 Assign links:

(2a) Given all links (i,j) with i in the tree and j outside the tree, compute L'(i,j) as:

$$L'(i,j) = \text{Cost}_{i,j} + \alpha(\text{Cost}_{i,c}) \tag{2.3}$$

where
$\text{Cost}_{i,j}$ is the cost of a link between nodes i and j
$\text{Cost}_{i,c}$ is the cost of a link between nodes i and c
α, a design parameter, is set to 1.

(2b) Sort $L'(i,j)$ values from low to high.

(2c) Check to see if all nodes have been connected into the network. If not, continue to the next step. If yes, then stop the algorithm. The solution has been found.

(2d) Select the lowest $L'(i,j)$ value and remove it from sorted list. Add node j and link (i,j) to the network. Check to see if cycling or capacity constraints have been violated. If the constraints have been violated, remove node j and link (i, j) from the solution. Return to Step (2a) above.

(2e) Update $L(i,j)$ to reflect the addition of a new link into the network. Return to Step (2a) above.

A sample problem illustrates Prim's algorithm. As stated above, the node locations, estimated traffic flow, and line costs must be collected before the algorithm can be initiated. We summarize the data for our sample problem in Table 2.1. Assume for this problem that the maximum traffic flow on a line is 10,000 units. Because the line capacity is very high relative to the traffic flows in this sample problem, the traffic constraint is not active. The design parameter, α, is set to 1 to encourage the

Table 2.1 Traffic and Cost Data for Star Topology Design Problem

Node Pair	Traffic Requirements	Line Cost
1,2	22	80
1,3	23	1307
1,4	6	1074
2,3	7	1387
2,4	10	1154
3,4	13	1528

Table 2.2   Node Weight Calculations for Sample Star Topology Problem

Node	Node Weight Calculation
1	Σ 22 + 23 + 6 = 51
2	Σ 27 + 7 + 10 = 44
3	Σ 23 + 7 + 13 = 43
4	Σ 6 + 10 + 13 = 29

selection of a star topology (recall that when α is set to 0, Prim's algorithm produces a tree topology).

Solution to Sample Star Topology Problem Using Prim's Algorithm:

Step 1   Find the network median:

(1a) Calculate a weight for each node i, as the sum of all traffic leaving and entering the node, using Equation 2.1.

(1b) Calculate a figure of merit for each node i using Equation 2.2.

(1c) Sort the figures of merit from low to high. See entries in Table 2.2 above, sorted from low to high.

(1d) Select the node with the smallest number as the network center, designated c. See Table 2.3 above, which shows that node 1 should be selected as the network center.

Step 2   Assign links:

(2a) Given links (i,j) with i in the tree and j outside the tree, compute: $L'(i,j) = \text{Cost}_{i,j} + \alpha(\text{Cost}_{i,c})$. Set the design parameter $\alpha = 1$. *See Table 2.4 for the first iteration calculations.*

(2b) Sort $L'(i,j)$ values from low to high. *Based on Table 2.4, this corresponds to the values 80, 1074, and 1307.*

(2c) Check to see if all nodes have been connected into the network. *Only the center node is in the network at this stage of the algorithm.*

Table 2.3   Figure of Merit Calculations for Sample Star Topology Problem

Node	Node Weight Calculation
1	90,867 = (44 * 80) + (43 * 1307) + (29 * 1074)
2	97,187 = (51 * 80) + (43 * 1387) + (29 * 1154)
3	171,997 = (51 * 1307) + (44 * 1387) + (29 * 1528)
4	171,254 = (51 * 1074) + (44 * 1154) + (43 * 1528)

Table 2.4   L'(i,j) Calculations for Sample Star Topology Problem

Node Just Brought into the Network	Distance from Node $(L'(i,j) = Cost_{i,j} + \alpha\ (Cost_{i,c}))$			
	1	2	3	4
1 (this is the center node, c)	—	80	1307	1074

Note: Use values from the first line of the cost matrix for $Cost_{i,j}$, for i = 2, 3, 4, because the $Cost_{i,c}$ = 0.

(2d) Select the lowest L'(i,j) value and remove from sorted list. Assign node j and link (i,j) as part of the network. If the constraints have been violated, remove node j and link (i, j) from the solution. Return to step (2a) above. The lowest L'(i,j) value is 80, corresponding to link (1,2). This link does not violate any constraints.

(2e) Update L'(i,j) to reflect the addition of a new link into the network. Compute the cost of bringing the remaining nodes into the network, either through the center node or by attachment to the end of the current tree built from the center. *These calculations are shown in Table 2.5.*

(2f) Sort L'(i,j) values from low to high. Based on Table 2.5, this corresponds to 1074, 1234, 1307, and 1467.

(2g) Check to see if all nodes have been connected into the network. *Nodes 1 and 2 have been connected to the network, but nodes 3 and 4 remain unconnected. So the algorithm proceeds to the next step.*

Table 2.5 L'(i,j) Calculations for Second Iteration of Sample Star Topology Problem

Node Just Brought into the Network	Distance from Node $(L'(i,j) = Cost_{i,j} + \alpha\ (Cost_{i,c}))$			
	1	*2*	*3*	*4*
1 (this is the center node, c)	—	80	1307	1074
2 (this was selected above)	—	—	1467	1234
	Checking (2,3) = 1387 + 1(80) = 1467 Checking (2,4) = 1154 + 1(80) = 1234			

Note: $Cost_{i,c} = 80$, because this is the cost of link (2c) which is now in the network.

Because both these new link costs are greater than a direct connection to the center node, they are rejected. Node 4 is brought into the network through a connection through the center.

Table 2.6 L'(i,j) Calculations for Third Iteration of Sample Star Topology Problem

Node Just Brought into the Network	Distance from Node $(L'(i,j) = Cost_{i,j} + \alpha\ (Cost_{i,c}))$			
	1	*2*	*3*	*4*
1 (this is the center node, c)	—	80	1307	1074
2 (selected in previous step)	—	—	1467	1234
4 (selected in previous step)	—	—	2602	
	Checking (3,4) = 1528 + 1(1074) = 2602			

Note: $Cost_{i,c} = 1074$, because this is the cost of link (4,c) that is now in the network.

Because this link cost is greater than a direct connection to the center node, it is rejected. Node 3 is brought into the network through a connection through the center.

> (2h) Select the lowest L'(i,j) value and remove from sorted list. Assign node j and link (i,j) as part of the network. *The lowest L'(i,j) value is 1074, corresponding to link (1,4). The L'(i,j) calculated for connecting node 3 to node 2 is 1467, and the value calculated for connecting node 4 to node 2 is 1237. Because these values are higher than the cost of connecting node 4 directly to the center, the links associated with them are not selected.*

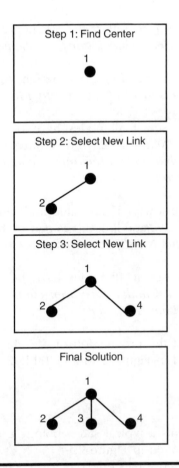

Figure 2.13   Prim's algorithm.

Table 2.7   Final Solution of Sample Star Topology Problem

Node Just Brought into the Network	Distance from Node $(L'(i,j) = Cos_{t,i,j} + \alpha\ (Cost_{i,c}))$			
	1	2	3	4
1 (this is the center node, c)	—	80	1307	1074
2 (selected in previous step)	—	—	1467	1234
4 (selected in previous step)	—	—	2602	
3 (final node selection)	—	—	—	—

(2i)   Update L'(i,j) to reflect the addition of a new link into the network. *These calculations are shown in Table 2.6.*

(2j)   Sort L'(i,j) values from low to high. *Based on Table 2.6, this corresponds to 1307, 1234, and 1467.*

(2k)   Check to see if all nodes have been connected into the network. *Nodes 1, 2, and 4 have been connected to the network, but node 3 remains unconnected. So the algorithm proceeds to the next step.*

(2l)   Select the lowest L'(i,j) value and remove from sorted list. Assign node j and link (i,j) as part of the network. *The lowest L'(i,j) value is 1307, corresponding to link (1,3).*

(2m) Check to see if all nodes have been connected into the network. *All nodes have been connected to the network, so the algorithm terminates.*

The final solution and the progression of the algorithm for this sample problem are illustrated in Figure 2.13 and Table 2.7.

2.2.3.3 Case Study

This case study presents a typical star topology used in WAN networks. This topology is illustrated in Figure 2.14.

In this example, a customer employs a star network for its terminal-to-host based applications as well as voice/fax communications between end-sites and the host location. The IBM 3174 Cluster Controllers serve to aggregate traffic from the various terminals/CRTs and coordinate communications with the front-end processor (FEP). The data communications protocol used is typically IBM's Synchronous Network Architecture (SNA), which operates in a master/slave arrangement whereby each station takes a turn communicating with the far-end host computer by way of the cluster controller.

2.2.4 Tree Networks

2.2.4.1 Overview

Tree WAN topologies are distinguished by their use of a multi-point line that is shared by many stations. A controlling device serves as a common,

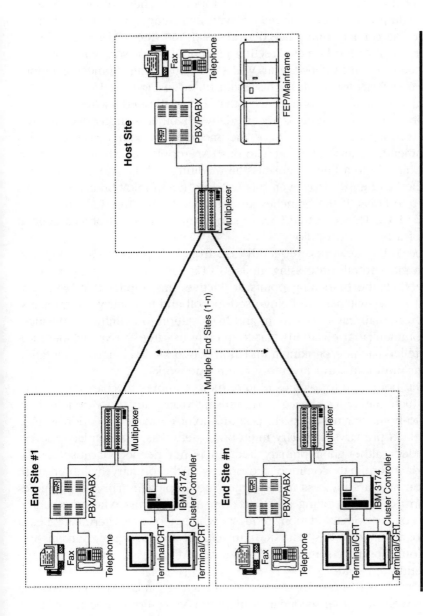

Figure 2.14   Star topology case study. (*Source:* B. Piliouras.)

central processor to which each station connects directly via a branching point. Transmission is usually restricted such that all communications between each node go through and are controlled by a central mainframe or node. Tree WANs are a form of star topology. They are either *passive,* with multi-point circuits, or *active,* with switching nodes at the branch points. The primary difference between star and tree WANs is that in the former, each station homes directly to a central processor, while in a tree they need not. In addition, star WANs do not contain branching points, as do tree WAN topologies. This is illustrated in Figure 2.15.

In Local Area Networks (LANs) that are configured as trees, a transmission from one station is received by all other stations because the signal propagates the length of the medium in both directions. This phenomenon is also exhibited with bus LAN topologies where the stations attach directly to a linear transmission medium, or bus, and communicate with the head-end.[36] However, bus LANs do not employ branching points as do tree LANs. If the branches are removed from a tree LAN, the result is a bus LAN. Hence, a tree LAN topology can be thought of as a general form of a bus LAN topology.

Tree WAN networks frequently use front-end processors (FEPs) and mainframe central processing units (CPUs) as the central networking device(s). If the branching points are active, this implies that data concentrators or switches are being used to poll and to control the transmissions from mainframe to station, and from station to mainframe. Business and commercial applications that frequently use tree WANs include automatic teller machine/banking transaction networks, stock market quotation systems, and parts and inventory control networks.

Multi-point tree topologies require more complex mechanisms to control which station or device may transmit data than those required by point-to-point star networks. In point-to-point networks, no addressing is needed. In primary-secondary multi-point networks, a secondary address is needed (although a primary address is not needed because all the secondary terminals communicate only with the primary mainframe). In addition, a shared access method must be employed. This usually takes the form of cyclic polling and selection. A *poll* occurs when the central processor, or primary station, requests whether or not a terminal (i.e., a secondary station) has data to transmit. In the opposite transmission direction, a primary station can issue a *select* to transmit data back to the requesting terminal.

Delays occur in tree topologies when each secondary station must wait its turn while the other secondary stations are being served. Because of these inherent delays, tree networks are usually restricted to data transmissions, and are generally not recommended for voice, fax, or video applications.

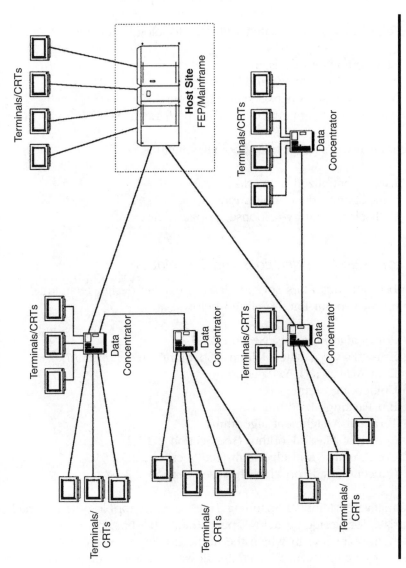

Figure 2.15   Tree network design. (*Source:* B. Piliouras.)

As compared to a star topology, the advantages of a tree topology include:

- Reduced facility costs
- Lower cost to expand
- Reduced number of hubbed facilities required

The disadvantages associated with tree topologies include:

- More complex to design
- More complex to expand
- More difficult to network management
- Number of drops affects network performance

Recommended uses of tree topologies include:

- Large, centralized networks
- Time-sharing data applications
- Poll-select (inquiry-response) applications

2.2.4.2 Design Considerations and Techniques

This subsection describes several classic heuristic design techniques for designing multi-point lines, as listed below:

- Prim's algorithm (see Section 2.2.3.2)
- Kruskal's Capacitated Minimal Spanning Tree algorithm
- Esau-William's (E-W) algorithm
- Unified algorithm
- Bin Packing algorithms
- Terminal Assignment algorithm
- Center of Mass algorithm (see Section 2.2.5.2)
- Concentrator Add algorithm (see Section 2.2.5.2)
- Concentrator Drop algorithm (see Section 2.2.5.2)

A Capacitated Minimal Spanning Tree[37] (CMST) topology is especially appropriate for designing a low-speed (i.e., 9.6 Kbps, 56 Kbps, or 128 Kbps) multi-point line in which the traffic flowing over the link is small compared to the link size. When traffic flows are comparable in magnitude to the smallest link in the network, it may be necessary to design several shared access multi-point lines that are, in turn, connected to concentrators that are part of a larger backbone network. When there is enough traffic flowing through the network to fill many smaller links, this leads to a choice between several lower speed lines or a single high-speed line. In the latter two cases, the optimal placement of the concentrators (or multiplexers, etc.) in the backbones can be determined using terminal assignment or concentrator location algorithms. The terminal assignment procedures are discussed later in this section, and concentrator location algorithms in Section 2.2.5.2.

The heuristic design techniques in this section are low in computational complexity yet they provide high-quality[38] results. They are all considered a form of greedy algorithm. The first three CMST algorithms (i.e., Prim's, Kruskal's, and E-W's) listed above are special cases of the Unified algorithm. CMST algorithms should be used when there are *no* constraints on the number of incoming multi-point lines that the host or concentrator can support. When the number of multi-point lines connecting to the host or concentrator must be minimized, bin-packing algorithms should be used instead of CMST algorithms. Terminal assignment problems are used to determine which terminals should be associated with a given concentrator or host location. Thus, the terminal assignment determines which terminals are grouped together on a multi-point line.

Sometimes the order in which terminals or links are considered can make a significant difference in the cost of the resulting design. However, with greedy algorithms, once a link assignment is made, it is not taken back. This is one of the ways the computational complexity of these procedures is kept under control. However, in some cases, one may wish to selectively re-order a chain of link assignments in the network. Branch and exchange algorithms are designed to do this, but they are *much more* complex and work only on limited problems. Branch and exchange algorithms work by exchanging specific link pairs to see if cost reductions can be found in the design solution. For a more detailed discussion of branch and exchange algorithms, the reader is referred to [KERS93].

After the multi-point lines are designed, the Add, Drop, and Center of Mass algorithms can be used to determine the concentrator or backbone node to which they should connect.

2.2.4.2.1 Kruskal's Capacitated Minimal Spanning Tree Algorithm

This algorithm begins with no links, only nodes in the network. The first step is to create a sorted list, from low to high, of all possible link costs. The algorithm then constructs a minimal spanning tree, link by link. As links are added to the network, they are checked to make sure that the capacity and cycling restrictions are not violated. If the constraints would be violated by a link addition, the link is ignored and another link is examined to see if it is suitable for inclusion in the network.

1. Sort all possible links in ascending order by cost and put in an ordered list.
2. Check to see if all the nodes are connected. This will occur when (n–1) links have been selected (this is true by the definition of a minimal spanning tree).
   - If all the nodes are connected, then terminate the algorithm with the message "Solution Complete."
   - If all the nodes are not connected, continue to the next step.
3. Select the link at the top of the list.
   - If no links are on the list, terminate algorithm. Check to see if all nodes are connected, and if not, terminate the algorithm with the message "Solution Cannot Be Found."
4. Check to see if the link selected creates a cycle in the network.
   - If the link creates a cycle, remove it from the list. Return to Step 2.
   - If the link does not create a cycle, check to see if it can handle the required traffic load.
     - If so, add the link to the network, and remove it from link list. Return to Step 2.
     - If not, remove the link from link list and return to Step 2.

An efficient implementation of this algorithm can be shown to be of O (m log n), where m is the number of edges tested, and n is the number of nodes in the network. [KERS93] A potential problem with this algorithm is that it may strand the last nodes connected to the network far from the center. The next algorithm — the Esau-Williams (E-W) algorithm — attempts to address this problem by starting with all the nodes connected directly to the center. Links to the center are only replaced when they can be cost justified.

We now present an example to demonstrate Kruskal's CMST algorithm. The line cost data for our sample problem is summarized in Table 2.8 and the traffic flow data in Table 2.9. If an entry in a table is missing, as indicated by "—," this means that there is no cost or traffic associated with this particular location. Assume for this problem that the maximum traffic a line can carry is ten (10) units. One wants to find the lowest cost multi-line network to connect terminals 2, 3, 4, and 5 to the host 1.

2.2.4.2.1.1 Solution of Sample Problem Using Kruskal's CMST

1. Sort all possible links in ascending order of cost and put them in a link list (Table 2.10).
2. Check to see if all the nodes are connected. *This will occur when (n − 1 = 4) links have been selected. No links have been added to the solution; therefore, proceed to the next step.*

Table 2.8   Cost Matrix for Sample CMST Problem

	Host 1	Node 2	Node 3	Node 4	Node 5
Host 1	—	6	3	4	5
Node 2	6	—	3	5	7
Node 3	3	3	—	3	5
Node 4	4	5	3	—	3
Node 5	5	7	5	3	—

**Table 2.9   Traffic Matrix
for Sample CMST Problem**

	Traffic Units Generated
Host 1	—
Node 2	5
Node 3	4
Node 4	3
Node 5	5

**Table 2.10   Sorted Link Costs
for Sample CMST Problem**

Link	Cost
1-3	3
2-3	3
4-3	3
4-5	3
1-4	4
2-4	5
3-5	5
1-5	5
1-2	6
2-5	7

3. Select the link at the top of the list. *This is link (1-3). Note that when ties occur in the sorted link costs, it does not matter which link is selected, although this may change the final solution.*
4. Check to see if the link selected creates a cycle in the network. *Link (1,3) does not create a cycle. It does not violate a capacity constraint, because only four (4) units of traffic are carried between nodes 1 and 3. So we add the link to the network.*
5. Check to see if all the nodes are connected. *Because only one of four required links have been selected, proceed to the next step.*
6. Select the link at the top of the list. *This is link (2-3).*
7. Check to see if the link selected creates a cycle in the network. *Link (2,3) does not create a cycle. It does not violate a capacity constraint, because only nine (9) units of traffic would be carried between nodes 2,3, and 1. So add the link to the network.*
8. Check to see if all the nodes are connected. *Because only two of four required links have been selected, proceed to the next step.*
9. Select the link at the top of the list. *This is link (4-3).*
10. Check to see if the link selected creates a cycle in the network. *Link (4,3) does not create a cycle. This link does violate the capacity constraint, because twelve (12) units of traffic would be carried between nodes 1, 2, 3, and 4. So do not add the link to the network.*
11. Check to see if all the nodes are connected. *Because only two of four required links have been selected, proceed to the next step.*
12. Select the link at the top of the list. *This is link (4-5).*
13. Check to see if the link selected creates a cycle in the network. *Link (4-5) does not create a cycle. It does not violate a capacity constraint, because only eight (8) units of traffic would be carried between nodes 4 and 5. So add the link to the network.*
14. Check to see if all the nodes are connected. *Because only three of four required links have been selected, proceed to the next step.*
15. Select the link at the top of the list. *This is link (1-4).*
16. Check to see if the link selected creates a cycle in the network. *Link (1,4) does not create a cycle. It does not violate a capacity constraint, because only eight (8) units of traffic would be carried between nodes 4,5, and 1. So add the link to the network.*
17. Check to see if all the nodes are connected. *Because four of four required links have been selected, terminate the algorithm.*

The results of this algorithm are illustrated in Figure 2.16.

2.2.4.2.2 Esau-Williams (E-W) Capacitated Minimal Spanning Tree Algorithm

This algorithm starts by connecting all terminals to the center node or concentrator. It then attempts to replace the links, one by one, with

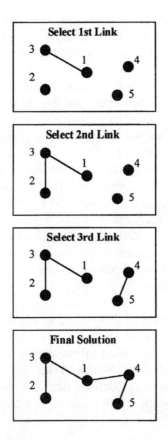

Figure 2.16   Kruskal CMST algorithm.

cheaper ones. When a new link is added, the corresponding connection to the host/concentrator is removed. The cycling and traffic constraints are checked as the links are selected, and no link replacements are allowed that would cause a violation of the constraints.

1. Connect all the nodes to the center.
2. Calculate a trade-off value, d_{ij}, representing the difference in cost C_{ij} between the connection (i,j) and the cost C_{i1} to connect terminal i to center node 1.

$$d_{ij} = C_{ij} - C_{i1} \qquad (2.4)$$

(calculate for all i,j, where 1 is the concentrator location)

3. Sort the links in increasing order of d_{ij}.

4. Remove all links with an associated d_{ij} that is equal to or greater than zero (0). If no links have a negative d_{ij} value, then the algorithm terminates and the network remains as it is.
5. Select the link with the lowest (i.e., the most negative) d_{ij} value. Check to see if the link selected creates a cycle in the network.
   - If the link creates a cycle, set the dij value $= +\infty$ (this is a very large positive number). Return to Step 4.
   - If the link does not create a cycle, check to see if it can handle the required traffic load.
     ■ If the link has sufficient capacity to carry the required traffic, replace link (i,1) with link (i,j). Recalculate affected d_{ij} values to reflect the fact that node i is now connected to node j, and not the center. Thus, replace C_{i1} with C_{ij} in the d_{ij} calculations. Return to Step 2.
     ■ If the link lacks sufficient capacity to carry the required traffic, do not replace the link. Set the link d_{ij} value $= +\infty$ and return to Step 2.

An efficient implementation of this algorithm can be shown as O $(n^2 \log n)$, where n is the number of nodes in the network. [KERS93]

We now present an example to demonstrate the E-W CMST algorithm using the same cost and traffic data as in our previous example (see Table 2.8 and Table 2.9). Assume, as before, that the maximum traffic constraint is ten (10) units. We want to find the lowest-cost multi-line network to connect terminals 2, 3, 4, and 5 to the host 1.

2.2.4.2.2.1 Solution of Sample Problem Using the E-W CMST Algorithm

1. Connect all the nodes to the center. This is shown in the first diagram of Figure 2.17.
2. Calculate a trade-off value, d_{ij}, representing the difference in cost C_{ij} between the connection (i,j) and cost C_{i1} to connect i to center node 1. The results of these calculations are summarized in Table 2.11.
3. Sort the links in increasing order of d_{ij} (see Table 2.12).
4. Remove all links with an associated d_{ij} that is equal to or greater than zero (0). This results in Table 2.13. Because there are still links with a negative d_{ij} value, the algorithm proceeds.
5. Select the link with the lowest (i.e., the most negative) d_{ij} value. Check to see if the link selected creates a cycle in the network. *This is link (2,3), which replaces link (2,1). The (2,3) link does not create a cycle. The combined traffic flows on links (2,3) and (3,1)*

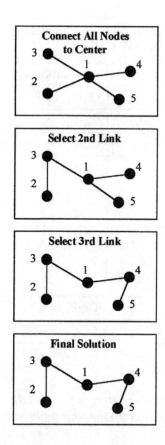

Figure 2.17 E-W CMST algorithm.

will equal nine (9) units, which does not exceed the traffic constraint of ten (10) units. So, the link is accepted as a replacement for (2,1). Set $d_{23} = C_{23} - C_{23} = 0$. Updating the other d_{ij}'s affected by the link replacement: set $d_{24} = C_{24} - C_{23} = (5 - 3) = 2$. Set $d_{25} = C_{25} - C_{23} = (7 - 3) = 4$. Because these values are positive, we do not need to further consider these d_{ij}'s as potential link replacements.

6. Returning to Step 4, remove all links with an associated d_{ij} that is equal to or greater than zero (0). *Because there are still negative d_{ij} values, proceed with the algorithm. (See Table 2.14)*

7. Select the link with the lowest (i.e., the most negative) d_{ij} value. Check to see if the link selected creates a cycle in the network. *As shown in Table 2.14, link (5,4) should be chosen to replace link (5,1). The (5,4) link does not create a cycle. The combined traffic flows on links (5,4) and (4,1) will equal eight (8) units, which does not exceed the traffic constraint of ten (10) units. So, the link is*

Table 2.11 d_{ij} Calculations for E-W CMST

d_{ij}	d_{ij} Calculations
d_{23}	$C_{23}\text{-}C_{21} = -3$
d_{32}	$C_{32}\text{-}C_{31} = 0$
d_{24}	$C_{24}\text{-}C_{21} = -1$
d_{42}	$C_{42}\text{-}C_{41} = 1$
d_{25}	$C_{25}\text{-}C_{21} = 1$
d_{52}	$C_{52}\text{-}C_{51} = 2$
d_{34}	$C_{34}\text{-}C_{31} = 0$
d_{43}	$C_{43}\text{-}C_{41} = -1$
d_{35}	$C_{35}\text{-}C_{31} = 2$
d_{53}	$C_{53}\text{-}C_{51} = 0$
d_{45}	$C_{45}\text{-}C_{41} = -1$
d_{54}	$C_{54}\text{-}C_{51} = -2$

accepted as a replacement for (5,1). Set $d_{54} = C_{54} - C_{54} = 0$. Updating the other d_{ij}'s affected by the link replacement: set $d_{53} = C_{53} - C_{54} = (5 - 3) = 2$. Set $d_{52} = C_{52} - C_{54} = (7 - 3) = 4$. Set $d_{45} = C_{45} - C_{45} = (3 - 3) = 0$. Because these values are greater than or equal to zero, we do not need to further consider these d_{ij}'s as potential link replacements.

8. Returning to Step 4, remove all links with an associated d_{ij} that is equal to or greater than zero (0). This results in Table 2.15. Because there are links with a negative d_{ij} value, the algorithm proceeds.
9. Select the link with the lowest (i.e., the most negative) d_{ij} value. Check to see if the link selected creates a cycle in the network. *This is link (4,3). This link would exceed the traffic constraint of ten (10) units, because nodes 2, 3, 4, and 5 would create a combined traffic load of 17 units. So, the link is rejected. Set $d_{43} = +\infty$.*
10. Returning to Step 4, remove all links with an associated d_{ij} that is equal to or greater than zero (0). Because there are no links with a negative d_{ij} value, the algorithm terminates.

The solution using the E-W algorithm is illustrated in Figure 2.17. Although the results for this sample problem are the same as those obtained using the Kruskal algorithm, in general, they may not be. The

Table 2.12   Sorted d$_{ij}$ Values for E-W CMST

d_{ij}	d_{ij} Calculations
d_{23}	$C_{23}\text{-}C_{21} = -3$
d_{54}	$C_{54}\text{-}C_{51} = -2$
d_{24}	$C_{24}\text{-}C_{21} = -1$
d_{43}	$C_{43}\text{-}C_{41} = -1$
d_{45}	$C_{45}\text{-}C_{41} = -1$
d_{32}	$C_{32}\text{-}C_{31} = 0$
d_{34}	$C_{34}\text{-}C_{31} = 0$
d_{53}	$C_{53}\text{-}C_{51} = 0$
d_{42}	$C_{42}\text{-}C_{41} = 1$
d_{25}	$C_{25}\text{-}C_{21} = 1$
d_{35}	$C_{35}\text{-}C_{31} = 2$
d_{52}	$C_{52}\text{-}C_{51} = 2$

Table 2.13   Negative d$_{ij}$ Values for First Link Selection

d_{ij}	d_{ij} calculations
d_{23}	$C_{23}\text{-}C_{21} = -3$
d_{54}	$C_{54}\text{-}C_{51} = -2$
d_{24}	$C_{24}\text{-}C_{21} = -1$
d_{43}	$C_{43}\text{-}C_{41} = -1$
d_{45}	$C_{45}\text{-}C_{41} = -1$

only time one can be certain the results will be identical (in cost) is when the capacity constraint is not active and does not affect the solution. This means that no link was rejected during the course of the algorithm due to insufficient capacity.

In general, the E-W algorithm produces very good designs for multi-point lines. It also does a good job configuring access routers on a shared T1 line. The algorithm performs well (i.e., it produces low-cost designs) when the traffic requirements vary considerably at each node. It also

Table 2.14   Negative d_{ij} Values for Second Link Selection

d_{ij}	d_{ij} calculations
d_{54}	$C_{54}\text{-}C_{51} = -2$
d_{43}	$C_{43}\text{-}C_{41} = -1$
d_{45}	$C_{45}\text{-}C_{41} = -1$

Table 2.15   Negative d_{ij} Values for Third Link Selection

d_{ij}	d_{ij} calculations
d_{43}	$C_{43}\text{-}C_{41} = -1$

usually does better configuring a network with a smaller number of sites on the line than it does configuring a network with a large number of sites on the line.

2.2.4.2.3 Unified Algorithm

Kershenbaum and Chou have shown that the Kruskal and E-W algorithms, and other CMST algorithms, are special cases of what they call the Unified algorithm. Not surprisingly, the Unified algorithm solution procedure is very similar to the E-W algorithm just presented. However, the Unified algorithm incorporates a weighting factor w that is defined for each terminal location. Variations on the definition of w yield specific heuristic CMST algorithms. For example:

■ If the weight w = 0, this reduces to the Kruskal algorithm.
■ If the weight w = C_{i1} = terminal distance from center, this reduces to the E-W algorithm.
■ If the weight w = 0 at center and w = ∞, this reduces to the Prim algorithm.

Other values for w are possible and may be useful, depending on the type of design desired. It can be shown that an efficient programming implementation of the Unified algorithm is of $O(n^2)$, where n is the number of nodes in the network. [KERSH93]

We present the Unified algorithm below. Begin by defining w_i as the weighting factor associated with terminal I, and C_{ij} as the link cost connecting terminals i and j.

1. Initialize w_i for i = 1 to N, where N is the number of terminals. Evaluate $d_{ij} = C_{ij} - w_i$ for all links (i,j) when C_{ij} exists and no constraints are violated with connection i-j.
2. Sort the d_{ij} and determine the minimum d_{ij} for all links (i,j), such that i does not equal j. If d_{ij} is equal to positive infinity ($+\infty$), terminate the algorithm.
3. Evaluate the constraints under connection (i,j). If any constraints are violated, set d_{ij} equal to positive infinity and return to Step 2. Otherwise, proceed to the next step.
4. Add link (i,j) and relabel one of the terminals i or j to correspond to each other. Reevaluate the constraints. Obtain new values for w_i and d_{ij} (for all i such that w_i has changed) and return to Step 2.

2.2.4.2.4 Bin-Packing Algorithms

If one needs to minimize the number of lines connected to the backbone node or central facility, CMST algorithms are not suitable. Minimizing the number of lines may be necessary if there are many groups of terminals in the same city with similar link costs. Presented below are three simple and effective bin-packing algorithms that are of computational complexity, when implemented efficiently, of O (N log N + NB), where N is the number of nodes in the network and B is the number of bins (or lines) to which the nodes are assigned. These algorithms are:

1. First Fit Decreasing
2. Best Fit Decreasing
3. Worst Fit Decreasing

These algorithms assume that the line capacities and traffic requirements at each site are known.

2.2.4.2.4.1 First Fit Decreasing Bin-Packing Algorithm

1. Sort terminal traffic requirements from high to low. Set the highest traffic requirement equal to t_{MAX}.
2. Check highest traffic requirement, t_{MAX}, against the line capacity. If the line capacity is less than the largest traffic requirement, terminate the algorithm with the message, "No solution possible due to insufficient line capacity." If the line capacity is sufficient, proceed to the next step.

Table 2.16   Sample Data for Bin Packing Algorithms

Terminal	Traffic requirement
1	17
2	10
3	9
4	15
5	6
6	8

3. Check to see if all the traffic flows have been assigned to a link. If not, continue to the next step. Otherwise, terminate the algorithm with the message, "Final Solution Found."
4. Put the first terminal requirement (i.e., the current t_{MAX}) on the *first* line with capacity. If a line does not have capacity for the requirement, install a new line and place the traffic requirement on it. Remove the current t_{MAX} from the sorted list. Return to Step 3.

An example is presented below to illustrate the First Fit Decreasing Bin-Packing algorithm. The data for the sample problem is summarized in Table 2.16. Assume that the maximum line capacity for the problem is 30 traffic units.

2.2.4.2.4.2 Solution to Sample Problem Using the First Fit Decreasing Bin-Packing Algorithm

1. Sort terminal traffic requirements from high to low *(See Table 2.17)*.
2. Check highest requirement against line capacity. *The line capacity is sufficient; therefore, proceed to the next step.*
3. Check to see if all traffic requirements have been assigned. *They have not, so we proceed to the next step.*
4. Put first terminal requirement on the *FIRST* line with capacity. *We select terminal 1 for placement on a new line A. Remove terminal 1 from Table 2.17 and proceed with the algorithm.*
5. Check to see if all traffic requirements have been assigned. *They have not, so we proceed to the next step.*
6. Put first terminal requirement on the *FIRST* line with capacity. *We select terminal 4 for placement on a new line B, because placement of terminal 2 on line A would exceed the maximum*

Table 2.17 Sample Data for Bin Packing Algorithms

Terminal	Traffic requirement
1	17
4	15
2	10
3	9
6	8
5	6

traffic constraint. Remove terminal 4 from Table 2.17 and proceed with the algorithm.

7. Check to see if all traffic requirements have been assigned. *They have not, so we proceed to the next step.*

8. Put first terminal requirement on the *FIRST* line with capacity. *We select terminal 2 for placement on line A. Remove terminal 2 from Table 2.17 and proceed with the algorithm.*

9. Check to see if all traffic requirements have been assigned. *They have not, so we proceed to the next step.*

10. Put first terminal requirement on the *FIRST* line with capacity. *We select terminal 3 for placement on line B. Remove terminal 3 from Table 2.17 and proceed with the algorithm.*

11. Check to see if all traffic requirements have been assigned. *They have not, so we proceed to the next step.*

12. Put first terminal requirement on the *FIRST* line with capacity. *We select terminal 6 for placement on a new line C, since it would exceed the maximum traffic constraint if placed on either line A or B. Remove terminal 6 from Table 2.17 and proceed with the algorithm.*

13. Check to see if all traffic requirements have been assigned. *They have not, so we proceed to the next step.*

14. Put first terminal requirement on the *FIRST* line with capacity. *We select terminal 5 for placement on line B. Remove terminal 5 from Table 2.17 and proceed with the algorithm.*

15. Check to see if all traffic requirements have been assigned. *They have, so we terminate the algorithm, having found the final solution.*

The final solution using this algorithm places terminals 1 and 2 on line A; terminals 3, 4, and 5 on line B; and terminal 6 on line C.

2.2.4.2.4.3 Best Fit Decreasing Bin-Packing Algorithm

1. Sort terminal traffic requirements from high to low. Set the highest traffic requirement equal to t_{MAX}.
2. Check highest requirement t_{MAX} against the line capacity. If the line capacity is less than the largest traffic requirement, terminate the algorithm with the message, "No solution possible due to insufficient line capacity." If the line capacity is sufficient, proceed to the next step.
3. Check to see if all the traffic flows have been assigned to links. If not, continue to the next step. Otherwise, terminate the algorithm with the message, "Final Solution Found."
4. Put current t_{MAX} requirement on the first feasible line with the *least* capacity. If no line has sufficient capacity for the requirement, install a new line and place the traffic requirement on it. Remove the current traffic requirement t_{MAX} from the sorted list. Return to Step 3.

We now illustrate the Best Fit Decreasing Bin-Packing Algorithm with an example that uses the same data as before (see Table 2.16). Continue to assume that the maximum line capacity for the problem is 30 traffic units.

2.2.4.2.4.4 Solution to Sample Problem Using Best Fit Decreasing Bin-Packing Algorithm

1. Sort terminal traffic requirements from high to low. See Table 2.18.

Table 2.18 Sorted Link Requirements for Best Fit Bin Packing Problem

Terminal	Traffic requirement
1	17
4	15
2	10
3	9
6	8
5	6

2. Check highest traffic requirement t_{MAX} against line capacity. *The line capacity is sufficient; therefore, proceed to the next step.*
3. Check to see if all traffic requirements have been assigned. *They have not, so we proceed to the next step.*
4. Put first terminal requirement on the line with the least capacity. *We select terminal 1 for placement on a new line A. Remove terminal 1 from Table 2.18 and proceed with the algorithm.*
5. Check to see if all traffic requirements have been assigned. *They have not, so we proceed to the next step.*
6. Put first terminal requirement on the line with the least capacity. *We select terminal 4 for placement on a new line B, because placement of terminal 4 on line A would exceed the maximum traffic constraint. Remove terminal 4 from Table 2.18 and proceed with the algorithm.*
7. Check to see if all traffic requirements have been assigned. *They have not, so we proceed to the next step.*
8. Put first terminal requirement on the line with the least capacity. *We select terminal 2 for placement on line A. Remove terminal 2 from Table 2.18 and proceed with the algorithm.*
9. Check to see if all traffic requirements have been assigned. *They have not, so we proceed to the next step.*
10. Put first terminal requirement on the line with the least capacity. *We select terminal 3 for placement on line B. Remove terminal 3 from Table 2.18 and proceed with the algorithm.*
11. Check to see if all traffic requirements have been assigned. *They have not, so we proceed to the next step.*
12. Put first terminal requirement on the line with the least capacity. *We select terminal 6 for placement on a new line C, because it would exceed the maximum traffic constraint if placed on either line A or B. Remove terminal 6 from Table 2.18 and proceed with the algorithm.*
13. Check to see if all traffic requirements have been assigned. *They have not, so we proceed to the next step.*
14. Put first terminal requirement on the line with the least capacity. *We select terminal 5 for placement on line B. Remove terminal 5 from Table 2.18 and proceed with the algorithm.*
15. Check to see if all traffic requirements have been assigned. *They have, so terminate the algorithm, having found the final solution.*

In the final solution, terminals 1 and 2 are on line A; terminals 3, 4, and 5 are on line B; and terminal 6 is on line C. This is the same solution obtained using the First Fit algorithm. In general, one would not necessarily expect the results to be the same.

2.2.4.2.4.5 Worst Fit Decreasing Bin-Packing Algorithm

1. Sort terminal traffic requirements from high to low. Set the highest traffic requirement equal to t_{MAX}.
2. Check the highest requirement t_{MAX} against the line capacity. If the line capacity is less than the largest traffic requirement, t_{MAX}, terminate the algorithm with the message, "No solution possible due to insufficient line capacity." If the line capacity is sufficient, proceed to the next step.
3. Check to see if all traffic flows have been assigned to links. If not, continue to the next step. Otherwise, terminate the algorithm with the message, "Final Solution Found."
4. Put first terminal requirement (i.e., the current t_{MAX}) on the first feasible line with the *most* capacity. If no line has sufficient capacity for the requirement, install a new line and place the traffic requirement on it. Remove the traffic requirement from the sorted list. Return to Step 3.

We now illustrate the Worst Fit Decreasing Bin Packing Algorithm using the same data as before. Continue to assume that the maximum line capacity for the problem is 30 traffic units.

2.2.4.2.4.6 Solution to Sample Problem Using Worst Fit Decreasing Bin-Packing Algorithm

1. Sort terminal traffic requirements from high to low. *The results are shown in Table 2.18.*
2. Check highest requirement against line capacity. *The line capacity is sufficient; therefore, proceed to the next step.*
3. Check to see if all traffic requirements have been assigned. *They have not, so proceed to the next step.*
4. Put first terminal requirement on the first feasible line with the *MOST* capacity. *We select terminal 1 for placement on a new line A. Remove terminal 1 from Table 2.18 and proceed with the algorithm.*
5. Check to see if all traffic requirements have been assigned. *They have not, so proceed to the next step.*
6. Put first terminal requirement on the line with the *MOST* capacity. *We select terminal 4 for placement on a new line B, because placement of terminal 2 on line A would exceed the maximum traffic constraint. Remove terminal 4 from Table 2.18 and proceed with the algorithm.*

7. Check to see if all traffic requirements have been assigned. *They have not, so proceed to the next step.*
8. Put first terminal requirement on the line with the *MOST* capacity. *We select terminal 2 for placement on line B. Remove terminal 2 from Table 2.18 and proceed with the algorithm.*
9. Check to see if all traffic requirements have been assigned. *They have not, so proceed to the next step.*
10. Put first terminal requirement on the line with the *MOST* capacity. *We select terminal 3 for placement on line A. Remove terminal 3 from Table 2.18 and proceed with the algorithm.*
11. Check to see if all traffic requirements have been assigned. *They have not, so proceed to the next step.*
12. Put first terminal requirement on the line with the *MOST* capacity. *We select terminal 6 for placement on a new line C, because it would exceed the maximum traffic constraint if placed on either line A or B. Remove terminal 6 from Table 2.18 and proceed with the algorithm.*
13. Check to see if all traffic requirements have been assigned. *They have not, so proceed to the next step.*
14. Put first terminal requirement on the line with the *MOST* capacity. *We select terminal 5 for placement on line C. Remove terminal 5 from Table 2.18 and proceed with the algorithm.*
15. Check to see if all traffic requirements have been assigned. *They have, so terminate the algorithm, having found the final solution.*

The final solution using this algorithm places terminals 1 and 3 on line A, terminals 2 and 4 on line B, and terminals 5 and 6 on line C.

2.2.4.2.4.7 Terminal Assignment Problem

The terminal assignment problem attempts to find the optimal connection of a terminal to a center or concentrator location. The Terminal Assignment algorithm is briefly summarized below and is nearly identical to the E-W algorithm. The only difference between this algorithm and the CMST version of E-W presented earlier is in the calculation of the trade-off function. Recall that the trade-off function is used to evaluate whether or not to replace a connection to the center node with the current link under consideration.

1. Calculate for each terminal the value associated with the trade-off function:

$$t_i = c_{i1} - \alpha\, c_{i2} \tag{2.5}$$

where t_i is the trade-off cost; c_{i1} and c_{i2} are the cost of connecting terminal to its nearest and second nearest neighbor respectively; and α is a parameter between 0 and 1 reflecting our preference toward "critical" terminals. As the α value approaches one, there is more of a penalty for failing to connect the terminal to its second nearest neighbor.

2. Find the minimum value trade-off value and assign the associated terminal to the cheapest *feasible* concentrator. If a terminal has lost its last feasible neighbor, stop the algorithm. If only one feasible neighbor remains, the terminal must be assigned to that concentrator.

3. Check and update trade-offs (setting them to $+ \infty$ as assignments are made) and remaining concentrator capacity affected by the last link assignment.

4. Continue until all t_i are greater than or equal to zero. At this point, terminate the algorithm.

2.2.4.3 Case Study

Figure 2.15 presents an example of a tree WAN. This network has several CRTs running IBM 3270 terminal emulation that are connected to 3174 data concentrators (which are also known as cluster controllers). These data concentrators either connect to other data concentrators or directly to the front-end processor (FEP), which communicates with the mainframe computer. Note that in many tree networks, the terminals connect directly to the FEP — typically a 3745 communications controller — or to a data concentrator. Many tree networks are actually a combination of one or more star networks interconnected with one or more tree networks. Because the costs of the links are usually mileage sensitive, it is sometimes economical to group various terminals into data concentrators via analog lines with modems, and then to connect the concentrators to the FEP using higher bandwidth digital facilities (such as 56 Kbps DDS or Nx64 Kbps Fractional T-1 links). The reduction in networking costs must be weighed against the decrease in reliability resulting from fewer links in the network. In a tree topology, if one of the higher bandwidth links fails, then multiple terminals are affected. In contrast, star networks are more reliable because each end terminal has a separate link to the central processor. However, these additional links add to the facility costs. The network designer must carefully weigh which terminals require direct links and which terminals should be aggregated via data concentrators.

2.2.5 Backbone Networks

2.2.5.1 Overview

A backbone network is comprised of high-capacity links and nodes (typically, routers or multiplexers) that consolidate traffic from smaller access networks. Backbone networks allow a diverse traffic mix to be carried on a single network infrastructure, irrespective of protocol. Backbone networks are attractive because they simplify the network and offer cost economies of scale.[39] For these reasons, many large U.S. and multinational corporations implement private backbone networks to support integrated and secure voice, data, and video applications across widely distributed locations.

Typically, backbone networks are hierarchically structured in three major ways, depending on the price and performance requirements:

1. Second-tier private access network(s) connecting into larger, private backbone
2. Second-tier private access network(s) connecting to local public network backbone
3. Second-tier private access network(s) connecting to large public backbone supplied by IEC carriers and service providers

Private backbone networks are commonly configured with T1 and T3 circuits. The traditional alternative to a private backbone, the public X.25 network, has a maximum bandwidth capacity of 56 Kbps, and thus is not suitable for high-speed inter-networking (such as would be required for LAN-to-WAN connectivity). Of the faster packet, high-bandwidth options (which include Frame Relay, ATM, and MPLS), Frame Relay is the most popular, with MPLS quickly gaining interest in the telecom community.

As companies merge to form larger corporate entities, many telecom managers are faced with the arduous task of integrating numerous external sub-systems into their existing corporate network infrastructure. As such, vast quantities of traffic flow and requirements data must be analyzed before a backbone network can be cost justified, designed, engineered, implemented, and maintained.

With the enormous influx of new telecom products from local, long-distance, and international carriers, network integrators, and value-added resellers (VARs), the network designer or manager can easily become inundated by the wide variety of networking solutions and alternatives. On the user side, software applications are growing in complexity and increasing the demands on the network. At the same time, management applies constant pressure to increase network utilization and user productivity while maintaining or cutting costs.

The telecom manager, therefore, must pursue many options to juggle the needs of users and upper management. In this climate, many large corporations place their networks up for bid every one to three years in the hopes of lowering costs and keeping pace with changing user demands. These bids often take the form of Requests for Information (RFIs), Requests for Quote (RFQs), and Requests for Proposal (RFPs) from telecom providers. When evaluating bids, the telecom manager should be guided by the business and technical requirements, and should avoid adopting new products and services for the sake of status or using the latest technology.

In summarizing the major advantages of private backbone topologies, they are:

- Used to integrate various sub-networks
- Relatively resistant to node and link failures
- Secure and private means of data transport
- Effective for certain time-sensitive applications
- Effective for distributed voice and data applications

The disadvantages of backbone networks, especially as compared to local access networks, include:

- More difficult and more complex to design
- More difficult to manage
- More redundant sites that need to be supported
- Can be costly if distributed rerouting and time-sensitive applications must be supported
- Increased redundancy needed for network reliability may come at the expense of idle bandwidth

Backbone networks are recommended to support:

- Large, multi-application networks
- Host-to-host connectivity
- Certain distributed voice and data applications
- Multiple host connectivity for disaster recovery
- Network availability for critical applications

2.2.5.2 Design Considerations and Techniques

Two major topological design problems must be solved when laying out a backbone network:

1. Node placement
2. Backbone connectivity (see Section 2.2.6.2)

Potential backbone nodes should be chosen at centers of major traffic volumes, in preference to small or remote sites. This reduces the potential for traffic bottlenecks, and it also increases the chances that the node will be located at a site where a full range of local and IEC services are available for supporting the backbone requirements.

Because backbone networks operate at high speeds, carrying lots of traffic between many nodes, the effects of network failures are magnified. For this reason, reliability is much more important when designing a backbone network than it is when designing a local access network. The performance and reliability demanded of backbone networks should be reflected in the selection and design of the network nodes and links.

With these practical guidelines in mind, four algorithms are useful for analytically deriving recommendations for locating potential backbone nodes. They are:

1. Center of Mass (COM) algorithm
2. Add algorithm
3. Drop algorithm
4. Mentor algorithm (see Section 2.2.6.2)

These algorithms are broadly applicable to many network design problems. This discussion reviews how they can be used in the context of identifying concentrator locations in centralized networks and backbone nodes in distributed networks.

The Center of Mass (COM) algorithm attempts to find natural traffic clusters based on estimated traffic between source and destination nodes. The COM algorithm is used when there are no candidate sites for the concentrator or backbone nodes. The node placements recommended by the COM algorithm must be checked to see if they are feasible.

2.2.5.2.1 Center of Mass Algorithm

The algorithm assumes that the following information is given:

- w_i traffic-based weights (calculated by totaling the traffic to and from each node).
- Desired maximum total weight W_M for a cluster (calculated by adding the w_i for each node used to construct a cluster center). This represents an upper limit on W that cannot be exceeded.

■ Desired minimum total weight W_m for a cluster (calculated by adding the w_i for each node used to construct a cluster center). This represents a lower limit on the cluster size.
■ Desired maximum total distance D_M between two clusters considered for a merge. This represents an upper limit on D_M that cannot be exceeded.
■ (x_i, y_i) coordinates for candidate i sites. These coordinates are used to calculate the distance between two sites, using a standard distance formula. The distance calculated is used as a proxy to estimate the cost of the link to connect the nodes. Therefore, in the discussion that follows, cost and distance are used synonymously.
■ Desired number of final clusters, C.

If during the course of the algorithm, there are conflicts enforcing the restrictions on W_M, W_m, and D, a trade-off function must be defined to resolve the differences. The problem must then be solved again, using the trade-off function.

1. Start with each (x_i, y_i) in a cluster by itself.
2. The cost to connect nodes i and j is assumed to be directly proportional to the distance between the two. The distance between each pair of nodes is calculated using a standard distance formula shown below:

$$Cost_{i,j} = \sqrt{\left[\left(x_i - x_j\right)^2 + \left(y_i - y_j\right)^2\right]} \qquad (2.6)$$

3. Sort the costs computed using Equation 2.6 from low to high for each node pair.
4. Find the two closest nodes as candidates for merging.
    −   If there are no merge candidates, check to see if the desired number of clusters, C, has been found. If not, terminate the algorithm with the message, "COM cannot find a solution." If the target for C and all other constraints are satisfied, terminate the algorithm with the message, "Final Solution Complete."
    −   If the constraints are violated, reject the cluster merge and remove it from the candidate list. Return to the beginning of Step 4.
    −   If the constraints are not violated, merge the two clusters (i.e., i and j) that are closest to each other to form a new cluster k. The new cluster k is chosen as the center of mass based on the traffic flowing at i and j.

The x coordinate of the new cluster k is:

$$x_k = \frac{(w_i * x_i) + (w_j * x_j)}{w_i + w_j} \qquad (2.7)$$

The y coordinate of the new cluster k is:

$$y_k = \frac{(w_i * y_i) + (w_j * y_j)}{w_i + w_j} \qquad (2.8)$$

5. Remove clusters i and j from further consideration. Add cluster k to the list of clusters to be merged. Return to Step 2 and calculate the distance between the existing nodes and the new cluster k.

The task of finding the two closest centers to merge is a major contributor to the algorithm's complexity. It can be shown that an efficient programming implementation of this algorithm has an overall complexity of $O(n^2)$, where n is the number of nodes considered. [KERS93]

2.2.5.2.2 Sample COM Problem

We now present an example to illustrate the COM algorithm. The data for the sample problem is given below:

See Table 2.19 for (x_i, y_i) coordinates and w_i traffic-based weights.

- Desired maximum total weight W_M for a cluster = 4.
- Desired minimum total weight W_m for a cluster = No Minimum.
- Desired maximum total distance D_M = 43 units. Any two clusters further apart than D_M cannot be merged.
- Desired number of final clusters C = 3.

1. Start with each (x_i, y_i) in a cluster by itself. *This means that initially each node is treated as a separate node center.*
2. Calculate the cost to connect nodes i and j. *The positional node coordinates are shown in Table 2.19. Using these coordinates and the distance formula, the cost to connect nodes i and j can be estimated.*
3. Sort the costs computed using Equation 2.6. *This is shown in Table 2.20, as are the results of the cluster merges.*
4. Remove clusters i and j from further consideration; add cluster k to the clusters to be merged. Return to Step 2 and calculate costs for distance between existing nodes and the new k cluster. *This*

Table 2.19 Sample COM Problem Data

Node	X_i	Y_i	Node weight
1	31	19	1
2	45	13	1
3	59	92	1
4	22	64	1
5	86	55	1
6	95	78	1
7	98	63	1
8	39	44	1
9	27	38	1
10	48	85	1

Table 2.20 Sample COM Calculations—First Iteration

Node pairs (original)	New node (merge-1)	X	Y	Cost	Weight
3,10	A	53.5	88.5	13	2
8,9	B	33	41	13.4	2
5,7	C	92	59	14.4	2
1,2	D	38	16	15.2	2
6,7	Reject: 7 already used	96.5	70.5	15.3	2
1,9	Reject: 9 already used	29	28.5	19.4	2
5,6	Reject: 5 already used	90.5	66.5	24.7	2
1,8	Reject: 8 already used	35	31.5	26.2	2
4,8	Reject: 8 already used	30.5	54	26.2	2
4,9	Reject: 9 already used	24.5	51	26.5	2
2,9	Reject: 9 already used	36	25.5	30.8	2
2,8	Reject: 8 already used	42	28.5	31.6	2
4,10	Reject: 10 already used	35	74.5	33.4	2
3,6	Reject: 3 already used	77	85	38.6	2

Table 2.20   Sample COM Calculations—First Iteration (continued)

Node pairs (original)	New node (merge-1)	X	Y	Cost	Weight
8,10	Reject: 10 already used	43.5	64.5	42	2
3,5	Reject: cost limit exceeded	72.5	73.5	45.8	2
1,4	Reject: cost limit exceeded	26.5	41.5	45.9	2
3,4	Reject: cost limit exceeded	40.5	78	46.4	2
6,10	Reject: cost limit exceeded	71.5	81.5	47.5	2
5,8	Reject: cost limit exceeded	62.5	49.5	48.3	2
5,10	Reject: cost limit exceeded	67	70	48.4	2
3,7	Reject: cost limit exceeded	78.5	77.5	48.6	2
9,10	Reject: cost limit exceeded	37.5	61.5	51.5	2
3,8	Reject: cost limit exceeded	49	68	52	2
7,10	Reject: cost limit exceeded	73	74	54.6	2
2,4	Reject: cost limit exceeded	33.5	38.5	56	2
2,5	Reject: cost limit exceeded	65.5	34	58.7	2
5,9	Reject: cost limit exceeded	56.5	46.5	61.4	2
7,8	Reject: cost limit exceeded	68.5	53.5	62	2
3,9	Reject: cost limit exceeded	43	65	62.8	2
4,5	Reject: cost limit exceeded	54	59.5	64.6	2
6,8	Reject: cost limit exceeded	67	61	65.5	2
1,5	Reject: cost limit exceeded	58.5	37	65.7	2
1,10	Reject: cost limit exceeded	39.5	52	68.2	2
2,10	Reject: cost limit exceeded	46.5	49	72.1	2
2,7	Reject: cost limit exceeded	71.5	38	72.9	2
4,6	Reject: cost limit exceeded	58.5	71	74.3	2
7,9	Reject: cost limit exceeded	62.5	50.5	75.3	2
4,7	Reject: cost limit exceeded	60	63.5	76	2
1,3	Reject: cost limit exceeded	45	55.5	78.2	2
6,9	Reject: cost limit exceeded	61	58	78.9	2
1,7	Reject: cost limit exceeded	64.5	41	80.2	2

Table 2.20 Sample COM Calculations—First Iteration (continued)

Node pairs (original)	New node (merge-1)	X	Y	Cost	Weight
2,3	Reject: cost limit exceeded	52	52.5	80.2	2
2,6	Reject: cost limit exceeded	70	45.5	82	2
1,6	Reject: cost limit exceeded	63	48.5	87	2

Table 2.21 Sample COM Calculations—Second Iteration

New & original node pairs	New node (Merge-2)	X	Y	Cost	Weight
C,6	E	62.3	65.3	19.2	3
B,D	F	35.5	28.5	25.5	4
B,4	Reject: B already used	29.3	48.7	35	3
A,4	G	43	80.3	39.9	3
A,6	Reject: A already used	67.3	85	42.8	3
A,C	Reject: cost limit exceeded	72.7	73.8	48.5	4
A,B	Reject: cost limit exceeded	43.3	64.8	51.7	4
B,C	Reject: cost limit exceeded	62.5	50	61.7	4
C,D	Reject: cost limit exceeded	65	37.5	69	4
C,4	Reject: cost limit exceeded	68.7	60.1	70.2	3
C,6	Reject: cost limit exceeded	93	65.3	72.2	3

is shown in Table 2.20. At the end of iteration 1, four (4) new nodes were created to reflect the merges of nodes A [3,10], B [8,9], C [5,7], and D [1,2]. Note that due to the constraints, nodes 4 and 6 are not merged with other clusters.

5. Calculate the cost to connect nodes i and j, and sort. *The updated cost data is computed in Table 2.21 and shows the costs to merge the remaining nodes A, B, C, D, 4, and 6. During the second iteration, the following merges were made: E [C, 6], F [8, 9, 1, 2], and G [3, 10, 4]. The final result is that three clusters were created, as specified in the original design objective. The final solution is shown in Figure 2.18. Note that the final solution provided by the COM algorithm must be checked for feasibility before it is accepted.*

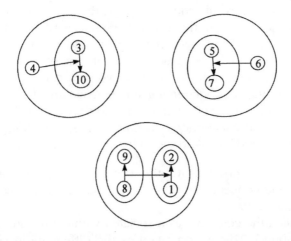

Figure 2.18   Sample COM problem: final solution. (*Source:* B. Piliouras.)

2.2.5.2.3 Concentrator/Backbone Placement

In this subsection, two new algorithms are introduced — the Add and the Drop algorithms. Both algorithms are used to select the best sites for concentrator/backbone nodes. These algorithms assume that the following information is available:

- Set of terminal locations i
- Set of potential concentrator/backbone locations j
- Cost matrix specifying c_{ij} to connect terminal i to concentrator/backbone node j for all i and j
- Cost matrix specifying d_j, the cost of placing a concentrator/backbone node at location j
- Maximum number of terminals that can be supported by each concentrator/backbone node j as MAX_j

2.2.5.2.3.1 Add Algorithm — The Add algorithm starts by assuming that no concentrator/backbone nodes are in the solution, and that all the terminals are connected directly to some central facility or node. It then proceeds to add concentrator/backbone nodes one at a time to lower the network costs. As each new node is added, some terminals are moved from the center or a previous node to a new node in order to obtain cost savings. The algorithm continues to add concentrator/backbone nodes until no more cost savings can be achieved.

Because the Add algorithm is a greedy algorithm, once a backbone/concentrator node is placed in the network it is neither reevaluated nor replaced, even if in subsequent steps all the terminals associated with the node are moved to another backbone/concentrator node. Thus, it is possible at the end of the algorithm to have added nodes that have no connections. This should be checked and if this situation exists, these nodes should be removed. The overall worst-case computational complexity of an efficient programming implementation of this algorithm is estimated at O (T^*B^2), where T is the number of terminals and B is the number of potential backbone/concentrator sites. [KERS93, p. 228]

The major steps of the Add algorithm are summarized below:

1. Connect all terminals to the central site. Do not include any backbone/concentrator sites in the initial solution. This is illustrated in Figure 2.19.
2. Calculate the cost savings associated with connecting terminal i to backbone j. This is computed as the difference in cost between connecting each terminal i directly to a potential backbone location j and a central site, minus the cost of the backbone node j. This can be expressed mathematically as:

$$Backbone \,/\, concentrator \; savings = S_j = \sum_{i=1}^{n}\left(c_{i,0} - c_{i,j}\right) - d_j \qquad (2.9)$$

where: $c_{i,0}$ is the cost to connect a terminal directly to a central node; $c_{i,j}$ is the cost to connect terminal i to backbone j; and d_j is the cost of the backbone node. This figure of merit must be calculated for each potential backbone node j with respect to all the terminals i. Sort the S_j values from high to low.

3. If all the S_j values are negative, no savings can be obtained by adding a backbone node. Therefore, the algorithm terminates with the message "Final Solution Obtained." Otherwise, pick the node associated with the largest positive S_j and add it to the solution. Move the terminals associated with the largest S_j from the center or the previous connection to the new backbone node.
4. Return to Step 2, recalculating the S_j for each remaining unassigned backbone/concentrator node.

2.2.5.2.3.2 Example of Add Algorithm

2.2.5.2.3.2 Example of Add Algorithm — Table 2.22 presents data for a sample problem that will be used to illustrate the Add algorithm. This table provides data on the set of terminal locations i, the set of potential concentrator/backbone locations j, and the cost matrix specifying c_{ij} to connect terminal i to concentrator/backbone node j. Assume that the d_j cost of placing a concentrator/backbone node in the network is equal to 50 units, and is the same for all nodes. Assume as well that the maximum

Table 2.22   Cost Data for Sample Concentrator/Backbone Location Problem

Terminal	Concentrator/backbone costs						
	Center 0	C1	C2	C3	C4	C5	C6
1	36	0	15	78	45	65	87
2	46	15	0	80	55	58	82
3	109	78	80	0	46	45	38
4	67	45	55	46	0	64	74
5	102	65	58	45	64	0	24
6	122	87	82	38	74	24	0
Total Cost:	482						

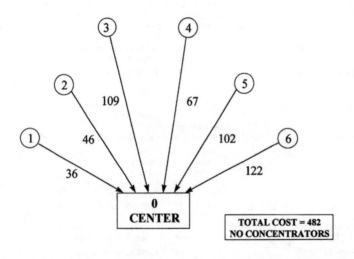

Figure 2.19   Initial solution for Add algorithm. (*Source:* B. Piliouras.)

number of terminals that can be supported by any concentrator/backbone node is equal to four (4) units. There are six (6) terminals, one (1) center location, and six (6) potential backbone sites in this sample problem. Note that in this example, there is no cost to connect terminals to co-located backbone nodes.

1. Connect all terminals to the central site. Do not include any backbone/concentrator sites in the initial solution. *This is illustrated in Figure 2.19.*

2. Calculate the cost savings associated with connecting terminal i to backbone j. This is computed as the difference in cost between connecting each terminal i directly to a potential backbone location j and a central site, minus the cost of the backbone node j. *For the sample problem, this involves calculating the possible cost savings associated with connecting Terminals 1–6 to Concentrators C1–C6. The Concentrator producing the largest savings is picked and added to the network design solution. This is shown in the first iteration calculations shown below.*

Savings Calculations for Iteration 1 of the Add Algorithm

CHECK C1

Cost Alternatives	Cost Savings (to Center vs. to C1)
C10-C11	36-0 = 36
C20-C21	46-15 = 31
C30-C31	109-78 = 31
C40-C41	67-45 = 22
C50-C51	102-65 = 37
C60-C61	122-87 = 35
Savings (4 largest savings)	139
Concentrator C1 Cost	-50
Total Savings using C1	**89**

CHECK C4

Cost Alternatives	Cost Savings (to Center vs. to C4)
C10-C14	36-45 = -9
C20-C24	46-55 = -9
C30-C34	109-46 = 63
C40-C44	67-0 = 67
C50-C54	102-64 = 38
C60-C64	122-77 = 48
Savings (4 largest savings)	216
Concentrator C4 Cost	-50
Total Savings using C4	**166**

CHECK C2

Cost Alternatives	Cost Savings (to Center vs. to C2)
C10-C12	36-15 = 21
C20-C22	46-0 = 46
C30-C32	109-80 = 29
C40-C42	67-55 = 12
C50-C52	102-58 = 44
C60-C62	122-82 = 40
Savings (4 largest savings)	159
Concentrator C2 Cost	-50
Total Savings using C2	**109**

CHECK C5

Cost Alternatives	Cost Savings (to Center vs. to C5)
C10-C15	36-65 = -29
C20-C25	46-58 = -12
C30-C35	109-45 = 64
C40-C45	67-64 = 3
C50-C55	102-0 = 102
C60-C65	122-24 = 98
Savings (4 largest savings)	267
Concentrator C5 Cost	-50
Total Savings using C5	**217**

CHECK C3

Cost Alternatives	Cost Savings (to Center vs. to C3)
C10-C13	36-78 = -42
C20-C23	46-80 = -34
C30-C33	109-0 = 109
C40-C43	67-46 = 21
C50-C53	102-45 = 57
C60-C63	122-38 = 84
Savings (4 largest savings)	271
Concentrator C3 Cost	-50
Total Savings using C3 (Best Savings; tied with C6)	**221**

CHECK C6

Cost Alternatives	Cost Savings (to Center vs. to C6)
C10-C16	36-87 = -51
C20-C26	46-82 = -36
C30-C36	109-38 = 71
C40-C46	67-74 = -7
C50-C56	102-24 = 78
C60-C66	122-0 = 122
Savings (4 largest savings)	271
Concentrator C6 Cost	-50
Total Savings using C6 (Best Savings; tied with C3)	**221**

Savings Calculations for Iteration 2 of the Add Algorithm

CHECK C1

Cost Alternatives	Cost Savings (to Center vs. to C1)	Cost Alternatives	Cost Savings (to previous C3 vs. to C1)
C10-C11	36-0 = 36	C33-C31	0-78 = -78
C20-C21	46-15 = 31	C43-C41	(Cheaper using C1) 46-45 = 1
C30-C31	109-78 = 31	C53-C51	45-65 = -20
C40-C41	67-45 = 22	C63-C61	38-87 = -49
C50-C51	102-65 = 37		
C60-C61	122-87 = 35		
	89	Savings (C11, C21, C41)	
	-50	Concentrator C1 Cost	
	-21	Remove savings from Concentrator 3, Iteration #1 (C40-C43)	
	18	**Total Savings using C1**	

CHECK C2

Cost Alternatives	Cost Savings (to Center vs. to C2)	Cost Alternatives	Cost Savings (to previous C3 vs. to C2)
C10-C12	36-15 = 21	C33-C32	0-80 = -80
C20-C22	46-0 = 46	C43-C42	46-55 = -9
C30-C32	109-80 = 29	C53-C52	45-58 = -13
C40-C42	67-55 = 12	C63-C62	38-82 = -44
C50-C52	102-58 = 44		
C60-C62	122-82 = 40		
	67	Savings (C12, C22)	
	-50	Concentrator C1 Cost	
	17	**Total Savings using C2**	

CHECK C4

Cost Alternatives	Cost Savings (to Center vs. to C4)	Cost Alternatives	Cost Savings (to previous C3 vs. to C4)
C10-C14	36-45 = -9	C33-C34	0-46 = -46
C20-C24	46-55 = -9	C43-C44 (Cheaper using C4) 46-0 = 46	
C30-C34	109-46 = 63	C53-C54	45-64 = -19
C40-C44	67-0 = 67	C63-C64	38-74 = -36
C50-C54	102-64 = 38		
C60-C64	122-77 = 48		
	67	Savings (C44)	
	-50	Concentrator C1 Cost	
	-21	Remove savings from Concentrator 3, Iteration #1 (C40-C43)	
	-4	**Total Savings using C4**	

CHECK C5

Cost Alternatives	Cost Savings (to Center vs. to C5)	Cost Alternatives	Cost Savings (to previous C3 vs. to C5)
C10-C15	36-65 = -29	C33-C35	0-45 = -45
C20-C25	46-58 = -12	C43-C45	46-64 = -18
C30-C35	109-45 = 64	C53-C55 (Cheaper using C5) 45-0 = 45	
C40-C45	67-64 = 3	C63-C65 (Cheaper using C5) 38-24 = 14	
C50-C55	102-0 = 102		
C60-C65	122-24 = 98		
	200	Savings (C55, C65)	
	-50	Concentrator C1 Cost	
	-141	Remove savings from Concentrator 3, Iteration #1 (C50-C53 and C60-C63)	
	9	**Total Savings using C5**	

Savings Calculations for Iteration 2 of the Add Algorithm (continued)

CHECK C6

Cost Alternatives	Cost Savings (to Center vs. to C6)	Cost Alternatives	Cost Savings (to previous C3 vs. to C6)
C10-C16	36-87 = -51		C33-C360-38 = -38
C20-C26	46-82 = -36		C43-C4646-74 = -28
C30-C36	109-38 = 71	C53-C56 (Cheaper using C6) 45-24 = 21	
C40-C46	67-74 = -7	C63-C66 (Cheaper using C6) 38-0 = 38	
C50-C56	102-24 = 78		
C60-C66	122-0 = 122		
	200	Savings (C56, C66)	
	-50	Concentrator C1 Cost	
	-141	Remove savings from Con. 3, Iteration #1 (C50-C53 and C60-C63)	
	9	**Total Savings using C6**	

3. Find the node associated with the highest savings S_j and add it to the solution. *The largest savings is obtained with the addition of either C3 or C6, which both yield a potential savings of 221 units. These results are shown in the first iteration calculations that follow. In cases of ties, either node can be selected for inclusion in the solution. We select node C3 for addition at site #3, and link terminals 3, 4, 5, and 6 to it. This is shown in Figure 2.20.*

4. Return to Step 2, calculating the S_j for each remaining unassigned backbone/concentrator node. *This is shown in the second iteration calculations below. In Iteration 2, check to see if any additional savings are possible with the Concentrators not used in Iteration #1 (e.g., C1, C2, C4, C5, and C6). We also verify that the original connections to Concentrator C3 are still the cheapest. For example, compare the savings with C3 versus the savings with C1, etc.*

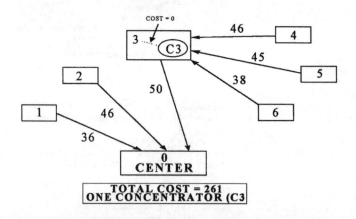

Figure 2.20 Solution obtained using Add algorithm after first iteration. (*Source*: B. Piliouras.)

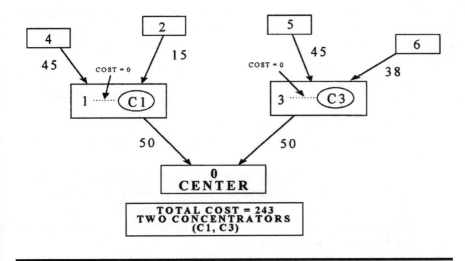

Figure 2.21   Final solution obtained using Add algorithm. (*Source:* B. Piliouras.)

5. Find the node associated with the highest savings S_j and add it to
the solution. *The largest savings is obtained with the addition of
C1, as shown in the savings calculations for Iteration 1. We select
node C1 for addition at site 3, and home terminals 1, 2, and 4 to
it. Note that in this iteration, terminal 4 has been moved from its
prior assignment at C3 to C1. This is shown in Figure 2.21.*

6. Because all the remaining S_j values are negative, the algorithm is
terminated with the message "Final Solution Obtained."

2.2.5.2.3.3 Drop Algorithm — The Drop algorithm starts by assuming
that all the concentrator/backbone nodes are in the solution, and that all
terminals are connected directly to the nearest backbone node. It then
proceeds to delete concentrator/backbone nodes one at a time to lower
the network costs. As a node is removed, some terminals are moved to
the center or to other nodes in order to obtain the savings. The algorithm
continues to delete concentrator/backbone nodes until no more cost
savings can be found.

Like the Add algorithm, the Drop algorithm is a greedy algorithm.
Despite apparent similarities between the two algorithms, there are also
substantial differences. Thus, one should not expect that they will produce
the same results, although it is possible that they might. The overall worst-
case computational complexity of an efficient programming implementation of this algorithm is estimated at O (T^*B^3), where T is the number of
terminals and B is the number of potential backbone/concentrator sites.

This is significantly worse than the computational complexity of the Add algorithm. Thus, for large problems, the Add algorithm might be preferable to the Drop algorithm, particularly because the Add algorithm produces solutions that are similar in quality to those produced by the Drop algorithm. [KERS93, p. 233]

The major steps of the Drop algorithm are summarized below:

1. Connect all terminals to the nearest backbone or concentrator site. Do not include any backbone/concentrator sites in the initial solution.

2. Calculate the cost savings associated with connecting terminal i to backbone j. This is computed as the difference in cost between connecting each terminal i directly to a potential backbone location j and a central site, and the cost of the backbone node j. This can be expressed mathematically as:

$$Backbone\ /\ concentrator\ savings = S_j = -\sum_{i=1}^{n}\left(c_{i,0} - c_{i,j}\right) - d_j \quad (2.10)$$

where $c_{i,0}$ is the cost to connect a terminal directly to a central node; $c_{i,j}$ is the cost to connect terminal i to backbone j; and d_j is the cost of the backbone node. This figure of merit must be calculated for each potential backbone node j with respect to all the terminals i. Sort the S_j values from high to low.

3. If all the S_j values are negative, no savings can be obtained by adding a backbone node. Therefore, the algorithm terminates with the message "Final Solution Obtained." Otherwise, pick the node associated with the largest positive S_j and drop it from the network. Move the terminals associated with the largest S_j from the previous connection to the center or to a new backbone node.

4. Return to Step 2, re-calculating the S_j for each backbone/concentrator node remaining in the solution.

2.2.5.2.3.4 Example of Drop Algorithm

2.2.5.2.3.4 Example of Drop Algorithm — The set of terminal locations i, the set of potential concentrator/backbone locations j, and the cost matrix specifying the cost c_{ij} to connect terminal i to concentrator/backbone node j for a new sample problem are given in Table 2.23. Assume that the d_j cost of placing a concentrator/backbone node is equal to two (2) units (and includes the cost of the node and the cost to connect the node to the center) and is the same value for all j. Assume that the maximum number of terminals allowed on each concentrator/backbone node, MAX_j, is equal to three (3) units, and is the same value for all nodes.

Table 2.23 Sample Cost Data for Drop Algorithm

T_i	C_0	C_1	C_2	C_3
1	2	1	2	4
2	1	0	1	2
3	4	1	2	2
4	1	2	1	2
5	2	3	2	0
6	4	4	3	2

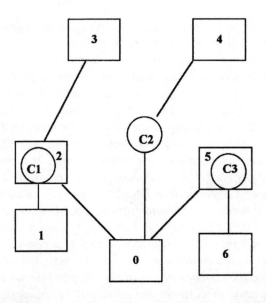

Figure 2.22   Initial solution to sample problem using Drop algorithm. (Source: B. Pilouras)

There are six (6) terminals, a C_0 center location, and three (3) potential backbone sites in this sample problem.

1. Connect all terminals to the nearest backbone site. Include all backbone/concentrator sites in the initial solution. This solution can easily be obtained by selecting the lowest row value in the cost matrix for each terminal. *The results of the initial solution are illustrated in Figure 2.22. The cost of the initial solution is calculated as the sum of the line costs to connect to concentrator plus*

Table 2.24   First Iteration Calculations for Drop Algorithm

Checking C1:	Checking C2:	Checking C3:
$c_{10} - c_{11} = 2 - 1 = 1$ (T1) $c_{20} - c_{21} = 1 - 0 = 1$ (T2) $c_{30} - c_{31} = 4 - 1 = 3$ (T3)	$c_{40} - c_{42} = 1 - 1 = 0$ (T4)	$c_{50} - c_{53} = 2 - 0 = 2$ (T5) $c_{60} - c_{63} = 4 - 2 = 2$ (T6)
Removing saves: -5	Removing saves: 0	Removing saves: -4
Total savings after concentrator cost considered: (-5+2) = -3	Total savings after concentrator cost considered: (0+2) = 2	Total savings after concentrator cost considered: (-4+2) = -2

the number of concentrators multiplied by the cost of each concentrator. For this problem, the initial solution cost = 1 + 0 + 1 + 1 + 2 + (3 * 2) = 11. Sites with an overlapping square and circle have a backbone/concentrator and terminal co-located at the same site.

2. Calculate the cost savings associated with disconnecting terminal i from backbone j. *For the sample problem, this involves calculating the possible cost savings associated with connecting terminals 1–6 to the center as opposed to concentrators 1, 2, or 3. The concentrator whose removal produces the largest savings is picked and deleted from the network design solution. This is shown in the first iteration calculations shown below. Recall that only three terminals, at most, can be assigned to a concentrator, based on a prespecified constraint. When a negative total savings is computed (as shown in the entries in Table 2.24), this means that it is cheaper to connect the terminal to the center directly than to maintain the current connection to the concentrator. Conversely, positive entries indicate that it is cheaper to connect the terminal to the concentrator being considered than to connect the terminal directly to the center.*

3. Find the node associated with the highest savings S_j and add it to the solution. *The largest savings is obtained with the removal of C3, which yields a potential savings of two (2) units. We select node C2 for removal, and home terminal 4 to the center. This is shown in Figure 2.23.*

4. Return to Step 2, calculating the S_j for each remaining backbone/concentrator node. *As shown in the first iteration calculation, no additional savings are possible by deleting concentrator 1 or 3. Because all the remaining S_j values are negative, the algorithm is terminated with the message "Final Solution Obtained." The solution remains the same as it was at the end of the first iteration.*

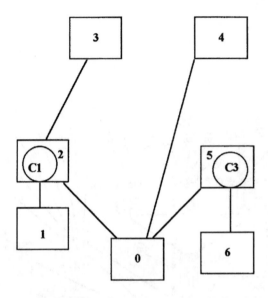

Figure 2.23   Solution to Sample Problem Using Drop Algorithm
(Source: B. Pilouras)

2.2.5.3 Case Study

As illustrated in Figure 2.24, large backbone networks are sometimes comprised of an inner backbone, referred to as a "core backbone," and multiple interconnecting semi-autonomous remote sub-systems. These sub-systems may be comprised of one or several regions or districts, where each region controls several smaller end-sites. Many topologies and designs can be used to interconnect the core backbone and the remote sub-systems. Which design is best depends on the needs of the users and the allocated telecom budget.

When the sub-systems are semi-autonomous, both intra-region and inter-region communications are needed. Examples of intra-regional communications include point-to-point voice, and client/server or peer-to-peer distributed data applications where sites share information within the district area. In comparison, inter-regional communications require data exchange across regions. This type of exchange might occur between a sales office and a district office, and from the district office to the core backbone. Examples of inter-regional applications include compressed broadcast video, point-to-point voice, and mainframe data applications. As LAN file servers replace mainframes, the corporate backbone infrastructure must provide increasing support for client/server applications.

In this case study, as shown in Figure 2.24, multiple high-speed private digital links are used to interconnect the company's core backbone. This

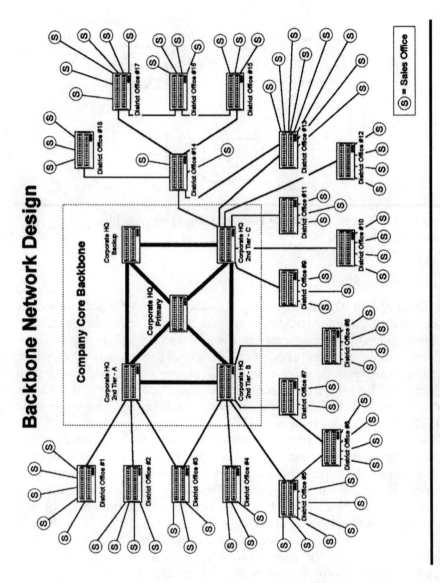

Figure 2.24 Backbone Case Study (Source: B. Piliouras)

network is comprised of corporate headquarter facilities, secondary main processing centers, and designated backup and disaster recovery location(s). In this scenario, the interconnecting links are usually dedicated circuits employing a range of speeds and mediums, including T-1 (1.544 Mbps), DS-3 (44.736 Mbps), or OC-n[40] SONET digital facilities. Note that within the core backbone network, a separate site has been designated as the corporate backup node, or disaster recovery site. This backup site may be an office of the corporation or it may be located within a company that provides disaster recovery services. To ensure true redundancy in the event of a major network failure, some companies install duplicate equipment (for example, voice and data switches, multiplexers, routers/bridges, mainframes, and file servers). In general, the telecom costs rise quickly when network failures cannot be tolerated and redundancy must be built into the network.

Many company backbones are a compilation of various separate network topologies overlaid to produce an overall asymmetrical corporate network. In this example, some district offices have only a single path to the core backbone, while other districts employ redundant rings, multidrop trees, or star topologies. Locations employing single paths are also called "single-threaded" sites. In this case study, this includes district offices 1, 2, 4, 8–12, and 18. These sites are relatively inexpensive to interconnect but do not have an alternate path back to the core backbone network. For mission-critical districts and sales offices, redundant paths may be needed, as is illustrated for district offices 3, 5–7, 13–14, and 15–17.

As revealed in Figure 2.24, redundancy can take many forms and depends on the types of applications supported, the urgency of the transmissions, and the associated networking costs. In the case of district office groups 5–7 and 14–17, for example, the company has installed TCP/IP routers using the Open Shortest Path First (OSPF) routing protocol. Therefore, if the path is disrupted from district office #6, through district office #7, to the core backbone, traffic could be rerouted through district office #5. Likewise, if the links from district offices 5 and 7 to the core backbone become inoperative, traffic can be rerouted in the opposite direction. However, these reroute procedures introduce additional node processing and facility propagation delays, and are usually recommended only for time-insensitive applications, such as large file transfers or noninteractive client/server data applications. As hardware and software vendors introduce products to compensate for the various delays, this type of network configuration can support more time-sensitive applications.

Network redundancy can take other forms, as shown in Figure 2.24 at district office #3. Suppose that at this location there are delay-sensitive voice, video, or data applications that are critical to the company's operations. At this site, there are two alternate paths providing redundant access to the core backbone and to two second-tier backbone nodes. Therefore, if a single link or backbone node fails, transmissions can be routed to the alternate facility or node. It should be stressed, however, that redundancy does not always equate to network infallibility. For example, if the second-tier-B node were out of service due to maintenance or failure, and the link from district office #3 to the second-tier-A node were to fail, then a re-route would prove futile. The redundant link would switch over and become the primary link, and retransmission would be attempted with the failed secondtier-B node.

The reader can easily deduce that when a given link fails and data is rerouted onto an active circuit carrying other user traffic, the utilization of the active link will increase. If full redundancy is required, the alternate link will need enough capacity to accommodate both the primary traffic and the rerouted traffic. Full link and equipment redundancy can be expensive and is usually reserved only for those sites that are considered mission critical by corporate headquarters.

A good compromise solution is illustrated in Figure 2.24 at district office #13. Instead of leasing a separate path to one of the second-tier nodes, the link from district office #13 to district office #14 can be used if necessary to reroute traffic. This scenario might occur when the district offices are closer to each other than to any one of the core backbone nodes, and the facilities are leased on a cost-per-mile basis. Thus, a partial backup path can be provided for some of the more critical applications at a relatively small incremental cost.

In addition to facility (or line) redundancy, separate hardware ports are needed if physical diversity is required. As discussed in later sections of this chapter, networks using X.25 packet, Frame Relay, Asynchronous Transfer Mode (ATM), and Multi-Protocol Label Switching (MPLS) can logically map multiple virtual circuits onto the same physical port. This approach can lead to substantial cost savings because the amount of hardware and the number of facilities ports can be reduced. These transport methods can be deployed on a private basis or procured from telecom providers through various public and semi-private product offerings.

2.2.6 Mesh Networks

2.2.6.1 Overview

A fully meshed topology allows each node to have a direct path to every other node in the network. This type of topology provides a high level of redundancy and reliability. Although a fully meshed topology usually facilitates support for all network protocols, it is not tenable for large packet-switched networks.[41] However, by combining fully meshed and star approaches into a partially meshed environment, telecom managers can improve network reliability without encountering the performance and network management problems usually associated with a fully meshed approach. A partially meshed topology reduces the number of routes within a region that have direct connections to other nodes in the region. Because all nodes are not connected to all other nodes, a non-meshed node must send traffic through an intermediary node acting as a collection point. There are many forms of partially meshed topologies.

Although a partial mesh diminishes the amount of equipment and carrier facilities required in the design, thus reducing networking costs, it may also introduce intolerable network delays. For example, voice/fax, video, and some data applications do not fare well with a partial mesh topology, due to the network latency introduced by circuit back-hauling, and nodal processing delays. To compensate for these delays, some legacy data protocols require local and remote "spoofing" to ensure that application timeouts do not occur. However, in general, partially meshed approaches provide the best overall balance for regional topologies in terms of the number of circuits, redundancy, and performance.

Many insurance companies, banking and financial institutions, and wholesale/retail operations employ mesh networks because their business transactions and order entry operations depend upon guaranteed connectivity to their processing centers. This is especially critical in situations where the value of the product or commodity varies widely with respect to time. As an example, the loss incurred in selling stock too late in a volatile market can more than compensate for the additional costs of building redundancy into the network.

Some of the advantages associated with mesh topologies include:

- Resistance to nodal and link failures
- Cost-effective method for distributed information computing
- Relatively easy to expand and modify

The disadvantages of mesh topologies include:

- More complex to design
- More difficult to manage
- Require more node site support
- Redundancy at the expense of idle, wasted bandwidth

Mesh topologies are recommended for:

- Large, multi-application networks
- Host-to-host connectivity

- Distributed application support
- User-initiated switching
- Redundant network support of mission-critical applications

2.2.6.2 Design Considerations and Techniques

As an alternative to maintaining a private mesh network, some companies use public switched networks. This involves attaching CPE[41] devices or nodes to the shared carrier network, usually by means of high bandwidth T-1 (1.544 Mbps) or DS-3 (44.736 Mbps) digital facilities that provide separate voice/fax and data channels. The digital facilities are usually de-multiplexed at the carrier's point-of-presence (POP) where they are interfaced to separate voice, data, and video networks owned and operated by the carrier. After traversing different networks, the channels are re-multiplexed and transmitted to the company's remote location over T-1 copper, DS-3 coaxial, or OC-n SONET fiber local loops. At the company's end-site, the high-bandwidth local loops are de-multiplexed back into the original voice, data, and video signals.

Using a public carrier reduces the number of I/O CPE ports and local access circuits needed. It may also be a cost-effective way to transmit each user application over the most appropriate transport network. For example, delay-sensitive and highly interactive voice, fax, and legacy data applications can be sent over dedicated circuit-switched connections with minimal delay. Conversely, less time sensitive client/server data applications can be sent over packet- or frame-switched networks where the facilities (and costs) are shared among several users. A public carrier network provides the benefits of a mesh topology at a cost that is often much less than the cost of a private network. The costs of a private mesh are often substantial due to the sheer quantity of circuits that must be dedicated to interconnect the various nodes.

When designing a network, the node traffic should be carefully analyzed to determine which nodes and applications require a full or partial mesh topology, and which ones might benefit from other networking architectures such as a star or tree topology. For most organizations, the cost of making a network completely fault tolerant is prohibitive. Determining the appropriate level of fault tolerance to be built into the network is neither a trivial matter nor an insignificant undertaking. As the complexity of the network increases (as it does with a mesh topology), it becomes increasingly important to use modeling and simulation tools to help design the network. The network decisions that must be made when planning a large mesh topology are too complex and numerous to manage without a systematic, analytical approach.

The Mentor algorithm was developed specifically for mesh topologies. Mentor is a heuristic, hierarchical design algorithm. As a first step, it selects the backbone sites to be used in the mesh topology. The algorithm then proceeds to design an "indirect" routing tree to handle traffic between nodes that is insufficient in volume to justify direct links. After an indirect routing tree is designed, the algorithm installs direct links between sites that have traffic above a prespecified threshold. Mentor is designed specifically to favor highly utilized, direct links to take advantage of the economies of scale offered by high-speed links. In addition to the traffic requirements, the designs produced by Mentor are determined by various preset design parameters, which are discussed in detail in the following paragraphs. These design parameters strongly influence the final design produced by Mentor. By making changes to the design parameters, it is possible to produce very different network topologies. Because the algorithm is so fast — it is of O (n^2) — it is possible to produce lots of network designs for analysis in a very short period of time.

The previous subsection presented several procedures — the COM, Add, and Drop algorithms — for selecting backbone nodes. Mentor also offers a procedure — Mentor Part I — for locating backbone nodes in a mesh network. All of these algorithms are suitable for locating backbone nodes and can be used to complement each other when exploring various design strategies. After the backbone nodes have been selected, they need to be interconnected to produce a final design. The Mentor algorithm Part II performs this task. An overview of the Mentor algorithm is provided in Figure 2.25 and Figure 2.26. For more details regarding the Mentor algorithm, the interested reader should refer to [KERSH91].

2.2.6.2.1 Mentor Algorithm Part I

The purpose of this part of the algorithm is to select backbone sites for the mesh topology. The algorithm examines all the potential node sites and separates them into two categories: end-nodes e_i and backbone nodes b_j. Mentor Part II ignores the end-nodes when developing the backbone mesh topology. End-nodes are connected to backbone nodes at a later time using the same techniques introduced earlier to solve the terminal assignment problem.

Mentor Part I takes as given a radius parameter, r, and a weight limit parameter, w. The radius parameter specifies the largest acceptable distance between a node and the nearest backbone node. The weight limit specifies the traffic threshold that must be satisfied to automatically qualify a node as a backbone site.

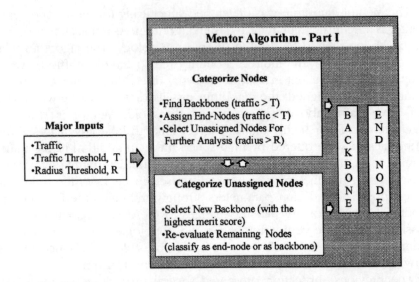

Figure 2.25 Overview of Mentor—Part I

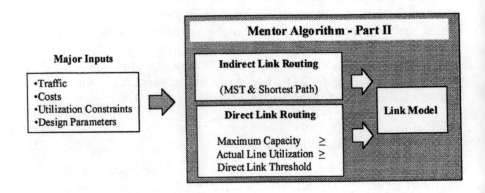

Figure 2.26 Overview of Mentor algorithm: Part II.

1. This step finds all the nodes that qualify automatically to become a backbone node because of the amount of traffic flowing through them. The remaining nodes are segregated for further analysis in the next step. Start by comparing the total traffic t_i flowing through each node i (i.e., the sum of all traffic entering and leaving the node) to the weight w:
   - If $t_i \geq w$, then assign terminal i as a backbone node, designated b_j.
   - If $t_i < w$, then assign terminal i as an unassigned node, designed u_i.
   - Find the maximum t_i, designated t_{MAX}.

2. This step examines all the u_i nodes that did not qualify, based on the previous calculations, to become a backbone. If a node u_i is within the allowed radius of a backbone site bj, the node is assigned to the nearest backbone and is redesignated as an end-node e_i. If the node u_i is too far from any currently designated backbone node, node is relabeled c_i to indicate that further clustering is necessary. This step begins by computing the distance d_{ij} between each node u_i and each b_j. The distance calculation provided in Equation 2.6 can be used for this purpose.
   - If $d_{ij} \geq r$, then designate terminal i as a node c_i that requires assignment to a closer backbone.
   - If $d_{ij} < r$, compare against all dij values calculated for terminal i. Find the smallest d_{ij} and assign terminal i as an end-site e_i associated with backbone b_j
   - Find the maximum d_{ij}, designated d_{MAX}.
3. This step examines all the nodes c_i that have not been assigned as either a backbone or an end-node. A trade-off function, or figure of merit, m_i is calculated below for each c_i. The node with the highest m_i represents the node with the best balance between central location and sufficient traffic flow to justify designation as a new backbone node. Compute m_i for each c_i as:

$$m_i = \frac{1}{2}\left(\frac{d_{max} - d_{ij}}{d_{MAX}} + \frac{t_{ij}}{t_{MAX}} \right) \tag{2.11}$$

   - Find the node with the largest m_i and designate it as a new backbone node b_j. Rename all c_i nodes as u_i and return to Step 2 (where the u_i nodes will be compared against the b_j just selected to see if they now qualify as end-nodes). Continue until all nodes are designated as either an end-node or a backbone node, at which time the algorithm is terminated.

2.2.6.2.2 Mentor Algorithm Part II

This part of the algorithm takes as given the following design parameters:

■ A design parameter, α, controls the topology of the indirect routing tree. This parameter is set somewhat arbitrarily, to a value between 1 and 0, inclusive. When α is set to 0, a minimal spanning tree is generated from the center. When α is set to 1, a star-like configuration is built around the center. Typically, values between 0.2 and 0.5 yield good mesh topologies.

■ A slack parameter, S, specifies the maximum percentage of the available bandwidth that can be used during the design process. For example, if S = 0.40, this means that only 40 percent of the available line capacity can be used. Any traffic levels up to and including this amount can be placed on the link. However, any traffic level exceeding this amount must be routed on other links.

■ A direct link threshold, D, specifies the level of traffic that is sufficient to justify a direct link between any two nodes.

1. Find the most central node. This is used to build an indirect routing tree. For each backbone node b_j, calculate a figure of merit as the sum over all nodes j of the distance — or cost, which is an approximation of distance — $dist_{ij}$ from i to j multiplied by the total traffic to and from j, w_{ij}:

$$f_j = \sum_{i=1}^{n} \left(dist_{ij} \times w_{ij} \right) \qquad (2.12)$$

   – Find the minimum f_{ij}, designated f_{min}. The node associated with this value is chosen as the center of the indirect routing tree, c.

2. Design the indirect routing tree. Start with node c, which is the center of the network as computed in Step 1. Initially, c is designated as part of the indirect routing network and all other nodes are considered outside of the network. For all links (i,j) with i in the tree and j outside the tree, we define L'(i,j) as:

$$L'(i,j) = (i,j) + \alpha \ (i,c) \qquad (2.13)$$

   where α is a prespecified design parameter between 0 and 1, inclusive.

   – Find the minimum computed value of L'(i,j). The node associated with this value is brought into the tree.
   – Update the L'(i, j) values affected by the addition of this node. Find the minimum computed value of L'(i,j). The node associated with this value is now brought into the tree. Continue this process, bringing in one node at a time, until all the nodes are connected.

3. Sequence the traffic requirements for each pair of nodes i and j.
   – Logically, we are trying to establish a sequence for loading traffic on the indirect routing tree in such a way that direct links will be encouraged. The indirect tree is designed to carry traffic that is insufficient to justify a direct link between two nodes. The indirect tree will also carry overflow traffic from

direct links that do not have sufficient capacity to carry all the traffic demand between two directly linked nodes. Overflow traffic can be routed on the indirect routing tree, up to the point that the traffic reaches the critical direct link threshold. (Note: at this point, a new direct link is installed on that portion of the indirect routing tree. This is done in Step 4 below.)

— We start by identifying all the pendent node pairs.[43] These pairs are considered first-level node pairs. Find the nearest neighbors of all the other nodes.

— Compute the cheapest route for detouring traffic through the nearest neighbors of the two nodes under consideration. These alternate routes represent dependencies that must be sequenced *after* the node pairs. Thus, these node pairs are considered second-level node pairs, and their dependencies are considered third-level node pairs, and so on. Although these dependencies *are not actually used* in routing the traffic during the course of the algorithm, we are attempting a routing that detours the traffic through the fewest number of hops, and that encourages optimal loading of traffic to justify direct links.

— The node pairs must be sequenced according the following rules: (1) node pairs at the highest sequence level (i.e., the pendent node level) can be considered in *any* order; and (2) node pairs at a lower level can sequenced only *after all* dependencies at higher levels have first been considered. Node pairs that are separated by the most number of hops must be sequenced before node pairs separated by fewer hops. For example, all the node pairs separated by four hops should be sequenced before all the node pairs dependencies separated by three hops, and so on. This sequencing is not necessarily unique, and it is possible that many valid sequences can be defined. Only one valid sequence is used in the fourth and final step of the Mentor algorithm.

4. Assign direct links. Consider each node pair in the sequence determined in Step 3.
   - If the direct link traffic threshold between the node pairs is equaled or exceeded, then assign a direct link between the two nodes.
   - If the traffic threshold is not reached, route the traffic between the nodes on the indirect routing tree.
   - If the traffic on the indirect routing tree reaches or exceeds the traffic threshold, install a new link on the indirect routing tree to handle the overflow. Conceptually, if an indirect link is loaded to the threshold level, the indirect is converted into a

direct link, and a new indirect link is inserted. In this way, the indirect tree is never overloaded beyond its usable link capacity. Once a direct link is assigned, no further attempt is made to put more traffic on the link, even if the link has excess capacity.
– After every node on the sequence list has been evaluated, the algorithm terminates.

Inherent in this presentation of the Mentor algorithm are several key assumptions, which are listed below:

■ *Traffic is bi-directional.* This is realistic for voice data, and not necessarily for packet-switched data. If this assumption is not true, the algorithm can be modified to explicitly consider the load on each link in each direction during the link selection.
■ *Only one link capacity at a time is considered.* The algorithm can be modified to account for multiple link capacities; however, this adds to the computational complexity of the procedure.
■ *Cost is represented by an increasing function of distance.* This assumption is made when distance and cost are treated synonymously. The cost data comes into play when the indirect routing tree is being constructed. If exact tariff data is available, it can be used in the cost matrix without changing the execution of the algorithm. The algorithm would have to be modified only if the distance approximation is altered and therefore the cost matrix used is affected.
■ *Cost is symmetric in that the line costs are the same, irrespective of the starting and ending node.* Thus, a link (i,j) will have the same cost as a link (j,i). This assumption is consistent with the use of virtual circuits in a TDM-based network. The algorithm can be modified to handle different costs in different directions. However, this adds to the computational complexity of the algorithm.
■ *The traffic requirement can be split over multiple routes.* If this is not true, the algorithm can be modified to use a bin-packing algorithm to assign traffic to the links.
■ *Each link is moderately utilized.* This assumption helps to account for the exceptionally low computational complexity of the Mentor algorithm. This assumption is enforced through the selection of the slack parameter, which ensures that no link is over utilized.

The version of Mentor presented here is best suited for designing TDM-based networks. If a router-based network needs to be designed, then modifications may be needed to ensure that reasonable designs are produced. For example, this might involve decreasing the usable line capacity assumed by the algorithm, as represented by the slack parameter. In addition, it may be advisable to modify the construction of the indirect routing tree to encourage a topology that minimizes the number of hops.[44] The interested reader should refer to [CAHN98], which presents numerous modifications and extensions to the Mentor algorithm for various types of networks.

In Mentor, the design of the indirect routing tree is controlled by an α parameter that governs the trade-off between tree length and path length. When the α parameter is set to 0, a minimal spanning tree is produced. This type of structure will tend to produce lower-cost designs. When the parameter is set to 1, a star is produced. This type of structure will tend to produce smaller path delays. With everything else held constant, increasing the direct link threshold — that is, the slack parameter — increases the available line capacity used during the design process. Because more line capacity is available, there will be a tendency to use fewer lines in the overall design. To increase the reliability of the networks generated using Mentor, try decreasing the link utilization threshold. Decreasing the link utilization will encourage more links in the final design, thus providing opportunities for more alternative routes. In addition, because the links are less heavily utilized with a lower link utilization threshold, the overall impact of a specific link failure will tend to be reduced correspondingly. However, if the relative traffic requirement is very small in relation to the line capacity, changing the slack parameter may have little overall impact on the final topology.

2.2.6.2.3 Sample Problem Using Mentor Algorithm

The Mentor algorithm is perhaps best understood by example. Data for a sample problem is presented in Table 2.25 and Table 2.26. The assumptions for this problem include:

- All nodes are potential backbone nodes (this assumption means that one does not need to use Mentor Part I to determine the backbone locations).
- Traffic can be split on the network as required to fill the links.
- All lines are full-duplex.

The design parameters selected for the sample problem are:

- $\alpha = 0.2$
- Maximum usable capacity = 19.2 units
- Direct link traffic threshold = 80 percent of the maximum usable capacity = 15.36 units

Table 2.25   Traffic Data for Sample Mentor Problem

Traffic requirements matrix					
	A	*B*	*C*	*D*	*E*
A	0	4	11	10	7
B	4	0	4	6	8
C	11	4	0	9	8
D	10	6	9	0	5
E	7	8	8	5	0

Table 2.26   Cost Data for Sample Mentor Problem

Cost requirements matrix					
	A	*B*	*C*	*D*	*E*
A	1	7	11	11	8
B	7	1	4	2	6
C	11	4	1	7	3
D	11	2	7	1	7
E	8	6	3	7	1

2.2.6.2.4 Solution to Sample Problem Using Mentor Algorithm

Step 1:   Calculate the network center.

– First add the traffic flows through each node:

Node A Traffic Flow:
AB + AC + AD + AE + BA + CA + DA + EA =
4 + 11 + 10 + 7 + 4 + 11 + 10 + 7 = 64

Node B Traffic Flow:
BA + BC + BD+ BE + AB + CB + DB + EB =
4 + 4 + 6 + 8 + 4 + 4 + 6 + 8 = 44

Node C Traffic Flow:
CA + CB + CD+ CE + AC + BC + DC + EC=
11 + 4 + 9 + 8 + 11 + 4 + 9 + 8 = 64

Node D Traffic Flow:
DA + DB + DC+ DE + AD + BD + CD + ED =
10 + 6 + 9 + 5 + 10 + 6 + 9 + 5 = 60

Node E Traffic Flow:
EA + EB + EC+ ED + AE + BE + CE + DE =
7 + 8 + 8 + 5 + 7 + 8 + 8 + 5 = 56 bps

- Now find the Figure of Merit, m_i, for each node using the traffic flows calculated above multiplied by the link costs.

Node a: (44 * 7) + (64 * 11) + (60 * 11) + (56 * 8) = 2120
Node b: (64 * 7) + (64 * 4) + (60 * 2) + (56* 6) = 1160
Node c: (64 * 11) + (44 * 4) + (60 * 7) + (56 * 3) = 1468
Node d: (64 * 11) + (44 * 2) + (64 * 7) + (56* 7) = 1632
Node e: (64 * 8) + (44 * 6) + (64 * 3) + (60 * 7) = 1388

- Because node b has the smallest figure of merit, it is selected as the network center.

Step 2: Find the indirect routing tree that is used to carry overflow traffic (where α = 0.2). Recall that:

$$\text{Distance to the tree} = (\text{distance}_{i,j}) + 0.2\ (\text{distance}_{i,\ center})$$

- Because D is the cheapest link to attach to the center node B, it is brought into the tree first.
- Next, the connection costs to bring A, C, and E into the network through attachment to D are computed. When these costs are compared to the cost of bringing a node in directly through the center node, the next node selected is C, which is homed to the center node. The costs to connect A and E into the network through an attachment to C are computed in the third line of Table 2.27. When these costs are compared to the cost

Table 2.27 Indirect Routing Tree Calculations for Mentor

Node just brought into tree	A	B	C	D	E
B (center node)	7	—	4	2	6
D (attached to center)	11 +.2(2) = 11.4	—	7 +.2(2) = 7.4	—	7 +.2(2) = 7.4
C (attached to center)	11 +.2(4) = 11.8	—	—	—	3 +.2(4) = 3.8
E (attached through C)	8 +.2(4+3) = 9.4	—	—	—	—
A (attached to center)					

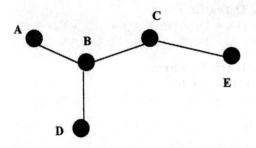

Figure 2.27   Indirect routing tree created by Mentor algorithm.

of bringing these nodes in directly through the center node, the decision is made to select node E, and to home it through node C. The final node A is attached to the network at the center node, because it is cheaper to attach node A through the center than it would be attach A to the end of the current tree (which consists of links from node B, to node E, to node C).

– The final indirect routing tree produced in this step is shown in Figure 2.27.

Step 3:   Find the sequencing requirements.

– For each potential node pair, the alternative traffic detour routes are examined. The alternative detour routes are formed by detouring each node's traffic to its nearest neighbor and connecting the nearest neighbor to the other endpoint of the node pair under consideration. This is done only for the purpose of sequencing the links in preparation for Step 4. This does not represent the actual traffic assignment and routing, which is done in Step 4.

– For the problem at hand, the possible node pairs in lexicographic order are:

(A,B), (A,C), (A,D), (A,E), (B,C), (B,D), (B,E), (C,D), (C,E), (D,E)

At this point, the lexicographic node pair ordering is not significant. This ordering is used solely to clarify the notation.

– By computing the cost of the alternative routes for each node pair, one can logically deduce the traffic overflow dependencies, and from this a link sequence for Step 4. Now determine the via-node for all nonadjacent nodes, as shown below. This is done by examining each node in a node pair to find its nearest neighbors. Each nearest neighbor is a potential via-point. From

Table 2.28 Dependency Matrix for Sample Mentor Problem

Node pair	Possible detour routes	No. of Hops between nodes	Cost of cheapest route
AB	Nearest neighbors	1	(Note: Since A and B are separated by only 1 hop and are nearest neighbors, an alternative path does not need to be considered.)
AC	AE to C	2	There is only one alternate path
AD	AB to D	2	There is only one alternative path
AE	AB to E EC to A	3	$7 + 6 = 13$ * $3 + 11 = 14$
BC	BD to C CE to B	1	$2 + 7 = 9$ $3 + 6 = 9$
BD	Nearest neighbors	1	(Note: Since B and D are separated by only 1 hop and are nearest neighbors, an alternative path does not need to be considered.)
BE	BD to E EC to B	2	$2 + 7 = 9$ $3 + 4 = 7$ *
CD	DB to C CE to D	2	$2 + 4 = 6$ * $7 + 3 = 10$
CE	Nearest neighbors	1	(Note: Since C and E are separated by only 1 hop and are nearest neighbors, an alternative path does not need to be considered.)
DE	DB to E EC to D	3	$2 + 2 = 4$ * $3 + 7 = 10$

this, list all possible routes for nonadjacent nodes. Choose the via-route with the cheapest path.

- Now list each node pair in the lexicographic list with the associated dependencies, thereby creating a dependency matrix as shown in Table 2.28.

- In sequencing the links, always start with the pendent node pairs. In this case, there are three pendent node pairs: (A,D), (A, E), and (D, E). Therefore, a possible sequence is:

AD, AE, DE, AC, AB, BE, CE, BC, CD, BD

This is not the only possible valid sequence.

Table 2.29 Direct Link Assignment for Sample Mentor Problem

Links	Traffic path	Traffic requirement	Load on direct links	Load on indirect links
AD	AB; BD; and AD	20	AD (19.2)	AB (.8) BD (.8)
AE	EC, CB, BA	14		EC (14) CB (14) AB (14.8)
DE	EC, CB	10	EC (19.2) CB (19.2)	EC (4.8) CB (4.8) BD (10.8)
AC	AC and AB, BC	22	AC (19.2)	BC (7.6) AB (3.6)
AB	AB	8		AB (11.6)
BE	BE	16	BE (16)	
CE	CE	16	CE (16)	
BC	BC	8	BC (15.6)	
CD	CD	18	CD (18)	
BD	BD	12	BD (19.2)	BD (3.6)

Step 4: Assign direct links.

- The direct link placements made by the Mentor algorithm are summarized in Table 2.29.
- The final design produced by Mentor, including both indirect and direct links, is provided in Figure 2.28.

2.2.6.2.5 Modification to Mentor Algorithm Using Half-Duplex Lines

To use Mentor to design a half-duplex network, the direct link assignments made in Step 4 above are calculated somewhat differently. In the case of half-duplex links, the direct link assignments are based on the *highest* traffic requirement in either direction. Recalling the previous sample problem, the maximum usable line capacity is 19.2 units, and the direct link traffic threshold is 15.36 units. The resulting half-duplex line placements for this revised problem are summarized in Table 2.30.

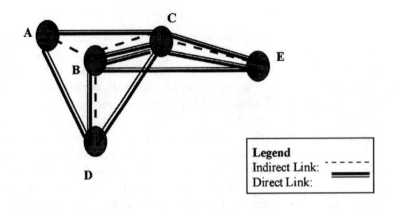

Figure 2.28   Final design created by Mentor algorithm.

Table 2.30   Direct Link Assignment Using Half-Duplex Links

Links	Traffic path	Traffic requirement	Load on direct links	Load on indirect links
AD	AB, BD	10		AB (10) BD (10)
AE	AB, BC, CE	7	AB(17)	BC (7) CE (7)
DE	EC, CB, BD	5		EC (12) CB (12) BD (15)
AC	AB, BC	11	BC (19.2)	BC (3.8) AB (11)
AB	AB	4		AB (15)
BE	BC, CE	8	CE(19.2)	BC (11.8) CE (.8)
CE	CE	8		CE (8.8)
BC	BC	4	BC(15.8)	BC (0)
CD	CB, BD	9	BD(19.2)	CB (9) BD (4.8)
BD	BD	6		BD (10.8)

2.2.6.3 Case Study

Figure 2.29 represents a typical mesh network. In this example, a company uses a two-tier network that is interconnected to two backup and disaster recovery sites. Time-division multiplexers (TDMs) are used as circuit switches to dedicate specific information channels between each node pair. Thus, all the network switching occurs within and between the PBXs, the DEC VAX computers, and the LAN/WAN routers. The reader should note that although poll-select mainframe applications can operate over mesh topologies, the applications should be routed on the most direct paths that are economically feasible. This is to minimize the delays inherent in this type of distributed computing environment. The factors that determine which users are connected to what sites include the following:

- Distance to the nearest site, Site A or Site B
- Location of application(s) requested by the users
- User security clearance issue
- Protocol interoperability and conversion/proxy requirements
- Delay and processing intolerance limits

It is easy to see why mesh networks can be very expensive. In a fully meshed network, there are $(N*(N-1))/2$ full-duplex links (where N is the number of nodes in the network). This quickly becomes a large number of links as the number of nodes added to the network increases. The input/output (I/O) hardware required grows as the square of the number of devices, because each device requires $(N-1)$ I/O ports for network connectivity. For example, when four multiplexer devices $(N = 4)$ are used in the network, $N-1$ (or $4-1 = 3$) I/O port cards and cables are needed by each multiplexer to connect to the other multiplexers. As the number of links and I/O ports and associated hardware increases, so does the network cost.

2.2.7 Value-Added Network Services

The network services discussed in this section have evolved from network topologies to standard product offerings from a wide range of service providers. Network managers should carefully compare and contrast the various features and functions, which vary considerably from one provider to another, to understand the respective strengths and weaknesses of each product offering. Often, Requests for Information (RFIs) and Requests for Proposal (RFPs) will contain matrices highlighting the attributes and corresponding benefits of each provider's offerings. As these services become more and more productized, every possible selling feature is highlighted

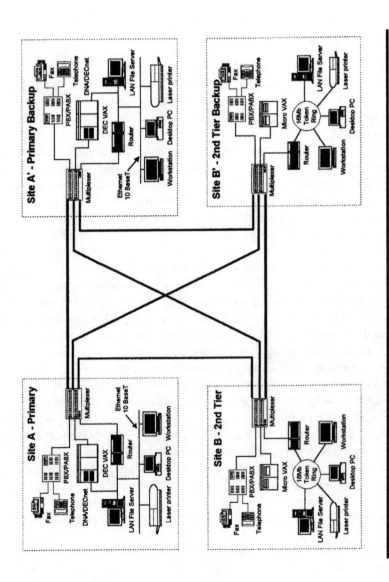

Figure 2.29 Mesh network case study. (*Source:* B. Piliouras)

by the vendor to its best advantage. Informed sleuthing on the part of the network manager is needed to identify the corresponding limitations or deficiencies, which providers are not as likely to point out.

2.2.7.1 Packet-Switched Networks

2.2.7.1.1 Overview

In 1976, the Consultative Committee for International Telegraphy and Telephony (CCITT)[45] adopted the ITU-T X.25 standard for packet switching. The X.25 standard specifies the interface between a host system and a packet-switched network via a dedicated circuit. Packet switching and its associated X.25 protocols were developed specifically to support data communications during the early days of analog networking. Analog lines are inherently noisy, creating signal distortion and transmission errors. Although a certain amount of error in voice communications is tolerable, in data communications — which involve transmission of precise binary bits — it is not. X.25 was designed to compensate for these reliability problems by providing extensive error control and handling to ensure that transmissions are correctly received. As the reliability of modern communications equipment steadily improved, and as digital and fiber lines have replaced analog lines, the need for network-level error control and handling has declined. The inherent latency and delay introduced by X.25 error checking limits the performance of packet-switched networks, especially in handling high-volume data transmissions. As discussed in Section 2.2.7.2, Frame Relay was developed to overcome some of the performance limitations of X.25 and has evolved as a popular successor to X.25.

Nonetheless, X.25 packet switching is still available domestically and internationally, providing dependable, economical networking for geographically dispersed sites. Packet switching is a flexible, scalable, and robust networking technology. Packet-switched networks are relatively easy to manage and readily accommodate the addition of new WAN links as the need arises. Packet-switched networks are particularly well suited to supporting small- and medium-scale data networking applications, such as terminal-to-mainframe access, electronic mail transmission, and data file transfer. Packet switching is also used to support low-volume internetworking.

Packet switching is implemented in two basic ways using either datagrams or virtual circuits. Datagrams, which are employed by the Internet community, are created by "chopping" a message into a series of minimessages. These mini-messages are, in turn, forwarded through a series of intermediate network nodes. The path each datagram takes through the network is independent of the path taken by the other datagrams.

The datagrams are then resequenced at the final destination into their original order. Virtual circuits, in contrast, establish a logical or "virtual" end-to-end path prior to the transmission and reception of data packets. Each approach has its own inherent advantages and disadvantages. They also introduce transmission delays in different ways.

Datagrams do not require preestablished transmission paths because the network determines the packet routing at the time of the transmission. This is a "connectionless" method of packet transport. This method is very robust in compensating for network failures. If a path on the packet-switched backbone fails, the datagram is simply sent over an alternate path. There is no need to reestablish the connection between the sender and receiver to continue the transmission. As is evident from this discussion, no call setup or call teardown procedures are required when using datagrams; this helps to control one aspect of transmission delay.

When using datagrams, there are no guarantees that a packet will be successfully transmitted and received. Instead, datagrams offer a "best-effort" packet delivery service. It is possible that packets will not be delivered successfully, or they may arrive out of sequence, because different paths may be taken by each datagram. Thus, the devices at the receiving end must resequence the packets into their original order. If the packets cannot be sequenced correctly, or if they are corrupted during transport, it is usually up to the end devices to request a retransmission. As the number of packets in the transmission increases, so does the chance that there will be an error in the transmission. Therefore, this method of error control is best suited to transmissions consisting of relatively few packets, because errors will not be corrected until the end of the transmission. If the transmission is very long, it will also take a long time to retransmit the message when an error occurs.

Datagrams are similar in concept to "message switching" in that both methods use connectionless packet transport. One major difference between the two, however, is that message switching sends the entire message as one continuous segment while datagrams are sent as a series of small packets.

When a traditional circuit switch is overloaded, "blocking" results. Any newcomers attempting to use the overloaded network are denied service. In contrast, datagram packet switching allows packets to be rerouted around congested areas using alternate paths. The trade-off against this resiliency is the increased transmission time needed to calculate alternative routes and to perform the necessary error checking and handling. The overhead included in the datagram packets (to perform such functions as addressing, routing, and packet sequencing) also reduces the possible end-to-end line utilization. These are significant factors in determining the performance limits of packet-switched networks.

Packet addressing, resequencing, and error control functions are performed by devices called packet assembler/dissemblers (PADs). Terminals and host devices send data in asynchronous (i.e., data is sent as individual characters) or synchronous (i.e., data is sent as blocks of characters) format. PADs convert these data streams into packets. Many PADs act as a concentrator or multiplexer, consolidating transmission streams from multiple devices and device types. Devices are typically connected to PADs in one of three ways: (1) through a direct, local connection; (2) through a leased line connection; or (3) through a public dial-up line. The PAD, in turn, can be connected to an X.25 network or some other non-X.25 device.

Virtual circuits provide another form of packet switching. In contrast to datagrams, virtual circuits are "connection" oriented. With virtual circuits, paths are set up once, using permanent virtual circuits (PVCs), or on a call-by-call basis, using switched virtual circuits (SVCs). PVCs require only one call setup (at the start of the transmission) and are permanently mapped to a particular end-site. However, PVCs require predetermined mappings specifying which sites can communicate with each other, and they do not support any-to-any communication, as do SVCs.

Although SVCs experience more call delay (relating to call setup and call teardown procedures) than PVCs, they offer more flexibility because virtually anyone can communicate with anyone else on the network. The open networking approach afforded by SVCs may, however, translate into increased CPU[46] and switch processing; and if the circuit becomes inoperative during a transmission, a new connection must be established. These factors contribute to the potential for transmission delays. However, once a path is established in a virtual circuit, each packet is transmitted over the same path. Because the same logical path is employed for the duration of the transmission, multiple packets can be sent before an acknowledgment (ACK) or negative-acknowledgment (NCK) is returned to the originating packet node. This significantly increases the efficiency of the link utilization. The number of packets that can be sent within a given ACK/NAK interval is known as the "sliding window." For X.25 networks, window sizes of eight (8) and one-hundred-twenty-eight (128) are common and are also referred to as Modulo 8 (MOD 8) and Modulo 128 (MOD 128), respectively. When packets are lost or corrupted during transport, the window "slides" down to a smaller number of frames per ACK/NCK. Similarly, when the quality of transmission improves, the window slides open. The sliding window provides an effective way for the network to adapt to changing traffic loads and conditions.

2.2.7.1.2 Design Considerations

A packet-switched network consists of the following components:

- *Non-X.25 devices:* the terminals and end-devices on the network.
- *PADs:* devices used to attach non-X.25 devices to the X.25 network via a packet switch.
- *Packet switch:* devices attached to PADs, other packet switches, and to the X.25 backbone.
- *Private/carrier facilities:* the links that comprise the access network(s) and the X.25 backbone.

A packet-switched network can be implemented as a public, a private, or a hybrid solution, as illustrated in Figure 2.30. All one needs to use a public X.25 network is a connection, usually via a packet switch. In a typical configuration, multiple end-devices are clustered into a single PAD, and in turn, multiple PADs are connected to a single packet switch. The packet switch can be connected to other PADs or to other packet switches. The major decisions when implementing a public X.25 network solution involve selecting the service provider and the packet switch. The packet switch should be chosen with the following in mind:

- Switch capacity
- Switching functions supported
- Type of access (public or private)
- Modularity required (memory, CPU, number of ports, etc.)
- SNMP support (for network management purposes)
- Interoperability with existing network equipment (routers, bridges, etc.)

Most packet switches are configured to support Frame Relay, T1/E3 speeds, and multiple protocols. This means it is fairly easy to transition from X.25 to Frame Relay. In addition, most packet switches are modular so they can be easily expanded to support growing traffic demands. This facilitates network scalability and helps to preserve the organization's investment in equipment. When selecting a packet switch, a word of caution is in order. Just because two packet switches comply with X.25 standards does not mean that they are compatible and can be used in the same network. X.25 specifies the standards for interfacing to an X.25 network, and it does not apply to equipment interfaces. Most packet switches use proprietary protocols and cannot be interconnected.

There are several compelling reasons for using a public switched X.25 network. First, X.25 may provide a cost-effective network solution, particularly if the sites to be connected are far apart. However, when transmitting high traffic volume over short geographic distances, it may be cheaper to use a private line WAN, due to the pricing strategies of the service providers. Second, this type of network is easily scalable and

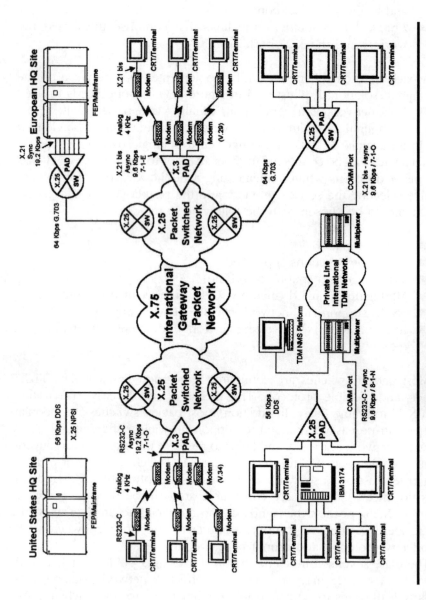

Figure 2.30 Packet-switched network. (*Source:* B. Piliouras.)

provides consistent, reliable performance. Third, it also requires minimal equipment and facilities to implement. Finally, X.25 provides a way to integrate many diverse protocols onto a single network infrastructure. This ability relates to the fact that X.25 networks operate at OSI layer 3 (or network layer). Because X.25 networks operate at the OSI network boundary level, they are generally compatible with all network architectures.

When a company owns a private X.25 network, it must be responsible for maintaining and operating the X.25 backbone through the use of leased line facilities. This type of network is usually more costly than a public X.25 network. According to [DINT94], to be truly cost effective, a private X.25 network should be used at or near its capacity. Toward this end, some organizations resell their excess network capacity, possibly during off-peak hours, to other organizations to help pay for the costs of the network. The need for network security, scalability, and control are the reasons most often cited for implementing a private X.25 network.

Hybrid solutions, consisting of both private and public X.25 facilities, are possible and may be justifiable, depending on circumstances.

2.2.7.1.3 Case Study

As shown in Figure 2.30, in this case study, X.25 PADs are used as port aggregators to reduce the number of network circuits and ports on the FEP/mainframes. In some cases, an X.25 PAD can take the place of a 3174 cluster controller in an SNA environment.

In general, charges for X.25 packet services typically include an access charge per location and a usage charge based on a certain price per kilobyte (i.e., 1000 bytes). Many packet services utilize "postalized" rates, whereby the infamous mileage-sensitive circuit charge is factored into the overall costs. To improve the cost effectiveness of the network, some organizations opt to use a small portion of the bandwidth to transport data from other networks. In our case study, the company has elected to utilize a low-cost X.25 link to connect to its higher-cost international private line multiplexer network. The X.25 line is used to support various network management functions.

As illustrated in Figure 2.30, many different types of packet service are available, including:

- *Dedicated access.* This provides a direct connection to serving points of presence through a dedicated private-line arrangement. Some of the available access options include X.25 direct, Bisync, SDLC, SNA, and Tymnet protocol connections. Access speeds are typically in the range of 9600 bps to 64 Kbps.

■ *Switched access.* This provides dial access, which in turn offers low-cost access to the X.25 packet network. Dial access provides secure remote access via the public switched telephone network (PSTN). The access options include public asynchronous dial-up and X.25 dial-up. The speeds supported by switched access usually range from 300 bps to 19.2 Kbps.

■ *X.75 gateway services.* Connectivity can be extended through the network to international postal telegraph and telephone (PTT) and other third-party packet networks via an extensive array of X.75 gateway arrangements. This is an important aspect of international packet services as these X.75 gateways provide companies with a cost-effective way to extend their global reach.

■ *X.25 to Frame Relay gateway.* This type of connectivity provides an economical way to bring small remote offices onto a backbone WAN that has been implemented with Frame Relay services. In countries where Frame Relay is not yet offered, X.25 can be used to access other networks that utilize Frame Relay.

■ *Protocol and rate conversion.* This involves a protocol or rate conversion function within the network. The following protocol conversion services are generally available:
  – Async to X.25
  – Async to SNA
  – Async to 2780/3780 HASP
  – Async to TCP/IP
  – Async to UTS (Universal Terminal Service — Unisys)
  – Async to 3270 Bisync/Host
  – X.25 to 3270 Bisync
  – X.25 to 3270 SNA (QLLC)
  – X.25 to SDLC
  – X.25 to TCP/IP
  – 3270 Bisync to 3270 SNA

A major advantage of packet-switched networks is that rate conversions are possible on each end of the transmission, thus allowing the use of different source and destination access speeds. As shown in Figure 2.30, some sites use 9.6-Kbps lines while other sites use 19.2-Kbps lines. Modems provide a cost-effective way to use an analog local loop[47] to connect remote locations into the network.

Many types of modulation techniques are available around the world. In the United States, the most typical modem in use is a V.90 modem, which is capable of supporting transmission speeds up to 56 Kbps. In some cases, the network will not support 28.8-Kbps transmission, and may only accommodate speeds up to 19.2 Kbps. This may be due to noise on the local analog circuit or to limitations on the port speeds available on the X.25 PAD/switch. Analog loops are inherently noisy and signal distortion may result in less than optimal modulation performance. Typically, the originating modem attempts to connect with the destination modem at the highest possible speed. If the receiving modem is capable of the same modulation scheme (e.g., V.90), and the local circuit is free of noise and distortion, then data will transmit up to the maximum rate of 56 Kbps. When the receiving modem uses a lower-speed modulation technique (i.e., V.29), the originating modem will usually lower its speed to match the receiving modem. Figure 2.30, for illustrative purposes, shows a connection to a European site using V.29 dial-up modem technology. This modem supports transmission speeds up to 9.6 Kbps.

2.2.7.1.4 Advantages, Disadvantages, and Recommended Uses of X.25 Packet-Switched Networks

This subsection summarizes the major advantages, disadvantages, and recommended uses of X.25 packet-switched networks.

The major advantages of X.25 packet-switched networks include:

- Easy to manage and scale in size
- Provides a way to share a link with multiple devices
- Available on a worldwide basis
- Supports some low-volume LAN-to-LAN internetworking
- Network level (OSI layer 3) operation supports data transmission across multiple protocols and network architectures
- For networks using datagrams:
  - Ability to bypass network congestion (which increases performance)
  - Ability to bypass network failures (which increases reliability)
- For networks using virtual circuits:
  - End-to-end transmission guaranteed

Some of the major disadvantages of packet-switched networks include:

- Delays and long response times due to "store-and-forward" transmission method
- Address, routing, and sequence packet overhead requirements
- Not good for time-sensitive applications
- Usually only allows data transport up to 64 Kbps
- For networks using datagrams:
  - "Best effort" transmission (no guarantees that data will be received)

- For networks using virtual circuits:
  - No ability to bypass network congestion (which decreases performance)
  - No ability to bypass network failures (which decreases reliability)

The recommended uses of packet-switched networks include:

- Terminal-to-host and host-to-host connectivity
- Distributed, international network topologies
- Low- and medium-speed data applications
- Low-cost networks (particularly those offering "postalized" rates)

2.2.7.2 Frame Relay Networks

2.2.7.2.1 Overview

X.25 packet services were developed to operate on high BER[48] analog copper circuits. To compensate for high BERs, X.25 packet networks use a substantial amount of overhead to ensure reliable transmission. With the widespread deployment of highly reliable fiber networks, this processing overhead is increasingly unnecessary. Frame Relay was developed to capitalize on these improvements in network technology. Frame Relay operates as a layer 2 data-link protocol interface between end users and networking equipment, and assumes that upper layer protocols, such as TCP/IP or SNA, will handle end-to-end error correction. If transmission errors are detected, Frame Relay discards the corrupted packets (or as they are known in Frame Relay networks, "frames").[49] It is then up to the customer premise equipment (CPE) and the higher layer protocols to request retransmission of the lost frames. This form of single error checking lowers network overhead and increases response time dramatically as compared to X.25.

Frame Relay is based on the Integrated Services Digital Network (ISDN) standards. Frame Relay protocols have been established by ANSI (T1.606, T1.617, and T1.618), the CCITT (I.233 and I.370), and various CPE vendors. For example, the Local Management Interface (LMI) was developed by the "gang of four"[50] to extend standard Frame Relay features and to help ensure some degree of compatibility between various Frame Relay products and vendor solutions. LMI extensions can be "common" or "optional" and include virtual circuit status messages (common), global addressing (optional), multicasting (optional), and simple flow control (optional). Other protocols — such as I.441 and Q.922 — support Frame Relay and can be referenced in the ISDN signaling standards.

Frame Relay is often called a "stripped-down" version of X.25 packet-switched technology. In some respects, Frame Relay is similar to X.25 and synchronous data-link control (SDLC) protocols — particularly because it is based on a form of packet switching — but it offers much better overall performance. The main difference between Frame Relay and X.25 packet networks is the way that the address and control bits in the packet are employed. Frame Relay addressing is based on either a two- or a four-octet address field specifying a permanent virtual connection (PVC). The two-byte default address field contains a ten (10) bit Data Link Connection Identifier (DLCI), which, in theory, represents an address range from 0 to 1023. The four-byte field, when used with an address field extension bit (EA), provides up to twenty-four (24) address bits. However, in practice, they are not all available for addressing purposes, because numerous DLCIs are reserved for control, maintenance, and housekeeping network tasks. Many companies provision Frame Relay technology over private or shared user-group networks using customized DLCI and PVC numbering plans.

Frame Relay uses PVCs to establish end-user connectivity. A Frame Relay permanent virtual connection is similar to an X.25 permanent virtual circuit, and is identified by a Logical Channel Number (LCN). Several hundred PVCs can be defined at a single access point that connects the customer premises equipment to the Frame Relay network. Thus, Frame Relay provides a convenient, single physical interface to support multiple data streams. Because applications can be mapped directly to a specific DLCI using static virtual circuit mapping or data encapsulation methods, the corresponding data streams can be segregated according to protocol type. This may be useful in improving the performance of time-sensitive applications or "chatty" protocols. The number of Frame Relay permanent virtual connections used in any given network is highly dependent upon the protocols in use and the traffic load that needs to be supported. In general, the number of DLCIs needed on each line depends on several interrelated factors, including line speed, static routes, routed protocols in use, amount of broadcast traffic, the size of the routing protocol overhead, and SAP[51] messaging.

The bandwidth provided by the Frame Relay service is based on the Committed Information Rate (CIR). The CIR is measured in bits per second. By definition, the CIR is the maximum *guaranteed* traffic level that the network will allow into the packet-switching environment through a specific DLCI. In practice, the actual usage can be up to the actual physical capacity of the connecting line. If a transmission requires more bandwidth than the CIR and there is sufficient capacity on the line, the transmission will likely go through. However, if the line becomes congested, Frame Relay will respond by discarding frames. The committed burst (Bc) size is the number of bits that the Frame Relay network is committed to accept and transmit at the CIR. A related metric, the committed burst excess (Be) size sets the upper bit limit for a given DLCI. Hence, Be represents the number of bits that the Frame Relay network will attempt to transmit after Bc bits are accommodated.

Traffic increases can lead to queuing delays at the nodes and congestion in the network. The CCITT I.370 recommendation for Frame Relay congestion control lists several objectives that need to be balanced while maintaining flow control in the network. These include minimizing frame discard, lessening QoS[52] variance during congestion, maintaining agreed upon QoS, and limiting the spread of congestion to other aspects of the network.

Frame Relay uses both congestion avoidance (which uses explicit messaging) and congestion recovery (which uses implicit messaging) methods to maintain control over traffic flows in the network. Two bits in the address field of the Frame Relay frame header are used to send an explicit signal to the user that congestion is present in the network. One of these is the Backward Explicit Congestion Notification (BECN) bit. The BECN bit signals when the network is congested in the opposite direction of the transmission stream. The Forward Explicit Congestion Notification (FECN) bit is used to notify the user when the network is congested in the direction of the transmission stream. A single Discard Eligibility (DE) bit is used to indicate which data frame(s) the network considers "discard eligible" and which ones are not. When the network becomes congested, the frame(s) marked DE are the first to be discarded. Until recently, these congestion control bits were not used by device manufacturers to signal congestion problems in the network to the end user. Although Frame Relay is designed to provide various kinds of congestion warnings, these warnings are not always fully utilized or implemented in network products and services.

In summary, Frame Relay was developed specifically to handle *bursty*[53] data network applications, and is commonly used to consolidate data transmissions from and to terminal-to-host or LAN-to-LAN applications onto a single network infrastructure. Generally, Frame Relay is not mileage sensitive, so it can be a cost-effective solution for connecting widely dispersed sites. Thus, organizations use Frame Relay to reduce network complexity and costs. Frame Relay can be used in private or public network solutions.

2.2.7.2.2 Design Considerations

Network managers should consider the cost, reliability, and performance characteristics of Frame Relay to decide if it is suited to their networking applications. Frame Relay is excellent at handling bursty data communications and internetworking functions. It provides bandwidth on-demand (although the bandwidth allocation is not guaranteed beyond the CIR), and supports multiple data sessions over a single access line. Frame Relay is generally capable of operating at speeds up to T1 (1.544 Mbps) or E1 (2.048 Mbps), and some service providers are offering Frame Relay at speeds up to DS-3 (45 Mbps). Because the majority of networking in place today uses DS-3 line speeds or lower, the bandwidth Frame Relay provides is more than adequate for the networking needs of most companies.

However, Frame Relay does exhibit transmission delays inherent in any packet service. While Frame Relay provides higher access speeds than X.25 and accommodates various data protocols, it is not meant to support delay-sensitive applications such as interactive voice and video. Therefore, some time-sensitive applications may be better suited to a private line network, because private lines offer a guaranteed quality of service (QoS) that Frame Relay does not. It is likely that with future advancements in the technology there will be fewer limitations on the type of traffic that can be carried successfully on a Frame Relay network.

A number of issues must be addressed before implementing a Frame Relay network. First, Frame Relay is an interface standard and not an architecture. Thus, a common network architecture must be in place to support end-to-end connectivity. TCP/IP is commonly used as a network architecture that supports Frame Relay. Second, the network design parameters must be estimated. This involves analyzing traffic patterns and requirements in order to select the appropriate Frame Relay service parameters (i.e., the CIR, Be, and Bc levels of service). Third, the impacts of Frame Relay on the (existing and planned) network components must be determined. Routers, bridges, multiplexers, FEPs, packet switches, and other devices that do not support Frame Relay must be upgraded to support it. This involves installing FRADs on the network equipment. FRADs[54] implement Frame Relay Interface (FRI) standards that establish how to connect a device to a Frame Relay network. Most routers are configured with a built-in FRAD; however, most bridges are not. Thus, the costs to upgrade to Frame Relay depend on the type of equipment being used in the network.

There are four major ways that Frame Relay networks can be implemented: as a public network, a private network, a hybrid, or as a managed service. Frame Relay is most commonly implemented as a public network solution. With public Frame Relay service, each network location gets a connection into the Frame Relay public network service through a port connection. The port connection is the node's gateway into the public Frame Relay service. Permanent virtual circuits (PVCs) are established between these port connections (note at this juncture that Frame Relay offers both PVS and SVCs. However, this discussion concentrates on PVCs because they are used far more often, particularly because not all the telecom providers offer SVC service). One big advantage of Frame Relay over private leased lines is that one connection into the network can support many PVCs to many other locations. This is illustrated in Figure 2.31. Compared with a private line, mesh topology with dedicated point-to-point connections, Frame Relay offers opportunities for substantial equipment savings. In addition, because Frame Relay pricing is generally not distance related, it is usually cheaper to connect widely separated sites with public Frame Relay than it is to use private leased lines that are priced by distance. Finally, public Frame Relay networks are usually easily modified and reconfigured by the service provider, providing more flexibility for change and growing requirements than dedicated, private line WANs.

In a private Frame Relay implementation, an organization leases private lines from a carrier. These private lines interconnect the various network nodes to corporate backbone switches. One of the primary reasons for using Frame Relay in a private network is that it is capable of supporting diverse traffic types over a single backbone (however, it is possible that organizations with low-volume traffic may find it more cost effective to use a TDM private line network using T1 or E1 lines that offers guaranteed levels of performance. In fact, in this situation, the most common solution is a private, leased line network). As the number of sites in the network grows, the rationale for using a private Frame Relay network over a private leased line network becomes more compelling.

The costs for Frame Relay are usually based on the Committed Information Rate (CIR) of each PVC, port costs for the backbone network, and the access link costs to connect the various sites to the Frame Relay network. Typically, these are fixed monthly fees (however, some service providers offer other pricing options, including capped usage based fees). Although fixed fees are appealing because they make it easier to budget and plan network expenses, they also present special challenges to the network manager. As telecom managers struggle to balance their communications budgets every year, it becomes increasingly important to establish an effective balance between network optimization and system overload. One of the dangers of using Frame Relay in a private network is that it might make it more difficult to manage these pressures because the Frame Relay pricing provides a powerful incentive to use the network facilities to the maximum extent possible.

A hybrid solution interconnects Frame Relay and other kinds of networks into a larger network structure. In general, a hybrid network is more difficult to manage because of the diverse traffic and equipment it must support. Because different portions of the network can use different

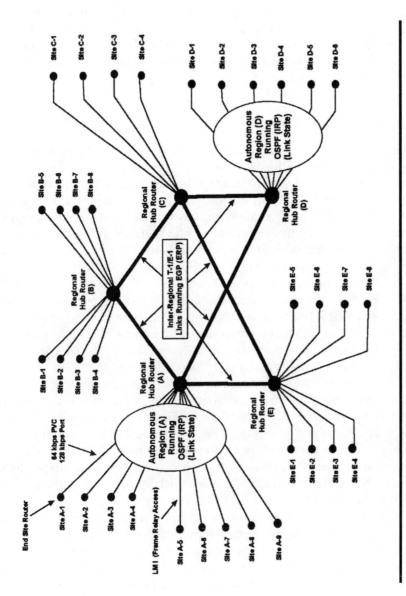

Figure 2.31 Frame Relay network. (*Source:* B. Piliouras.)

technologies and protocols, there is also more potential for difficulties in the end-to-end transmission. However, the business and technical requirements may be such that a hybrid network solution is necessary.

In an effort to control the complexity and cost of increasingly sophisticated and larger networks, many organizations are outsourcing their Frame Relay networks entirely. The service provider is responsible for all aspects of the local access provisioning, the customer premise equipment, disaster recovery arrangements, and network management. This is one of the easiest ways to implement a Frame Relay network.

2.2.7.2.3 Case Study

As illustrated in Figure 2.31, hierarchical meshed Frame Relay networks can be used to localize traffic by region. This segmentation is accomplished by placing regional hub routers at network mesh points. The regional hub routers also reduce the number of DLCIs needed at each physical interface to the network. This configuration is used to help manage the network complexity. This configuration is not always suitable if there is a lot of traffic on the network, because it may result in too much packet broadcast and replication, thus overloading the DLCIs.

Packet broadcasts are intrinsic to Frame Relay routing protocols. OSPF and RIP are two commonly used Frame Relay routing protocols. OPSF, derived from OSI's Intermediate System to Intermediate System (IS-IS) routing protocol, is a link state intra-domain hierarchical routing protocol developed for large, heterogeneous IP internetworks. Link state protocols typically employ a "shortest path first" algorithm that *floods* all network nodes, but only broadcasts routing information about the state of its links. The earlier Routing Information Protocol (RIP), in contrast, is a distance vector routing protocol that uses "Bellman-Ford" type shortest-path algorithms to determine packet routing. OSPF broadcasts a partial routing table to all nodes, while RIP sends out an entire routing table, but only to its neighbors in a localized region of the network. Both schemes can produce a significant amount of broadcast traffic in a large network. However, of the two protocols, RIP generates considerably more broadcast traffic.

While differing in their methods of updating routing information, both OSPF and RIP are considered interior routing protocols (IRPs), or intra-domain routing protocols. In contrast, exterior routing protocols (ERPs) such as the Border Gateway Protocol (BGP) provide inter-domain routing to support communications between core backbone routers. Like OSPF and RIP, BGP is a dynamic routing protocol that adjusts in real-time to changing network conditions.

Gateway routers are expensive; however, when used in conjunction with both intra- and inter-routing protocols, they provide a relatively straightforward and easy way to interconnect large networks. In the case study design, gateway routers are used to separate autonomous systems (AS) or regions, running the OSPF IRP from the meshed T-1/E-1 backbone running the EGP ERP.

Frame Relay can be used to implement star, partially meshed, or fully meshed topologies for regional and inter-regional networks. Star topologies are frequently used to minimize the number of DLCIs required in the network, thus reducing network costs. However, a "single-threaded" star topology does not offer the fault tolerance needed by many networking solutions. In a fully meshed Frame Relay network, every routing node is logically linked, via an assigned DLCI, to every other node in the network. This topology may not be practical for large Frame Relay internetworks because of its potential costs (which is related to the number of DCLIs needed) and performance problems (which is related to the packet broadcast scheme that Frame Relay uses, which can cause severe congestion in larger networks). A partial mesh Frame Relay network is a common compromise solution because it combines the cost savings of a star topology with the redundancy of a fully meshed topology.

2.2.7.2.4 Advantages, Disadvantages, and Recommended Uses of Frame Relay

This section summarizes the major advantages, disadvantages, and recommended uses of Frame Relay networks.

The major advantages of Frame Relay networks include:

- Good support for bursty data traffic
- Easy to manage and scale in size
- Allows resource sharing
- Improved service provided by network rerouting capabilities
- Widely available and rapidly growing utilization on a worldwide basis
- Supports medium-volume internetworking of LAN-to-LAN applications
- Provides faster network transport (using OSI Layer 2) than X.25 packet technology
- May reduce costs and complexity as compared to a comparable private line, leased network

The disadvantages of Frame Relay networks relate to the following:

- Not designed to support interactive voice and video applications
- Congestion control mechanisms are not fully implemented by providers of Frame Relay products and services

- Wide fluctuations in QoS make it more difficult to justify end-user chargebacks
- Does not support connectionless-oriented applications
- Inadequate bandwidth support for large, steady data file transfers

Frame Relay is recommended for use in:

- Small- to large-scale data applications
- Autonomous and inter-regional networking
- Peer-to-peer, terminal-to-host, and host-to-host connectivity requirements
- Distributed, international network topologies with wide connectivity
- Low-, medium-, and medium-high-speed data applications
- Low- to medium-cost networks (particularly when "postalized" rates are available)

2.2.7.3 ATM Networks

2.2.7.3.1 Overview

Until the mid-1990s, most public and private networks relied on time division multiplexing (TDM) data transport techniques, using DS0, DS1, DS2, and DS3 circuits. Until recently, this was adequate for most voice and data applications. However, as data communications have grown and new data applications have emerged, there has been an increasing need for more speed and greater bandwidth. Current packet-switching technology is simply not adequate in speed or sophistication to handle such emerging broadband applications as multimedia, high-speed digital transports of medical images, television service, and color facsimile. The telcos/PPTs responded to this need by developing ATM (Asynchronous Transfer Mode). They designed ATM to support a variety of tailored services for their customers, based on a new form of cell relay switching technology.

ATM transports data in small fixed-length 53-byte cells analogous to packets. From queuing analysis, it is known that segmenting data into smaller units reduces the overall transmission times in the network. The ATM cell size was designed from a queuing perspective specifically to optimize the transport of many types of voice and data applications with minimum delay. In addition, ATM makes use of advances in fiber optic and switching technology so that high-bandwidth applications can be transported at very high speeds.

One of the major benefits of ATM is that it provides integrated transport on a single network for many types of data, while optimizing the *simultaneous* transmission of each data type at a guaranteed Quality-of-Service level. Like X.25 and Frame Relay, ATM provides multiple logical connections over a single physical interface. However, ATM can transport data on the order of several hundred megabits per second (Mbps), whereas X.25 and Frame Relay can typically accommodate speeds up to 128 Kbps and 12 Mbps (occasionally 45 Mbps), respectively. In short, ATM combines the flexibility of packet and frame switching with the (desirable) steady transmission delays of circuit switching at new, higher levels of service. In addition, ATM is backward compatible with packet, Frame Relay, and SMDS switching, and can support these protocols on the same network.

The basic concept of ATM is analogous to an endlessly circulating set of box-cars. Although each box-car travels synchronously with respect to the other box-cars, the rate at which the contents of the box-cars are loaded or removed is asynchronous. When a data cell is ready for transport, it is loaded into a particular slot (or box-car). In ATM, each slot can accommodate, at most, one cell. Other data cells waiting for transport are queued until a slot is available. When there are no data cells awaiting transport, there may be empty slots (i.e., box-cars) circulating. The transmission rate is, in essence, the rate at which the slots, or box-cars, are moving. The queuing method ATM uses for data awaiting transport guarantees that transmission delays and QoS can be maintained consistently throughout the network.

ATM is an outcome of the Broadband Integrated Services Digital Network (B-ISDN) standards. B-ISDN is based on Common Channel Signaling (CCS).[55] In the United States, CCS commonly takes the form of Signaling System #7 (SS7) while throughout Europe and other parts of the world, Common Channel Signaling System #7 (CCSS7) is used. Because differences exist between the two CCS methods, international signaling may require conversion in each direction before B-ISDN transmission can occur. B-ISDN is designed to support switched, semi-permanent, and permanent broadband connections for point-to-point and point-to-multi-point networks. Channels operating at speeds of 155 Mbps and 622 Mbps are possible under B-ISDN. B-ISDN is based on cell switching and is intended to work with ATM. B-ISDN uses SONET as the physical networking conduit for ATM transmissions.

ATM standards define three operating layers that correspond roughly to OSI layer 1 (physical) and layer 2 (data link). The ATM physical layer defines the physical means of transporting ATM cells. Although ATM cells may be transported over many different types of media, their transport is optimized by Synchronous Optical NETwork (SONET) fiber optic technology.

SONET, a high-speed transport medium based on single-mode and multi-mode fiber optic technology, provides a very reliable layer 1, physical backbone infrastructure. Furthermore, because SONET can operate at gigabit per second (Gbps) data speeds, larger bandwidth applications are possible over ATM networks that employ SONET technology.

The second layer, the ATM layer, defines the switching and routing functions of the network. Located within the ATM five-byte header are various UNI[56] fields, which include the Virtual Channel Identifier (VCI) and the Virtual Path Identifier (VPI). The VCI serves a function similar to that of an X.25 virtual circuit. The VCI field is used to define a logical connection between two ATM switches. Thus, the VCI serves as an access point to the "virtual circuit" connection established over one or more virtual paths (VPs). The VPI field, eight bits (UNI) or twelve bits (NNI)[57] in length, is used to aggregate VCIs corresponding to multiple virtual channels (VCs). When a VPI value is assigned, a virtual path is established, and when a VPI value is removed, a virtual path is terminated. As their names imply, both VPIs and VCIs are virtual and coexist within the same ATM interface, and therefore switches can act upon one or the other or both the VPI and VCI fields. For example, in a straight-through connection, the entire VPI is mapped across the ATM input and output ports, and there is no need to process the VCI addresses. However, in other types of connections, a VCI can enter the ATM switch via one VPI and can exit by way of another VPI. Thus, a virtual circuit is fully identified using *both* the VCI and VPI fields (because two different virtual paths may have the same VCI value).

The Cell Loss Priority (CLP) bit is used to control congestion in ATM networks. This bit indicates whether a cell should be discarded. ATM's method of congestion control is similar in some respects to Frame Relay's use of the Discard Eligible (DE) bit. ATM attempts to balance the discard of incoming data against a preestablished traffic threshold while steadily "leaking" data into the network. A "leaky bucket" algorithm is often used to estimate and control how much congestion is present in the network based on the number of cells traversing the network with the CLP bit turned on. ATM also uses Operations, Administration, and Maintenance (OAM) cells to carry network management information between the ATM switches.

ATM's third layer, the ATM adaptation layer (AAL), provides the mechanism for handling different types of simultaneous user applications. To accommodate dissimilar transmission requirements, ATM separates the incoming data into the following categories:

- *AAL-1*: constant bit rate (CBR) connection-oriented transport for time-sensitive type applications, including voice, full-motion video, and TDM data.

- *AAL-2*: variable bit rate (VBR) connection-oriented mechanism for time-sensitive applications, such as compressed H.261 video, that do not require permanent connectivity.
- *AAL-3/4*: unspecified bit rate (UBR) connectionless-oriented transport for LAN-to-LAN data such as Switched Multi-megabit Digital Service (SMDS).
- *AAL-5*: available bit rate (ABR) connection-oriented method for bursty data traffic such as Frame Relay.

These groups are processed through a single network entry point referred to as the user-to-network interface (UNI). The network-to-network interface (NNI) controls the internetworking between the various ATM switches. The Private NNI (PNNI) protocol is designed specifically to support private ATM networks. ATM's traffic grouping promotes an integrated approach to network management, eliminating the need for separate local access facilities for different types of transmissions.

The AALs (or ATM adaptation layers) reside "logically" above the ATM 53-byte common switching fabric and below the transport data layer. The AAL functions are organized into two logical sublayers: the Convergence Sublayer (CS) and Segmentation and Reassembly (SAR) sublayer. The CS resides logically below the transport protocols, such as TDM voice and Frame Relay, and above the SAR. CS components include the Common Part (CP), and either a Service Specific Part (SSP) or a Service Specific Convergence Procedure (SSCP). The CS components are associated with the AALs and are used to create protocol data units (PDUs).[58] The PDUs consist of the transported data, a beginning header, an end tag, and a field length trailer. The SAR functions are responsible for segmenting (going down the OSI stack from the CS) and reassembling (going up the OSI stack from the ATM switching fabric) the payload (i.e., the data to be transmitted) of the ATM cells. While the ATM payloads are 48 bytes in length, actual usable payloads average about 44 bytes due to AAL overhead. When coupled with a 5- byte header and 48-byte payload, the ATM Cell Relay structure totals 53 bytes. One of the downsides of ATM, therefore, is the relatively high amount of cell overhead ($9/53 \cong 17$ percent) that it requires.

2.2.7.3.2 Design Considerations and Techniques

ATM is a technology that integrates voice, data, video, and imaging traffic across a WAN backbone at gigabit speeds. However, ATM technology has been surpassed by Mult-Protocol Label Switching (MPLS) as the WAN infrastructure of choice due to its compliance with the open systems-based TCP/IP protocol suite.

Although ATM works on a variety of media, fiber optic cabling is needed to achieve its full bandwidth potential. However, one must have right of way or access to install fiber optic cabling, and this is often impractical over long distances. This means that a public ATM network solution is often the only viable solution for many types of WANs.

There are a number of ways that ATM is being implemented in the marketplace:

- *ATM LAN.* The most common solution today for internetworking LANs it to use intelligent hubs operating on FDDI.[59] This type of solution supports communication speeds up to 100 Mbps. An ATM star-configured LAN, in contrast, supports speeds up to 155 Mbps, using standard twisted-pair copper wiring. Thus, ATM provides better support for high-bandwidth applications than traditional FDDI LAN hubs.

- *ATM LAN-to-LAN.* It is possible to interconnect LANs into an ATM switch. ATM switches are more intelligent than LAN hubs and operate at faster speeds, allowing data to pass from one port to any other without blocking. In this solution, ATM-based routers and hubs are used to form a LAN backbone.

- *ATM Private WAN.* In this solution, customer premise-based ATM switching and multiplexing devices are interconnected via leased lines. This type of network might be used in preference to a more traditional Frame Relay or packet-switched network to reduce the number of lines needed, to increase the number of applications supported by the network, or to increase network throughput. In general, ATM is not widely deployed in private WANs.

- *ATM public network.* In this solution, devices are connected into a single ATM backbone that is, in turn, connected to a public carrier. The CPE is connected to a LEC or service provider, most typically through a microwave or fiber optic connection. Hand in hand with the selection of a public cell relay service is the selection of a provider for the access portion of the network that connects into the public ATM backbone. There are a number of access network providers, including (1) local exchange carriers (LECs) or Regional Bell Operating Company (RBOC), (2) dedicated LEC/RBOCX provided facilities to an IXC's POP, where the IXC's service is used, or (3) alternative access provider (AAP). A public ATM network is the easiest to implement and involves the fewest decisions of the choices listed here.

Once the decision is made to deploy ATM technology, the network must be designed. This involves evaluating choices and requirements relating to the following:

- Design and selection of the network topology
- Identification and selection of traffic to be supported on the network
- Identification and selection of protocols to be supported by the network
- Design and selection of access network type (e.g., IEC fast packet service or dedicated line)
- Selection of public or private facilities (including cabling and switching needed)

2.2.7.3.3 Case Study

As shown in the diagram for this case study, ATM networks can support various types of simultaneous applications by segregating traffic according to AAL category. A few of the many types of applications that ATM can support are listed below:

- Compressed H.261 video operating at 384 Kbps (AAL-2)
- Supercomputer 8 Mbps SMDS data with High-Speed Serial Interface (HSSI) and Data Exchange Interface (DXI) frame format (AAL-3/4)
- TDM voice/fax and serial data (AAL-1)
- Frame Relay "routed" data (AAL-5)

As shown in Figure 2.32, customer A communicates with site #1 and site #2 through an ATM UNI logical interface that runs over a SONET OC-n physical fiber optic local access loop. For illustrative purposes, customers A and B can communicate with other sites, but only through the appropriate backbone technology. For example, customer A/site #2 can communicate with customer A/site #3 to exchange data relating to AAL-1 voice and fax TDM applications. Customer A/site #2 can also communicate with customer A/site #4 to exchange AAL-3/4 SMDS traffic, and with customer A/site #5 to exchange data handled by a Frame Relay router. The same connectivity principles hold for customer B applications. Note that customer C is connected through a Switched Multi-megabit Data Service (SMDS) backbone. Using SMDS' E.164 public addressing scheme, customers A and B can communicate with customer C with the appropriate provisioning.

2.2.7.3.4 Advantages, Disadvantages, and Recommended Uses of ATM Networks

This section summarizes the major advantages, disadvantages, and recommended uses of ATM networks.

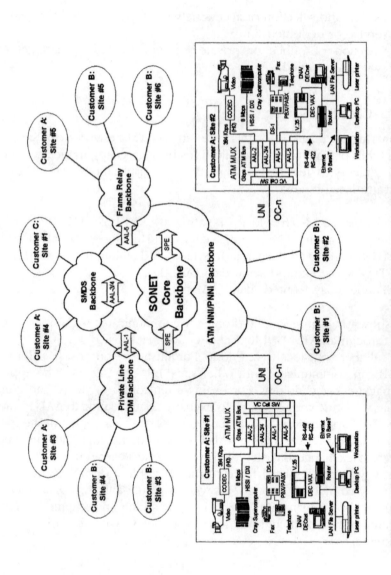

Figure 2.32 ATM network design. (*Source:* B. Piliouras.)

The advantages of ATM networks include:

- Good for integrating different types of user traffic
- Allows resource and application sharing
- Improved service availability and network bandwidth (particularly when using a SONET infrastructure)
- Supports high-volume and high-speed internetworking

Some of the disadvantages associated with ATM networks include:

- The technology is being replaced by Multi-Protocol Label Switching (MPLS)
- Congestion control mechanisms not fully developed
- Excessive overhead (ranging from 9 to 20 percent) required by the network
- Requires new hardware platforms
- High cost
- Does not support wireless or broadcast communications

ATM networks are recommended for:

- Integrated voice/fax, video, and data applications
- Large-scale internetworking backbones
- Single point for network management access
- High-speed bandwidth requirements, both bursty and continuous
- B-ISDN transport mechanism for SONET-based core backbones
- Medium- to high-cost networks

2.2.7.4 VPN Networks

2.2.7.4.1 Technical Overview

Public network access, whether by dial-up, ISDN, xDSL, or cable modem, is widely available and relatively inexpensive. Depending on the type of connection, speeds of kilobits and even megabits per second can be achieved, rivaling the performance of a private corporate or education networks. With the proliferation of WiFi-compatible access points in airports, hotels, bookstores, and coffee shops, access is available in more places than ever.

However, public networks have some disadvantages when compared to private networks. Because the organization does not have control over the network equipment employed on a public network, applications that require a certain Quality-of-Service (QoS), in terms of reliability or throughput, may not work reliably over a public network. Despite the work that has gone into making the Internet as reliable as possible, there is always a chance of a service interruption that is beyond the organization's control. In addition, communications on a public infrastructure are inherently less secure and prone to unauthorized viewing and/or modification.

Even with these shortcomings, in some situations (to support a mobile or global workforce), it may make sense to combine a public network with a private network. The combined flexibility of a public network with the security of a private network is called a *virtual private network* (VPN).

2.2.7.4.2 Basic Terminology

There are two basic types of VPN systems: a *secure VPN* and a *trusted VPN*. A secure VPN uses encryption technology to protect data across a public network. A trusted VPN relies on exclusive routing of one's traffic over a public network, so that the communication channel one is using is guaranteed to only carry one's traffic. A secure VPN is typically less expensive than a trusted VPN because it uses a public network versus expensive ATM or Frame Relay lines.

2.2.7.4.2.1 Secure VPN — The most common type of VPN is a secure VPN, which uses data encryption, such as IPSec, to secure data on an otherwise insecure network. A secure VPN also provides tunneling, making a client machine think it is part of a remote network, using a protocol such as L2TP (Level 2 Tunneling Protocol).

L2TP (specified in RFC 2661) combines Cisco's L2F (Layer 2 Forwarding) protocol, (specified in RFC 2341) with Microsoft's PPTP (Point-to-Point Tunneling Protocol) specified in RFC 2637. Because L2TP occupies level 2 — the data-link layer of the ISO OSI seven-layer networking model — the virtual circuit can provide all the functionality, except the actual physical connection, of a permanent wired connection.

2.2.7.4.2.2 Trusted VPN — A trusted VPN can guarantee network traffic will follow a specific path at a certain QoS. A variety of technologies can be used to create a trusted VPN. A layer 2 trusted VPN, which operates at the data-link layer, can be used with ATM (Asynchronous Transfer Mode), Frame Relays or the transport of layer 2 frames over MPLS (Multi-Protocol Label Switching). A layer 3 trusted VPN, which operates at the network layer, can be realized with MPLS and constrained distribution of routing information through a BGP (Border Gateway Protocol).

2.2.7.4.3 Design Considerations

To implement a secure VPN, equipment must be installed on the edge of the corporate network to interface with the public IP infrastructure. This involves making sure the following elements are in place:

- **VPN Device** — A secure VPN can be implemented on several different types of network and computing devices. Software to implement a VPN can reside within a network router, a firewall, a VPN concentrator, or a server operating system with the appropriate security modules.
- **Address Translation** — No matter what piece of equipment is used to provide VPN services, some level of NAT (network address translation) should be used so that the VPN network device is separated from the network to be accessed by an explicit mapping of one or more external IP addresses to one or more internal IP addresses. In no case should the network to be accessed be placed directly on the Internet.
- **Data Encryption** — Because a VPN typically operates over a public network, some form of data encryption should be used. One security protocol, IPSec, offers many features that protect the integrity and privacy of data going across a network. A primary concern is that unprotected data could be viewed by unauthorized parties. IPSec permits the use of many different encryption algorithms, the most recent being AES (Advanced Encryption Standard), which can support key lengths from 128 to 256 bits and beyond. Older encryption algorithms, such as DES (56 bits) and Triple DES (168 bits) can also be used; but in general, the more bits, the better the security.
- **Data Integrity** — IPSec also offers data integrity in the form of a hash algorithm. A hash algorithm takes the contents of a block of data, performs a calculation, and generates a hash value. If this is done both before and after a packet of data is sent over a network, the values can be compared to make sure the data has not been intentionally or unintentionally scrambled. Some of the more popular hash algorithms are SHA-1 (Secure Hash Algorithm) and MD5 (Message Digest).

2.2.7.4.4 Case Study

Cisco offers a popular suite of VPN solutions, with associated client software.

A VPN client can use many methods of authentication, starting with a simple username and password. For those who require more security, a certificate (i.e., a public key signed with a private key) can be used. Cisco offers the following VPN authentication methods: [CISCO04]

- Shared key authentication
- VPN group name and password authentication
- RADIUS server authentication
- SecurID authentication

Depending on the connection one is using, the form of transport tunneling may need customization. The Cisco client can offer network services using either UDP or TCP. For those working from home or a remote location, a VPN can provide all the services that the organization supports over a LAN, as long as the protocols used are compatible with the layer 3 VPN devices used.

As another example, Apple's latest operating system, Mac OS X 10.3, otherwise known as Panther, contains a VPN client that can connect to a VPN offered by Mac OS X 10.3 server, and perhaps other standards-based VPN products. Apple's client supports a VPN connection using either L2TP over IPSec, or PPTP.

The VPN configuration can be further customized. Instead of a password, an RSA SecurID device can be used to provide authentication information. A SecurID device generates an unpredictable code every 60 seconds, and must be in the possession of the user who would like to log in.

2.2.7.5 Access Technologies

2.2.7.5.1 Lit Buildings (Fiber): SONET and DWDM

2.2.7.5.1.1 Technical Overview — Synchronous Optical Network (SONET) is an international standard developed by the Regional Bell Holding Companies (RBHCs), Bell Communications Research (Bellcore), and various standards bodies. SONET formally became an ANSI (American National Standards Institute) standard — as ANSI T1.105 — in June 1988, with the last major update to the standard being made in 2001. There are also standards, designated T1.105.01 through T1.105.09, that deal with specific aspects of SONET, with the most recent updates made in 2003. Prior to the development of SONET, there had been a proliferation of proprietary fiber optic equipment that was largely incompatible across vendors. SONET was designed to establish a common standard for building fiber optical networks that would interface with existing electrical protocols and asynchronous communication devices.

The SONET standards we have today are the result of the cooperative effort of many interested groups from both private industry and other standard bodies. For example, in 1984, MCI proposed to the Interexchange Carrier Compatibility Forum (ICCF) the concept of standardizing the "mid-span meet," the handover point of an optical transmission system. This proposal was passed on to the Exchange Carriers Standards Association (ECSA) for discussion and elaboration. In 1985, Bellcore proposed the "SONET" concept, which was expanded on within the T1X1 committee of the ECSA. The resulting documentation was submitted to ANSI (American National Standards Institute) for endorsement and release in 1988 as the ANSI T1.105 standard in use today. Today, many service providers are installing fiber optic backbone networks based on SONET transport technology. New bandwidth-intensive services — including cable and DSL modems, as well as local area network (LAN) links — are being developed and are fueling the need for large-capacity SONET backbone networks. As broadband services have become more readily available and telecommunications requirements have become more time sensitive, the SONET backbone providers have continued to develop and implement "survivability" techniques to minimize the possibility of network downtime. The result is that SONET networks are extremely fault tolerant and reliable, and can recover from even catastrophic fiber cuts in a matter of seconds or less (the average recovery time in a SONET network for this type of fault is 50 milliseconds).

SONET is based on a worldwide standard for fiber optic transmission, modified for North American asynchronous rates. In the rest of the world, this technology is known as Synchronous Digital Hierarchy (SDH). Standards that have had a major impact on SONET development are summarized in Table 2.31.

The United States SONET standards were modified to permit internetworking with the networks of other countries. These modifications were incorporated into the CCITT standards covering SDH (Synchronous Digital Hierarchy), which is the international equivalent of SONET. These standards, designated CCITT 707-709, define the operation of network elements down to the line signal formats, interfaces, security features, and signal structures for transmission speeds from 51.84 Mbps (OC-1) to 9.95 Gbps (OC-192) to 40 Gbps (OC-768) and higher. SONET is the first international transmission standard that eliminates the differences between American, European, and Japanese multiplexing schemes. Table 2.32 provides a summary of the SONET signaling categories.

The lowest-level SONET signal is called the Synchronous Transport Signal level 1 (STS-1). This is the base signal used for SONET multiplexing. The STS-1 has a signal rate of 51.84 Mbps. The optical equivalent of the STS-1 is the Optical Carrier level 1 signal (OC-1), which is obtained by a

Table 2.31 SONET Standards

Organization	Standard	Description
Bellcore	TR-253	Generic requirements
Bellcore	TR-233	Wideband and broadband digital cross-connect systems (DCS)*
Bellcore	TR-303	Loop carriers
Bellcore	TR-496	Add/drop multiplexing (ADM)
Bellcore	TR-499	Transport system generic requirements
Bellcore	TR-782	SONET digital trunk interface
Bellcore	TR-917	SONET regenerator equipment
ANSI	T1.195xx	Rates, formats, jitter, etc.
ANSI	T1.106	Optical interface
ANSI	T1.119	OAM&P communications
ANSI	T1.204	Operations, administration, provision
ANSI	T1.231	In-Service performance monitoring

* DCS — Digital cross-connect systems are commonly referred to as DCSs, DXCs, and DACs, depending on the origin of the switch and the technology employed. For our purposes, we assume that all these acronyms relate to the same switching function. A digital cross-connect system electronically maps a DS0 (64 Kbps) channel into a DS1 (1.544 Mbps) line, allowing each DS0 channel to be routed and configured into different DS1 lines.

Table 2.32 SONET Signal Categories

Synchronous Transport Signal	Optical Carrier (OC-n)	Line Rate (Mbps)
STS-1	OC-1	51.84
STS-3	OC-3	155.52
STS-12	OC-12	622.08
STS-24	OC-24	1,244.16
STS-48	OC-48	2,488.32
STS-192	OC-192	9,953.28

direct electrical to optical conversion of the STS-1 signal. The STS-1 frame consists of 90 columns of bytes. The first three columns of data are dedicated to network management functions, while the remaining 87 columns and 9 rows are used to carry the data, that is, the Synchronous Payload Envelope (SPE). The STS-1 frame consists of 810 bytes, and is transmitted at the rate of one frame per 125 microseconds. The section overhead is made up of a block of 9 bytes (3 rows by 3 columns) that is transmitted 8000 times a second for a total transfer rate of 576 Kbps. The section overhead is used to perform various functions between Section Terminating Equipment (STE) — for example, a SONET regenerator or other network element — including framing, error monitoring, and STS identification. Similarly, the line overhead is a block of 6 rows by 3 columns (18 bytes) transmitted 8000 times a second for a 1.152-Mbps transfer rate. The line overhead is used to indicate when a line between two LTEs (line terminating equipment) has gone bad. Line overhead data is also used to perform such functions as synchronization, multiplexing, and automatic protection switching. The path overhead is computed in a similar manner, and is used to support and maintain transport of the STS-1 Synchronous Payload Envelope (SPE) between the terminating network elements, and to perform end-to-end error and connectivity checking. The section, path, and line overheads perform functions analogous to the functions of the section, path, and line overhead segments of the public telephone network. Layering the network functions allows different types of equipment to be built to perform specifically tailored functions.

The SONET STS-1 Synchronous Payload Envelope (SPE) has a channel capacity of 50.122 Mbps, and has been designed specifically to provide transport for a lower-speed DS3 tributary signal. Transport for a tributary signal with a signal rate lower than a DS3, such as a DS1, is provided by a Virtual Tributary (VT) frame structure. VTs support the transport and switching of payload capacities that are less than those provided by STS-1 SPE. By design, the VT frame structure fits neatly into the STS-1 SPE to simplify the VT multiplexing requirements. A range of different VT sizes is provided by SONET, including:

- *VT1.5.* Each VT1.5 frame consists of 27 bytes, structured as 3 columns of 9 bytes. At a rate of 8000 frames per second, these bytes provide a transport capacity of 1.728 Mbps and will accommodate the mapping of a 1.544-Mbps DS1 signal. A total of 28 VT1.5 signals can be multiplexed into the STS-1 SPE.
- *VT2.* Each VT2 frame consists of 36 bytes, structured as 4 columns of 9 bytes. At a rate of 8000 frames per second, these bytes provide a transport capacity of 2.304 Mbps and will accommodate the mapping of a 2.048-Mbps CEPT E1 signal. A total of 21 VT2 signals can be multiplexed into the STS-1 SPE.

■ *VT3*. Each V3 frame consists of 54 bytes, structured as 6 columns of 9 bytes. At a rate of 8000 frames per second, these bytes provide a transport capacity of 3.456 Mbps and will accommodate the mapping of a DS1C signal. A total of 14 VT3 signals can be multiplexed into the STS-1 SPE.

■ *VT6*. Each VT6 frame consists of 108 bytes, structured as 12 columns of 9 bytes. At a rate of 8000 frames per second, these bytes provide a transport capacity of 6.912 Mbps and will accommodate the mapping of a DS2 signal. A total of 7 VT3 signals can be multiplexed into the STS-1 SPE.

The concept of transporting tributary signals intact across a synchronous network has resulted in the term "synchronous transport frame" (STF) being applied to such synchronous signal structures. A synchronous transport frame (51.84 Mbps) is comprised of two distinct parts: the synchronous payload envelope and the transport overhead.

■ *Synchronous payload envelope (SPE):* 50.122 Mbps (includes 576 Kbps path overhead). This consists of all the user data, including the path overhead bytes. Individual tributary signals (such as a DS-3 signal) are arranged within the SPE, which is designed to transverse the network end-to-end. This signal is assembled and disassembled only once although it can be transferred from one transport system to another many times on its route through the network.

■ *Path overhead:* the overhead contained within the SPE and allows network performance to be maintained from a customer service end-to-end perspective (576 Kbps channel). Path overhead, which contains error detection, tracing, path status, and connection establishment, is terminated by STS/OC-N end equipment.

■ *Line overhead:* allows the network performance to be maintained between transport nodes and is used for most of the network management reporting (1.152-Mbps channel). Line overhead, which contains error detection, pointer justification, automatic protection information (rings), alarms, and provisioning commands, is terminated by all SONET equipment capable of OC-N/STS level switching and termination.

■ *Section overhead:* used to maintain the network performance between the line generators or between a line regenerator and a SONET network element (NE) and provide fault localization (576-Kbps channel). Section overhead, which contains error detection, framing, STS identification, and other information, is usually terminated by SONET equipment that converts the optical signal into an electrical signal.

SONET lightwave terminating equipment (LTE) accepts standard optical signals and converts them into lower-level electrical signals. The electrical signal is then de-multiplexed to the component STS-1s where signal processing occurs. Signal processing may consist of checking the overhead payload for the purpose of signal performance testing, distance alarm monitoring, signal switching, and other related tasks. Once processed, STS-1 signals can be routed to various electrical or optical tributaries, depending on the configuration of the LTE.

The SONET LTE is used to provide elementary connectivity between synchronous and asynchronous transmission equipment. In the most basic form, the SONET LTE will mimic many currently deployed asynchronous LTEs. However, when SONET support systems are integrated within the network, SONET LTEs offer added features and improved performance in: signal performance, advanced monitoring, broadband transport capabilities (OC-N), improved cable management (optical versus coaxial), and a more compact cable and equipment footprint.

The basic building block of a SONET network is the Add/Drop Multiplexer (ADM). The "synchronous" functionality provided by ADMs is one of the driving forces behind the development of SONET technology. The ADM offers many features, including but not limited to reduced cost, increased network reliability, automated ring protection, full use of SONET overhead, performance monitoring, and automated provisioning. The ADM allows selected payload channels (i.e., DS3s, STS-1s, DS1s, and other rates as allowed by vendor) to be "added" or "dropped" at a given site while allowing the rest of the signal to pass on to the next site. Due to the synchronous nature of SONET, these payload channels are mapped to higher rate signals at specific locations in the payload. In this way, the channel's exact location is always known and can be read out, or written over without de-multiplexing the entire higher rate signal. The ability to locate specific channels within the data stream is the basis for synchronous transmission. With the deployment of ADMs, the back-to-back LTEs and manual DSX-3 patching procedure in the asynchronous environment is greatly reduced or eliminated.

The add, drop, and throughput capabilities of the ADM are controlled by a time slot assignment (TSA) feature within the ADM that allows local and remote traffic provisioning. For example, a DS3 signal coming into the ADM from the west on time slot 4 may be dropped to time slot 2 at the ADM drop site. Similarly, a DS3 signal may be added at the ADM drop site to time slot 4 and be routed to the east on time slot 7. Thus, the ADM offers flexibility in provisioning the traffic. In addition, the ADM's TSA feature increases bandwidth utilization and signal performance as compared to conventional asynchronous technology.

ADMs provide the IECs (interexchange carriers) and LECs (local exchange carriers) a great deal of flexibility in completing SONET mid-span meets. Although mid-span meets are not new to these backbone providers, SONET offers capabilities that have never before been possible. Some of these capabilities include:

- Differentiation of vendors
- Elimination of electrical demarcation/cross-connect points
- Improved signal performance

Ideally, ADMs should be used at drop and re-insert (DREI) sites, and at other facilities that require small amounts of traffic grooming and add/drop functionality. ADMs or DREI facilities provide interconnects with LECs, CAPs, and other IECs while creating a seamless network. Additionally, ADMs can be used in metropolitan and possibly regional rings where SONET capabilities and self-healing systems are desired.

The Broadband DXC (BBDXC) is used to terminate synchronous optical, synchronous electrical, and asynchronous electrical signals. With this capability, the BBDXC can serve as a gateway device between synchronous and asynchronous transmission systems. A BBDXC performs six primary functions:

1. Signal termination
2. Grooming
3. STS-1 cross-connection
4. Access to asynchronous network
5. Performance monitoring
6. Restoration

The BBDXC is used primarily at major switch and relatively large fiber junction sites, where, in addition to restoration functions, it performs OC-12 grooming functions. A BBDXC can consolidate STS-1s at OC-12 endpoints, creating express trunks with high utilization rates.

Wideband DXCs (WBDXCs) have the capability of interfacing with BBDXCs at synchronous optical levels OC-3 or OC-12. The WBDXC performs the following four primary functions:

1. Signal termination
2. Grooming
3. VT cross-connection
4. Performance monitoring

The WBDXC is used primarily at switch sites, relatively large fiber junction sites with large amounts of terminating traffic, and other sites with moderate to large amounts of access/egress traffic. The WBDXC performs asynchronous to synchronous signal conversion, grooming for local express trunks, and performance monitoring of DS1 and VT level signals.

The narrowband DXC (NBDXC) — which is essentially the same as the DXC 1/0 in service today — allows SONET equipment to be interfaced at synchronous optical levels OC-3 or OC-12. The NBDXC performs four primary functions:

1. Signal termination
2. Grooming
3. DS0 cross-connection
4. Performance monitoring

The NBDXC is used primarily at switch sites, relatively large fiber junction sites with large amounts of terminating traffic, and other sites with moderate to large amounts of access and egress traffic.

OC-192 is an established transmission rate equating to 9.95328 Gbps, supporting 192 STS-1s. This is four times the capacity of an OC-48. This very high-speed lightwave system provides a means to alleviate fiber constraints and congestion on network backbone routes. The use of OC-192 will better position SONET backbone providers to provide OC-n trunks for broadband-based services and to accommodate unforeseen high-bandwidth traffic. It will also significantly reduce the transmission costs on a DS-3 mile basis, freeing capital dollars for other projects. To accommodate OC-192 LTE equipment and increasing traffic loads, there will be a need for OC-48 express pipes. Most of these would be local in scope and would be implemented using the same guidelines used when creating OC-12 trunk express pipes for OC-48 links. OC-768 is the next step in the OC hierarchy, being four times the capacity of an OC-192 line.

SONET technology is being developed primarily for use on fiber transmission routes, but digital radios benefit from SONET as well. Currently, many manufacturers are developing SONET digital radio products. Some manufacturers have proposed upgrades to their existing asynchronous radio systems to ease the transition into the synchronous platform. A high percentage of synchronous radio manufacturers have focused their efforts on capturing the European market, which is based on synchronous digital hierarchy (SDH) standards, which constitute the European version of SONET.

There is a need to accommodate SONET radios in the IEC network for a number of reasons, which include:

- Extending fiber routes into smaller markets often cannot be economically justified.
- Customers may require OC-n services at locations served only by digital radios.
- Some digital radio routes serve as backups for fibers and cannot perform this function without SONET capability.

The main options for extending SONET transmission to access areas of the network currently served via digital radio are to:

- Overbuild digital radios with SONET radios to extend the requirements for SONET bandwidth transmission.
- Build a new or diverse fiber route to these sites.
- Lease SONET transmission from an alternate carrier.

Although the second and third options listed above may prove technically feasible in some areas, they frequently are not economically feasible. The reader should note that SONET radios may be more susceptible to failure than fiber routes. A "burst error" hit to the payload pointer bits can cause immediate payload loss. Because radios can be prone to multipath fading and dispersing conditions, which manifest themselves as burst errors, SONET radio manufacturers have been working on different schemes to alleviate this problem. One solution to the pointer vulnerability is to use a forward error correction (FEC) technique. The higher speed rates necessary for SONET transport and the FEC feature creates a need for modulation formats above 64 QAM. Some vendors are offering 128, 256, and 512 QAM schemes. It is expected that any tributary interface to a synchronous radio must be done at the OC-3 optical level.

2.2.7.5.1.2 DWDM — Within the past few years, there has been an increase in the use of *dispersion compensator* technology in the SONET industry. Dispersion compensators are devices used on the line side of SONET LTEs to compensate for the inherent dispersion characteristics of fiber optic media. Wave division multiplexing (WDM) offers an alternative method to increase the transmission capacity of an installed fiber path by utilizing the other regions of the optical spectrum resident on each fiber. This process multiplexes optical signals from two or more transmission systems operating at different wavelengths so that they can be transmitted simultaneously over a single fiber pair. When used on single-mode fiber, WDM allows multiple optical signals to be transmitted in the same direction (uni-directional) or in opposite directions (bi-directional).

Table 2.33   Fiber Optic Cable Classifications

Classification	Frequency (nm)
VSR-1	850
SR-1	1310
IR-1	1310
IR-2	1550
LR-1	1310
LR-2	1550

To understand dense wavelength division multiplexing (DWDM), it is helpful to know about the different types of fiber. There are several ways to classify types of fiber. One classification is based on how far the signal must go and the frequency of light used. Table 2.33 provides a summary of different fiber types based on this classification. There is Very Short Reach (VSR-1), which operates at 850 nanometers; Short Reach (SR-1) Intermediate Reach (IR-1) and Long Reach (LR-1), which operate at 1310 nanometers; and Intermediate Reach (IR-2) and Long Reach (LR-2), which both operate at 1550 nanometers. These values are defined in Telcordia (formerly Bellcore) Standard GR-20 "Generic Requirements for Optical Fiber and Optical Fiber Cable."

Another way to classify fiber is by type. There are two major types today: single mode and multimode fiber. Single-mode fiber has a relatively thin core, typically ten microns or less, and carries a single light beam. Multimode fiber has a relatively thick core, typically in the tens or hundreds of microns, and carries multiple light beams. Single-mode fiber is more expensive but can carry signals farther without requiring a repeater.

DWDM is a standard that greatly increases the capacity of a fiber optic network, while using existing fiber infrastructure. It is described in standard ITU-T G.692, dated October 1998. Until DWDM, traditional fiber optic networks would transmit a single wavelength of light, typically between 850 and 1550 nanometers, over a single strand of fiber. Using DWDM, up to 40 bi-directional (80 uni-directional) or more channels can be sent through the same piece of fiber at up to OC-192 (9.6 Gbps) per channel yielding DWDM system capacities of 1 Tbps (Teri-bits per second). Compare this to other techniques like time division multiplexing (TDM), which breaks down at high speeds. DWDM solves this problem using the frequency domain of the fiber rather than time.

For more information on DWDM, the reader is referred to the following sources:

IEC DWDM Tutorial: http://www.iec.org/online/tutorials/dwdm/.

Cisco—Introduction to Optical Fibers, dB, Attenuation and Measurements.

Cisco—Introduction to DWDM Technology (OL-0884-01).

References: VPN Consortium, http://www.vpnc.org/vpn-technologies.html.

L2TP: http://www.ietf.org/rfc/rfc2661.txt.

MPLS: http://www.ietf.org/rfc/rfc3031.txt.

BGP-4: http://www.ietf.org/rfc/rfc1771.txt.

IPSec: http://www.ietf.org/html.charters/ipsec-charter.html.

2.2.7.5.1.3 Other Advances — Optical amplifiers boost the optical signal emerging from the line side of the LTE and allow it to travel further before requiring regeneration. There are currently several types of optical amplifiers. Transmit and receive amplifiers have increased the spans between repeaters from about 75 km to more than 140 km (i.e., about 45 miles to over 85 miles). Line amplifiers allow spans in excess of 500 km (300 miles) and more. Optical amplifiers operate at the optical level, are completely independent of bit rates, wavelengths, or transmission protocols, and are able to handle rates up to OC-768 and more. They eliminate the need for entire regeneration sites. This significantly reduces operations, maintenance, administration, and provisioning costs.

An optical cross-connect, which is also known as a *remotely controlled optical patch panel*, performs the same functions of a BBDXC. Optical cross-connect devices have been used to provide broadcasting services over multimode fiber transmission and to restore the network in the event of a catastrophic fiber failure. Optical cross-connect technology is still in the early stages of development; however, single-mode OPT-X devices are being deployed with single-mode fiber transmission systems in some SONET networks.

Many of the technical hurdles in fiber optics (i.e., cross-talk, alignment of cross-connections, and "directivity") have been overcome, but not by all manufacturers. Many SONET backbone providers today are considering the use of an OPT-X as a vehicle for network disaster recovery, network reconfiguration, facility testing, network management, and new service offerings. Because the unit is completely photonic, no electrical to optical signal conversion is required. It contains an MxN matrix, and it cross-connects incoming light at any one of the M ports on side A, to any of the N ports on side B. This type of device does not affect the direction of the photon beam.

2.2.7.5.1.4 Design Considerations — The survivability of telecommunications networks is becoming an increasingly important issue. In response, telecommunication service providers are building more "robustness" into their networks. Survivability is defined as the ability of a network to maintain or restore an acceptable level of performance during failure conditions using various means of network restoration. Currently, network survivability is provided by circuit diversity, equipment protection switching, and automated restoration techniques using DXC and control systems. The SONET architecture is designed specifically for survivability using self-healing ring techniques. These self-healing techniques are based on either DXC-based restoration or SHR (Self-Healing Ring) restoration. The SHR techniques provide an alternative traffic path in the event of fiber optic link failure, restoring service on average within 50 milliseconds (and 2.5 seconds in the worst case). The restoration happens so quickly that most communications functions will not be affected by the line failure and end users may not even know that there was a problem and it has been corrected.

A spur site is defined as any site where there is only one network transmission route to the fiber or radio backbone. When a fiber outage occurs on a single threaded spur route, several sites can be affected, impacting a large number of users. Repair times in hours are not acceptable to many companies today because their networks play an increasingly important role in their day-to-day operations. "Spur enhancement" is the terminology used to describe an alternate transmission facility at a single threaded site. Fiber is the most reliable alternate transmission facility for spur enhancement.

Having multiple backbone sites for spur enhancement is the most reliable means to reduce the risk associated with node outages. However, it is often prohibitively expensive to use multiple backbone sites. Therefore, SONET providers usually utilize a single backbone site. When a single backbone site is used, the necessary precautions should be taken at the site to ensure that the alternate route equipment is located separately from the existing equipment (i.e., by locating equipment at opposite ends of the building utilizing separate power distribution frames [PDFs], installing uninterruptable power supplies [UPS], etc.).

Because ADMs can transmit and receive traffic from two different directions simultaneously, they are ideally suited to supporting a self-healing ring topology. Two methods have evolved for implementing "self-healing" rings (SHRs) within the ADM: uni-directional (USHR) and bi-directional (BSHR) rings. The type of SHR used in any given type of network depends on the normal traffic routing pattern. Uni-directional SHRs are usually configured with one fiber pair. One fiber is the working channel and the other is the protection channel. Under normal conditions, all working traffic on the ring is transferred to one fiber in the same direction around the ring. Although the second fiber is reserved for protection, this does not mean it is idle until a failure occurs. The USHR uses a 1+1, or "dual-fed" configuration. This means that the working traffic is also bridged to the protection fiber and is routed in the opposite direction. Thus, each fiber is carrying a copy of the working traffic, but in different fibers and opposite directions of transport. BSHRs can be implemented using two or four fibers. In a two-fiber BSHR, traffic origi-nating at A and destined for B is transmitted to B on one fiber, and the responding traffic from B destined for A is returned via the other fiber in the opposite direction. In each ring type, half the channels are defined as "working," and the other half as "protection." Therefore, the working throughput of the ring is actually OC-n/2. For example, a BSHR/2 ADM operating at OC-48 has 24 working and 24 protection STS-1 time slots on both the transmitting and receiving fibers. However, the protection time slots can be reassigned when not needed.

Protection switching uses one of two SONET overhead levels: the line or the path. A typical example of line protection occurs when normal traffic is disrupted on a working fiber and all the traffic is switched to the protection fiber. The communication involved in making this switch is coordinated using the K1 and K2 bytes of the SONET line overhead. All or none of the working traffic is switched to the protection channel (as defined in ANSI T1.105). Path protection switching uses the path layer of the SONET overhead and avoids the complexity of the line communication process. Using path protection switching, a node can switch individual STS-1s or VTs in and out of the payload based on the path layer indications. In general, ANSI does not deal explicitly with path switching requirements or standards. SHRs generally implement protection switching in different ways. USHRs favor path switching, while BSHRs favor line switching. A USHR can, in fact, use line switching, but this configuration may create problems due to SS#7 transport constraints. The USHR is usually imple-mented using path switching as described in Bellcore document TR-496. USHR with path protection switching can be designated in several ways. Uni-directional Path Switched Ring (UPSR) and Uni-directional Self-Healing Ring-Path Switching (USHR/P) are two examples, and are essentially equivalent designations.

BSHRs can support path or line switching. While BSHR path switching is relatively new, several vendors are currently demonstrating product lines that offer it. On the other hand, line switching, which uses the line layer of the SONET overhead to initiate switching action, has been modified in ANSI T1.105 to support BSHRs. This document assumes that all BSHRs, both two and four fiber, use line switching. Line switching is further divided into ring and span switching. BSHR/2s supports only ring switching. When a ring switch occurs, the time slots that carry the working channels are switched to the corresponding empty time slots in the protection channels traveling in the opposite direction.

2.2.7.5.1.5 Case Study — As illustrated in the SONET Core Network Design, Figure 2.33, BSHR/4s support ring switching. This is similar to the ring switching already discussed, except that all the time slots are placed on the protection fiber and transmitted in the opposite direction with the span switching. Span switching occurs almost exactly as it does in point-to-point 1:1 systems. In the event of a working fiber failure, the traffic is switched to the protection fiber. BSHR/4 systems support several span switches in the same ring, providing full protection from multiple failures.

The Uni-directional Path Switched and Bi-directional Line Switched rings have subtle differences. These differences are primarily due to the adaptation of the bi-directional rings to a point-to-point, line layer based protection, as specified in ANSI T1.105. Freedom from the line layer affords USHR/P rings several advantages, including:

- *Vendor independence.* Each node of a USHR/P can be implemented with equipment from different vendors. Because vendors implementing BSHR/L use proprietary solutions to overcome protection-switching shortcomings at the line layer, each node must be from the same vendor.
- *Technology integration.* USHR/Ps can be designed to include asynchronous and synchronous components.
- *Line speed independence.* The link speed between nodes of a BSHR/L must be at the same speed. USHR/P rings, however, can employ different speeds link-by-link, node-by-node.
- *Number of nodes* BSHR/L rings are typically limited to 16 to 20 nodes while the number of nodes in a USHR/P ring is theoretically unlimited. Some vendors claim USHR/P rings in excess of 100 nodes.
- *Better services.* Because USHR/P rings switch at the path layer endpoints, switching of all the tributaries on the line due to a single failed tributary is avoided. For example, if the ring is an OC-12 and one STS-1 has failed, the entire OC-12 might appear bad, resulting in a line switch. Therefore, a "hit" on adjacent clean tributaries also occurs. Path switching substitutes the good STS-1 on the protection channel corresponding to the failed STS-1 on the working channel at the endpoint of the circuit. Therefore, end-to-end service performance can be provided to the customer.

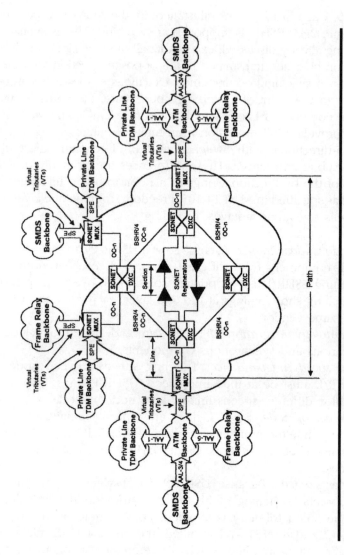

Figure 2.33 SONET network design. (*Source:* B. Piliouras.)

However, USHR/P rings also have disadvantages, including:

- *Bandwidth utilization.* Once bandwidth on the USHR/P ring is assigned, it cannot be reused. For example, if using an OC-12 USHR/P with four nodes (i.e., A through D) and nine STS-1s are allocated from A to B, there are only three STS-1 time slots available for all additional traffic on the ring.
- *Difference in path lengths.* Different physical lengths between the working and protection fibers may result in time-outs of some data applications in the event of a protection switch. Not only do some applications need to see a response within a certain amount of time, but they also require consistent amounts of time for each response. Switching to protection may result in a longer or shorter path, thus changing the delivery time of the data packet.
- *Short range design.* Vendor implementation of the USHR/P is based on a "virtual ring" concept and is usually installed on a customer-by-customer basis.

In addition to or in lieu of the transmission-level restoration techniques, there is also a need to look at how to restore application-level functionality, in the event of a network failure. The application level, or service layer, is defined as the individual network that provides the specific voice and data services, such as Asynchronous Transfer Mode (ATM). By using a grid network, and employing various application diversity and restoration techniques, a single fiber cut should not adversely impact the network. Application restoration and survivability should also incorporate equipment diversity. Specific restoration strategies for each service type/application should be developed, reflecting priorities that have been established for restoring each level of service in the event of a network failure.

Self-healing metro rings are used to provide survivable trunks. The optimal connection to the ring should occur at two separate locations to protect against local outages. ADMs using TSA functionality to switch traffic in the event of an outage provide an effective means of restoration. The decision to implement any or all of these restoration and survivability methods must be based on financial and business criteria.

Perhaps the most important criterion for determining which type of ring to use is the distribution of the traffic between nodes on the ring. If the traffic on all the nodes is destined for a single or centralized node on the ring and the traffic from each node is nearly equal, the path USHR/P is the most suitable method. If there is a high concentration of traffic between nodes and the traffic is fairly heavy, the BSHR/L is the preferred choice.

Photonic based restoration is available in two basic methods. Both methods require photonic elements at the node or junction sites. One method, which is based on the use of optical switches, is called the optical domain restoration (ODR) method. The second method uses optical cross-connects (OPX-T). Both methods provide vehicles to maintain and restore an acceptable level of performance during network failures, particularly in the case of catastrophic fiber cable cuts. These methods define a simplified network architecture called "photonic domains." This architecture breaks the network into small sub-networks to help simplify the network management and restoration mechanism. Both methods require interoperability between the system and physical restoration layers. In either case, these methods need to use BBDXCs to efficiently perform the circuit-level grooming needed for optimal interconnection of OC-12 logical facilities. The optical domain architecture applies to small independent sub-networks or optical domains (photonic domains). Therefore, the complexity of network management and restoration is reduced. ODR provides protected optical access into a specific domain and makes use of optical switches.

The optical cross-connection restoration architecture functions similarly to endpoint restoration, and parallels the methods used for DXC-based real-time restoration (RTR). It can be used in small independent or dependent sub-networks, or photonic domains. Optical cross-connects use the restoration capacity available in the network to provide alternate routes for failed circuits. These restoration capacities are determined on either a dynamic or preplanned basis. The optical cross-connect is used as a passive cross-connect to perform network restoration at the optical level or photonic domain without converting signals to an electrical rate. The OPT-X is used in meshed networks with evenly distributed traffic. If a fiber outage takes place, the disabled section is optically switched within the domain to the planned restoration capacity. Restoration over multiple photonic domains requires more complex restoration control systems, and increases the time to restore a failed link. Network restoration using optical cross-connects is applicable to single grid or multiple grid domains.

The most challenging aspects relating to the use of OPT-Xs are the control network management functions that are needed to restore failed sections of the network. Photonic-based restoration capacity is preplanned and is determined during the provisioning process. Fault correlation is another major network management process, and can be specified independently of the network topology based on a business, a service, or a network element orientation. Restoration methods can be centralized or distributed. Both types of methods require a mediation device to generate alarms indicating a catastrophic fiber cut failures via the detection of AIS, LOS, or BER threshold crossings. Another possibility is to use the K1/K2 bytes defined by ANSI to identify catastrophic failures.

Centralized restoration can be used in individual photonic domains, while distributed restoration can be used for restoration across domain boundaries. In a centralized restoration, an operations system (OS) controls the rerouting of traffic around the failure. Control of the restoration process is centered at the OS workstation(s). In distributed restoration systems, the photonic device controls rerouting of traffic. An algorithm controlling the restoration is programmed into each optical cross-connect. At the time of a failure, the optical cross-connects distribute control exchange messages (via signaling control channels) and coordinate activities of each device to reroute traffic. Control of the process is shared by the optical cross-connects in the network. This is essentially a sequential control process. Therefore, a sufficient amount of restoration capacity must be designed into the network for the distributed algorithms to work effectively in restoring traffic flows. This method increases the complexity of the optical cross-connect requirements and operations.

Almost all distributed algorithms use the restoration capacity available in the network to provide alternate routes for the failed circuits. When a dynamic algorithm is used, it resides in the cross-connect's system controller. In contrast, a preplanned algorithm "knows" about internal routes via "hooks" (indirect and indirect offset addressing) to preestablished connection look-up tables. A preplanned approach usually reduces the algorithm's execution time but it increases the complexity of the network provisioning and requirements planning process.

Analysts within the industry are discussing design techniques for transmission systems operating at 40 Gbps and higher, along with plans for OC-768 solutions. With the advent of new optical technologies such as dense wave division multiplexing (DWDM), ultra-long haul DWDM, dispersion-shifted fiber (DSF), optical amplifiers, and optical cross-connects (OPT-X), these very high network speeds are becoming less of a future vision and more of a reality every day.

2.2.7.5.1.6 Advantages, Disadvantages, and Recommended Uses of SONET — This section summarizes the major advantages, disadvantages, and recommended uses of SONET technology.

Some advantages of SONET networks include:

- Fiber optic transmission systems (FOTs) equipment interoperability
- Operation support systems (OSS) network management capabilities
- Simpler "mid-span meet" handover
- Support for future high-bandwidth services
- High network survivability and available bandwidth
- Excellent core backbone technology for integrating multiple network infrastructures

- Rapid worldwide growth for both SONET and SDH
- Supports very-high-volume and very-high-speed internetworking

The disadvantages of SONET include:

- Requires expensive initial outlay of capital
- Relatively high overhead (more than 4 percent) compared to T-1 1.544 Mbps (about 1 percent)
- Difference in path lengths between working and protection fibers may result in time-outs of some data applications in the event of a protection switch
- Standards are still emerging and evolving
- Requires new hardware platform

SONET networks are recommended for:

- Backbone infrastructure for LECs, IECs, PTTs, and large corporate users
- Bandwidth-intensive services such as Broadband ISDN, high-speed computer links, local area network (LAN) links, full-motion video, and high-definition TV (HDTV)
- Telecommunications requirements that necessitate "survivability" techniques to minimize network downtime
- Multiple, integrated, large-volume voice/fax, video, and data applications
- Very large-scale internetworking backbones
- Support for SDN and B-ISDN protocols
- Medium-high to very-high-cost networks

2.2.7.5.2 Small Office/Home Office (SOHO): DSL, Cable, Dial-Up

2.2.7.5.2.1 Technical Overview — Institutions that need to provide Internet access to a large group of users (numbering in the thousands or more) should look to an established Tier 1 global connectivity provider. These providers can manage all aspects of establishing and maintaining a connection to the Internet for a large group of people. However, there is a class of users called SOHOs (small office/home office) that has more modest requirements, needing connectivity for tens or hundreds of users. Fortunately, the technology required to offer this connectivity is commoditized, and the only significant issue is that high-speed Internet connectivity is not available in all areas. There are two major steps required to establish this connectivity. One is to select a type of technology and service provider. The second is to select a device to share this connectivity with others.

Specialized companies — called Internet service providers (ISPs) — are available to provide low-cost solutions for accessing the Internet. ISPs pay for a relatively expensive connection to the Internet, which they make accessible to others through the installation of high-performance servers, data lines, and modems. Acting as middlemen, the ISPs rent time to other users who want to access the Internet. An ISP might be your local telephone or cable TV company.

Once an ISP service provider has been selected, the SOHO must install a device that makes the connection between the ISP and the end-user's computer. This device is referred to as a modem, which stands for MOdulator/DEModulator. The term "modem" used alone usually refers to a telephone modem, which connects to an RJ-11 phone connector on one end and plugs into a PC serial port on the other end. A telephone modem sends sounds over the telephone network, converting these sounds to digital data before entering the computer. Because the phone networks were designed to carry human voice data, and not computer data, getting the maximum data rate advertised by a modem manufacturer may be difficult. Higher-speed units, such as cable and DSL modems, typically have either a cable or DSL connection on one end and a type of Ethernet connection to connect to the computer. Because these modems utilize a medium that has been designed to carry computer data, much higher speeds are possible when compared to a telephone modem. To support communication speeds of 10 Mbps (10 BaseT) or 100 Mbps (100 BaseT), a cable type known as Category 5 can be used. To support Gigabit Ethernet, also known as 1000 BaseT, one should use Category 5e cable. Category 6 cable is highly recommended to "future-proof" your installation.

In a SOHO environment, one typically wants to share this connection among the local network, so another type of networking device will be required. This device is commonly referred to as a cable/DSL router. One plugs the Ethernet cable that comes out of the cable modem into the cable/DSL router, and then plugs any additional network equipment, such as other computers, printers, etc., into the extra ports in the cable/DSL router. If there is a plan to use some type of wireless technology, either instead of or in addition to wired cables, there are cable/DSL routers that also offer some combination of 802.11a/b/g wireless connectivity.

For a small group of users, another option is to purchase a simple hub or switch, and use the connection sharing feature of the operating system. Most major operating systems include some capability to share the connection of one computer with other computers connected to the hub or switch.

Two important decisions must be made when deciding what type of Internet connection is most appropriate. The first decision is the company budget allocated for Internet connectivity, and the second is the Internet connection speed needed to support the business requirements. Both decisions are interrelated. ISPs offer a variety of options for connecting to the Internet, ranging from a simple dial-up account over phone wires to high-speed leased lines or cable. Unless one is really on a budget, a high-speed permanent connection, such as cable or DSL, makes the most sense. Dial-up accounts are typically available for a low, flat monthly fee, and are generally much cheaper than cable or DSL connections. However, the cable or DSL connection is faster and more reliable than the dial-up connection.

When a dial-up account is used, a modem and a phone line are used to call and log into the ISP server (or computer), which in turn acts as the doorway to the Internet. The transmission speed of the connection is limited by the speed of the modems used by the user and the ISP. DSL can offer throughput typically in the range of 384 Kbps to 6 Mbps downloads, and up to 384-Kbps uploads. Cable can have download speeds of up to 10 Mbps and upload speeds of up to 1 Mbps. This compares very favorably with more expensive solutions, such as ISDN (which support transmission speeds from 56 Kbps to 128 Kbps), T1 (transmitting at speeds up to 1.544 Mbps), and T3 (transmitting at speeds up to 45 Mbps).

If a SOHO user only needs to make an occasional connection to the Internet — for example, less than 20 to 50 hours per month for all users — a dial-up account should be sufficient. The costs for a basic dial-up account with unlimited access are in the range of approximately $250 per year. However, if a company needs faster data transfer speeds or has several users who must access the Internet for substantial periods of time over the course of a month, a DSL, cable or leased line connection should be considered. A business-class DSL connection can range from $35 to $160 per month, depending on the download and upload speeds required. A business-class cable connection can be had for around $100 a month, typically offering upload and download speeds in the megabit-per-second range.

Typical monthly charges and special equipment fees for a leased line connection might average $200 to $1200 per month for an ISDN connection; $1000 to $3000 per month for a T1 connection; and $2000 to $10,000 per month for T3 connection. Note that the actual line costs will vary according to the service provider and specific options the company has chosen to implement.

Principles of queuing analysis can be applied to the problem of sizing the links needed to support the Internet access, whether or not that access is to an ISP or to a direct Internet connection. The reader is referred to Chapter 8 for specific techniques on how to estimate the throughput and performance characteristics associated with using different-sized link capacities. This analysis can be used to determine whether or not a dial-up or leased line connection is sufficient to support the bandwidth requirements with tolerable transmission delays.

2.2.7.5.2.2 Case Study — A case study is now presented for a single user with a wired desktop and printer, and a wireless laptop. The concepts are easily generalized to a SOHO network with more users. In this example, there is a cable modem connection with Optimum Online, a popular cable provider, along with a cable/DSL router from Linksys (specifically, the WRT54G). This device offers four ports for expansion, and also offers high-speed 802.11g wireless connectivity.

One thing to keep in mind when designing a SOHO network with a cable/DSL router is that the router creates two virtual "sides" of the network. One is a public side, which is connected to a cable or DSL modem and is accessible to anyone on the Internet. Fortunately, the router also creates a "private" side of the network, into which one can plug a multitude of network devices that are assigned "private," nonroutable Internet addresses. These addresses, defined in RFC 1918, are in three ranges: 10.0.0.0 to 10.255.255.255, 172.16.0.0 to 172.31.255.255, and 192.168.0.0 to 192.168.255.255. Most devices use the 10.X.X.X range of addresses because it has the most addresses available.

The WRT54G has a public and private side of the network. Each device that connects to the router, which has a private address of 10.0.1.1, can do so either by DHCP (which is most appropriate for temporary clients, such as a wireless-enabled portable) or by a static address, in the case of a desktop machine. To guarantee sufficient addresses for both types of clients, the convention is to start assigning IP addresses for temporary clients at 10.0.1.100, and for permanent clients to start addressing at 10.0.1.200. The wired Ethernet printer is a permanent device, so it will take the next IP address available in the 10.0.1.200 range, which is 10.0.1.201.

To summarize, a SOHO network can easily handle both wired and wireless client computers and devices. Because wireless clients are temporary, it makes the most sense to have their addresses assigned by a dynamic protocol such as DHCP. For permanent, wired clients, the best strategy is to assign a permanent address, in order to allow clients to reliably connect and interact with the computer or other device. Of course, DHCP can also be used for permanent clients but this requires additional effort to make sure that the same DHCP address is always assigned to the same device.

2.2.7.5.5 ISDN Networks

2.2.7.5.5.1 Overview — The Integrated Services Digital Network (ISDN) supports voice, data, text, graphics, music, video, and other communications traffic over twisted-pair[60] telephone wire. ISDN was originally conceived by the telcos as a way to integrate voice, data, and video communication services for their customers. Although a key benefit of ISDN is the ability to carry simultaneous applications over the same digital transmission links, ISDN provides many other attractive services, including "intelligent" options and features, and network management and maintenance functions. B-ISDN was developed in the mid-1990s to enable the public networks to provide switched, semi-permanent, and permanent broadband connections for point-to-point and point-to-multipoint applications. B-ISDN is considered the second-generation version of ISDN. The B-ISDN standard defines a cell relay technology that is designed to work in conjunction with ATM and SONET. Channel speeds of 155 Mbps and 622 Mbps are available through B-ISDN and ATM services. The developers of ISDN and B-ISDN have attempted to develop an infrastructure for worldwide networking solutions based on comprehensive standards and services. The integration provided by a common infrastructure facilitates the provisioning of services and a more comprehensive approach to network management.

The CCITT is an international organization responsible for the development of ISDN standards. The ISDN standards are known as the I-series recommendations. By convention, all CCITT standards are referred to as recommendations. The signaling aspects of ISDN, including the Link Access Protocol D (LAP-D) specification, are defined in the CCITT Q-series recommendations. The B-ISDN standard is specified in ITU-T Recommendation I.121.

Some of the more commonly available ISDN services include:

- 64 Kbps (1B+D)
- 128 Kbps (2B+D)
- 384 Kbps (H0)
- 1472 Kbps (H10: 23B+D)
- 1536 Kbps (H11: 24B)
- 1920 Kbps (H12: 30B+D)

Bearer or B channels, as listed above, operate at 64 Kbps and are used to carry digitized voice, data, or video traffic over circuit-switched, packet-switched, or semi-permanent connection facilities. The D channels carry signaling information that is used to control the circuit-switched B channels. They can also carry low-speed, packet-switched data. H channels are aggregates of B channels.

The ISDN architecture delineates two service classes: BRI and PRI. Basic rate interface (BRI) circuits employ a 2B+D structure, which consists of two full-duplex 64-Kbps B channels and one full-duplex 16-Kbps D channel. Overhead bits are also carried on the channel and are used to perform framing, synchronization, and management functions. BRI facilities are the ones most often used for corporate networking applications.

When the traffic requirements exceed 64 Kbps per channel, multiple channels are grouped together or *bundled* to provide the necessary bandwidth. These bundled channels are typically digital 1.544-Mbps (T-1) circuits in the United States and in a few other countries, such as Canada and Japan. In the rest of the world, 2.048-Mbps (E-1) digital circuits are used. Access facilities operating at these speeds are known as primary rate interface (PRI) circuits and are formally offered in H channel structures beginning at 384 Kbps (H0).

Many times, PRI circuits include a separate 64-Kbps D channel for each PRI facility. When there are multiple PRI circuits to the same customer location, it is possible to control all transmissions through a single D signaling channel. In this case, all the PRI channels can be used to transmit data, except for the one PRI reserved for signal control purposes. Signaling overhead is greatly reduced when only one D channel is used to control multiple PRI circuits. However, there is the risk that if the D channel fails, no transmissions can be sent.

The actual call processing is performed by a Signaling System 7 (SS#7) network that controls the call setup and call teardown procedures for each ISDN call. This is the same SS#7 signaling network used by the U.S. local exchange carriers (LECs) and the interexchange carriers (IXCs) to place standard voice calls. In other parts of the world, the Common Channel Signaling System 7 (CCSS#7) signaling technology infrastructure is used. While many of the basic concepts are the same, there are differences between the two signaling methods that necessitate signal conversion equipment. Therefore, many popular switch manufacturers support both signaling formats so their equipment can be used to handle both domestic and international ISDN calls.

2.2.7.5.5.2 Design Considerations

— Although ISDN was slow to take off in the U.S. marketplace, its popularity has been growing in recent years. One of the reasons for this is the increased availability of ISDN products and services. Until fairly recently, ISDN was not widely available throughout the United States, in part because telephone carriers have been slow to upgrade their equipment to support ISDN. However, even when ISDN service is available, there is often a long lead time of weeks or months to install the lines after they are ordered.

The limited availability of ISDN contributed to the reluctance of vendors to offer ISDN-ready products. As ISDN has become generally available, so have ISDN and ISDN-compatible products. The development of IDSN device interface standards has also encouraged the development of new ISDN product offerings. ISDN protocols governing communication between devices in the network are based on the notion of "functional devices" and "reference points." ISDN equipment standards define several types of device classes. In these "functional device standards," each device type is characterized by certain functional or logical features that may or may not actually be present in the device. Device-to-device interfaces are defined through "reference point standards" that are based, in turn, on the functional device types. Reference points allow different ISDN functional devices to operate under different protocols on the ISDN network. The ISDN standards are designed to promote modular product development so that equipment manufacturers can make hardware and software improvements to their product lines without compromising ISDN compatibility. From the customer's perspective, this means that ISDN equipment can be purchased from a variety of vendors, thus offering more freedom of choice in the selection of ISDN networking solutions. As illustrated in Figure 2.34, several ISDN functional groups and reference points are currently defined, including:

- **TE1** *(subscriber terminal equipment type 1)*: devices that support the standard ISDN interface, such as digital telephones and digital (type IV) facsimile equipment.
- **TE2** *(subscriber terminal equipment type 2)*: devices that support non-ISDN equipment, such as video CODECs, terminals, and mainframe computers.
- **TA** *(terminal adapter)*: devices that convert non-ISDN equipment into ISDN formatted signals. The ISDN TA can often be either a stand-alone device or combined with a TE2 type device.
- **NT1** *(network termination 1)*: devices that include functions associated with the physical and electrical aspects (OSI layer 1) of ISDN termination on a customer's premises.
- **NT2** *(network termination 2)*: device with switching and concentration functions up to OSI layer 3 (network level). Some examples include digital PBXs, LAN routers, and terminal controllers.
- **NT1/2** *(combination network termination 1 and 2)*: not shown in diagram. Device that combines NTI and NT2 functions into a single piece of telecom equipment. Often used by foreign postal telegraph and telephone (PTT) ISDN providers.
- **R** *(reference point R — rate)*: provides a non-ISDN interface between non-ISDN-compatible equipment and a terminal adapter (TA).
- **S** *(reference point S — system)*: separates individual ISDN user equipment from network-related telecom functions.
- **T** *(reference point T — terminal)*: consistent with ISDN network termination, it separates end-user equipment (up to NT2) from network provider equipment (NT1).
- **U** *(reference point U — user-to-network interface)*: reference point between NT1 devices and line-termination equipment in the carrier network. The U reference point is typically employed in North America because the carrier network does not usually provide NT1 functionality.

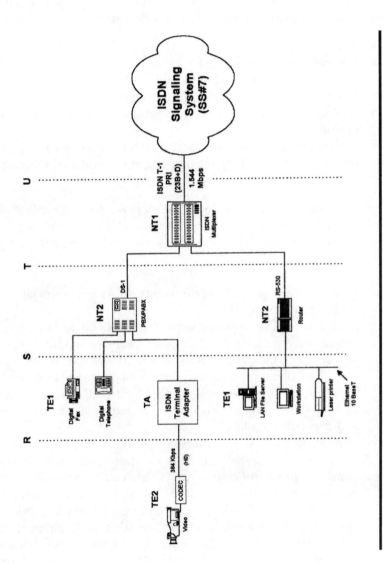

Figure 2.34   ISDN network design. (*Source:* B. Piliouras.)

Although ISDN continues to be more expensive than regular phone service, its costs have declined as service providers have developed a larger customer base. This, too, has encouraged ISDN use. A primary driver behind the market acceptance of ISDN is the suitability of ISDN for certain niche applications. In particular, ISDN provides a very cost-effective and convenient networking solution for the following types of applications:

- Telecommuting
- Internet connectivity to ISP[61]
- Long-distance videoconferencing
- Dial-up lines for backup and disaster recovery
- LAN-to-LAN connectivity (especially for small business and organizations)
- Branch-office connectivity between remote sites having low-volume traffic (particularly for banking and automated teller machine applications)

Not surprisingly, ordering ISDN service is more complicated than ordering regular telephone service because of the various service and circuit options available. The basic steps involved in implementing ISDN involve the following:

1. *Determine the network requirements.* (This is true for all network design problems.)
2. *Determine the appropriate service.* In addition to bearer services, mentioned earlier for data transport, there are supplementary ISDN services and teleservices. Supplementary services provide options for voice communication and include such things as caller ID and call waiting. Teleservices is a sort of catchall service category for various enhanced offerings, such as teleconferencing, videotext, and message handling.
3. *Determine the type and number of circuits needed.* B, D, and H channels are available with various options for ISDN service.
4. *Order the lines and equipment.* As indicated previously, there can be significant lead-times before the ISDN lines are installed. This means that the ISDN service should be planned in advance. The network designer must also make sure that the equipment selected matches the ISDN services ordered, and vice versa, because not all equipment is compatible with ISDN service. For example, most analog modems do not support speeds above 56 Kbps. IDSN transmissions, on the other hand, begin at 56 Kbps and can exceed 1 Mbps. In addition, not all ISDN-compliant equipment supports all ISDN functions.

5. *Install the equipment.* The supplier of the ISDN equipment and services must be selected. ISDN service providers, for example, are working with many of the major equipment vendors to develop value-added reseller (VAR) relationships, authorizing the VARs to sell and install ISDN lines. This will increase the number of service options available to the organization. A complaint businesses have had in the past about ISDN is that it is difficult to incorporate different products and vendors into an integrated ISDN environment. The ISDN Forum and various standards bodies are working to alleviate this problem.

2.2.7.5.5.3 Advantages, Disadvantages, and Recommended Uses of ISDN — This section summarizes the major advantages, disadvantages, and recommended uses of ISDN networks and standards.

Some of the immediate benefits that can be achieved using ISDN networks include:

■ Charges are based on the duration of the call, and not according to a predetermined rate
■ Easy to design, manage, and implement an ISDN solution
■ Consolidates multiple "circuits" onto a single access facility
■ Improves end-to-end service quality versus private line (OSI layer 1), frame-switching (OSI layer 2), and packet-switching (OSI layer 3) network infrastructures

Some of the disadvantages associated with ISDN include:

■ Relatively high overhead (more than 4 percent) compared to T-1 private lines (about 1 percent)
■ Vendor compatibility varies widely, depending on ISDN function and reference point
■ Uncertain overall implementation by numerous worldwide carriers and equipment providers

Recommended ISDN uses include:

■ Support for a variety of end-user telecom requirements, including voice, data, text, graphics, music, video, and other communications traffic
■ Telecom applications where the end user pays only for the duration of the call

- Special communications needs that require high network reliability and performance
- Large-scale internetworking requiring any-to-any "connectivity"
- Low- to high-cost networks

Notes

1. In Chapter 8, considerations involved in the selection of node equipment are presented. This selection is based on the functionality that must supported at a particular network location. For instance, a firewall may be needed at a strategic network access point, and therefore a router may be selected as a note type. Once the functionality needed at a particular site has been determined, specific equipment options can be evaluated based on their performance and cost characteristics.
2. DDS, Digital Data Service, a transmission service that supports line speeds up to 56 Kbps.
3. DS1, Digital Signal 1, a transmission service that supports 24 voice channels multiplexed onto one 1.544-Mbps T1 channel.
4. DS3, Digital Signal 3, a transmission service that supports 44 Mbps on a digital T3 channel.
5. IEC, the acronym for interexchange carrier.
6. Mesh: a mesh network provides point-to-point connectivity between various nodes in the network.
7. Terminal: any device capable of sending or receiving information on a communications line. It includes such devices as a workstation, CRT, keyboard, and remote printer.
8. Router: a protocol-specific device that finds the "best" route (i.e., the cheapest, fastest, or least congested route available) to send data from sender to receiver. Routers are commonly used in large networks with low bandwidth connections, such as those found in legacy SNA networks.
9. Cluster controller: specialized computer that offloads data communications tasks from the mainframe which allows a single, high-speed line to be shared between several lower-speed interactive terminals. A cluster controller is also known as an establishment controller. A cluster controller is considered a front-end processor if it is co-located with the mainframe. It is considered a switching node if it is located remotely from the host and acts as a stand-alone switch. The network control program (NCP) is operating system software that resides in the communication controller to manage communication and operation of links with the SNA network. Cluster controllers are also referred to as peripheral nodes and operate at the terminal, low end of the SNA hierarchy.
10. Front-end processor (FEP): a dedicated device that intercepts data before it is processed by the host computer, to perform such functions as error control, line control, message handling, and code conversion. Along with the mainframe, the FEP is known as a sub-area node. These nodes form the top level of the SNA hierarchy.

11. Broadcast traffic: the data sent to *all* devices on a network. Broadcast traffic adds to the network overhead and bandwidth requirements. In contrast, unicast traffic is sent only to a single node in the network, and multicast traffic is sent only to a selected set of nodes on the network.

12. NCP, which is resident in the communications controller, interfaces with ACF/VTAM, which is resident in the host mainframe, to manage and control network communications.

13. Token Ring: a local area network in which devices are connected in a ring or star topology. A token is passed between devices, granting them permission to transmit on the line. This scheme prevents possible data collisions that might occur if other devices were to try to transmit at the same time. IBM's Token Ring led to the development of the IEEE 802.5 Token Ring protocol. The Fiber Distributed-Data Interface (FDDI) is also based on a Token Ring protocol.

14. AppleTalk: This is Apple Computer's seven-layer stack communication protocol, which is similar to the OSI stack. This protocol allows interconnection of Macintosh computers and printers on local area networks. AppleTalk is compatible with Ethernet and Token Ring protocols. AppleTalk uses dynamic addressing, which allows plug-and-play device operation on the network.

15. TCP/IP: protocol suite for networking occupying layers 3 and 4 of the OSI Reference Model.

16. Bandwidth: range of frequencies that can be supported on a communication link. High bandwidth supports higher rates of data transmission.

17. AS/400: a distributed processor that supports many host processor functions, except for network management. This IBM product has been discontinued.

18. Bridge: a device to interconnect networks using the same protocols.

19. Dedicated line: a leased line usually supplied by a telephone company that permanently connects two or more locations, and is for the sole use of the subscriber.

20. Switched line: a line through the public telephone network that is connected to a switching center.

21. TDM (time division multiplexing): a way of transmitting multiple data streams in a single signal. It does so by decomposing the signal into many unique, small time slots. At the destination end, the signals are reconstituted. A multiplexer combines signals at the transmission source and prepares them for transport. At the destination end, a de-multiplexer separates the signals for delivery to the end-user. Two-way communication circuits use multiplexers and de-multiplexers at each end of long-distance, high-bandwidth cable.

22. Backhauling is a technique in which the user payload is carried past a switch that has a line to the user and to another endpoint. After the traffic destined for the endpoint is dropped off, the first user's payload is then sent back to the switch where it is relayed to the first user. In many existing configurations, grooming eliminates the need for back-hauling, but it also requires expensive configurations (such as back-to-back multiplexers that are connected with cables, panels, or electronic cross-connect equipment).

23. Internet2: a consortium of hundreds of universities working in partnership with industry and government to develop and deploy advanced network applications and technologies. The goal is to investigate real-time multimedia and high-bandwidth applications for teaching and collaborative research, and to implement the network infrastructure needed to conduct these studies. Internet2 is not intended as a replacement for the Internet, but as a means to augment its future development. [INTE03]
24. Lit-building refers to a building in which fiber optic cable has been installed.
25. Dark fiber: fiber optic cable that is not attached to light termination equipment.
26. NAS: consists of storage devices connected to a LAN with specialized network interface cards.
27. SAN: a separate network from a LAN or a WAN. It is used exclusively to connect servers to storage devices. SANs are used for disk mirroring, backup and recovery, data archiving, etc.
28. T1: United States digital communication line operating at 1.544 Mbps.
29. T3: United States digital communication line operating at 44 Mbps.
30. OC-1: the base SONET signal rate of 51.84 Mbps for transmitting digital signals on fiber optics. Thus, OC-n translates into a signal rate of (51.84 Mbps * n).
31. Teleradiology: high-quality medical imaging transmitted over communications links.
32. Multiplexer: also known as a "mux," this device allows more than one signal to be sent over a channel simultaneously.
33. Concentrator: a device that consolidates traffic coming from a number of circuits onto a smaller number of circuits, to reduce costs.
34. PDA (personal digital assistant): refers to a mobile hand-held device used for ready reference and entry of calendar, address, and note information. This term is used synonymously with "hand-held." A popular example is 3Com's PalmPilot. New PDAs offer additional telephony and paging functions. An example of this is RIM's (Research In Motion) Blackberry, which supports Internet connectivity, e-mail, and paging.
35. IRC (Internet Relay Chat): allows people to start or join an online meeting (a.k.a. a chat or channel) and to exchange messages in real-time. It allows one to discover existing chat groups and members who are identified by nicknames. Its use requires the installation of an IRC client program. Some popular chat clients include: mIRC for Windows, Ircle for Macintosh machines, and irc2 for UNIX machines. After you have installed an IRC client, you must connect to an IRC server. Some of the largest IRC networks include EFnet, IRCnet, Undernet, and Dalnet. To join an IRC chat, type: "/join #channelname." Two very popular channels are #hottub and #riskybus. On UNIX using irs2, you can list channels with 20 members, along with nicknames and topics by typing: "/list -min 20."
36. Head-end: also known as the central processor.
37. Capacitated Minimal Spanning Tree (CMST): a topology representing a minimal spanning tree that has been designed to conform to link capacity constraints.

38. For the cases they tested, Chandy and Russell have shown that heuristic CMST techniques are usually within 5 to 10 percent of optimal. [AHUJ82, p. 130]

39. Backbone networks offer economies of scale because higher-speed links carry more traffic and cost less than an equivalent number of lower-speed lines.

40. OC-n: a SONET-related term for the optical carrier level n. This is the optical signal that results from an STS-n signal conversion. STS-1 is the basic SONET signal that operates at a transmission rate of 51.84 Mbps. An STS-n signal has a transmission rate of n times the basic STS-1 transmission rate.

41. Packet-switched networks are not suited to full mesh topologies due to the large number of virtual circuits required (one for every connection between nodes), the problems associated with the large number of packet/broadcast replications required, and the resulting configuration complexity in the absence of multicast support in non-broadcast environments.

42. CPE (customer premise equipment): the terminal equipment connected to the public switched network. The CPE can be provided by the common carrier or some other supplier.

43. Pendent node: is characterized as having only one link into the node. A pendent node pair, as the name implies, consists of two pendent nodes.

44. Hop: the number of hops in the network refers to the number of nodes that the data must traverse in the network when going from source to destination.

45. CCITT: since 1993, this organization has been known as the ITU-T. It is an advisory committee dedicated to the development of standards and recommendations for the telecommunications industry.

46. CPU (central processing unit): a computer.

47. Local loop: the part of the communications circuit that connects the subscriber's equipment to the equipment of the local exchange.

48. BER (bit error rate): a measure of the line transmission quality that specifies the probability that a given bit will have an error in transmission. The BER of a typical analog circuit is on the order of 1 in 10^6, while the BER of a typical fiber circuit is on the order of 1 in 10^9.

49. Frame Relay packets: in Frame Relay, the packets transmitted over the network are called "frames." A frame is a block of data consisting of a flag sequence, an address field, a control field (which is not used by Frame Relay but is defined for compatibility with other protocols), an FECN bit, a BECN bit, an EA expansion bit, a DE delete eligibility bit, a Frame Relay information field containing user data, and a frame checking sequence field used to detect transmission errors. The data-link layer (layer 2) creates Frame Relay packets using this standard format.

50. Gang of Four: refers to a group of four telecom vendors: Cisco Systems, Northern Telecom, StrataCom, and Digital Equipment Corporation.

51. SAP (Service Advertising Protocol): the protocol used to broadcast messages to ascertain the connectivity between various end-users and processing nodes.

52. QoS (Quality-of-Service): this metric is used by the service provider or carrier to define the guaranteed level of quality over the transmission facilities.

53. Bursty traffic is characterized by intermittent high-volume traffic spikes, interspersed with moderate levels of traffic.
54. FRAD (Frame Relay Assembler Disassembler): a device similar in concept to the PAD used in packet switching. It enables asynchronous devices, such as personal computers, to send data and communicate over the Frame Relay network. FRADs are usually implemented as an integral part of the network device. In addition to Frame Relay, many FRADs support X.25, ISDN, and other protocols in the same device.
55. CCS: a signaling protocol developed by the ITU-T for high-speed digital networks.
56. UNI (user network interface): the point at which the user connects to an ATM network.
57. NNI (network-network interface): the interface between two devices in an ATM network. There is more discussion on NNIs later in this section.
58. PDU (protocol data unit): an OSI term for the data to be transmitted, and the headers and trailers that are appended to the data by each OSI layer for processing purposes.
59. FDDI (Fiber Distributed Data Interface): a LAN standard specifying LAN-to-LAN backbone transmission at 100 Mbps over fiber optic cabling.
60. Twisted pair: consists of two insulated wires twisted together without an outer covering. This is the standard wiring used in telephone installations.
61. ISP (Internet service provider): a company providing public access to the Internet, usually through a leased line ISDN connection via a modem.

References

[AHUJ82] Ahuja, V., Design and Analysis of Computer Communication Networks, McGraw Hill, New York, 1982.
[AXNE03] Axner, David, Gigabit Ethernet Goes on MAN Hunt, IT World.com, http://www.nwfusion.com/edge/research/2000/1204edgefeat.html
[CAHN98] Cahn, R., *The art and science of network design,* Morgan Kaufmann, 1998.
[CISCO01B] Cisco Data Systems, *Schroder, Salomon Smith Barney:Citigroup Global IPMulticasting Network uses Cisco IOS ® Multicast Technology to Deliver Real-Time Trading Data,* Cisco Systems, 2001. http://www.cisco.com/warp/public/732/press/docs/citigroup_cp_final1.pdf
[CISC02] Cisco Systems, IBM Communications Controller Replacement Design Guide, copyright 1999-2002, pp. 1-20. http://www.cisco.com/warp/public/cc/so/neso/ibso/data/datacent/ibmcm_wp.pdf
[CISC004] Cisco Systems, *Establishing a VPN Connection,* 2004, http://www.cisco.com/en/US/products/sw/secursw/ps2308/products_user_guide_chapter09186a008015cffc.html.
[DINT94] Dintzis, M., X.25 packet and Frame Relay switches, *Datapro*, McGraw-Hill, Inc., March 1994.
[DOBR96] Dobrowski, George and Humphrey, M., ATM and its critics: separating fact from fiction, *Telecommunications,* November 1996, pp. 31–37.

[FEDE03] Feder, Barnaby, Information On-Ramp Crosses a Digital Divide, *New York Times*, July 8, 2003, p. C1, C6.

[FOUN02] Foundry Networks, Telecom Ottawa selected Foundry to power its Metropolitan Area Network, one of the world's first standards-based 10-Gigabit Ethernet MANs, July 24, 2002, http://www.foundrynet.com/about/newsevents/releases/pr7_24_02b.html

[IEC03] International Engineering Consortium, MultiProtocol Label Switching (MPLS), http://www.iec.org/cgi-bin/acrobat.pl?filecode=94

[IEEE02] IEEE 802.16 Working Group, *Broadband Wireless Access: An Introduction to the Technology Behind the IEEE 802.16 WirelessMAN™ Standard*, May 24, 02, http://grouper.ieee.org/groups/802/16/pub/backgrounder.html

[INTE03] About Internet2®, www.internet2.edu

[KERS91] Kershenbaum, A., Kermani, P., and Grover, G., Mentor: an algorithm for mesh network topological optimization and routing, *IEEE Transactions on Communications*, Volume 39, No. 4, April 1991, pp.503-513,

[KERS93] Kershenbaum, A., *Telecommunications network design algorithms*, McGraw-Hill, Inc., New York, 1993.

[[MCDY95] McDysan, D. and Spohn, D., Future Directions of ATM, *ATM Theory and Application,* McGraw-Hill, New York, 1994, pp. 579–595.

METC00] Metcalfe, Bob, From the Ether: Quality multimedia in real time is the impetus for Enkido's 768 service, *InfoWorld*, April 28, 2000, http://archive.infoworld.com/articles/op/xml/00/05/01/000501opmetcalfe.xml

[MFN99] MFN, Metromedia Fiber Network Leases Dark Fiber Infrastructure To America Online, *MetroMedia Fiber Press Release*, New York, February 2 1999. http://www.mfn.com/news/pr/19990202_Lease_AOL.shtm

[MPLS02] MPLS Forum, *GMPLS Interoperability Event Test Plan and Results*, copyright 2002,http://www.mplsforum.org/interop/GMPLSwhitepaper_Final1009021.pdf

[NORT03] Nortel Networks, *Enabling Technologies*, http://www.nortelnetworks.com/corporate/technology/oe/protocols.html

[RESI01] Resilient Packet Ring Alliance, An Introduction to Resilient Packet Ring Technology, October 2001, www.rpralliance.org

[RYAN97] Ryan, Gerald, Editor, Dense Wavelength Division Multiplexing, Ciena Corporation, copyright 1997.

[SCHO02] Schoolar, Daryl, Plenty of Light Left in Optical MAN Services, *Cahners In-Stat Press Release,* June 12, 2002, http://global.mci.com/news/marketdata/network/

[SEAR00] SearchNetworking.com, Glossary: OC-768, August 9, 2000, http://search-networking.techtarget.com/gDefinition/0,294236,sid7_gci294572,00.html

[SPIR02] Spirent Communications, BGP/MPLS Virtual Private Networks Performance and Security over the Internet, Copyright 2002. http://www.spirentcom.com/analysis/index.cfm?WS=27&D=2&wt=2

[WELS03] Welsh, William, Deficits don't deter new CIOs, *Washington Technology*, May 12, 2003.

Chapter 3

Local Area Network Design and Planning Techniques

3.1 Management Overview of Local Area Networks (LANs)

In contrast to wide area networks (WANs), local area networks (LANs) operate on smaller geographic distances and support higher data transmission rates. Typically, LANs connect computers that reside in the same building or campus, and are usually limited to a few city blocks. LANs provide connectivity between devices operating on a private network, including file servers, printers, host processors, and personal computers. LANs are commonly used to support a wide range of applications, including client/server[1] applications, file sharing, and centralized file backup and retrieval.

3.1.1 IEEE LAN Standards

The IEEE 802.1-802.12™ standards define LAN protocols.[2] LANs are usually classified in accordance with transmission schemes (baseband[3] and broadband[4]), transmission media (twisted pair, coaxial cable, fiber, and wireless), transmission techniques (balanced and unbalanced), topologies (star, ring, bus, and tree), and access control techniques (random control such as CSDA/CD,[5] slotted ring,[6] and register insertion ring;[7] distributed control such as Token Ring; and centralized control such as polling, circuit switching, and time division multiplexing). The three main *de facto* LAN access protocols are: Ethernet,[8] Token Ring,[9] and FDDI.[10] The access protocols determine how the devices on the LAN share the cabling system that connects them. On a LAN, usually only one device at a time can transmit data over the network. The access protocol determines which and how each device takes turns transmitting data. Examples of typical LAN configurations are shown in Figure 3.1.

3.1.2 Ethernet LANs

Ethernet is by far the most prevalent type of local area network. Devices are typically connected to an Ethernet LAN using coaxial cable, twisted-pair wire, or wireless transmission media, and compete for access using the Carrier Sense Multiple Access with Collision Detection (CSMA/CD) protocol (as defined by the IEEE 802.3™ standard).

Ethernet supports transmission at a variety of speeds. Ethernet LAN transmission at speeds of 10 Mbps includes:

- *10Base5:* Ethernet using thicknet[11] coaxial cable.
- *10Base2:* Ethernet using thinnet[12] cable.
- *10BaseT:* Ethernet over unshielded twisted-pair[13] cable.
- *10BaseF:* Ethernet over fiber optic cable.

Ethernet LAN transmission at speeds up to 100 Mbps includes:

- *100VG-AnyLan:* Ethernet using voice-grade unshielded twisted-pair cable and is capable of supporting cascading hubs.
- *100BaseT:* "Fast Ethernet," which typically uses two-pair Category 5 unshielded twisted-pair cable and is used to support LAN backbones.

Gigabit Ethernet achieves transmission speeds up to 1 Gbps, while 10-Gigabit Ethernet, as its name implies, supports transmission speeds up to 10 Gbps. Token Ring networks support transmission speeds up to 10 Mbps[14] and 16 Mbps, respectively. Fiber Distributed Data Interface (FDDI) LANs support transmission speeds up to 100 Mbps using fiber optic cable.

3.1.3 Wireless LANs

Wireless technology is having a profound impact on the LAN market. The IEEE 802.11™ wireless networking standard — which is also known as Wi-Fi (for wireless fidelity) or WLAN — is being used to provide high-speed

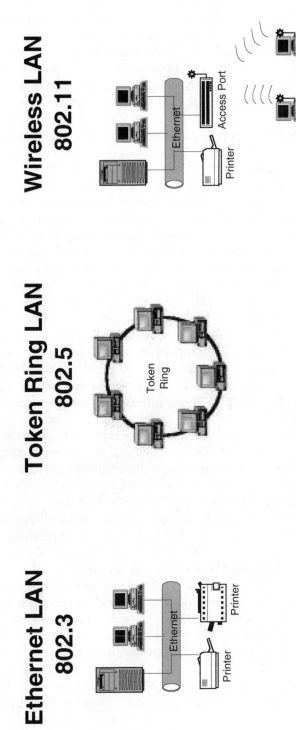

Figure 3.1   Types of local area networks.

Internet access in "airports, cafes, bars, and restaurants — anywhere one finds a surfeit of laptop-toting customers and a scarcity of telephone jacks. ... Offices and factories also use Wi-Fi technology to create private networks. With new phones from Cisco and Motorola, people are even using it to make cellular calls." [LAND03] This phenomenon is worldwide, with wireless Internet access available from "Amsterdam to Mount Everest to the Dead Sea." [HEIN03] "By the year 2004, the value of WLAN installations is expected to top $34 billion on sales of 12 million units per year. When you consider that WLAN devices were almost nonexistent until the year 2000, that growth is nothing short of astonishing." [CIAM03]

The IEEE 802.11a™ wireless standard — also known as Wi-Fi5 — has a maximum rated speed of 54 Mbps and, in non-standard mode, can support speeds up to 108 Mbps. This standard operates at a different frequency than other 802.11 standards and is not compatible with them. The IEEE 802.11b™ standard supports wireless transmission speeds of 11 Mbps and is accepted worldwide. It is the most widely used, particularly in an office environment. This trend is recognized by Microsoft, which provides built-in support for IEEE 802.11b™ in Windows XP, including automatic detection of wireless NICs (network interface cards) and installation of external software drivers for wireless network installation. IEEE 802.11g™ was approved as a standard in June 2003, and it supports transmission speeds up to 54 Mbps. IEEE 802.11b™ is an older, slower protocol, which also means that it is cheaper and more readily available in devices than IEEE 802.11g™. Because the total available bandwidth is shared among users — for example, in an airport or library offering 11 Mbps — the transmission capacity can be quickly gobbled up if lots of users send large numbers of files simultaneously. With the increasing popularity of wireless computing, there will be added incentive to adopt the new 802.11g™ standard because it offers an almost five-to-one improvement in bandwidth over 802.11b. IEEE 802.11g™ is backward compatible with 802.11b™, meaning that 802.11g™ devices can fall back to 802.11b™ speeds if need be, so that both types of devices can coexist on the same network.

A subset of the IEEE 802.11i wireless security specification — the Wi-Fi Protected Access (WPA) protocol — has been recently introduced to strengthen the weak Wired Equivalent Privacy (WEP) encryption standard — and its relatively easily hacked keys — previously built into Wi-Fi products. It is intended to secure 802.11 devices and to ensure interoperability. To be Wi-Fi certified, all devices must incorporate WPA. Cisco's wireless encryption scheme — LEAP — also offers significant improvements over the original version of WEP. These developments will further encourage adoption of wireless LANs. Wireless transmission will eventually become ubiquitous as it is embedded in an increasing array of new consumer and office products.

3.1.4 Similarities and Differences between WAN and LAN Design and Planning

WANs and LANs operate on different geographic scales and use different protocols, equipment, and facilities. LANs are usually private networks, while WANs can be either private or public. Designing a WAN, as compared to designing a LAN, is more complex and requires the consideration of more options (with respect to protocols, equipment, etc.).

One area where LAN and WAN design is converging is on the issue of how to interconnect the two. Bridges, routers, and brouters are typically used to interconnect LANs and WANs. *Bridges* are relatively inexpensive, as well as faster and easier to maintain than most routers. However, they are not suitable for large, complex networks because they can only interconnect networks running the same protocols. They also lack the ability to reroute traffic in the event of line failures and to filter traffic that may create security or performance problems. In contrast, *routers* interconnect networks running multiple protocols and support dynamic routing schemes to redirect traffic in response to changing network conditions. Routers can also be used to implement firewalls[15] to prevent unauthorized transmissions in a network. A *brouter* is a hybrid of a bridge and router, and performs the functions of both. When it cannot perform the function(s) of a router, it defaults to the function(s) of a bridge.

3.2 LAN Design and Planning

Until recently, the design, planning and implementation of LANs has been, to a large extent, an intuitive process. The purchase price of LAN equipment and bandwidth has been low in comparison with the cost of modeling and design packages, and an analyst's time spent using the tools. Connectivity and compatibility were of much greater importance than properly sizing servers and bandwidth requirements for satisfactory network performance. However, as traffic loads increase, LAN topologies are becoming more complex, and LAN performance evaluation is becoming more critical.

LAN design, visualization, asset management, planning, and design tools are a fairly new product area. Obstacles that organizations might encounter when considering these tools include a *lack* of required:

- Budget for tool acquisition
- Computer and other equipment needed to use the tools
- Data (e.g., performance statistics, etc.) for analysis
- Monitoring instruments installed on the network
- Know-how to implement and use tools
- Staffing and infrastructure to keep up with very rapid changes in the technology, which may make frequent updates necessary
- Product integration, necessitating separate products for the logical and physical LAN design
- Integration between monitoring and modeling tools

3.2.1 Criteria for Evaluating LAN Design and Modeling Products

To evaluate LAN design and modeling products, the following criteria should be considered. [TERP96]

3.2.1.1 System Requirements

The processing power should be sufficient to support the simulation techniques required. This can be very extensive for large segments or interconnected LANs.

3.2.1.2 Input Data Requirements

This involves determining what data is needed to conduct a "what-if" network evaluation. Input parameters can be grouped by LAN segment and by interconnecting parameters. Modeling parameters can be classified into two groups:

1. LAN segment parameters
2. Internetwork parameters

LAN segment parameters include:

- Sizes: average packet and measure sizes
- Protocols: lower, middle, and high level
- Application and network operating systems
- Measurements of the LAN capacity relative to the average background load
- Number of workstations on a remote LAN
- Speed of the segment

Internetwork parameters include:

- Network architecture
- Bridge, router, switch, and gateway transfer rates
- Lower level protocol on the interlink.
- Number of hops between two LANs
- Background load on links between LANs
- Throughput rates

3.2.1.3 Control Parameter Extensions

Users may want to change or extend the network model to examine the potential effects of new operating systems, unsupported protocols, and new transmission media. This criterion is used to evaluate the openness and programmability of the modeling process.

3.2.1.4 Technology Used

This impacts the accuracy that can be achieved in the modeling results. Queuing equations provide quick evaluations of expected performance ranges. Complex simulation allows modeling in greater detail, with higher accuracy. Some products support both techniques.

3.2.1.5 Applicability of Live Data

Once the LAN is running, LAN analyzers can be used to collect actual traffic data. Some performance models can read and import collected data to augment the modeling capabilities. This is useful in model calibration and validation, and in modeling the effects of traffic growth on the network's performance.

3.2.1.6 Post-Processing

Visualization and presentation of the results are essential for a proper interpretation of the findings. It is extremely important to be able to examine the modeling results from different perspectives, without completely rerunning the simulation. Graphics and colors are extremely useful in analyzing results, and should be part of the visualization tool package.

3.2.1.7 Animated Displays

This capability allows designers and planners to get a feel for the impact of certain modeling parameters, such as queuing delays at congestion points or collisions in certain LAN segments. Some products provide both a step mode and an automatic mode to support this type of visual display. This type of graphical support accelerates the evaluation process by highlighting potential performance bottlenecks.

3.2.1.8 Product Maturity

It useful to discuss the experiences that other users have had with the product (at user group meetings, etc.) to get a better idea of how easy it is to learn and use, and how well the product is supported. The integration of existing solutions for LAN segments and interconnected LANs is a positive indicator of product maturity.

3.2.2 LAN Performance Indicators and Modeling Parameters

Building models prior to implementing a real network helps save time and money. In complex and interconnected LAN environments, modeling and rapid prototyping are a necessity. In this context, management platforms are often used to integrate modeling tools and monitoring instruments to enable rapid prototyping. The management platform plays the role of a broker. The broker is a standard feature of the management platform. Performance management applications are also key components of network management platforms. Figure 3.2 gives an overview of the functionality provided by a management platform.

The performance indicators used to model actual and hypothetical LANs are the same, and are summarized in Table 3.1.

3.2.2.1 LAN Modeling Parameters

LAN performance indicators can be grouped into fixed, variable, and performance measurement metrics. [TERP96]

3.2.2.1.1 Fixed Metrics

■ *Transmission capacity.* The transmission capacity is normally expressed in terms of bits per second. Although the bit rate is fixed, the total capacity can be divided into multiple, smaller capacities to support different types of signals. One of the common myths regarding LAN transmission capacity is that Ethernet is saturated at an offered load (the actual data carried on the channel, excluding overhead and retransmitted bits) of 37 percent. Many studies have shown that Ethernet can offer a 10-Mbps data rate at distances of less than one kilometer with the CSMA/CD protocol.

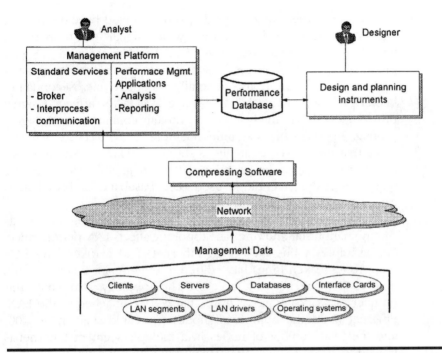

Figure 3.2   Network management platform.

Table 3.1   Overview of LAN Modeling Parameters
Fixed metrics:
Transmission capacity
Signal propagation delay
Topology
Frame/packet size
Variable metrics:
Access protocol
User traffic profile
Buffer size
Data collision and retransmission
Performance metrics
Resource usage
Processing delays
Throughput
Availability
Fairness of measured data
Communigram

■ *Signal propagation delay.* Signal speed is limited by the speed of light; the longer the distance over which signals are propagated, the longer the delay associated with their transmission. Signal propagation time is defined as the time required to transmit a signal to its destination. It is generally about five microseconds per kilometer. Therefore, cabling distance is a factor affecting signal propagation delay. In the case of satellite communication, signal propagation delay plays an influential role, as the distance between an earth station and the satellite is about 22,500 miles. Within LANs, the internodal signal propagation delay is negligible. However, the type of signaling technique used (i.e., baseband or broadband) can produce different levels of delay.

■ *Topology.* A LAN can be configured as a star, tree, ring, bus, or as a combination star and ring. The topology affects LAN performance. For example, a bus LAN (e.g., Ethernet) and a token ring LAN (e.g., IBM's Token Ring) have different built-in slot times (i.e., the time to acquire network access). The topology also limits the number of workstations or hosts that can be attached to the LAN. Ethernet limits the number of nodes per cable segment to 100, and the total number of nodes in a multiple-segment Ethernet is limited to 1024. A single IBM Token Ring supports 260 nodes. The higher the number, the greater the performance impact because all network traffic is generated from these nodes.

■ *Frame/packet size.* Most LANs are designed to support only a specific, fixed size of frame or packet. If the message is larger than the frame size, it must be broken into smaller sizes occupying multiple frames. The greater the number of frames per message, the longer the delay a message will experience. Like every LAN, Ethernet, for example, has a minimum packet size requirement which must not be shorter than the slot time (51.2 microseconds) to allow time for collision detection. This limit is equivalent to a minimum length of 64 bytes, including headers and other control bytes. Similarly, Ethernet has a maximum upper boundary in the packet size of 1518 bytes, in order to minimize access time.

3.2.2.1.2 Variable Metrics

■ *Access protocol.* The type of access protocol used by a LAN is an extremely influential metric affecting performance. IBM's Token Ring uses a proprietary token access control scheme in which a circulating token is passed sequentially from node to node to grant transmissions. A node must release a token after each transmission and is not allowed to transmit continuously on the single ring architecture. Ethernet, on the other hand, employs the 1-persistent CSMA/CD access control, in which a node that waits for a free channel can transmit as soon as the channel is free with a probability of 1 (i.e., 100 percent chance to transmit).

■ *User traffic profile.* A computer system and network is useless without users. There are many factors constituting a user's traffic profile: message/data arrival rate (i.e., how many key entries a user makes per minute); message size distribution (i.e., how many small, medium, and large messages are generated by a user); type of messages (to a single user, multiple users, or all receivers); and the number of simultaneous users (e.g., all active, 50 percent active, or 10 percent active).

■ *Buffer size.* A buffer is an area of memory allocation to receive, store, process, and forward messages. If the number of buffers used is too small, transmissions may suffer delays or be discarded. Some LANs have a fixed number of buffers, and some use a dynamic expansion scheme based on the volume of messages and the rate of processing on the LAN. In particular, LAN internetworking devices are likely sources of buffer problems.

■ *Data collision and retransmission:* Data collision is inevitable, especially in a bus LAN, unless the transmission is controlled in an orderly manner. Two factors must be considered: how long it takes nodes to detect a data collision and how long it takes to transmit messages. Various detection schemes are used by different topologies. For example, Ethernet employs a *jam time*. This refers to the time needed to transmit 32 to 48 more bits after a transmitting station detects a collision. This gives other stations time to reliably detect the collision. The time required to transmit data after a collision occurs is a significant determinant of LAN contention and delay. Many LANs use a binary exponential back-off scheme to avoid a situation in which the same two colliding nodes collide again at the next interval. Both collision detection and retransmission contribute delays to the overall processing delay. Generally, waiting time depends on network load and may become unacceptably long in extreme cases.

3.2.2.1.3 Performance Measurement Metrics

A single dimension cannot quantify or properly express the performance of a LAN. It is very difficult to interpret LAN metrics without knowing what applications (users) are involved. The following measurement metrics are recommended and are generally readily obtainable:

■ *Resource usage.* The processor, memory, transmission medium, and in some cases, peripheral devices all contribute to the processing of a user request (e.g., open a file, send a message, or compile a program). This metric measures how much of their respective capacities are used and how much reserved capacity remains. These values should be evaluated in conjunction with processing delay information, and compared with service level goals.

■ *Processing delays.* A user request is likely to suffer delays at each processing point. Both host and network introduce processing delays. Host delays relate to system processing and application processing delays. Network delays can be viewed as a combination of delays due to hardware and software processing. However, from the end-user perspective, the total processing delay (or response time) is the only meaningful performance metric.

■ *Throughput.* Transmission capacity can be measured in terms of throughput (i.e., the number of messages or bytes transmitted per unit of time). In LAN measurement, throughput is an indication of the fraction of nominal network capacity actually used to carry data. In general, packet headers are considered useful in estimating throughput, if no other measurement facilities are available, because the header contains the number of bytes in a frame. A metric related to throughput is *channel capacity*. Each transmission medium has a given maximum capacity (e.g., bits per second) that is a function of message volume and message size.

■ *Availability.* From an end user's point of view, *the network availability and its consistency determine service availability.* A network can be operational, but if a user suffers long delays, as far as the user is concerned, the network is virtually unavailable and is perceived as unreliable. Most LAN measurement tools are only able to measure availability (up- and downtime) because delay measurement may add several orders of magnitude to the measurement tool complexity.

■ *Fairness of measured data.* Because network traffic tends to be sporadic, the measured period and the internal data-recording rate are quite important. For example, an hourly averaged measured data rate may not reveal any performance bottlenecks, which a one-second recording rate might reveal. However, this recording rate will also generate an enormous amount of data, with commensurately large processing and storage requirements. As a general practice, a peak-to-average ratio is used when high activity data can be measured and collected in short time intervals. The ratio between the high activity periods and the average periods can be established for studying network capacity requirements.

■ *Communigram.* Traffic volume is used to quantify the traffic between communication partners. This measurement is greatly affected by the measured and reported intervals. An hourly average rate may not reveal performance bottlenecks that a smaller one may reveal. However, as discussed earlier, the recording interval should reflect an appropriate trade-off between collecting the required level of data without excessive detail. The ratio between the high activity periods and the average periods can be used for sizing resources supporting the communication between partners.

3.2.2.2 LAN Hubs as Source of Modeling Parameters

Hubs play a very important role in managing local area networks. This role is emphasized when hubs are used to collect and preprocess performance-related information that is used to build baseline models. This section lists some key indicators for Ethernet and Token Ring segments.

3.2.2.2.1 Ethernet Indicators

Hub level indicators provide information about the data transmission results collected on a logical hub. These statistics consist of multiple physical-level port statistics. Typical statistics required for diagnostics and performance analyses include:

■ Peak traffic in the segment within a specified time window
■ Average traffic in the segment for a specified period of time
■ Current traffic at the time of the last sample
■ Total packets received
■ Total bytes received
■ Missed packets
■ Number of cyclic redundancy check errors on the segment
■ Frame alignment errors on the segment
■ Collision rate in the segment for a specific period of time

3.2.2.2.2 Token Ring Indicators

Token Ring indicators consist of hard and soft error indicators, and general performance statistics. Hub vendors usually support most of the indicators below:

■ Hard error indicators:
  – Ring purges by the active monitor
  – Number of times input signal is lost
  – Beacons in the ring for a specified period of time

- Soft error indicators:
  - Number of line errors
  - Number of burst errors
  - Number of AC (access control) errors
  - Number of abort sequences
  - Number of lost frames
  - Number of receive data congestion errors
  - Number of frame copied errors
  - Number of token errors due to token loss or frame circulation
  - Number of frequency errors
- General indicators:
  - Cumulative number of bytes on the ring for a specified period of time
  - Frame count
  - Average and peak utilization levels
  - Average and peak frame rates
  - Bytes per second on average or peak
  - Average frame size for a specified period of time
  - Current and peak number of stations on the ring
  - Current operating speed of the ring

Modeling, design, and planning of LANs are based on monitoring all or at least some of these indicators. Principal data collection sources include:

- LAN analyzers and monitors
- SNMP MIBs
- RMON MIBs
- Device-dependent applications from equipment vendors

3.2.2.3 Modeling Objects in LANs

The targets or managed objects (MOs) are the same for performance analysis, tuning, modeling, and design. This section describes the major elements (a.k.a. elements or managed objects) used to model and analyze a LAN:

- LAN servers (especially the CPU and I/O devices they use)
- LAN drivers
- LAN interface cards
- LAN operating systems
- Workstations
- Peripherals

3.2.2.3.1 LAN Servers

LAN servers have a major impact on network performance. These impacts relate to whether the servers are high-end or low-end, and how they are loaded (i.e., how they are equipped with systems management, automatic server recovery, remote maintenance, predictive diagnostic software, etc.). Servers use Complex Instruction Set Computing (CISC) or Reduced Instruction Set Computing (RISC) chips and industry-standard PC-I/O buses, such as Extended Industry Standard Architecture (EISA), Micro Channel Architecture (MCA), and Peripheral Component Interconnect (PCI). Despite technological advances in CPUs and I/O devices, they are often the source of system bottlenecks.

Increasing the speed or number of CPUs provides only a partial solution to the problem of optimizing LAN performance. When it is not working, the CPU is in a *waiting state*. To minimize this time, high-performance servers use special cache designs, bus designs, memory management, and other architectural features to keep the CPU as busy as possible. These enhancements apply to servers that run in asymmetric multiprocessing (ASMP) mode (which dedicates individual CPUs to independent tasks), SMP mode (which enables multiple CPUs to share processing tasks and memory), and clustering mode (which enables CPUs on multiple servers to work in ASMP mode). To prevent the CPU from accessing main memory too often, high-speed cache memory can be installed. Measurement tests confirm that this results in CPU efficiency improvements of about 10 to 20 percent.

Another technique, called *pipelining*, is also used to prevent unnecessary CPU wait time while data is being transported from memory. This transport process requires a cycle of time on the CPU-to-memory bus. In a non-pipelined architecture, a second cycle is not started until the first one completes, and there is a time delay before the second cycle starts. In pipelined bus architecture, the second cycle begins before the first cycle completes. This way, the data from the second cycle is available immediately after the completion of the first cycle.

The vendor usually determines the size of the cache, and the LAN analyst is expected to set the systems parameters accordingly.

Cache systems are designed to keep the CPU supplied with instructions and data by managing access to main memory. However, bus controlled I/O devices also contend for access to main memory, and they run at a much slower speed than the CPU. Therefore, CPU-to-memory operations should take priority over I/O-to-memory operations, but not at the expense of interrupting these I/O operations. This is why most high-performance servers are engineered to let the CPU and I/O devices simultaneously access main memory by maximizing concurrency and minimizing contention.

Placing buffers between high-speed system buses and the I/O-to-memory bus achieves maximizing concurrency. These buffers capture data reads and writes between buses to prevent one device — such as a CPU or I/O card — from waiting for another to finish. Vendors use these buffers in segmented bus architectures that segregate different devices on various buses.

Disk I/O bottlenecks are often associated with high-volume transaction processing (because it involves moving many small transactions between the CPU and disks) and decision support systems (with many record moves). In these and similar cases, performance is affected by disk speed, the number of disk drives, and the intelligence and speed of drive array controllers. The number of disk drives in the system has a greater effect on server performance than the speed of individual drives. This is because of reduced latency in the positioning of drive heads and because more than one set of read/write heads may be active at a given time. As the number of drives increases, so does the performance of the drive array. There are still other ways to improve disk I/O performance. Intelligent array controllers support more than one disk channel per bus interface and implement support for multiple Redundant Array of Inexpensive Disks (RAID) hardware levels.

3.2.2.3.2 LAN Drivers

Other elements affecting LAN performance are software routines called *drivers*. Drivers accept requests from the network and provide the interface between the physical devices (e.g., disk drives, printers, and network interface cards) and the operating system. The drivers also control the movement of data throughout the network and verify that the data has been received at the appropriate address.

Drivers plan a critical role, and can severely impact the performance of the overall network when they have problems. Traditionally, drivers are supplied by LAN vendors and are tailored and sized according to the operating system in use.

Third-party software developers also provide customized drivers for networks. If a driver takes up too much RAM, other applications will have insufficient room in which to operate, causing them to alter their normal operating procedures to reduce memory use. Also, if the customized driver is not programmed compactly or efficiently, it may introduce network delay, increasing the effects of competition with other network requests (such as requests for printer services or requests from other users processing jobs). For these reasons, care should be taken when selecting and installing customized drivers.

Interface cards also affect LAN performance. Memory management is crucial to speed and performance. Factors such as DMA (direct memory access) versus shared memory, onboard processors, and buffers can make a significant difference in the actual network throughput. The performance difference, for example, between Ethernet cards can be as much as 50 percent.

3.2.2.3.3 LAN Interface Cards

When data from the CPU is sent to the network port of a disk, this can create bottlenecks and limit the number of users that can simultaneously make server requests. When the network is expected to support high-volume file and print services, video servers, and imaging systems, this difficulty should be anticipated, and high-performance, low-utilization network interface cards (NICs) should be used. The performance of server-to-LAN channels are affected by NIC driver optimization, the bus mastering capabilities of the controller, concurrent access to server memory, and the number of LAN channels per bus interface.

In cases where CPU utilization is an issue, a critical limit can be reached where placing additional NICs in the server will not improve the performance. This is due to the overhead associated with routing and servicing the NIC. The practical threshold is around three NICs per server. Vendors are working to improve server connections to high-speed technologies, such as ATM, FDDI, Frame Relay, and 100 Mbps Ethernet. They are also addressing the concept of placing servers on a dedicated high-speed LAN and of using switching to keep the server and end users from waiting on the network.

3.2.2.3.4 LAN Operating Systems

The LAN operating system is a prominent factor affecting the performance of a LAN. The LAN operating system performs many functions, including communication with the operating system of servers and clients, support of inter-process communication, maintenance of networkwide addressing, data movement within the network, file management, and control of input and output requirements of interconnected LANs. The more efficiently the LAN operating system is able to perform these tasks, the more efficiently the LAN overall will operate.

Network managers and IT procurers want high performance combined with low purchase costs and operating expenses, something that is not satisfied by many vendors. The LAN operating system market is concentrated on a few powerful products (such as NetWare, Windows NT, LAN Manager, Vines, and LANtastic).

3.2.2.3.5 Workstations

The performance of a workstation (a.k.a. client) has considerable impact on both the perceived system performance and actual system performance. Workstations frequently use the same operating system as the servers. When this is true, LAN coordination between the two is easier. In other cases (for example, a powerful file server operating on a 10-Mbps LAN in conjunction with an older PC running DOS with limited RAM), bottlenecks may occur because the workstation cannot accept or display data as fast as the file server and network hardware can supply it. At times, it is cheaper and more practical to upgrade the workstation, rather than the LAN itself. Adding more RAM or a co-processor to the workstation might improve the overall performance substantially. The protocol software used also affects the workstation performance. A full seven-layer OSI stack protocol implementation requires considerable resources to run. Even user-friendlier protocols can have significant effects on performance, depending on the packet sizes, transfer buffers, and address translation techniques they use.

Like the server and its operating system, the network workstation also affects the overall performance of the LAN. The workstation executes the network's protocols through its driver software; thus, a faster workstation will add to the performance of the LAN. One factor to consider is whether or not the workstation should contain a disk drive of its own. Obviously, a diskless workstation will ease the budget and improve security. But diskless workstations have their own set of costs. For one, these workstations depend on shared resources. If the work being performed at the station does not involve resource sharing, a workstation with its own disk may be more appropriate. Moreover, diskless workstations add to the traffic load on the LAN. This could be significant, especially if the workstations are for programmers who typically do not need to share files but who often work on files that are extremely large.

3.2.2.3.6 Peripherals

Printing requirements affect LAN performance in a variety of ways. Modern printers provide much more advanced printing capabilities than were available just a few years ago. Complete pages are transmitted all at once with improved fonts and high-end graphics. These printing capabilities, however, if not handled properly, can degrade network performance. If the user runs into such a performance problem, and if enough printers are available, redirecting the printing job to a local printer may help. It may also help to introduce another server dedicated to handling printer functions.

Another way to avoid bottlenecks caused by demanding printing requirements is to use a network operating system that incorporates a spooler to control these requirements. Spoolers are designed to accept a printing request from the network, and to logically, and in order, complete print requests without additional help from workstations.

Additionally, effective filtering across bridges and routers helps to reduce traffic volume on the network. If consistently high volumes indicate the need for LAN partitioning, bridges, routers, or virtual LANs can be used to alleviate the network congestion.

These examples relate to a client/server-oriented network. Other approaches might be needed in a peer-to-peer LAN and in interconnected LANs. In the latter, routers assume many responsibilities while directing traffic to targeted destinations. However, routers have inherent limitations on the number of packets, frames, and messages they can process and forward before they become overloaded and unable to perform their router functions well.

Figure 3.3 illustrates three possible operating states of routers:

1. Below A: normal performance
2. B: stress point
3. Beyond B in the area of C: degraded performance

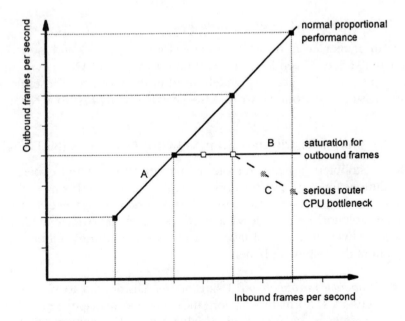

Figure 3.3   Router performance as a function of inbound traffic.

Other types of LAN performance bottlenecks may be related to:

■ Too many message collisions, due to too many workstations on the LAN or too many messages being sent
■ Bandwidth of the LAN is too low relative to the traffic
■ Excessive message transmissions due to high error rates
■ Message storms[16]
■ An uncontrolled chain of confirmations and reconfirmations in the LAN[17]

3.2.3 Information Sources for Baseline LAN Models

The LAN designer and planner should not start their efforts from scratch. In most cases, networks are running and can give measurement data about their actual performance. This could be used to perform a network optimization analysis, or to create a new design, or some combination of both. There are many information sources available to the LAN designer and planner, including monitors, SNMP-based managers, RMON-based information collection, and device-specific management applications. Each of these can provide the data needed to validate a baseline model for the network designer and capacity planner.

3.2.3.1 LAN Analyzers and Monitors

There are numerous instruments available for LAN analysis and monitoring through a variety of vendors. This section reviews two LAN analysis/monitoring products that are very widely used in the industry. These include Sniffer Distributed from Network Associates and *nGenius* from NetScout.

3.2.3.1.1 Sniffer Distributed from Network Associates [NETW04]

Sniffer Distributed provides remote monitoring (RMON) and expert analysis solutions for proactive network management and problem resolution. Sniffer Distributed is an enterprisewide network performance and management solution which allows a network manager to look at detailed network activity at each OSI layer, via remote monitoring. Sniffer Distributed offers the following benefits:

■ *Proactive management.* Using Sniffer Distributed to baseline the network's normal behavior, the network manager can identify incremental changes and problems in traffic congestion. Sniffer Distributed has built-in troubleshooting, alarm generation, and problem reporting capabilities to help network managers solve underlying issues before they have larger impacts on end-user productivity.

■ *Fast problem resolution.* Sniffer Distributed targets the underlying cause of problems on Ethernet and token ring LANs as well as bridged and routed internetworks. The network manager can look at detailed interpretations of more than 140 protocols. Sniffer Distributed identifies network problems and provides LAN managers with suggestions for corrective action to help expedite problem solving.

■ *Maximizes performance.* Sniffer Distributed reports a number of statistics from all seven OSI layers in real-time. This information helps keep the network running, enabling operators to solve problems before they cause downtime. Using historical statistics, Sniffer Distributed can help provide recommendations on server placement and segmentation to maximize network performance.

■ *Reduces network management costs.* Sniffer Distributed can be used to automate software distribution and to provide out-of-band serial support. This allows centralized monitoring and analysis network performance and quicker problem resolution.

The Sniffer Distributed products include:

■ *Sniffer Distributed Expert:* provides troubleshooting, fault management, and network performance management support for the network core. Expert Analysis provides three types of information: symptoms, diagnoses, and explanations to help quickly resolve performance issues and bottlenecks; protocol violations and nonstandard protocol activity; internetwork link problems; and physical layer problems (such as Ethernet and Token Ring lower-layer problems, including congestion and ring beaconing[18]).

■ *Sniffer Distributed RMON:* provides remote RMON monitoring capabilities. To protect investments in other RMON technologies, the Network Associates standards-based monitoring solution fully supports all the RMON MIB groups for Ethernet and Token Ring LANs.

■ *nPo Manager:* provides centralized management, administration, and security functions from a console and the collection of data from Sniffer Network Protection Platform components. It is a browser-based system.

■ *nPo Visualizer:* a Web-based network analysis and trend reporting tool for LANs and WANs based on ART (Application Response Time), Frame Relay, Expert, and RMON data collected by individual or multiple Sniffer Distributed instruments.

- *s400 Series Appliances:* These are modular, high-performance hardware platforms for collecting data for the purposes of monitoring and troubleshooting high-speed core networks. These appliances are designed to work in conjunction with the nPo Visualizer and nPo Manager.

- *Sniffer Portable/Sniffer Voice:* Sniffer Voice is a value-added package that integrates with the Sniffer portable product to monitor voice and video traffic. It is designed for use in a VoIP network. It is used to identify problems with jitter, packet loss, packet sequencing problems, and latency, and to generate the appropriate alarms.

- *Sniffer Wireless:* It is used to monitor and troubleshoot wireless mobile networks.

3.2.3.1.2 NetScout [NETS04]

NetScout is based on an architecture that provides an integrated structure for mapping data so it can be presented in a consistent context, regardless of where or how data is collected in the network. NetScout's *nGenius* solution is a network and application performance solution comprised of *nGenius* Performance Manager and *nGenius* Probes:

- *nGenius Performance Manager:* This is a software application that provides "enterprisewide perspectives on the interrelationship between the many applications and complex technologies deployed across the network." It also encapsulates business intelligence to help in the interpretation of and response to network data. *nGenius* Manager supports application monitoring, network monitoring, capacity planning, troubleshooting, fault prevention, and service level management.

- *nGenius Probes:* These are used to gather comprehensive network data and to perform network monitoring functions in the core, distribution, access, and storage areas of the network. NetScout offers specialized probes, such as the *nGenius* Gigabit Ethernet Multiport Probes, which are used to monitor critical core segments and switches. Probes are also available for the distribution and access network layers (such as Fast Ethernet networks, T3/E3 WAN/Frame Relay, and T3/E3 WAN/ATM networks). *nGenius* Flow Directors are available to provide security and network monitoring for Fast Ethernet, Gigabit Ethernet, ATM, POS (Packet-over-SONET), and other WAN technologies.

3.2.3.2 SNMP MIBs as Information Source

The Simple Network Management Protocol (SNMP) originated in the Internet community as a means for managing TCP/IP and Ethernet networks. SNMP's appeal broadened rapidly beyond Internet, attracting waves of users searching for a proven, available method of monitoring multivendor networks. SNMP's monitoring and control transactions are actually completely independent of TCP/IP. SNMP only requires the datagram transport mechanism to operate. It can, therefore, be implemented over any network media or protocol suite, including OSI.

Chapter 7 discusses SNMP in more detail. However, a brief summary of its key concepts, which include the manager, agents, and the management information base (MIB), is presented here and illustrated in Figure 3.4.

- An *agent* is a software program housed within a managed network device (such as a host, gateway, bridge, router, brouter, hub, or server). An agent stores management data and responds to the manager's request for this data.
- A *manager* is a software program housed within a Network Management Platform. The manager has the ability to query agents using various SNMP commands.
- The management information base (MIB) is a virtual database of managed objects, accessible to an agent and manipulated via SNMP to achieve network management.

SNMP's major advantages include:

- Its simplicity eases the vendor implementation effort.
- It requires less memory and fewer CPU cycles than CMIP, an alternative network management approach that implements the entire OSI stack.
- It has been used and tested on the Internet.
- SNMP products are widely available and are affordable.
- Development kits are available free of charge.
- It offers the best direct manager–agent interface.
- It is robust and extensible.
- Its polling approach is good for LAN-based managed objects.

SNMP has several disadvantages, including:

- It has weak security features.
- It demonstrates a lack of global vision.

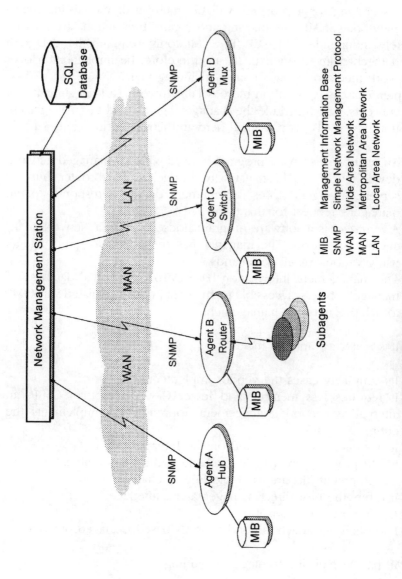

Figure 3.4 Architecture of SNMP-based network management.

- There are limitations associated with the Trap command (see Chapter 7 for details) relating to the type and method of data collection supported.
- Unique semantics exist that make integration with other approaches difficult.
- It has a high polling overhead.
- It allows too many private MIB extensions, thus leading to vendor incompatibilities.

3.2.3.3 RMON Probes and RMON MIBs as Information Sources

The Remote MONitoring (RMON) MIB helps bridge the gap between the limited services provided by management platforms and the rich sets of data and statistics provided by traffic monitors and analyzers. RMON is the definitive standard for network monitoring, and provides more comprehensive network fault diagnosis, planning, and performance tuning features than any other current monitoring solution.

The design goals for RMON include: [STAL93]

- *Offline operation.* To reduce overhead over communication links, it may be necessary to limit or halt polling of a monitor by the manager. In general, the monitor should collect fault, performance, and configuration information continuously, even if a manager is not polling it. The monitor should continue to accumulate statistics that may be retrieved by the manager at a later time, and may also notify the manager if an exceptional event occurs.
- *Preemptive monitoring.* If the monitor has sufficient resources and the process is not disruptive, the monitor can continuously run diagnostics and log performance data. In the event of a failure somewhere in the network, the monitor may be able to notify the manager and provide useful information for diagnosing the failure.
- *Problem detection and reporting.* Preemptive monitoring involves active probing of the network to determine the consumption of network resources and to check for error and exception conditions. Alternatively, the monitor can passively — without polling — recognize certain error conditions and other conditions, such as congestion and collisions, on the basis of the traffic that it observes. The monitor can be configured to continuously check for such conditions. When one of these conditions occurs, the monitor can log the condition and notify the manager.
- *Value-added data.* The network monitor can perform analyses specific to the data collected on its sub-networks, thus offloading the manager of this responsibility. The monitor can, for example,

observe which station generates the most traffic or errors in network segments. This type of information is not otherwise accessible to a manager that is not directly attached to the segment.

■ *Multiple managers.* An internetworking configuration needs more than one manager to provide network management capabilities to different units within the organization. Monitors can be configured to deal with more than one manager concurrently.

Chapter 7 discusses RMON MIB groups for Ethernet segments and RMON MIB groups for Token Ring segments in detail. At the present time, there are just a few monitors that can measure both types of segments using the same probe.

RMON is very rich on features, and there is the very real risk of overloading the monitor, the communication links, and the manager when all the details are recorded, processed, and reported. The preferred solution is to do as much of the analysis as possible locally, at the monitor, and send just the aggregated data to the manager. This requires powerful monitors. In other applications, the managers may reprogram monitors during operations. This is very useful when diagnosing problems. Even if the manager can define specific RMON requests, it is still necessary to be aware of the trade-offs involved. A complex filter will allow the monitor to capture and report a limited amount of data, thus avoiding overhead on the network. However, complex filters consume processing power at the monitor; if too many filters are implemented, the monitor will become overloaded. This is particularly true if the network segments are busy, which is probably the time when measurements are most valuable.

The most widely used RMON-1 is basically a MAC standard. It does not give LAN managers visibility into conversations across the network or connectivity between various network segments. The extended standard — RMON-2 — targets the network layer and higher. It gives the network manager the ability to peer into the network across the enterprise, up and down the protocol stack. With remote access and distributed workgroups, this can result in substantial inter-segment traffic. Functions supported by RMON-2 include:

■ Protocol distribution
■ Address mapping
■ Network layer host table
■ Network layer metrics table
■ Application layer host table
■ Application layer matrix table
■ User history
■ Protocol configuration

With RMON, more and more complete information can be collected for performance analysis and capacity planning. Remote monitoring allows continuous monitoring of LAN segments so that problems and performance bottlenecks can be highlighted in real-time or in near-real-time. An RMON implementation includes the following components: clients and servers. Clients consist of hardware/software components and are called either a *monitor* or a *probe*. The server is responsible for centrally collecting and processing monitored data. The server is usually implemented on a management platform and communicates with the clients using SNMP.

RMON offers multiple benefits, including:

- *Reduction in problem resolution time.* Monitors can be programmed for various problem conditions. If conditions are met and thresholds are violated, the management station is immediately notified. Trouble tickets are opened and dispatched to the workforce.
- *Reduction in outage time for the network and its components.* Actions to problem resolution are initiated earlier because information is available in real-time. Triggering is automatic, and support personnel do not need to be on-site.
- *Reduction in traveling expenses.* Portable monitors require technicians and engineers to travel to the LAN site. Using RMON probes, this is not necessary. This is very beneficial in networks with many distributed LAN segments.
- *Better scheduling for expensive engineers and technicians.* By allowing centralized support of multiple LAN segments, technical personnel can be more economically utilized. In addition, the quality of problem resolution is improved.
- *Availability of history data.* RMON probes provide detailed information on multiple indicators. At the central site, data can be compressed and maintained for future performance analysis and network design.

Certain considerations should be borne in mind when RMON is used:

- *It requires the TCP/IP protocol suite.* Most RMON products are based on TCP/IP. When legacy protocols (such as SNA, DSA, DNA, or Novell) are used, RMON probes cannot be used without protocol converters. However, protocol converters increase the complexity and cost of the network.
- *RMON-probes should be installed in each LAN segment.* To guarantee the best results, each LAN segment should be continuously monitored. Rotating RMON probes to take measurements on the LAN segments offers cost savings (on probes), but do not guarantee

complete data sets with which to manage and tune the network. RMON-1 is segment oriented, while RMON-2 offers cross-segment statistics. Additional RMON probes are needed if continuous performance measurement of LAN switches is also required.

■ *Cost of probes.* The price/performance ratio is continuously improving but the probes cost money. In the case of LANs with many single segments, dedicated probes are required. In general, probes can be used to continuously measure up to four segments in any combination of Ethernet and Token Ring.

There are three basic ways of implementing RMON probes in LAN segments, as shown in Figure 3.5:

1. Probe as stand-alone monitor
   – Advantages:
     ■ Excellent performance
     ■ Full functionality
     ■ Available in stackable or rack-mountable forms
   – Disadvantages:
     ■ High expenses for an average segment
     ■ Many probes are required for LAN switches with probe ports
     ■ The vendor must maintain stand-alone hardware
     ■ Advanced LAN technology support is slow
2. Probe is embedded in hubs and routers
   – Advantages:
     ■ Not much individual investment is required
     ■ Lower costs as compared to stand-alone probes
     ■ Switch integration is less expensive than installing probes on each switched segment
   – Disadvantages:
     ■ Hubs and routers should use the latest version to accommodate probes
     ■ Upgrades are not always economical
     ■ Performance is not always good
     ■ Standards conformance is not guaranteed
     ■ Probes may cause breakdowns and performance bottlenecks
     ■ To reduce performance impacts, not all RMON indicator groups are supported
     ■ Probes may come from different vendors
     ■ Integration with management platforms is not intuitive
     ■ Functionality is usually limited

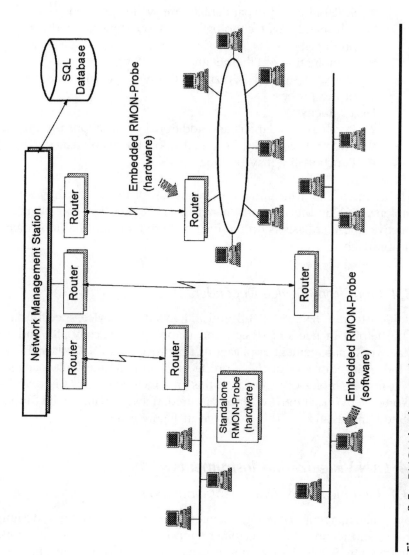

Figure 3.5   RMON implementation.

3. Probe is implemented as software module on a UNIX or personal computer workstation
   - Advantages:
     - Much lower cost in comparison to other alternatives
     - Performance is comparable with stand-alone probes
     - Scalability and expandability are well-supported
     - Advanced LAN technology is supported more rapidly in the products
     - Combination of Ethernet and Token Ring support is possible
     - Outband support is possible through special configuration of the probes
   - Disadvantages:
     - Purchase of adapters and additional workstations is required
     - The end user must be relied upon for maintenance
     - Functionality may be limited

It is assumed that a meaningful combination of these alternatives will be implemented by the network manager. From a design and planning perspective, the emphasis is on the data gathered, not on the implementation approach.

3.2.3.4 Input from Device-Dependent Applications

In addition to raw SNMP (MIB-II and RMON) data, device specific data may be helpful in network design and planning. Practically all vendors offer network management applications for processing SNMP-data. However, the main emphasis of these products is on real-time fault management. A few examples of products in this area include Cisco's CiscoWorks and Nortel Network's Optivity. Once captured, network data can be further processed and analyzed. This is shown in Figure 3.6.

3.2.4 LAN Visualization Instruments

3.2.4.1 Graphical LAN Documentation Tools

Change management, ongoing maintenance, troubleshooting, design, planning, and technical and customer support all need accurate network documentation. Tools with graphical capabilities are very useful in visualizing networks and preparing design alternatives. NetViz is a widely used example of this type of product. However, these tools are *not* considered design and planning tools, but rather visualization and documentation tools.

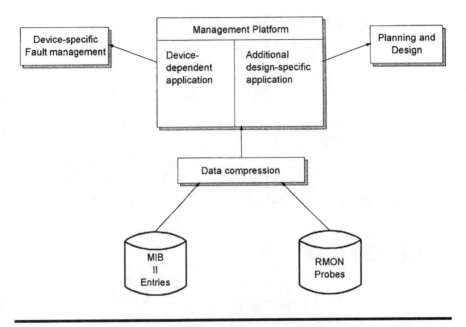

Figure 3.6   Data collection for network design and planning.

3.2.4.2 *netViz [NETVO4]*

To accurately and thoroughly document a network, information must be collected on the various technologies, geographical access, networking services, business applications, and users it supports. netViz supports complex data visualization, which reflects underlying relationships at different levels of abstraction and granularity. Traditional drawing programs and database management programs, even when used together, are often inadequate for documenting networks because they fail to effectively capture these interrelationships. netViz can be used for diagramming and documenting networks, systems, and processes. Networks can be divided into pieces and linked together to improve their manageability. A diagram demonstrating how netViz can be used to visualize multiple dimensions in a graphical representation is shown in Figure 3.7.

In summary, netViz provides a broad array of capabilities:

- Drag-and-drop graphics
- Graphics that are completely under user control
- Multi-level documentation capabilities
- Integrated data manager

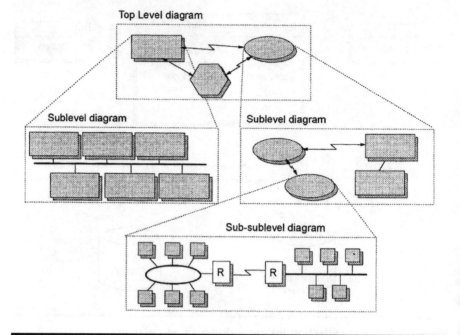

Figure 3.7 netViz: multi-dimensional data representation.

- On-screen data display
- Graphics export capability

This product is distinguished by:

- Object-oriented business graphics plus integrated dynamic data management
- Drag-and-drop simplicity in the creation or change phase
- Supports multi-level network topologies; by double-clicking on any node, subnetworks can be viewed in unlimited depth
- Imports text files so the user can use data stored in existing databases and spreadsheets
- Imports map and floor plan graphics for backgrounds; imports custom node symbols
- Exports diagrams and data for use in presentations and word processing; exports data for use in other programs

3.2.5 LAN Planning and Design Tools

Modeling, design, and capacity planning always start with the intelligent interpretation of measurement results. Performance visualization tools and

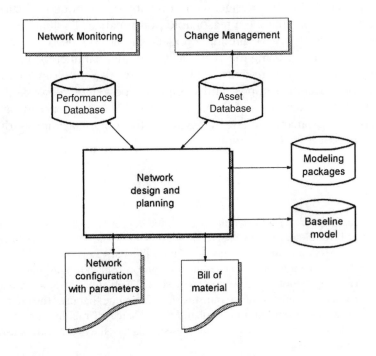

Figure 3.8   Integrated data collection for LAN planning and design.

topology layouts are often used to aid in understanding how the current network is functioning. Configuration management data, which deals with the maintenance of assets, often provides a source of data to help with network topology planning. Network configuration data collection is an especially demanding task due to the very dynamic nature of LANs and interconnected LANs. Tools of this category also encompass asset management, change management, and network monitoring. Figure 3.8 shows the interfaces of the network design and planning process to these other network management processes.

3.3 Summary

The number of alternatives for designing, sizing, and interconnecting LANs is continuually increasing. LAN design and planning should start with connectivity and compatibility considerations. Certain technologies and equipment cannot be combined with each other. Unsophisticated design instruments can filter out incompatible solutions.

After completing the design with compatible components, sizing the components helps guarantee reasonable performance. To quantify performance from the very beginning, models should be used to compute multiple alternatives using the "what-if" technique. Parameters that should be investigated in a sensitivity analysis include the topology, load estimates, configuration settings, and computing power of networked components. Network managers demand validated models that can be used to assist change management by assessing the potential impacts of new and/or modified LAN components.

Notes

1. Client/server: paradigm that defines clients (e.g., a program or computer) and how they are to be "served" by other network devices (e.g., a printer, etc.). A client can potentially be served by many servers.
2. IEEE 802 Standards: for more information on IEEE 802 LAN standards, the interested reader is encouraged to visit the Web site: http://www.ieee.org.
3. Baseband: a short-distance transmission technique that uses the entire cable bandwidth to send an unmodulated digital signal. Time division multiplexing is used to interleave multiple signals so they can be transmitted simultaneously.
4. Broadband: a transmission technique that is used to send a wide spectrum of frequencies simultaneously over long distances.
5. CSDA/CD (Carrier Sense Multiple Access/Collision Detection): an Ethernet protocol for transmission access specified in the IEEE 802.3 standard. In an Ethernet network, a device can transmit at any time. If another device attempts to transmit at the same time, a collision may occur, and the message frames are discarded. Each device waits for a random period of time before attempting to retransmit.
6. A slotted ring network supports unidirectional data transmission between nodes, using predefined circulating slots. Transmissions start at the beginning of a slot.
7. A registered insertion ring is a type of LAN topology in which each node contains a register that can temporarily hold a circulating packet. When a node is transmitting, it can hold other oncoming packets until it has completed its transmission.
8. Ethernet: local area networking protocol developed by Xerox and supported by a variety of vendors, including Hewlett-Packard, Intel, and others. It is defined by the IEEE 802.3 standard.
9. Token Ring: defined by the IEEE 802.5 standard, this protocol is derived from the IBM Token Ring protocol and is based on a token passing scheme that guarantees that message collisions cannot occur on the network. A small frame, called a token, is passed around the network. If a node receives a token, it is granted permission to transmit. If it has nothing to transmit, it passes the token on to the next node on the ring.

10. Fiber Distributed Data Interface (FDDI): a local area networking protocol for transmitting data at 100 Mbps speeds over optical fiber.
11. Thicknet: a 0.4-inch coaxial cable generally used for Ethernet backbone wiring.
12. Thinnet (a.k.a. ThinWire or Chapernet): a 0.2-inch diameter coaxial cable used for Ethernet wiring. It has the same impedance as Thickwire, and is cheaper and easier to install.
13. Twisted pair: a commonly used form of copper wiring for home and business use. Higher-grade twisted-pair wiring is used in LAN installations because it is cheaper than coaxial cable.
14. Mbps = 1,000,000 bytes per second
15. Firewall: router or access server used to filter transmissions for a private network for security purposes.
16. A message storm can arise when a broadcast message results in propagating a stream of responses, which in turn also propagates more responses. A severe message storm can essentially block out all other network traffic. Networks should be configured to block illegal broadcast messages to help avoid this problem. Illegal message storms are also called "denial of service" or DoS.
17. For example, in an electronic trade application, a user's transaction might be processed on his or her workstation, by a local LAN server, the server at the broker site, and by an SQL server that updates the user's database records. This may involve numerous message exchanges related to authentication, reconfirmation, and requests for user input.
18. *Ring beaconing* refers to the warning signal a Token Ring station sends to all other stations when it detects a hard error on the ring.

References

[CIAM02]   Ciamp, M., *Guide to Wireless Communications*, Course Technology, 2002, p. 181.

[HEIN03]   Heingartner, D., Roving the Globe, Laptops Alight on Wireless Hot Spots, *New York Times*, June 5, 2003, p. G4.

[LAND03]   Landler, M., With Wireless, English City Reaches across the Digital Divide, *New York Times*, Business Day, May, 31, 2003, p. C1, C3.

[NETV04]   NetViz Corporation, http://www.netviz.com/home/index.asp.

[NETW04]   Network Design and Analysis Corporation, 2004; http://ndacorp.com.

[TERP96]   Terplan, K., *Effective Management of Local Area Networks, 2nd Edition*, McGraw-Hill, New York, 1996.

Chapter 4

Intranets and Extranets

4.1 Management Overview

An *intranet* is a company-specific, private network based on Internet technology; and as such, it is a form of local area network (LAN). However, one of the major distinctions between traditional LANs and intranets is the reliance of the latter on TCP/IP, packet switching, and Internet technologies. In the case of the Internet, the technology is deployed over a public network; while in the case of intranets, the technology is deployed within a private network. An *extranet* blurs the line between the Internet and an intranet by making an intranet available to select external customers via the Internet. Unless otherwise noted, the term "intranet" is used to encompass both intranet and extranet deployments.

According to George Eckel, author of *Intranet Working*, one of the important benefits of intranets is that they provide a cost-effective vehicle for communication, because the expense of reaching one person or one million people is essentially the same. Intranets are the corporate world's equivalent of a town hall where people can meet, chat, and exchange information.

The emergence of intranets has changed the way companies communicate information throughout the organization. Information that used to be communicated via bulky paper documents or expensive satellite links can now be made available via a centralized server.

4.1.1 Benefits of Intranets

Intranets offer many potential benefits, including:

- Reduced operating costs
- Improved employee productivity
- Streamlined processing flows
- Improved internal and external communication
- New and improved customer service
- Cross-platform capability

Some of the ways these benefits can be achieved are discussed in the sections that follow.

4.1.1.1 The "Less-Paper" Office

Although a paperless office will probably never come to pass, one can certainly create an environment in which less paper is used — thanks to the intranet. Many companies find that intranets simplify corporatewide communications, and reduce printed material costs by eliminating the need for many paper-based processes. For example, some organizations offer complete manuals on their corporate Web site in electronic form, instead of distributing the information in printed form. Companies can benefit immediately from an intranet by replacing their printed materials, little by little, with electronic versions. Electronic media is cheaper to produce, update, and distribute than printed material. Oftentimes, printed material is out-of-date by the time it is distributed. Electronic documents, however, can be easily modified and updated as the need arises. Paper forms can also be distributed electronically, with a popular cross-platform standard like Adobe's PDF (Portable Document Format). Better yet, some intelligence can be embedded in the form, thus helping to reduce data entry errors. Although there are some tasks for which paper remains well-suited, paper usage can be greatly reduced by replacing paper-based forms with electronic forms.

4.1.1.2 Improved Customer Service

For many organizations, having the right information at the right time can make a significant difference in their ability to close a sale or meet a deadline. In today's competitive business environment, companies are under constant pressure to improve productivity while reducing costs. To achieve these productivity gains, companies must constantly improve their relationships with employees, customers, vendors, and suppliers. Intranets provide an important avenue for making these improvements.

Using an extranet, vendors, employees, and customers can access information as it is needed, thus alleviating the delays associated with mailing or distributing printed materials. For example, extranets have been used to:

- Distribute software updates to customers, thus reducing the need to distribute printed materials and floppy diskettes or CD-ROMs
- Process customer orders online
- Process and respond to customer inquiries and questions about products and services
- Provide customer self-service by making a FAQ (Frequently Asked Questions) file available, to help reduce expensive call center time spent answering the same questions again and again.
- Collect customer and survey data

Using an extranet, all these activities can be completed electronically in a matter of minutes.

4.1.1.3 Improved Help Desks

Intranets have been used to augment help desk services. For example, when someone in the organization learns about a new technology or how to perform a new task (e.g., running virus software), he or she can put information and instructions for others on a personal Web page. Others within the organization, including help desk staff, can then access this information as needed. In an organization empowered by an intranet, every employee can leave the imprint of his or her expertise. One method of communicating one's experience to a group of people is known as a "blog," which is short for "web log." There is software specifically designed to capture one's thoughts and make them available in this format.

4.1.1.4 Improved Corporate Culture

Intranets help to cultivate a corporate culture that encourages the free flow of information. Intranets place information directly in the hands of employees, promoting a more democratic company structure. The danger of "information democracy" is that once it is in place and taken for granted, management cannot easily revert to older, more controlled forms of communication without seriously damaging employee morale and cooperation. Every individual in an intranet environment is empowered to access and distribute information — both good and bad — on a scale heretofore unknown in the corporate environment.

Intranets dissolve barriers to communication created by departmental walls, geographical location, and decentralized organizations. Placing information directly in the hands of those who need it allows organizations to decentralize and flatten decision-making and organizational processes, while maintaining control over the information exchange. Individuals and groups can distribute ideas freely without having to observe traditional channels of information (i.e., an individual, a printed document, etc.) that are far less effective in reaching geographically dispersed individuals.

Some companies post periodic communications from executive leadership, helping employees understand high-level strategy, the corporate mission, and other important issues. They may even solicit and answer employee questions, further breaking down the walls that previously separated the executives from regular employees.

4.1.1.5 Cross-Platform Compatibility

Since the early 1980s, organizations with private networks have struggled with connecting and disseminating information between different types of computers — such as PCs, Macintoshes, and UNIX-based machines. To help manage potential barriers to electronic communication posed by hardware and software incompatibilities, many companies have instituted strict standards limiting corporate users to specific hardware and software platforms. This may have unforeseen consequences because a single platform is rarely the best at all tasks that an organization needs to perform. Macs still have a foothold in the graphic design and publishing arenas, disproportionate to their relatively small market share. UNIX machines have been known to run for months and years without requiring a reboot.

Intranets provide a means to overcome many of these software and hardware incompatibilities because Internet technologies (such as TCP/IP) are platform independent. Thus, companies using intranets no longer need to settle on one operating system because users working with Macintosh, PC, or UNIX-based computers can freely share and distribute information. The sections that follow explain why this is so. Care should be taken with respect to the use of browser-specific technologies, such as ActiveX controls, plug-ins, or scripting languages that are limited to a specific platform or browser, as this will limit the organization's ability to migrate to or support other platforms in the future.

4.1.2 Intranet Planning and Management

To implement an intranet, a company needs a dedicated group of Web servers, communications links to the intranet, and browser software. *Unfortunately, intranets do not come prepackaged and fully assembled.*

They require careful planning and construction if they are to be effective in meeting the needs of the organization. The subsections that follow discuss recommendations for planning and implementing an intranet.

4.1.2.1 Gaining Support

The first step toward a successful intranet implementation is to obtain companywide support for the project, including support from upper management. A quality presentation should be made to both management and staff to explain the benefits of the intranet project. Some of the benefits of the intranet are tangible and easy to measure, while others are intangible and difficult to measure. To gain widespread support for the intranet project, decision makers must be shown what an intranet is and how it will benefit the organization. There are many resources (including complete presentations) available on the World Wide Web to help promote the intranet in a corporate environment.

4.1.2.2 Planning The Intranet Strategy

After selling upper management on the idea of an intranet, the next step is to define the goals, purpose, and objectives for the intranet. This is an essential part of intranet project planning.

The intranet project plan should include an overview of the organizational structure and its technical capabilities. The current communication model used to control information flows within the organization should be examined with respect to its strengths and weaknesses in supporting workflow processes, document management, training needs, and other key business requirements. It is important to understand and document existing systems within the organization before implementing the intranet. Legacy systems may need to be combined with some type of middleware in order to be accessible via the intranet.

The intranet plan should clearly define the business objectives to be achieved. The objectives should reflect the needs of the potential users of the intranet. Conducting interviews with employees and managers can help identify these needs. For example, the Human Resources department may wish to use the intranet to display job opportunities available within the organization. If this need is to be satisfied, the intranet should be designed to display job information and job application forms on a Web server, so that applicants can apply for positions electronically. The Human Resources department might also want to offer employees the ability to change their 401(k) information using the intranet. Each goal identified shapes and defines the functionality that the intranet must support. An employee survey is also an excellent way to collect ideas on how to employ the intranet within the organization.

In summary, the following questions are helpful in defining the requirements of the intranet project:

- Will intranet users need to access existing (legacy) databases?
- What type of training and support will intranet users require?
- Who will manage, create, and update the content made available through the intranet?
- Will individual departments create their own Web pages autonomously?
- Will there be a central authority that manages changes to the content offered on the intranet?
- Do users need a way to access the intranet remotely?
- Will the intranet need to restrict access to certain users and content?
- Will a Web master or a team of technicians/managers be assigned to coordinate and manage the maintenance of the intranet?
- Will the intranet be managed internally, or will it be outsourced?

4.1.2.3 Selecting the Implementation Team

After developing and approving the intranet project plan, the implementation team should be assembled. If the organization does not have an infrastructure in place that is capable of implementing the intranet, additional staff and resources will need to be hired or the project will need to be outsourced to a qualified vendor.

It is important that the intranet team assembled has the requisite skills to successfully execute the project plan. Skill assessment checklists are provided below to help evaluate the resources available within an organization and their ability to successfully support the intranet implementation.

4.1.2.3.1 Technical Support Skills Checklist

The intranet project will require staff with the technical skills needed to solve network problems, understand network design, troubleshoot hardware and software compatibility problems, and implement client/server solutions (such as integrating network databases). Thus, the following network skills are required to support an intranet:

- Knowledge of network hardware and software
- Understanding of TCP/IP and related protocols

- Experience in implementing network security
- Knowledge of client/server operations
- Experience with custom programming
- Knowledge of database management

4.1.2.3.2 Content Development and Design Checklist

A typical organization has many sources of information: Human Resource manuals, corporate statements, telephone directories, departmental information, work instructions, procedures, employee records, and much more. To simplify the collection of information that will be made available through the intranet, it is advisable to involve people familiar with the original documentation and also those who can author content for intranet Web pages. If possible, the original authors of the printed material should work closely with the intranet content developers to ensure that nothing is lost in translation.

The following technical skills are needed to organize and present information (content) in browser-readable format:

- Experience in graphic design and content presentation
- Basic understanding of copyright law
- Knowledge of document conversion techniques (to convert spreadsheet data, for example, into a text document or PDF file for browser access)
- Experience in page layout and design
- Experience with Web browsers and HTML document creation
- Knowledge of image-conversion techniques and related software
- Knowledge of programming languages and programming skills
- CGI, PERL, ASP, or JSP programming and server interaction
- Human factors expertise, to make sure content is organized sensibly
- Familiarity with search and index facilities

4.1.2.3.3 Management Support Skills Checklist

As previously discussed, the company's management should be involved in the planning and implementation of the intranet. Ideally, management should have a good understanding of the intranet benefits, and the expected costs and timeframes needed for the project completion. Managers with skills relating to quality-control techniques, process-management approaches, and effective communication are highly desirable. Thus, the following management skills are recommended:

- Understanding of the organization's document flow
- Experience with reengineering processes
- Knowledge of quality-control techniques
- Awareness of the company's informal flow of information
- Experience with training and project coordination

4.1.2.4 Funding Growth

The initial cost of setting up a simple intranet is often quite low and may not require top-management's approval. In fact, all major operating systems offer some sort of Web server, so that any machine, with or without management approval, can serve content to others on the network. However, when complex document management systems are needed to integrate database access, automate workflow systems, and implement interactive training and other advanced features, the intranet should be funded with the approval of top management. To gain approval for the project, upper management must be convinced that the intranet is an integral part of the company's total information technology deployment strategy. This involves quantifying the tangible benefits of the intranet to the organization. Management also needs to understand how the intranet will change the way people work and communicate. As shown in Appendix A, an ROI analysis is an important step in justifying the economic benefits of the intranet project.

4.1.2.5 Total Quality Management

Effective deployment of an intranet often involves reengineering current process flows within the organization. Employees are usually most receptive to changes that make their jobs easier. To avoid perceptions that the intranet is an intimidating intrusion of yet another technology, it is advisable to involve staff as early on as possible in the deployment planning. This will facilitate the transition to the intranet and encourage employee participation in the intranet's success.

After migrating the company's work processes to the intranet, it is up to managers and employees to adhere to the procedures that have been put in place to improve productivity and teamwork. Management should not assume that because employees have a new tool — the intranet — this alone is sufficient to ensure that the desired attitudes and service levels will be attained. Instead, managers should view the intranet as one aspect of their quest for *Total Quality Management* (TQM) and *Six Sigma* methodology.

TQM involves creating systems and workflows that promote superior products and services. TQM also involves instilling a respect for quality throughout the organization. TQM and the successful deployment of intranets represent a large-scale organizational commitment fostered by upper management.

4.1.2.6 Training Employees

If employees are expected to contribute to the intranet, they will need to be given tools and training so they can author HTML documents. Many organizations have dedicated intranet staff for collecting, authoring, and distributing content requests, and adding them to the intranet as appropriate.

If one chooses to have employees or end users contribute content, they should be surveyed to determine if the tools they have been provided satisfy their needs. Many users find that creating HTML documents is difficult. If the users have difficulty creating HTML documents, the training efforts may need to be improved or modified. For example, this might involve training one person in each department to assume the responsibility of training the rest of the department. Patience and diligence are needed when introducing employees to new HTML authoring skills.

In summary, the following actions are recommended to help develop an effective program for training employees to author high-quality HTML documents, or to submit requests to those who can author the documents:

- Conduct a survey to assess user training needs and wants.
- Train users on tools and methods to develop HTML content. This includes providing users with HTML authoring tools that complement what they already know (e.g., the Internet Assistant for Microsoft Word is a good choice for users already familiar with Microsoft Word).
- Review the design and flow of material that will be "published" on the intranet.
- Provide feedback to HTML authors on ways to improve the site appearance and ease of use.

4.1.2.7 Organizational Challenges

In addition to technological challenges, companies may also face these organizational challenges after the initial release of an intranet:

- Offering a method for employees, technical and otherwise, to contribute content and useful information to the intranet
- Marketing the intranet within the organization so that all employees will support its growth and continued use

▪ Obtaining funding on an ongoing basis to implement new capabilities

▪ Encouraging an information-sharing culture within the company

▪ Merging a paper-based culture with the new culture of electronic documentation

▪ Ensuring that the content on the intranet is updated on a regular basis

▪ Preventing one person or group from controlling (monopolizing) the content of the intranet

▪ Training employees to author HTML content so they can contribute material to the intranet

▪ Training employees on intranet etiquette, thereby facilitating online discussion forums and other forms of user interaction on the intranet

▪ Using the intranet as an integral part of working with customers and vendors

▪ Measuring the intranet's overall effectiveness and contribution to the organization

As is the case when introducing any information technology to an enterprise, intranet deployment requires careful planning, effective implementation, and employee training. In the short term, most of the organizational focus is usually on the technical aspects of the intranet deployment. But as time goes on, organizational issues relating to how the intranet is used within the organization must be managed. When organizations actively examine and work toward resolving these issues, they are more likely to achieve a culture of teamwork and collaboration.

4.1.3 Intranets versus Groupware

An intranet can allow a group to collaborate and interact using standard protocols and low-cost, bundled or free tools. There is another class of tools called *groupware* that provides similar functionality. However, groupware is based on proprietary client and server tools. Some of the more popular products in the arena include Lotus Notes from IBM and Exchange from Microsoft. These tools typically integrate e-mail, scheduling, task management, instant messaging, and audio- and videoconferencing in a single package. These products can be expensive, and the cost of the solution is usually tied to the total number of users. While one can piece together a solution that contains free, standards-based software to provide these same services, groupware solutions are more tightly integrated, and easier to maintain and administer.

In light of the increased popularity and lower cost of standards-based solutions, groupware vendors have opened up their products. For example, both Notes and Exchange offer browser-based access to most, if not all, of their products' functionality. While a browser-based implementation may lack some functionality when compared to the proprietary client, it does allow easier access for home users, or those who are frequently on the road. Groupware also offers the benefit of centralized administration, which helps keep support costs down. When the groupware must support hundreds or thousands of users, the benefit of free standards-based solutions may be quickly offset by increased support costs. Another benefit of groupware is a centralized database infrastructure, which is easier to maintain.

4.1.4 Management Summary

This section summarizes key management considerations surrounding the use of intranets:

- An intranet is a company-based version of the Internet. Intranets provide an inexpensive solution for information sharing and user communication.
- An extranet is an extension of an intranet to select business partners. Current technology can allow one to share select internal content with authorized external users.
- Companies that have installed intranets have found that installation costs are generally low and system versatility is high.
- An intranet provides an easy way for users to communicate and share common documents, even if they are using different machines, such as IBM compatibles and Macintosh personal computers.
- Some organizations have expanded their intranets to allow customers to access internal databases and documents.
- Many companies can establish a functional intranet using in-house personnel with a minimal amount of new equipment.
- Intranet solutions are open and are shaped by competitive forces, whereas groupware products tend to be closed and proprietary. However, groupware solutions may offer lower support costs because they are more tightly integrated and easier to centrally manage.

Personal computers have moved computing power away from the Information Services department and into the hands of users. Today, intranet technology is taking this one step further, giving users even more control over the creation and distribution of information throughout the enterprise.

Internet technology adheres to open standards that are well-documented. This, in turn, encourages the development of cost-effective and easy-to-implement intranet solutions. As the popularity of intranets has increased, so has the demand for new tools and Web-based solutions. This demand has fueled competition among software manufacturers, which in turn has resulted in better and cheaper intranet products. Groupware products are already dropping in price because of the increased popularity of intranets.

In summary, intranets can be used to improve productivity, simplify workflows, and gain a competitive advantage over those who have yet to learn how to capitalize on them.

4.2 Technical Overview

4.2.1 Internet Basics

4.2.1.1 Packet Switching

Packet switching was introduced in the late 1960s. In a packet-switched network, programs break data into pieces, called packets, that are transmitted between computers. Each packet contains the sender's address, the destination address, and a portion of the data to be transmitted. For example, when an e-mail message is sent over a packet-switched network, the e-mail is first split into packets. Each packet intermingles with other packets sent by other computers on the network. Network switches examine the destination address contained in each packet and route the packets to the appropriate recipient. Upon reaching their destination, the packets are collected and reassembled to reconstitute the e-mail message.

4.2.1.2 TCP/IP

The U.S. Advanced Research Projects Agency (ARPA) was a major driving force in the development and adoption of packet-switched networking. The earliest packet-switched network was called the ARPAnet. The ARPAnet was the progenitor of today's Internet. By the early 1980s, ARPA needed a better protocol for handling the packets produced and sent by various network types. The original ARPAnet was based on the Network Control Protocol (NCP). In January 1983, NCP was replaced by the Transport Control Protocol/Internet Protocol (TCP/IP). TCP/IP specifies the rules for the exchange of information within the Internet or an intranet, allowing packets from many different types of networks to be sent over the same network.

4.2.1.3 Connecting to the Internet

One way to connect to the Internet is to install a link from the company network to the closest computer already connected to the Internet. When this method is chosen, the company must pay to install and maintain the communications link (which might consist of a copper wire, a satellite connection, or a fiber optic cable) to the Internet. This method was very popular with early adopters of the Internet, which included universities, large companies, and government agencies. However, the costs associated with installing and maintaining the communications link to the Internet can be prohibitive for smaller companies.

Fortunately, specialized companies — called Internet service providers (ISPs) — are available to provide a low-cost solution for accessing the Internet. ISPs pay for the (expensive) connection to the Internet, which they make accessible to others through the installation of high-performance servers, data lines, and modems. Acting as middlemen, the ISPs rent time to other users who want to access the Internet. Figure 4.1 illustrates how small offices and a schoolhouse might connect to the Internet using an ISP.

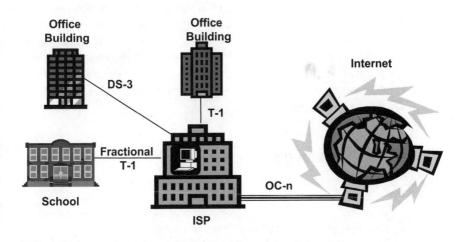

Figure 4.1   Internet access through an ISP.

Two important decisions must be made when deciding what type of Internet connection is the most appropriate. The first decision is the company budget allocated for Internet connectivity, and the second is the Internet connection speed needed to support the business requirements. Both decisions are interrelated. ISPs offer a variety of options for connecting to the Internet, ranging from a simple dial-up account over phone wires to high-speed leased lines from the company to the ISP. Dial-up accounts are typically available for a low, flat monthly fee, and are generally much cheaper than a leased line connection. However, the leased line connection is usually much faster than the dial-up connection.

When a dial-up account is used, a modem and a phone line are used to call and log in to the ISP server (or computer), which in turn acts as the doorway to the Internet. The transmission speed of the connection is limited by the speed of the modems used by the user and the ISP. A modem is not needed when a leased line connection is available to the ISP. Leased lines are available in many different configurations with a variety of options. The most common link types are ISDN, T1 (transmitting at speeds up to 1.54 Mbps), and T3 (transmitting at speeds up to 45 Mbps).

If a small company only needs to make an occasional connection to the Internet (e.g., less than 20 to 50 hours per month for all users), a dial-up account might suffice. However, if a company needs faster data transfer speeds or has several users who must access the Internet for substantial periods of time over the course of a month, a leased line connection should be considered. Line costs will vary according to the service provider and specific options the company has chosen to implement.

Principles of queuing analysis can be applied to the problem of sizing the links needed to support the Internet access, whether or not that access is to an ISP or to a direct Internet connection. The reader is referred to Chapter 8 for specific techniques on how to estimate the throughput and performance characteristics associated with using different size link capacities. This analysis can be used to determine whether or not a dial-up or leased line connection is sufficient to support the bandwidth requirements with tolerable transmission delays.

4.2.2 Basic Terminology

This section defines basic Internet-related terminology and nomenclature.

4.2.2.1 The World Wide Web

The World Wide Web — or Web — is a collection of seamlessly interlinked documents that reside on Internet servers. The Web is so named because it links documents to form a web of information across computers worldwide. The "documents" available through the World Wide Web can support text, pictures, sounds, and animation. The Web makes it very easy for users to locate and access information contained within multiple documents and computers. "Surfing" is the term used to describe accessing (through a Web browser) a chain of documents through a series of links on the Web.

4.2.2.2 Web Browsers

To access and fully utilize all the features of the Web, special software — called a Web browser — is necessary. The main function of a Web browser is to allow one to traverse and view documents on the World Wide Web. Browser software is widely available for free. The two most commonly used browsers are Netscape Navigator and Microsoft Internet Explorer. The same Web browser software used for accessing the Internet can also be used for accessing documents within an intranet.

4.2.2.3 Uniform Resource Locator (URL)

The Web consists of millions of documents that are distinguished by a unique name called a URL (Uniform Resource Locator), or more simply, a Web address. The URL is used by Web browsers to locate Internet information. Examples of URLs include:

- http://www.netscape.com
- ftp://ftp.microsoft.com

A URL consists of three main parts:

1. A service identifier (such as http)
2. A domain name (such as www.ups.com)
3. A path name (such as www.ups.com/tracking)

The first part of the URL (i.e., the service identifier) tells the browser software which protocol to use to access the file requested. The service identifier can take one of the following forms:

- *http://* — This service identifier indicates that the connection will use the Hypertext Transport Protocol (HTTP). HTTP defines the rules that software programs must follow to exchange information across the Web. This is the most common type of connection. Thus, when Web addresses start with the letters "http," it indicates that the documents are retrieved according to the conventions of HTTP.

■ *ftp://* — This service identifier indicates that the connection will use the File Transfer Protocol (FTP). This service identifier is typically used to download and copy files from one computer to another.

■ *gopher://* — This service identifier indicates that the connection will utilize a gopher server to provide a graphical list of accessible files.

■ *telnet://* — This service identifier indicates that a telnet session will be used to run programs from a remote computer.

The second part of the URL, the domain name, specifies which computer will be accessed when running server software. An example of a domain name is: www.tcrinc.com.

The final part of the URL, the pathname, specifies the directory path to the specific file to be accessed. If the pathname is missing from the URL, the server assumes that the default page (typically, the homepage) should be accessed. Large, multi-page Web sites can have fairly long pathnames. For example, the following URLs request specific pages within a given Web site:

■ http://www.microsoft.com/downloads/search.aspx?display-lang=en&categoryid=7
■ http://story.news.yahoo.com/news?tmpl=story&cid=514&e=5&u=/ap/20040419/ap_on_re_mi_ea/iraq_coalition

4.2.2.4 Home Pages

Information published on the Internet is usually organized into "pages." A home page is the first page that users see when they access a particular Web site. The home page should be appealing, concise, informative, and well-organized to succeed in maintaining the reader's interest. The home page is usually used to convey basic information about the company and what it is offering in the way of products or services.

Many companies publish the Internet address (known as a Uniform Resource Locator, or URL) of their home page on business cards, television, magazines, and radio. To access a Web site, a user has merely to type the URL into the address area on the Web browser screen.

4.2.2.5 Client Programs and Browsers

Across the Internet, information (i.e., programs and data) is stored on the hard disks of thousands of computers called servers. The servers are so named because, upon request, they serve (or provide) users with information. A server is a remote computer that can be configured to run several different types of server programs (such as Web server, mail server, and ftp server programs).

A client program is used to initiate a session with a server. Client programs are so named because they ask the server for service. In the case of the Web, the client program is the Web browser. All client/server interactions take the same form. To start, the client connects to the server and asks the server for information. The server, in turn, examines the request and then provides (serves) the client with the requested information. The client and server may perform many request-response interactions in a typical session.

The browser may launch other software when specialized content is required from an audio, video, or animation file. There are three major streaming media formats: Windows Media, Real Player, and QuickTime. One may need to install a special player to properly experience this type of content.

4.2.2.6 Where Web Documents Reside

When users publish Web pages, they store the pages as files that are accessible through a file server. Typically, Web pages reside on the same computer on which the server program is running, but this is not necessarily true. For security reasons, it may be necessary to limit accessibility to various files on the Web server. Obviously, it might be disastrous if internal documents and data were made available to competitors or to the general public. To prevent this type of security risk, a Webmaster (or systems administrator) can configure the Web server so it only allows specific clients to access confidential information, based on a need to know. The Webmaster can control access to the server by requiring users to log in with a username and password that has predetermined access privileges. This configuration is also known as an extranet.

4.2.2.7 HTML: The Language of the World Wide Web

The European Particle Physics Laboratory at CERN (in Geneva, Switzerland) developed HTML in the late 1980s and early 1990s. The Hypertext Markup Language (HTML) is the language of the World Wide Web. Every site on the Web uses HTML to display information.

Each Web document contains a set of HTML instructions that tells the browser program (e.g., Microsoft Internet Explorer) how to display the Web page. When you connect to a Web page using a browser, the Web server sends the HTML document to your browser across the Internet. Any computer running a browser program can read and display HTML, regardless of whether that computer is a Windows-based system, a UNIX-based system, or Macintosh.

If word processor formatted files — such as Microsoft Word — were used to create Web pages, only users with access to Microsoft Word would be able to view the Web page. HTML was designed to overcome this potential source of incompatibility. All users can access Web pages from their browser because all Web pages conform to HTML standards. An HTML Web page is a plaintext file (i.e., an ASCII text file) that can be created and read by any text editor. There are many software programs to convert document files to HTML equivalents, and many standard presentation and word processing packages offer built-in routines to convert a standard document into a Web-ready HTML file. This type of conversion might be helpful, for example, if you wanted to convert a Microsoft PowerPoint presentation into a set of HTML files for display on the Web. After HTML files are transferred to a Web site, anyone with a browser can view them.

HTML provides the browser with two types of information:

1. "Mark-up" information that controls the text display characteristics and specifies Web links to other documents
2. "Content" information consisting of the text, graphics, and sounds that the browser displays

The functionality of a browser can be extended by a scripting language such as JavaScript or VBScript. For example, a script could be used to verify the integrity of data on a Web form before it is submitted to a database. The scripting language and other extensions should be compatible with the entire user population.

4.2.2.8 Hypertext and Hyperlinks

Documents on the Web can be interconnected by specifying links (called hyperlinks) that allow the user to jump from one document to another. The Hypertext Markup Language (HTML) code, which drives all Web pages, supports Hypertext. Hypertext, in turn, supports the creation of multimedia documents (containing pictures, text, animation, sound, and links) on the Web.

Hyperlinks (or simply, links) are visually displayed on the Web pages as pictures or underlined text. When a user clicks on a hyperlink displayed on the browser screen, the browser responds by searching for and then loading the document specified by the hyperlink. The document specified in the hyperlink might reside on the same computer as the Web page on display or on a different computer on the other side of the world. Much of the Web's success has been attributed to the simplicity of the hyperlink point-and-click user interface.

4.2.2.9 FTP: File Transfer Protocol

The FTP (File Transfer Protocol) is a standard protocol for transferring and copying files from one computer to another. Depending on the configuration of the FTP server program, one may or may not need an account on the remote machine to access systems files. In many cases, one can access a remote computer with FTP by logging on with a username of "anonymous," and by entering one's e-mail address as the password. This type of connection is referred to as "anonymous FTP session."

After logging in to the remote FTP server, it is possible to list a directory of the files that are available for viewing or copying. The systems administrator determines which files can be accessed on the remote server, and who has access privileges. When system security is a major concern, the system administrator may require a specific username and password (as opposed to allowing an anonymous log-on procedure) to gain access to system files.

FTP is very useful in accessing files on the World Wide Web. Most browsers have built-in FTP capabilities to facilitate downloading files stored at FTP sites. To access an FTP site using a browser, type in the FTP site address, which is very similar to a Web address. For example, to access the Microsoft FTP site, the address "ftp://ftp.microsoft.com" would be entered into the browser address window.

4.2.2.10 Java

Java is a programming language released by Sun Microsystems that closely resembles C++. Java is a platform-independent language that can be used to create both client and server software. A client-side Java program (a.k.a. an applet) can be accessed via a Web browser, but offers much more functionality than a simple Web page. All a client needs is a version of a JVM (Java Virtual Machine) to execute Java code. JVM software is available with nearly all modern operating systems, including Windows, UNIX, and Mac OS. To check out the latest Java offerings, the interested reader should visit the following Web site: http://java.sun.com/.

Java can run on multiple platforms, but compatibility can still be an issue when one is working in a multi-platform or multi-OS environment. Even on the same platform and operating system, there could be compatibility issues if there are different versions of the JVM running. For an internal application, one can dictate which JVM should be used. For applications that are exposed to external users, whose platform and operating system are unknown, additional testing of the Java application must be done to ensure compatibility.

A server-side Java application has the advantage of being able to run on almost any platform, including Windows, UNIX, and Mac OS. This allows content providers to run an application on any available hardware. Server-side Java code can integrate with a Web page using JSP (Java Server Pages). Tomcat is a program that allows a Web server to understand both HTML and embedded Java code.

4.2.3 Intranet Components

This section provides an overview of the components necessary to create an intranet and their associated costs. The final selection of intranet components depends on the company's size, level of expertise, user needs, and future intranet expansion plans. A detailed cost analysis of an intranet deployment project is provided in Appendix A.

An Internet requires these basic components:

- A computer network for resource sharing
- A network operating system that supports the TCP/IP protocol suite
- One or more server computers that can run Internet server software
- Server software that supports HTTP requests from browsers (clients)
- Desktop client computers equipped with network software capable of sending and receiving TCP/IP packet data
- Browser software installed on each client computer

Note that if a company does *not* want to use an internal server, an Internet service provider (ISP) can be used to support the intranet. It is very common for organizations to use an ISP, especially when they have very little interest in maintaining a corporate-operated intranet server. ISPs are also used when the organizational facilities cannot support the housing of an intranet server.

In addition to installing the software and hardware components listed above, HTML documents must be developed to display information content on the intranet. The creation and conversion of documents to HTML format is very easy using commercial software packages, such as Macromedia Dreamweaver, Adobe GoLive, or Microsoft's FrontPage. Numerous third-party sources are also available to provide this service at a reasonable cost.

4.2.3.1 Network Requirements

An intranet is a computer network. For the purpose of this discussion, assume that a basic computer network is in place. Thus, the discussion that follows focuses on the hardware and software modifications that may be needed to support an intranet.

Most computer networks are local area networks (LANs). LANs are based on a client/server computing model that uses a central, dedicated computer — called the server — to fulfill client requests. The client/server computing model divides the network communication into two sides: a client side and a server side. By definition, the client requests information or services from the server. The server, in turn, responds to the client's requests. In many cases, each side of a client/server connection can perform both client and server functions.

Network servers are commonly used to send and receive e-mail, and to allow printers and files to be shared by multiple users. In addition, network servers normally have a storage area for server programs and to back up file copies. Server applications provide specific services, such as corporate e-mail.

A server application (or server process) usually initializes itself and then waits for a request from a client application. Typically, a client process will transmit a request (across the network) for a connection to the server, and then it will request some type of service through the connection.

A network interface card (NIC) is needed to physically connect a computer to the network. The NIC resides in the computer and provides a connector to plug into the network. Depending on the network, twisted-pair wiring, fiber optic, coaxial cable, or wireless access can be used to physically connect the network components. The network interface or wireless adapter card must be compatible with the underlying network technology employed in the network (e.g., Ethernet, Token Ring, or 802.11b).

The principles of queuing analysis and minimal cost design techniques are applicable to the design of intranets. These principles and specific guidelines on how to design and size the computer network and network components are described in detail in Chapter 8.

4.2.3.2 Network Operating Systems

The Internet supports connectivity between different hardware platforms running a variety of operating systems. In theory, there is no reason why an organization must stay with one type of machine or operating system when implementing an intranet. However, in practice, many organizations use only one network operating system to simplify the task of managing the network overall. One must balance several factors when selecting a network operating system, such as cost, maturity, effort required to install and maintain, and scalability. The primary choices for network operating systems include UNIX and Microsoft Windows.

4.2.3.2.1 UNIX

Many larger companies use UNIX-based machines as their primary business application server platform. UNIX is a proven operating system that is well-suited to the Internet's open system model. Unfortunately, learning how to use UNIX is not easy (although this matter is the subject of debate among some UNIX aficionados). Also, using a UNIX-based machine limits the choices available for developing interactive intranets and other software applications. Many programmers, for example, prefer to develop applications using Windows-based machines and programming languages, such as Microsoft's Visual Studio suite of tools.

4.2.3.2.2 Windows

Many companies choose Windows over UNIX because Windows is easier to install and administer. Windows provides a high-performance, multi-tasking workstation operating system. It also supports advanced server functions, including Web and FTP services, and communications with clients running under Windows, UNIX, or Macintosh operating systems. The latest version of Windows Server includes Internet Information Server (IIS) and the Internet Explorer Web browser. Microsoft designed the Internet Information Server so that it can be installed quickly and easily on a Windows computer. The Windows Server also comes with a built-in Remote Access Services feature that supports remote access to the intranet through a dial-up phone or network connection.

4.2.3.3 Server Hardware

Server machines run the network operating system and control how network computers share server resources. Large businesses with thousands of users typically use high-speed UNIX-based machines for their servers. Small and medium-sized companies normally use less-expensive, Intel-based machines running Windows or Linux.[1] A recent newcomer is Mac OS X, which runs on Apple hardware but has a UNIX core.

If UNIX is chosen on a non-Intel platform, such as HP PA-RISC or Sun UltraSPARC, one will typically pay more for the hardware, due to the huge economies of scale that come with Intel-based systems.

The decision to use a UNIX-based machine versus an Intel-based Windows machine as the intranet server is also influenced by maintenance costs. A non-Intel UNIX-based machine typically requires more resources to maintain a Windows-based machine. Hardware upgrades for Intel-based Windows machines are also cheaper than hardware upgrades for UNIX workstations.

One serious consideration when choosing between UNIX and Windows is reliability. Because UNIX is a relatively mature operating system, it is generally more stable than Windows, and can literally run for years without crashing. Windows is a relative newcomer, and due to constant product revisions, it tends to be less stable. Many Windows system administrators are quite familiar with the "blue screen of death" which appears when Windows is unable to function and crashes.

Principles of queuing analysis, presented in Chapter 8, can be applied to estimate the server capacity needed to support the anticipated traffic at the required throughput levels.

4.2.3.4 Web Server Software

A working intranet requires server software that can handle requests from browsers. In addition, server software is needed to retrieve files and to run application programs (which might, for example, be used to search a database or to process a form containing user-supplied information).

For the most part, selecting a Web server for an intranet is similar to selecting a Web server for an Internet site. However, Internet servers must generally handle larger numbers of requests and must deal with more difficult security issues. The performance of the Web server has a major impact on the overall performance of the intranet. Fortunately, it is fairly easy to migrate from a small Web server to a larger, high-performance Web server as system usage increases over time.

In general, stand-alone Web servers have been replaced by servers that are bundled with the operating system. Web servers handle many tasks that require custom programming such as seamless connection to databases, video and audio processing, and document management.

4.2.3.4.1 UNIX Web Servers

Currently, the most popular Web server is Apache. Apache was initially developed on UNIX but is available for other operating systems, including Windows. According to the Netcraft Web Server Survey, as of July 2003, Apache is being used on 63 percent of Web servers on the Internet, and is more popular than all other Web servers combined. Apache started as a continuation of the NCSA Web server, but eventually it took on a life of its own. The source code for Apache is freely available, and everyone is invited to fix bugs and add new features to benefit the entire Apache community. Apache is highly configurable and extensible with third-party modules, and can be customized by writing "modules" using the Apache module API.

4.2.3.4.2 Windows Web Servers

If cross-platform Web server software is not needed, Microsoft's Internet Information Server (IIS) should be considered. This server is used to drive Microsoft's Internet site, and it works well for a large organization. Currently, Microsoft offers this server free of charge. It also comes bundled with Microsoft's Windows Server software. IIS is easy to install and allows new users to be added to the intranet with minimal effort. IIS comes with an FTP server.

4.2.3.4.3 Macintosh Web Server

Apple's latest operating system, Mac OS X, is based on UNIX, and therefore it can support Apache. Apache is bundled with OS X.

4.2.3.5 Desktop Clients Running TCP/IP

In the early days of the World Wide Web, one had to install a TCP/IP stack on desktop computers because it was not included with the operating system. These days, all modern operating systems come bundled with some form of TCP/IP. Users of older operating systems should seriously consider upgrading to benefit from built-in TCP/IP and additional features that are a standard part of modern desktop operating systems.

4.2.3.6 Web Browsers

The final component needed to make a functional intranet is a Web browser. The operating system will determine what browsers can be used. On Windows, the standard choice is Internet Explorer, because it is bundled free of charge with the operating system. Other popular browsers include Netscape, Mozilla, and Opera. For Macintosh users, Internet Explorer is available but Microsoft's support of their Mac product line is inconsistent. Apple now bundles its own browser, Safari, which is based on open source code. Netscape, Mozilla, and Opera are available for Mac, Linux, and some other flavors of UNIX.

If the intranet supports only internal users, then IT staff can control the browser end-users use to access content. However, if the network must support access by outside vendors and/or customers, it may be difficult or impossible to control the type of browser used to view the site's content. This can have significant implications on the Web site design and the mechanisms employed for information display and retrieval.

Microsoft integrates its browser with the Windows operating system. Outside of the Windows world, a browser is always a stand-alone program. Time will tell if other operating systems will take the Microsoft lead and fold the browser into the operating system.

The following is a list of popular browsers, and where one can get more information on and download them:

- Internet Explorer: http://www.microsoft.com/windows/ie/default.asp
- Netscape Navigator: http://channels.netscape.com/ns/browsers/default.jsp
- Mozilla: http://www.mozilla.org/
- Opera: http://www.opera.com/
- Safari: http://www.apple.com/safari/

4.2.3.7 Intranet Component Summary

This section summarizes the basic components of an intranet, recapitulating some of the key concepts discussed hertofore.

- Intranets are based on a client/server network computing model. By definition, the client side of a network requests information or services and the server side responds to a client's requests.
- The physical components of an intranet include network interface cards (NICs), cables, and computers.
- Suites of protocols, such as TCP/IP and IPX/SPX, manage data communication for various network technologies, network operating systems, and client operating systems.
- Windows-based intranets are easier and less expensive to deploy than those that are UNIX based.
- Netscape and Microsoft provide both low- and high-end server software products designed to meet the needs of large, medium, and small organizations.
- Netscape's Navigator and Microsoft's Internet Explorer provide advanced browser features for intranet applications.

4.2.4 Intranet Implementation

4.2.4.1 Information Organization

After the physical components of the intranet are in place, the next step is to design the information content of the intranet or Internet Web pages.

This task involves identifying the major categories and topics of information that will be made available on the intranet. Information can be organized by department, function, project, content, or any other useful categorization scheme. It is advisable to use cross-functional design teams to help define the appropriate informational categories that should be included on the corporate Web site. The following types of information are commonly found on corporate intranet home pages:

- What's new
- Corporate information (history and contacts)
- Help desk and technical support
- Software and tools library
- Business resources
- Sales and marketing information
- Product information
- Human Resources-related information (benefits information, etc.)
- Internal job postings
- Customer feedback
- Telephone and e-mail directory
- Quality and system maintenance records
- Plant and equipment records
- Finance and accounting information
- Keyword search/index capability

4.2.4.2 Content Structure

After identifying the main topics of information to be displayed on the corporate Web page(s), the flow and manner of presentation on the intranet must be developed. Four primary flow models are used to structure the flow of presentation on an intranet Web site: linear, hierarchical, nonlinear (or Web), and combination information structures. These are illustrated in Figures 4.2, 4.3, and 4.4.

A linear information structure is similar in layout to a book, in that information is linked sequentially, page by page. When a linear layout is used, the Web pages are organized in a "slide show" format. This layout is good for presenting pages that should be read in a specific sequence or order. Because linear layouts are very structured, they limit the reader's ability to explore and browse the Web page contents in a nonsequential (or nonlinear) manner.

When a hierarchical layout is used to structure the information, all the Web pages branch off from the home page or main index. This layout is used when the material in the Web pages does not need to be read in any particular order. A hierarchical information structure creates linear

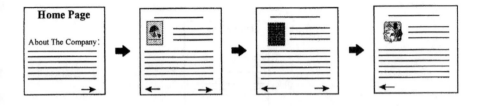

Figure 4.2   Linear hyperlink sequence.

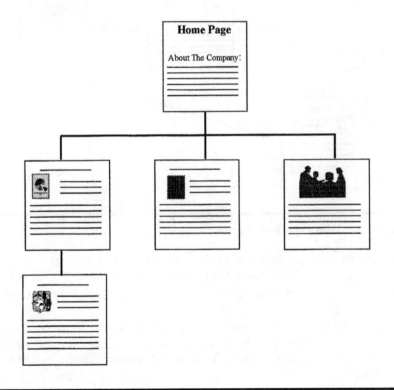

Figure 4.3   Hierarchical hyperlink sequence.

paths that only allow up and down movements within the document structure.

A nonlinear, or Web, structure links information based on related content. It has no apparent structure. Nonlinear structures allow the reader to wander through information spontaneously by providing links that permit forward, backward, up and down, diagonal, and side to side movement within a document. A nonlinear structure can be confusing,

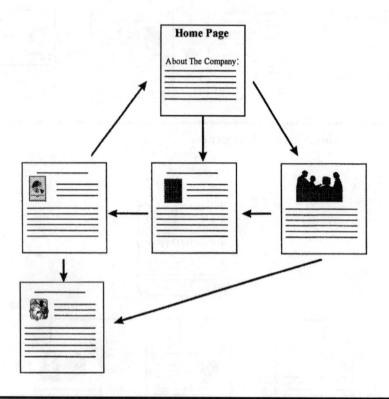

Figure 4.4 Web hyperlink sequence.

and readers may get lost within the content, so this structure should be chosen with care. The World Wide Web uses a nonlinear structure. The advantage of a nonlinear structure is that it encourages the reader to browse freely.

The combination Web page layout, as the name implies, combines elements of the linear, Web, and hierarchical layouts. Regardless of the type of flow sequence employed, each Web page typically has links that allow the user to move back and forth between pages and back to the home page.

Over the lifetime of the intranet, it is likely that the layout and organization of information on the corporate Web pages will change many times. It is often helpful to use flowcharting tools to help manage and document the updated information flows. Visio for Windows by Visio Corp. and ABC Flowcharter by Micrografx are two excellent tools for developing flowcharts. In addition, some of the Web authoring tools offer flowcharting and organizational tools to help design and update the information structure on the Web pages.

4.2.4.3 Interface Design

After defining the intranet's structure, the next step is to define the functionality and user interface. The intranet design should be consistent with the organization's corporate image. For example, items such as corporate images, logos, trademarks, icons, and related design themes add a familiar look and feel to the content. Where possible, they should be included in the Web page design. It is also advisable to work with the marketing department when designing the Web page layouts to ensure that a consistent theme is maintained in all company communications that will be viewed by the outside world.

A technique called "storyboarding" is frequently used to design the Web page layout. Storyboards are used by film producers, story writers, and comic strip artists to organize the content and sequence of their work. A storyboard depicts the content, images, and links between pages of the intranet in the form of a rough outline.

Software — such as Microsoft PowerPoint or another similar presentation program — can be used to develop a storyboard and sample Web pages. It is a good idea to test the interface design to ensure that the icons, buttons, and navigational tools are logical and intuitive. An intranet without intuitive navigational tools is like a road without signs. Just as it would be difficult for drivers to find their way from one city to another without the aid of signs, street names, and directional information, intranet users will find it difficult to retrieve information without easy-to-follow categories, buttons, and links. It is often helpful to employ graphic designers and marketing-communications staff to create effective graphics and images for the Web site.

4.2.4.4 Free Clip Art and Images

Many icons and navigational signs are available as clip art that comes with word processing, page layout, and presentation software programs. In addition, many Web sites offer images and clip art, that can be downloaded free of charge. However, the licensing agreements for downloading free images may have restrictions and requirements that should be observed.

4.2.4.5 Intranet Functionality

The required functionality of an intranet dictates many of the design and user interface features. One of the goals in designing an intranet should be to improve existing systems and infrastructures. After examining the current information structure, it may become clear which aspects of the structure work well and which ones need improvement.

Workflow processes, document management, and work collaboration are areas that the organization should strive to improve through the use of an intranet. A workflow analysis should consider ways in which the intranet can automate various organizational tasks and processes. For example, if a company has a geographically dispersed project team, the intranet might be used to post and update project information as various tasks are completed. Other team members could then visit the intranet page at any time to check the project status.

The following checklist is helpful when developing a list of the functions that an intranet must support:

Functionality Checklist

- The user interface must be intuitive.
- The intranet's design should support continuous updates.
- The intranet may need to be integrated with database management systems to allow users to access information (such as customer and product data).
- The intranet should support existing (legacy) applications, as needed.
- The intranet should have built-in directories, such as corporate telephone numbers and e-mail addresses.
- The intranet should incorporate groupware applications, if needed.
- Support (or future expansion) for online conferencing should be considered.
- The intranet should provide division-specific and corporatewide bulletin boards for electronic postings.
- The intranet should be designed with a document sharing and management process in mind.
- The intranet should foster teamwork and collaboration by enhancing channels of information distribution.
- Search engines — which simplify a user's ability to locate and access information — should be made available.
- The intranet should support electronic mail.
- The intranet should support (for future expansion) multimedia applications that use text, images, audio, and video.
- The intranet should be designed so it can interface, at least potentially, with factory equipment, other manufacturing devices, or other critical legacy systems.
- The intranet should support the automation of organization workflows.

4.2.4.6 Content Management

Many organizations struggle with the tasks of information creation, management, and dissemination. They are time consuming and difficult to control. The intranet alone cannot solve information management problems unless specific intranet solutions are implemented that directly address the need for document management. The following list identifies content-management tasks that should be considered in the intranet plan:

- Users must have the ability to easily add or update content on a regular basis.
- Users must have the ability to protect their content from changes by other users.
- A content-approval process should be defined and in place. This process should encompass ways to manage and control document revisions, especially changes to shared documents. It should also specifically address security-related concerns.

As policies and procedures relating to content management are formulated, it is important to designate responsibilities to specific individuals to ensure that they are put in place and followed. An intranet style guide should be developed that provides page layout, design elements, and HTML code guidelines. The style guide will help the organization to maintain a consistent look and feel throughout the intranet's Web pages. The style guide should contain information on where to obtain standard icons, buttons, and graphics, as well as guidelines on page dimensions and how to link to other pages. As part of the style guide, it is helpful to create Web page templates. These templates consist of HTML files, and are used to provide a starting point for anyone interested in developing Web pages or content for the intranet. Although it is very easy to create a working Web page and to publish it for mass viewing, the real challenge is in producing a well-conceived Web page. Large web-based projects can benefit from a source control system, such as Microsoft's SourceSafe, to help keep track of the various documents used to create a Web site.

4.2.4.7 Training And Support

After the intranet is up and running, efforts should focus on how to maintain the information content and on employee training. Part of the document-management strategy should encompass the selection of content stakeholders. Content stakeholders are individuals in different departments or workgroups who are responsible for the creation and maintenance of specific content. Stakeholders can be department managers, team leaders, or content authors and publishers.

Some organizations create a position called *Webmaster*. This position is responsible for maintaining and supporting the content published on the intranet. A good Webmaster should have the following skills:

- Basic Internet skills, including an understanding of e-mail, FTP, and Telnet
- A thorough understanding of HTML document creation
- Experience with CGI programming, or a similar scripting language
- Programming experience with languages such as Perl, PHP, or C/C++
- Skill using rich content creation tools such as Flash
- Familiarity with media technology such as Windows Media Player, Real Player, or QuickTime
- Experience with content creation and the conversion of text and images
- Knowledge of client/server processing
- Experience with server setup and maintenance
- Knowledge of the organization's structure and inner workings
- Organizational and training skills

It is possible that the organization may choose to decentralize the maintenance of the information content. In this case, individuals from various departments might be selected to maintain the content relating to their respective departments. These individuals should be trained to handle a variety of maintenance issues. A decentralized approach depends on having more than one individual with the necessary skills available to maintain the Web pages. A decentralized support structure gives authors and content owners direct control and responsibility for publishing and maintaining information. This can help prevent bottlenecks in making information available in a timely fashion.

Training for stakeholders, Webmasters, and intranet users is an important part of an intranet strategy. Intranet customers and content stakeholders should be trained to understand the intranet and how it will improve the organization and the way the company does business. They should also be given training on how to create, utilize, and maintain content on the Web page(s). Companies that invest in the education and training of their employees will have a better chance of creating and maintaining a successful intranet.

4.2.5 Intranet Deployment

Because intranets are easy to set up, many companies are not fully cognizant of the true resource requirements needed to maintain the intranet with up-to-date information. The goal of this section is to provide a realistic perspective on how organizations are most likely to achieve long-lasting benefits from the intranet.

4.2.5.1 Technological Considerations

The major technological challenges facing the organization after the initial implementation of an intranet include:

- Converting existing paper documents into electronic documents that employees can access electronically via the intranet
- Connecting existing databases to the intranet so they are accessible by a wide range of computing platforms (such as Windows- and Mac-based systems)
- Coordinating the use of multiple servers across departmental lines
- Continuously enhancing the intranet's features and capabilities
- Installing security features within the intranet to prevent unauthorized access to confidential or sensitive information

Intranet technology, and information technology in general, is changing so fast that keeping up with the latest software and hardware solutions requires a substantial ongoing organizational commitment.

4.2.5.1.1 Conversion of Paper Documents into Electronic Form

A key issue facing companies is how to convert large numbers of existing paper documents into electronic format ready for distribution on an intranet. One popular product for converting documents into a common format is Adobe Acrobat, which creates PDF documents. PDF readers and web browser plug-ins are freely available from Adobe. Microsoft's Internet Assistant for Microsoft Word can also be used to easily convert existing Word documents into HTML documents. After paper documents have been converted to PDF or HTML and placed on the intranet, the next challenge is to keep the documents up-to-date.

Obsolete information can frustrate intranet users and may encourage them to revert to old ways of information gathering (i.e., calling people, walking to various offices, and writing memos). One way to minimize this problem is to create a database containing the document title, date of last change, and frequency of update in a database. Other useful information that can be used to track the status and nature of documents on the intranet is shown in Table 4.1. A program can then be written to search the intranet for documents that have not been updated recently. The program

Table 4.1   Intranet Document Tracking Information

Data for Tracking Intranet Documents
Name of document
Document description
Page owner
Type of document (i.e., official, unofficial, personal)
Confidentiality status (i.e., confidential, non-confidential, etc.)
Original publish date
Date document last modified
Frequency of update (i.e., daily, weekly, monthly, etc.)

can then issue an e-mail to the document owner to request an update and to confirm the status of its content.

4.2.5.1.2 Interface to Legacy Database(s)

Companies may need to connect the intranet to legacy databases to access:

- Financial reports (regarding project costs, product costs, the overall financial health of the enterprise, etc.)
- Document-management systems
- Human Resources information (e.g., so employees can review information on health care and benefits)

Connecting databases to the intranet is not an easy task, and may require additional staff or reassignment of current programming staff. Legacy database vendors are currently working on various intranet solutions to facilitate the implementation of this requirement. A popular API used to allow interoperability between different databases is called ODBC (Open Data Base Connectivity). If a database offers an ODBC connector, any modern Web server can interact with it, using a scripting language such as Microsoft's ASP (Active Server Pages) or Java JSP (Java Server Pages).

4.2.5.1.3 Use of Multiple Servers

As the intranet becomes more complex, multiple servers may be needed. This is especially true for companies that have a large number of divisions and business units using the intranet. For example, a product development group may need to provide team members with the ability to search project-specific databases, submit forms to various databases, and to use a private online discussion group. The Webmaster may find it impossible to support these service needs in a timely manner. When this happens, companies frequently relegate the task of server maintenance to each respective department.

Fortunately, some form of Web server is included with all modern operating systems. When each department is responsible for maintaining its own Web server, it is particularly important to choose server software that is easy to install and maintain. A Pentium-class machine running Windows Server software and Microsoft's Internet Information Server is a good choice for small departments. Another way to provide departments with their own domain name and disk space is to use a virtual domain name. Companies use virtual servers to reduce hardware costs. In the case of the Web, an HTTP-based server runs on a server computer. For example, a company may need two types of Web servers, one that allows easy access and one that requires usernames and passwords. In the past, the company would have purchased two different computers to run the Web server software. Today, however, the company can run both servers on the same system — as virtual servers.

4.2.5.1.4 Standardizing Hardware and Software

To avoid supporting multiple hardware and software components, it is important to standardize the server software, hardware, HTML editing tools, and browser software. This will help minimize the potential for unexpected network errors and incompatibilities.

4.2.5.2 Maintaining the Information Content on the Intranet

One of the major challenges organizations must face is how to transition from paper-based systems to computer-based systems, while keeping information up to date.

4.2.5.2.1 Automating HTML Authoring

After establishing a policy for the distribution of intranet documents, it is advisable to develop a set of guidelines that clearly specifies who is responsible for keeping them current. Inaccurate information greatly reduces the effectiveness of the intranet. If employees lose confidence in the accuracy of the online information, they will revert to calling people to find information. Unfortunately, many people tend to ignore the need to update information, irrespective of its form (i.e., electronic or print).

In some cases, the intranet will contain information that employees must update daily, weekly, or monthly. Spreadsheets can be used to capture highly time-sensitive data. Major spreadsheets such as Microsoft Excel have a feature that will export data directly to HTML, ready for placement on a Web site.

4.2.5.2.2 Managing Document Links

In a traditional document-management system, documents often reference one another. In most cases, authors list the applicable references at the top of each new document. Intranets, unfortunately, create a situation in which organizations cannot easily control the accuracy of links in documents.

HTML document developers can use links freely and, in many cases, without checking the accuracy of those links. Even if employees test the initial accuracy of their document links, it is difficult to maintain and check the accuracy of those links after the document is released. If you have ever encountered a "broken" link when surfing the Web, you know that it can be frustrating. People depend on links within a Web document to find information. Today, however, there are a few mechanisms available to ensure the accuracy of document links. Employees must understand that other people can link to their pages, and that they should not freely move the location of their documents. Regular review of the log files created by the Web server can reveal any bad links or other pages that do not work as expected.

4.2.5.3 Centralized versus Distributed Control

The implementation of an intranet is a major change for any organization. Although change is not easy, people are more inclined to modify their behavior when leaders have a clear sense of direction and are able to demonstrate how the intranet will positively affect the employees' well-being. Managers should work with their employees to show that intranets can free them from routine aspects of their job. This, in turn, will allow employees to spend more time learning and being productive for the corporation.

Some of the benefits that can be obtained from a distributed model of intranet control include:

- Employees can tap into the knowledge of everyone in the organization, and thus empowering them.
- The span of control of any one Webmaster to dictate the intranet's form and function is limited.
- It empowers departments to create their own information databases, and facilitates new communication vehicles.

4.2.6 Intranet Security Issues

By their very nature, intranets encourage a free flow of information. This means that it is also very easy for information to flow directly from the intranet to the desktops of those who might seek to gain access to information they should not have. To guard against this situation, adequate security measures should be in place when the intranet is deployed. The discussion that follows reviews various security techniques to protect an intranet from unauthorized external and internal use.

4.2.6.1 Firewalls

The Internet was designed to be resistant to network attacks in the form of equipment breakdowns, broken cabling, and power outages. Unfortunately, when using the Internet today, a variety of security measures are needed to ward off attacks against user privacy and company security. One of the most essential security tools is the *firewall*. A firewall is a collection of hardware and software that interconnects two or more networks and, at the same time, provides a central location for managing security. It is essentially a computer specifically fortified to withstand various network attacks. Network designers place firewalls on a network as a first line of network defense. It becomes a "choke point" for all communications that lead in and out of an intranet. By centralizing access through one computer (which is also known as a *firewall-bastion host*), it is easier to manage the network security and to configure appropriate software on one machine. The bastion host is also sometimes referred to as a server.

The firewall is a system that controls access between two networks. Normally, installing a firewall between an intranet and the Internet is a way to prevent the rest of the world from accessing a private intranet. Many companies provide their employees with access to the Internet long before they give them access to an intranet. Thus, by the time the intranet is deployed, the company has typically already installed a connection through a firewall. In addition to protecting an intranet from Internet users, the company may also need to protect or isolate various departments within the intranet from one another, particularly when sensitive information is being accessed via the intranet. A firewall can protect the organization from both internal and external security threats.

Most firewalls support some level of encryption, which means data can be sent from the intranet, through the firewall, encrypted, and sent to the Internet. Likewise, encrypted data can come in from the Internet, and the firewall can decrypt the data before it reaches the intranet. Using encryption, geographically dispersed intranets can be connected through the Internet without worrying about someone intercepting and reading the data. Also, a company's mobile employees can also use encryption when they dial into the company system (perhaps via the Internet) to access private intranet files.

In addition to firewalls, a router can be used to filter out data packets based on specific selection criteria. Thus, the router can allow certain packets into the network while rejecting others.

One way to prevent outsiders from gaining access to an intranet is to physically isolate it from the Internet. The simplest way to isolate an intranet is to not physically connect it to the Internet. However, this is not practical or reasonable for people who must use the Internet.

Even without a connection to the Internet, an organization is susceptible to unauthorized access. To reduce the opportunity for intrusions, a policy should be implemented that requires frequent password changes and keeping that information confidential. For example, disgruntled employees, including those who have been recently laid off, can be a serious security threat. Such employees might want to leak anything from source code to company strategies to the outside. In addition, casual business conversations, overheard in a restaurant or other public place, may lead to a compromise in security. Unfortunately, a firewall cannot solve all these specific security risks.

It should be noted that a low-level firewall cannot keep viruses out of a network. Viruses are a growing and very serious security threat. Preventing viruses from entering an intranet from the Internet by users who upload files is necessary. To protect the network, everyone should use and update their anti-virus software on a regular basis. A server-based solution that scans all incoming e-mail may make sense for larger organizations.

The need for a firewall implies a connection to the outside world. By assessing the types of communications expected to cross between an intranet and the Internet, one can formulate a specific firewall design. Some of the questions that should be asked when designing a firewall strategy include:

- Will Internet-based users be allowed to upload or download files to or from the company server?
- Are there particular users (such as competitors) who should be denied all access?
- Will the company publish a Web page?
- Will the site provide Telnet or FTP support to Internet users?
- Should the company's intranet users have unrestricted Web access?
- Are statistics needed on who is trying to access the system through the firewall?
- Will dedicated staff be used to monitor firewall security?
- What is the worst-case scenario if an attacker were to break into the intranet? What can be done to limit the scope and impact of this type of scenario?
- Do users need to connect to geographically dispersed intranets?

There are three main types of firewalls: network level, application level, and circuit level. Each type of firewall provides a somewhat different method of protecting the intranet. Firewall selection should be based on the organization's security needs.

4.2.6.1.1 Network, Application, and Circuit-Level Firewalls

Network-Level Firewall — A network-level firewall is typically a router or special computer that examines packet addresses, and then decides whether to pass the packet through or to block it from entering the intranet. The packets contain the sender and recipient IP addresses, and other packet information. The network-level router recognizes and performs specific actions for various predefined requests. Normally, the router (firewall) will examine the following information when deciding whether to allow a packet on the network:

- Source address from which the data is coming
- Destination address to which the data is going
- Session protocol, such as TCP, UDP, or ICMP
- Source and destination application port for the desired service
- Whether the packet is the start of a connection request

If properly installed and configured, a network-level firewall will be fast and transparent to users.

Application-Level Firewall — An application-level firewall is normally a host computer running software known as a proxy server. A proxy server is an application that controls the traffic between two networks. When using an application-level firewall, the intranet and the Internet are not physically connected. Thus, the traffic that flows on one network never mixes with the traffic of the other because the two network cables are not connected. The proxy server transfers copies of packets from one network to the other. This type of firewall effectively masks the origin of the initiating connection and protects the intranet from Internet users.

Because proxy servers understand network protocols, they can be configured to control the services performed on the network. For example, a proxy server might allow FTP file downloads, while disallowing FTP file uploads. When implementing an application-level proxy server, users must use client programs that support proxy operations.

Application-level firewalls also provide the ability to audit the type and amount of traffic to and from a particular site. Because application-level firewalls make a distinct physical separation between an intranet and the Internet, they are a good choice for networks with high-security requirements. However, due to the software needed to analyze the packets and to make decisions about access control, application-level firewalls tend to reduce network performance.

Circuit-Level Firewalls — A circuit-level firewall is similar to an application-level firewall in that it also is a proxy server. The difference between them is that a circuit-level firewall does not require special proxy-client applications. As discussed in the previous subsection, application-level firewalls require special proxy software for each service, such as FTP, Telnet, and HTTP. In contrast, a circuit-level firewall creates a circuit between a client and server without needing to know anything about the service required. The advantage of a circuit-level firewall is that it provides service for a wide variety of protocols, whereas an application-level firewall requires an application-level proxy for each and every service. For example, if a circuit-level firewall is used for HTTP, FTP, or Telnet, the applications do not need to be changed. One simply runs existing software. Another benefit of circuit-level firewalls is that they work with only a single proxy server. It easier to manage, log, and control a single server than multiple servers.

4.2.6.2 Firewall Architectures

Combining the use of a router and a proxy server in the firewall can maximize the intranet's security. The three most popular firewall architectures are the dual-homed host firewall, the screened host firewall, and the screened subnet firewall. The screened-host and screened-subnet firewalls use a combination of routers and proxy servers.

4.2.6.2.1 Dual-Homed Host Firewalls

A dual-homed host firewall is a simple, yet very secure configuration in which one host computer is designated as the dividing line between the intranet and the Internet. The host computer uses two separate network cards to connect to each network. When using a dual-home host firewall, the computer routing capabilities should be disabled, so the two networks do not accidentally become connected. One of the drawbacks of this configuration is that it is easy to inadvertently enable internal routing.

Dual-homed host firewalls use either an application-level or a circuit-level proxy. Proxy software controls the packet flow from one network to another. Because the host computer is dual-homed (i.e., it is connected to both networks), the host firewall can examine packets on both networks. It then uses proxy software to control the traffic between the networks.

4.2.6.2.2 Screened-Host Firewalls

Many network designers consider screened-host firewalls more secure than dual-homed host firewalls. This approach involves adding a router and placing the host computer away from the Internet. This is a very effective and easy-to-maintain firewall. A router connects the Internet to your intranet and, at the same time, filters packets allowed on the network. The router can be configured so that it sees only one host computer on the intranet network. Users on the network who want to connect to the Internet must do so through this host computer. Thus, internal users appear to have direct access to the Internet, but the host computer restricts access by external users.

4.2.6.2.3 Screened-Subnet Firewalls

A screened-subnet firewall architecture further isolates the intranet from the Internet by incorporating an intermediate perimeter network. In a screened-subnet firewall, a host computer is placed on a perimeter network that users can access through two separate routers. One router controls intranet traffic and the second controls Internet traffic. A screened-subnet firewall provides a formidable defense against attack. The firewall isolates the host computer on a separate network, thereby reducing the impact of an attack on the host computer. This minimizes the scope and chance of a network attack.

4.2.6.3 CGI Scripting

Some Web sites provide two-way communications using CGI (Common Gateway Interface) scripting. For example, if one fills in a Web form and clicks the mouse on the form's Submit button, the browser issues a request to the server computer to run a special program, typically a CGI script, to process the form's content. The CGI script runs on the server computer, which processes the form. The server then returns the output to the browser for display.

From a security perspective, the danger of CGI scripts is that they give users the power to make a server perform a task. Normally, the CGI process works well, providing an easy way for users to access information. Unfortunately, it is also possible to use CGI scripts in ways that were never intended. In some cases, attackers can shut down a server by sending potentially damaging data through the CGI scripts. It is important to make sure that users cannot use CGI scripts to execute potentially damaging commands on a server. If security is a great concern, a more restrictive scripting solution, such as VB Script or JavaScript, may be appropriate.

4.2.6.4 Encryption

Encryption prevents others from reading your documents by "jumbling" the contents of your file in such a way that it becomes unintelligible to anyone who views it. You must have a special key to decrypt the file so its contents can be read. A key is a special number, much like the combination of a padlock, which the encryption hardware or software uses to encrypt and decrypt files. Just as padlock numbers have a certain number of digits, so do encryption keys. When people talk about 40-bit or 128-bit keys, they are simply referring to the number of binary digits in the encryption key. The more bits in the key, the more secure the encryption and less likely an attacker can guess your key and unlock the file. However, attackers have already found ways to crack 40-bit keys.

Several forms of encryption can be used to secure the network, including link encryption, document encryption, Secure Socket Layer (SSL), and Secure HTTP (S-HTTP). The following sections describe these encryption methods in more detail.

4.2.6.4.1 Public Key Encryption

Public key encryption uses two separate keys: a public key and a private key. A user gives his public key to other users so anyone can send him encrypted files. The user uses his private key to decrypt the files (which were encrypted with a public key).

A public key only allows people to encrypt files, not to decrypt them. The private user key (designed to work in conjunction with a particular public key) is the only key that can decrypt the file. Therefore, the only person that can decrypt a message is the person holding the private key.

4.2.6.4.2 Digital Signatures

A digital signature is used to validate the identity of the file sender. A digital signature prevents clever programmers from forging e-mail messages. For example, a programmer who is familiar with e-mail protocols can build and send an e-mail using anyone's e-mail address, such as Bill-Gates@microsoft.com.

When using public key encryption, a sender encrypts a document using a public key, and the recipient decodes the document using a private key. When a digital signature is used, the reverse occurs. The sender uses a private key to encrypt a signature, and the recipient decodes the signature using a public key. Because the sender is the only person who can encrypt his or her signature, only the sender can authenticate messages. To obtain a personal digital signature, one must register a private key with a certificate authority (CA), which can attest that one is on record as the only person with that key.

4.2.6.4.3 Link Encryption

Link encryption is used to encrypt transmissions between two distant sites. It requires that both sites agree on the encryption keys that will be used. It is commonly used by parties that need to communicate frequently with each other. Link encryption requires a dedicated line and special encryption software. It is an expensive way to encrypt data. As an alternative, many routers have convenient built-in encryption options. The most common protocols used for link encryption are PAP (Password Authentication) and CHAP (Challenge Handshake Authentication Protocol). Authentication occurs at the data-link layer and is transparent to end users.

4.2.6.4.4 Document Encryption

Document encryption is a process by which a sender encrypts documents that the recipient(s) must later decrypt. Document encryption places the burden of security directly on those involved in the communication. The major weakness of document encryption is that it adds an additional step to the process by which a sender and receiver exchange and receive documents. Because of this extra step, many users prefer to save time by skipping the encryption. The primary advantage of document encryption is that anyone with an e-mail account can use document encryption. Many document encryption systems are available on the Internet free of charge or for little cost.

4.2.6.4.5 Secure Socket Layer (SSL)

The Secure Socket Layer (SSL) was developed by Netscape Communications to encrypt TCP/IP communications between two host computers. SSL can be used to encrypt any TCP/IP protocol, such as HTTP, Telnet, or FTP. SSL works at the system level. Therefore, any user can take advantage of SSL because the SSL software automatically encrypts messages before they are put onto the network. At the recipient's end, SSL software automatically converts messages into a readable document.

SSL is based on public key encryption and works in two steps. First, the two computers wishing to communicate must obtain a special session key (this key is valid only for the duration of the current communication session). One computer encrypts the session key and transmits the key to the other computer. Second, after both sides know the session key, the transmitting computer uses the session key to encrypt messages. After the document transfer is complete, the recipient uses the same session key to decrypt the document.

In contrast with SSL, which creates a secure connection, S-HTTP creates a secure message transmission.

SSL VPNs are gaining dominance over traditional IPSec VPNs for remote access and extranet use because of its greater flexibility and scalability. IPSec was designed for site-to-site VPNs, while SSL was designed for site-to site connectivity anywhere over the Internet. SSL VPNs provide "client-less access to Web, client/server, and legacy applications, and to files from a single solution, and secure, anywhere access from any location. It seamlessly works over any network and traverses any firewall."[2] As Black-berries, PDAs, and other very thin-client applications become increasingly prevalent, this type of solution becomes very appealing. SSL VPNs also offer "significant savings in user administration and support costs because you don't have to deploy and manage IPSec clients and a higher level of security by eliminating direct network connections and providing granular access control."[2] Aventail Corporation specializes in SSL VPN solutions that are secure, easy to implement, clientless, and accessible anywhere with an Internet browser. For more information, the reader is referred to the Aventail Web site at http://www.aventail.com/company/vision.asp.

4.2.6.4.6 Secure HTTP (S-HTTP)

Secure HTTP is a protocol developed by the CommerceNet coalition. S-HTTP is less widely supported than Netscape's SSL. Because S-HTTP works only with HTTP, it does not address security concerns for other popular protocols, such as FTP and Telnet.

S-HTTP works similarly to SSL in that it requires both the sender and receiver to negotiate and use a secure key. Both SSL and S-HTTP require special server and browser software to perform the encryption. In contrast with SSL, which creates a secure connection, S-HTTP creates a secure message transmission.

4.2.6.5 Intranet Security Threats

This section examines additional network threats that should be considered when implementing intranet security policies.

4.2.6.5.1 Source-Routed Traffic

Packet address information is contained in the IP packet header. When source routing is used, an explicit routing path for the communication can be chosen. For example, a sender could map a route that sends packets from one specific computer to another, through a specific set of network nodes. The roadmap information contained in the packet header is called "source routing," and it is used mainly to debug network problems. It is also used in some specialized applications. Unfortunately, clever programmers can also use source routing to gain (unauthorized) access into a network. If a source-routed packet is modified so that it appears to be from a computer within your network, a router will obediently perform the packet routing instructions, permitting the packet to enter the network, *unless special precautions are taken*. One way to combat such attacks is simply to direct the firewall to block all source-routed packets. Most commercial routers provide an option to ignore source-routed packets.

4.2.6.5.2 ICMP Redirects (Spoofing)

ICMP (Internet Control Message Protocol) defines the rules routers use to exchange routing information. After a router sends a packet to another router, it waits to verify that the packet actually arrived at the specified router. Occasionally, a router may become overloaded or may malfunction. In such cases, the sending router might receive an ICMP-redirect message that indicates which new path the sending router should use for retransmission.

It is fairly easy for knowledgeable "hackers" to forge ICMP-redirect messages to reroute communication traffic to some other destination. The term "spoofing" is used to describe the process of tricking a router into rerouting messages in this way. To prevent this type of unauthorized access, it may be necessary to implement a firewall that will screen ICMP traffic.

4.3 Summary

Intranets are being used to improve the overall productivity of organizations. Important intranet concepts discussed in this chapter are summarized below:

- TCP/IP was created because of the need for reliable networks that could span the globe. Because of its reliability and ease of implementation, TCP/IP has become the standard language (or protocol) of the Internet. TCP/IP defines how programs exchange information over the Internet.
- An intranet is based on Internet technology. It consists of two types of computers: a client and a server. A client asks for and uses information that the server stores and manages.
- Telnet, FTP, and gopher are widely used network programs that help users connect to specific computers and to transfer and exchange files.

- The World Wide Web (or Web) is a collection of interlinked documents that users can access and view as "pages" using a special software program called a browser. The two most popular browser programs are Netscape Navigator and Microsoft Internet Explorer.
- HTML (Hypertext Markup Language) is used to describe the layout and contents of pages on the Web.
- Java is a new computer programming language that allows users to execute special programs (called applets) while accessing and viewing a Web page.
- A network computer is a low-cost, specialized computer designed to work in conjunction with the Internet and Java application programs.

To be effective, the intranet must deliver quality information content. To ensure this, management must play a proactive role in assigning staff who will keep the corporate information reservoirs on the intranet current and relevant. The following is a checklist of some of the ways to encourage the development of a high-quality intranet:

- Give users access to document-management systems and various corporate databases.
- Distribute the responsibility of maintaining the intranet to increase the number of staff involved in developing and enhancing intranet content.
- Create a corporate culture based on information sharing.
- Include employee training as part of the intranet deployment strategy.
- Design and implement appropriate security measures as soon as possible to minimize the potential for security breaches.
- Use firewalls to control access to the network.
- Use anti-virus software and update it regularly.
- Implement a security plan that controls the access that employees and outsiders have to the network and internal systems and data.
- Design and implement CGI scripts and applets with security in mind.
- Encourage users to encrypt files before sending confidential data across the Internet or intranet.

Note

1. An entry-level server-class machine can be purchased for a few thousand dollars.
2. Aventail Corporation, http://www.aventail.com/company/vision.asp.

Chapter 5

Client/Server and Distributed Networking

5.1 Overview of Client/Server and Distributed Object Networking

Client/server networking distributes system processing across multiple computing devices. The ultimate client/server network is the Internet, which supports millions of distributed processes, operating across disparate time zones, geographic locations, and organizational boundaries. This chapter presents the major forms of client/server architecture, including two-tier, three-tier, and n-tier distributed network models.

In legacy mainframe networks, there is a tightly controlled and synchronized master–slave relationship between the central computer and end-user devices. In contrast, client/server networks distribute processing functions between clients, which initiate and request services, and servers, which wait for and respond to client requests. Clients and servers are more loosely coupled, interacting through message exchanges. Thus, servers can be modified without disrupting clients, provided that the messaging interface between the two is not altered, and vice versa. This loosened coupling through a standardized, predetermined interface also provides the basis for allowing various software and hardware platforms to "peacefully" coexist in a client/server network. The major client/server networking standards (i.e., CORBA, COM+, and Web Services) define message interface conventions, and are distinguished by the degree to which they couple client and server implementation.

Distributed networks are a specialized form of client/server network. In a distributed network, no distinction is made between the client or server program. No longer is the server a passive participant that must wait for a client's initiative. Rather, the clients and servers are interacting peers. In a distributed environment, peer programs can initiate, use, or provide services to other peer processes operating on the network.

Middleware is essential in maintaining the operation of high-volume, mission-critical applications in a distributed network environment. Middleware sits between and manages interactions between processes or components on a distributed network. Middleware can be purchased from a variety of vendors, or it can be implemented as a proprietary, custom software solution. Except in very specific, demanding applications, commercial solutions are preferable to proprietary solutions, due to the inherent complexity of the tasks that middleware must support. Middleware is commonly used to (1) provide communication control relating to network performance, security, fault management, recovery, and other network management functions; (2) manage high-volume transaction processing, such as database updates, recovery, and data conversion; and (3) support messaging services to allow interoperability and communication between different applications running on the network. CORBA, COM+, and Web Services represent alternative types of middleware solutions, each with differing philosophical approaches and operational repercussions. A discussion of the trade-offs and impacts involved in the selection of middleware appears in more detail later in this chapter.

In theory, client/server applications and networks scale well and are suitable for both small and large user environments. For example, a small business or doctor's office might use a client/server application on a single machine. As scope of the system usage increases, the office might want to link together many — perhaps even thousands of — computers to access and process information. This scalability is supported in a client/server network, which easily permits changes in the numbers and types of clients as long they do not violate or exceed the network design specifications. Similarly, the number and types of servers used on the network can be upgraded or increased with relatively minimal impact on other network components.

Migration from centralized, legacy systems to a client/server network environment generally involves either *upsizing* or *downsizing*. Upsizing refers to the migration of single-user applications operating on a personal computer to a multi-user client/server platform. Upsizing is often prompted when the scope of an application expands from a single user to multiple users, and a single workstation cannot support the requirements for data/file access and transmission. For example, some applications left over from the early days of the personal computer fall into this category. The development tools of the time — such as the DOS version of Clipper — did not support a multi-user user environment, and had very inefficient file and data handling capabilities. Many thousands of stand-alone departmental systems developed outside the domain of the mainframe or MIS department have since become candidates for client/server conversion.

Given Microsoft's dominance in the industry, it is not surprising that many companies have invested heavily in Windows desktop applications created using such tools as Visual Basic. When migrating these applications to a client/server environment, these companies are likely to favor Microsoft solutions as a way to preserve their investment. However, this is but one approach among many. There is considerable diversity in the types of client applications in use, ranging from custom to prepackaged solutions running on a variety of operating systems and network protocols. Similarly, there are numerous custom and COTS (commercial-off-the-shelf) products for developing and running server applications (such as SQL database servers, Web servers, groupware servers, TP monitors, and object application servers).

Downsizing refers to the migration of applications from a mainframe to a cheaper client/server platform, such as a personal computer. Downsizing involves separating the application functionality into independent but interacting client and server programs. The three major approaches to downsizing are:

1. *Platform migration.* This involves moving a centralized mainframe application to a client/server platform, while preserving as much of the legacy code as possible. Oftentimes, the application software and user interface must be extensively reengineered to support the client/server paradigm, so this approach may involve substantial new program development and network infrastructure.
2. *Wrappering.* This involves encapsulating existing legacy application code inside distributed objects or client/server programs that are, in turn, installed on various remote computers.
3. *New application development.* Unlike the previous two downsizing approaches, which seek to preserve at least some portion of the existing application code, this approach involves scrapping old code and writing all new software specifically tailored to the application and client/server networking requirements.

Although there are numerous advantages to client/server networking, as evidenced by its widespread adoption, it is not without its attendant drawbacks. Perhaps foremost among these is the potential for security breaches. Security is inherently more difficult in a distributed environment, simply because there are more potential entry points from which the system might be compromised. Similarly, as more users access system resources from different types of devices and locations, the performance of the system becomes more difficult to manage. The entire gamut of network management — encompassing fault management, security management, configuration management, performance management, and asset management — is especially vital in a distributed environment and should be planned for accordingly.

Until the past decade, relatively few vendors dominated the market. Thus, the selection of a particular vendor (e.g., IBM, etc.) effectively sealed the organization's fate with regard to its use of network equipment, protocols, operating systems, and software. With the advent of client/server networking, organizations are freer to choose an optimal mix of platforms, network equipment, protocols, and software tailored to their specific requirements. However, with this freedom of choice comes the added responsibility of keeping up with new technology and understanding how it can best be utilized to serve the business objectives. This is not always easy, particularly because there are so many choices and a preponderance of vendor "hype." There are a number of important new standards in various stages of evolution for client/server networking. Understanding what can be realistically achieved in the present and what is needed to set the stage for future infrastructure development is a very challenging task, particularly as different client/server standards and their implementations provide varying levels of interoperability. The advantages and limitations of each client/server approach should be carefully considered before a final selection is made to assess the impacts on the organization's infrastructure and possible future technology migrations. For example, the open CORBA standard provides interoperability between a heterogeneous mix of platforms and programming languages. However, various vendor implementations of CORBA may not be interoperable, and thus the organization may be tied to a specific vendor with this approach. The proprietary COM standard developed by Microsoft is also a language-neutral specification. However, COM is not compatible with other competing client/server protocols — such as EJB (Enterprise Java Beans),[1] RMI (Remote Method Invocation), and CORBA — and is best suited to a Windows-based operating environment.

5.2 Motivations for Distributed Networking

Total cost of ownership[2] (TCO) reduction is a major motivation behind many networking/IT initiatives. Client/server technology offers many economies of scale, and is often easier and cheaper to deploy and maintain than traditional mainframe computer system architectures. These advantages are achieved through modularized, reusable, standardized plug-and-play hardware and software. This is a key reason why client/server architectures have become the pervasive standard for new application development and network implementation.

Other characteristics of client/server networking include:

- *Scalability.* Additional clients and servers can be added as needed in a distributed environment, without necessitating major changes in the existing application code or networking infrastructure.
- *Fault tolerance.* Because it is relatively straightforward to add additional servers and network components in a distributed environment, load balancing and failover configurations can be utilized to ensure continuous operations as needed.
- *Concurrency.* Unlike a single workstation system, multiple users can simultaneously access and share data, files, and other network resources supported by the client/server network.
- *Transparency.* The interaction between clients and servers is seamless and transparent to the end user, who need not be aware of the location of remote network devices.

These features provide a basis for a flexible and adaptable infrastructure that can accommodate significant change in the number and location of end users and computing/networking devices. In a corporate era where an international presence and mergers and acquisitions are prevalent, this flexibility is extremely useful for the integration of disparate systems and networks.

In addition to providing a cost-effective, adaptable infrastructure, client/server technology offers rich functionality to support new applications. For example, using XML, client/server networks can process multimedia documents and attachments across a multitude of platforms. Intelligent, self-adapting technology is also being used in client/server networks to provide dynamic system responses to changing conditions for a wide range of applications. For example, this technology is being used to implement intelligent self-managing network entities, distributed over the network to assist with network management tasks. As shown in Figure 5.1, intelligent agent software is also being used by marketers to learn and model customer preferences, needs, and perceptions, and to match them against company products and services. Technology for intelligent mobile agents that can move freely within the network is also currently in development. All these capabilities, when implemented in a client/server network, offer an unprecedented opportunity to perform instantaneous decision making and information processing that harnesses the computational power of the Internet.

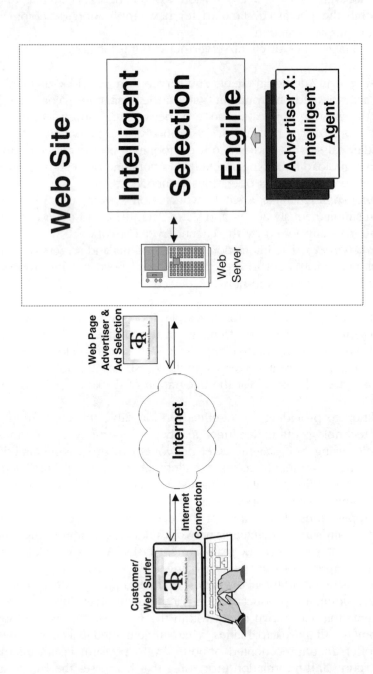

Figure 5.1 Intelligent Web agent for customized advertising.

5.3 Two-Tier and Three-Tier Client/Server Architectures

Client/server architecture is based on the decomposition of application processing into logical processes or layers, as defined below:

- *Presentation.* This layer is responsible for the collection, presentation, and display of data and information to and from an end user. It provides the "look and feel" of the application via a GUI (graphical user interface), non-GUI, or object-oriented interface.
- *Application processing.* This layer implements the application functions and activities.
- *Data management.* This layer performs and manages system database functions.

Specific types of client/server architecture (i.e., thin client, fat client, two-tier, three-tier, and n-tier) differ largely in how and how many processing layers are implemented in the client and server programs. The partitioning of the client and server functionality is a critical design issue with weighty impacts on the scalability of the network. The split should properly balance network performance and cost trade-offs in accordance with the size and scope of the business functions supported.

In simple two-tier client/server networks, there may be a one-to-one, many-to-many, or many-to-one mapping of clients to servers. Although generally there are more clients than servers, the actual network configuration will depend on the application requirements. In a two-tier network, clients must have knowledge of the available servers and the services they provide. However, servers need not be aware of clients until a request is made of the server.

Many of the earliest two-tier client/server networks were based on a *fat-client* model. With this approach, the client program implements both the presentation and application processing layers. The server is responsible for handling database processing functions. Thus, most of the processing load is relegated to the client. This is illustrated in Figure 5.2. Database (a.k.a. Database Management Servers or DBMS) and file servers are an example of this type of network, as shown in Figures 5.3 and 5.4, respectively. For example, a fat client might be appropriate when a personal computer application program (e.g., computer-aided design tools, statistical software, etc.) is used to process data and create files that are then sent to a central file repository. A fat-client approach is suitable for computationally intensive applications, which if run on a server might result in excessively heavy network processing and traffic. However, when updates are made to the client software that interfaces with the application, they must be applied to all the clients in the network. Developing a

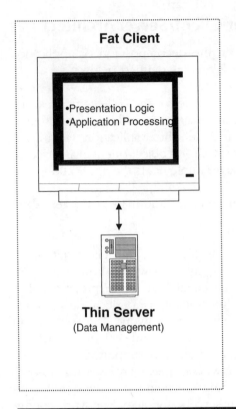
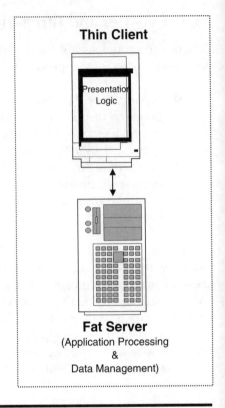

Figure 5.2   Fat versus thin clients and servers.

mechanism for keeping track of all the clients and for updating them can require significant effort, particularly when there are many clients and lots of changes. This is a drawback to network scalability, and is one reason that fat-client architectures are not as widely used as they were in the past, when application volatility was not as great as it is today. Therefore, this approach works best in a relatively static network and development environment.

In contrast, a *thin-client* architecture offloads the application and data processing to the server. The client is only used to run user interface software. This type of approach is appropriate, for example, with an application that must handle a heavy volume of queries but not much actual processing of data. Thin-client networks are inherently easier to upgrade than fat-client networks. This is because major application changes and updates need only be applied to the servers, which are generally centrally managed and fewer in number than clients. However, a thin-client approach increases the processing load on the server and the network, and may necessitate higher performance network components to satisfy the user

Note:
- Client passes SQL requests in message format to database server
- Results of SQL commands are returned over network.
- In contrast to file severs (which pass all data back to client), the software processing the SQL requests and data reside on same machine (i.e., the DBMS server).
- DBMS server code is provided by vendor (such as Oracle); however, programmers must create SQL tables and populate them with data. The application code resides within the client.
- This approach is used for decision support systems, ad-hoc queries, etc.

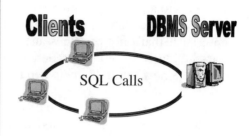

Figure 5.3   Fat client: database server configuration.

Note:
- Client passes request for file records over the network to a file server via a message exchange.
- This approach is used for file sharing.

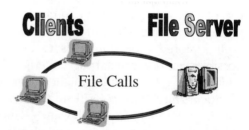

Figure 5.4   Fat client: file server configuration.

and application demands. This, in turn, can result in significant increases in the cost of the networking equipment and software.

In a two-tier client/server network, when the client is thin, this implies the server is fat (see Figure 5.2). This means the server performs more functions than the client. Transaction servers are a good example of fat servers used in two-tier networks. These servers minimize message interchanges by creating abstract, higher levels of service through the use of *remote* or *stored procedure* calls (or, in an object-oriented environment, method invocations). Remote procedure calls (RPC) are defined by the American National Standards Institute as part of the SQL Data Definition Language (DDL), Data Manipulation Language (DML), and Data Control Language (DCL) standards.[3] A stored procedure consists of a sequence of SQL commands that are precompiled and stored in the system tables of a database management system (such as Oracle, Ingress, DB2, Sybase, Informix, etc.). A client application can initiate the execution of a stored procedure by calling the procedure name and passing the appropriate parameters to a properly configured transaction server. A stored procedure effectively consolidates the message exchanges between the client and server, such that only one message from the client is needed to trigger the execution of an entire predefined set of messages. This significantly reduces the network traffic between the client and server. It also promotes software manageability, because predetermined business processes and rules can be enforced by the implementation logic contained in the remote procedure. For example, a stored procedure to process a customer payment might include a sequence of SQL statements to update various database tables (such as the customer history, accounts receivable, general ledger accounts, billing history, and other application database tables), to verify customer status, and to prepare various reports, customer correspondence, etc. The encapsulation of business logic within a series of interrelated processing steps performed by the SQL statements generated by the stored procedure is called a *complex transaction unit*. The processing of complex transaction logic on a transaction server through the use of remote procedures contrasts with that of database servers, which process only one SQL statement per client request.

If a stored procedure fails to execute completely or properly, it is possible that some database tables might be left only partially updated, while other related tables might not be updated at all. Because this type of processing failure could lead to inconsistent, incomplete, or erroneous data in system tables, middleware — such as a transaction server — must enforce the requirement that all SQL statements in a stored procedure must either all succeed or else they must all fail. In the event of a failure, the transaction server must also roll back all the database tables to the state that they were in prior to the start of the stored procedure. Figure 5.5 provides an overview of the role of transaction servers in supporting transactional and database integrity.

It should be noted that DBMS (database management system) vendors support stored procedures using their own proprietary formats and conventions. Thus, stored procedures are generally not portable between different vendors without at least some form of modification. This is an area where the "plug-and-play" objective of client-server network has not been fully achieved because it is in the vendors' interest to try to maintain a proprietary advantage over their competitors in their DBMS offerings at the expense of compatibility.

A groupware server is another example of a fat server and is depicted in Figure 5.6. A groupware server manages and executes software — such

Note:
- *Online Transaction Processing (OLTP)* applications consist of a client (which provides a GUI) and a server (which issues a SQL transaction against a database).
- When a client issues a *Remote Procedure* (which is a type of server service), this initiates the execution of a series of grouped SQL statements called *Transactions* on a SQL server.
- In contrast with a database server (which issues one client request and receives one server reply per SQL statement), the client and server exchange consists of only one request/reply message, thereby reducing network traffic. The SQL statements either all succeed or all fail as a unit.
- Used for mission critical applications that require high reliability, database integrity, fast response time, and high security.

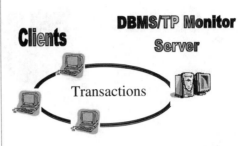

Figure 5.5   Fat server: transaction server configuration.

Note:
- Groupware typically operates on a corporate Intranet or LAN.
- Groupware is designed to support collaborative processing by helping end-users manage their e-mail, documents, meeting schedules, conferencing, file distribution, and other functions.
- Lotus Notes and Netscape are examples of commercial groupware products.
- Typically, groupware vendors provide scripting languages and form based interfaces to allow end-users to extend the groupware functionality.

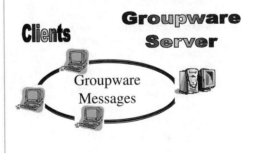

Figure 5.6   Fat server: groupware configuration.

as IBM's Lotus Notes — which is designed to promote collaborative processing between individuals working in a virtual environment.

Three-tier client/server architectures provide an even greater level of functional decomposition, as each processing layer can be executed on separate computers and can be updated independently. A three-tier client/server configuration splits the processing load between clients running a

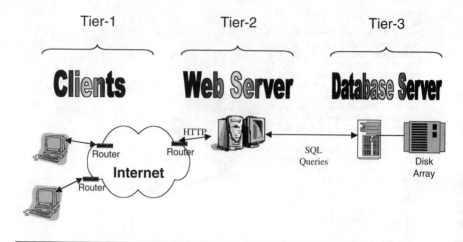

Figure 5.7   Example of three-tier client/server architecture.

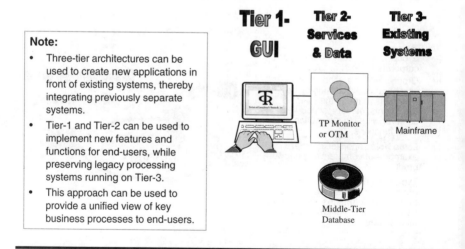

Figure 5.8   Three-tier architecture for integrating new and old applications.

GUI, the application server running business logic, and the database application. This is depicted in Figure 5.7.

This approach is more scalable and easier to manage than either thin- or fat-client two-tier architectures. A three-tier architecture allows hardware and software components to be phased in or out as needed. As diagrammed in Figure 5.8, this can be very useful in integrating new and legacy applications, while incrementally enhancing the infrastructure. This approach is used to avoid taking on a massive wholesale conversion effort all at one time. In a three-tiered network, thin clients can be downloaded as Java applets or beans, making it easier to manage and distribute changes in the application code to the end user. As the number of clients or end users increases, servers can be added to the network with relative ease to handle and distribute the increased processing load.

A three-tier infrastructure provides rich new message paradigms to support new services. Like two-tier architectures, it supports remote procedure calls and synchronous messaging. In addition, it also provides asynchronous messaging for *publish/subscribe* and *broadcast services*. Publish/subscribe services allow a client to publish (i.e., send) a topically categorized message to a message server. The message server, in turn, posts the message for viewing by other clients that have subscribed to that particular topic. The message server manages the message queue and keeps the posted messages until all subscribers have viewed them. This communication model is also referred to as *push technology*. Broadcast messaging refers to a communication model that supports sending a particular message to all clients on the network.

Three-tier client/server networks are often used to integrate heterogeneous platforms and applications in a single networking infrastructure. This architecture is also employed to implement a robust infrastructure (i.e., it is scalable and supports failover configurations, load balancing, and hardware/software upgrades) for high-volume, low-latency mission-critical applications. However, the design of this type of network is considerably more complex than the design of a corresponding two-tier network. It also requires a greater level of staff expertise and training to properly implement and maintain.

5.4 n-Tier Distributed Object Architectures

Generally, when a network is built using more than three tiers, it is referred to as an *n-tier network*.[4] In addition to the client, application, and database layers found in a typical three-tier network, an n-tier network may also include:

■ Web server layer supporting Internet services (such as S-HTTP servers, etc.)
■ Backend host system layer
■ Storage Area Network (SAN) layer for additional data and file handling
■ Backup network layer for failsafe configurations
■ Network management layer (such as SNMP servers, etc.)

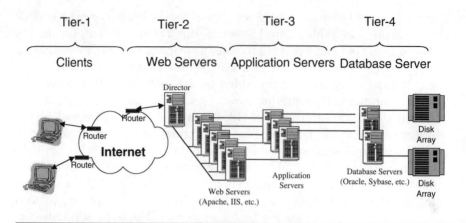

Figure 5.9   Example of n-tier architecture.

The list above is not exhaustive. The actual number associated with "n" in a particular network depends on how the processing functionality is partitioned across the network. A major distinguishing feature of n-tier architectures is that the processes are smaller and more highly compartmentalized and specialized than either two- or three-tier architectures. Middleware (e.g., CORBA, COM+, etc.) provides the "glue" that binds the tiers together and allows them to interoperate. An example of an n-tier network is shown in Figure 5.9.

Many n-tier networks are also distributed object networks. In distributed object architectures, interacting objects replace the notion of interacting clients and servers. Objects communicate and request the services of each other through a middleware component called an object request broker (ORB). This approach leverages object-oriented technology with the intent of achieving a plug-and-play component network environment that is extremely open and flexible. Object-oriented approaches support incremental development through the use of reusable, building block components that are used to create larger, more complex systems.

As shown in Figure 5.10, in a distributed object network, the application is written as a set of communicating objects, which can reside anywhere within the network. Objects are referenced by a unique name, and possess attributes and an interface that controls the object behavior. A group of similar objects offering the same services is referred to as a *class*. Objects composed of other objects are referred to as *composite objects*. A composite object is not like a simple object (consisting of methods, interfaces, attributes, and data). Instead, a composite object contains only references to its constituent objects. Objects provide services through *methods,* which are accessed by remote clients via method invocations. Methods operate on private instance (or state) data contained within the object, which, in turn,

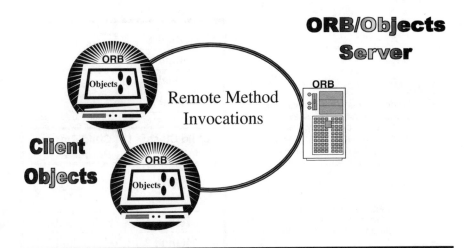

Figure 5.10 **Example of distributed object architecture.**

determines the specific response to the invocation. The object's public interface defines its functions, what can be done to the object, and how other objects can access it. This interface is independent of the actual object implementation. Middleware platforms (such as CORBA, COM+, or EJB) use these interfaces to create a cohesive object/component-based infrastructure.

Inheritance, encapsulation, and polymorphism are key properties of objects. *Inheritance* provides the means to augment or modify the methods of a parent class through an associated child class. However, the parent method still controls and establishes child inheritance rules and is not affected by the child methods. Some object models support only single inheritance. In this case, a class can have only one parent class. Other object models support multiple inheritance. With this model, a class can have multiple parents. The property of inheritance is used to create reusable units of code. Figure 5.11 presents a diagrammatic overview of this concept.

Encapsulation is used to delineate the object's private and public data. The object uses a published external interface to expose its public instance data and methods to other objects. The external interface is an immutable contract for client and object communication. Thus, even if changes are made to the object, these methods and data must continue to be supported. However, the internal methods and data that constitute the object's inner workings are hidden from public view, so they can be modified as needed without disrupting the object's ability to communicate and interact with others. In this way, encapsulation provides object modularity and flexibility in making internal changes to the object data and methods. This is shown in Figure 5.12.

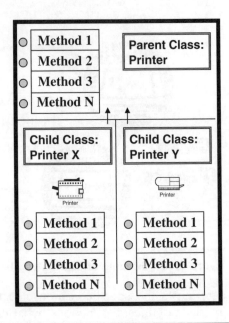

Note:
- Inheritance is a mechanism for creating new child classes (*subclasses or derived classes*) from an existing parent class.
- Child classes inherit their parent methods and data structure. This is indicated by the arrows pointing upwards to the parent.
- The child class can add new methods and override their inherited methods without affecting the parent methods.

Figure 5.11   Overview of object inheritance.

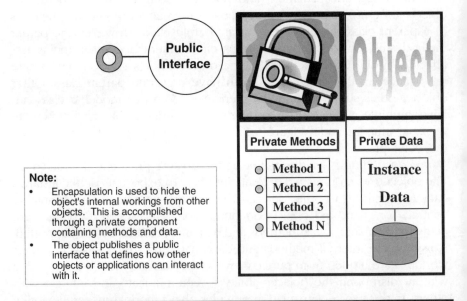

Note:
- Encapsulation is used to hide the object's internal workings from other objects. This is accomplished through a private component containing methods and data.
- The object publishes a public interface that defines how other objects or applications can interact with it.

Figure 5.12   Overview of object encapsulation.

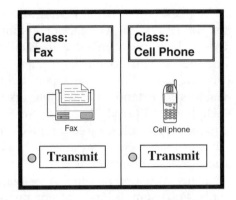

Note:

- Polymorphism allows the same method to have a different expression, depending on the class that implements it.
- Objects in different classes receive the same message but respond differently.
- *Overloading* is a variant of polymorphism used to define different versions of a method, each with different parameter types.

Figure 5.13 Overview of object polymorphism.

Polymorphism is a way of handling similar, but not identical, objects with a common interface. Polymorphism allows the same method to have a different expression, depending on the class that implements it. This means that objects in different classes receive the same message but respond to it differently. This property is useful in building object-oriented user interfaces and is shown in Figure 5.13.

Objects can be used to build *components*. Components are reusable, self-contained, and self-descriptive software that is language, application, and hardware neutral. Once built or purchased from a commercial vendor, components can be combined to construct new applications running on different machines and platforms. The scope of the component function depends on its implementation. Interfaces to components can be generated by compiling component objects and their associated classes, methods, and data into binary executables (having a .DLL or .EXE file extension) using a framework such as COM+ or CORBA.

Microsoft's OLE (Object Linking and Embedding)/Active X is a leading desktop component standard with widespread vendor support. ActiveX components are called *controls*. They provide a standard application programming interface to allow the embedding and execution of multiple applications within a single compound working document. OLE/ActiveX is built into the Windows operating system on top of the DCOM (Distributed Component Object Model) ORB. ActiveX controls can be automatically downloaded and executed in Internet Explorer, providing a means to distribute components over the Internet. ActiveX is also designed to support live display of OLE documents (such as Excel spreadsheets, PowerPoint presentations, and other Windows applications) from the Internet Explorer Web browser. Visual J++, Visual C++, and Visual Basic are among the tools offered by Microsoft for ActiveX development.

ActiveX is an integral part of Microsoft's strategy to promote enterprise use of DCOM as a component framework. Traditionally, ActiveX and DCOM have been linked with Windows and WinNT on Intel platforms, and they perform optimally in these environments. However, Microsoft is working with TOG (The Open Group)[5] and a host of other industry vendors to extend DCOM's interoperability to other operating systems, including UNIX and Macintosh. As part of its efforts to promote vendor support, TOG offers free detailed documentation on the COM/DCOM ActiveX Core Technology at its Web site: http://www.opengroup.org/pubs/catalog/ax01.htm. [OPEN04]

Many vendors provide cross-platform ActiveX/DCOM development tools. For example, in a recent press announcement, IBM and Taligent announced a new product — the JavaBeans Migration Assistant for ActiveX in Taligent's VisualAge WebRunner and VisualAge for Java development tools. The JavaBeans Migration Assistant for ActiveX is a tool for converting ActiveX controls into Javabeans. [JAVA04]

Although IONA Technologies boasts that its CORBA-based object request broker — Orbix — has the largest installed base of any ORB, it also recognizes the need to coexist with DCOM. As part of its development tools for building distributed networking applications, it offers a CORBA/DCOM bridge that allows Windows NT applications to be ported to UNIX, and other platforms. [IONA03] This is yet another example of the deep base of vendor support and user demand for ActiveX and DCOM.

Although theoretically, ActiveX controls can be run in a cross-platform environment, as noted by Cornelius Willis, Microsoft's group product manager of Internet developer marketing, "... any time you have an application (or Web page) that depends on a platform-specific component like an OpenDoc/Mac component or an Active/X/Windows component you build that platform dependency into your app." [SHOF97]

ActiveX controls should be used with care, particularly in an extranet/Internet environment. ActiveX controls have security flaws that have been exploited by malicious hackers to spread worms and viruses, and to gain unauthorized access to data and machine execution. Using ActiveX controls, it is possible for a hacker to build a Web page that, when viewed by Internet Explorer, can initiate destructive actions without the end user being aware of it. According to the SANS (SysAdmin, Audit, Network, Security) Institute, ActiveX vulnerabilities are among the top-20 Internet security threats. [SANS03] To help protect against these vulnerabilities, the SANS Institute recommends that Microsoft patches be applied to Internet Explorer on an ongoing basis and, if possible, the Web browser should be upgraded to the latest possible version. In addition, the SANS Institute recommends the following Internet Explorer (IE) security settings: (1) disable downloads of unsigned Active X controls, (2) disable initialize and script ActiveX controls not marked as safe, (3) under Microsoft VM, select High safety for Java permissions to prevent inappropriate access to your system, and (4) disable access to data sources across domains (under Miscellaneous) to avoid cross-site scripting attacks. [SANS03]

The ActiveX security model is based on *signed code*. Microsoft's Authenticode technology is a tool to allow software publishers to digitally sign executable (EXE), Microsoft ActiveX controls (OCX), cabinet (CAB),[6] and dynamic-link library (DLL) files. This provides a means to identify the author of the ActiveX control and to verify that the control has not been modified since it was signed.

Before a control can be signed, the first step is to obtain a certificate from a certificate authority, such as VeriSign. Authenticode offers class 2 certificates for individual software publishers and class 3 certificates for commercial software publishers. Class 2 certification requires that individuals submit their names, addresses, e-mail addresses, dates of birth, and social security numbers to the certificate authority for verification. If the individual passes the verification process, the certificate authority issues a certificate. Similarly, a commercial software publisher must provide its company name, location, and contacts, and have a Dun and Bradstreet rating for verification purposes.

When end users download Web pages that attempt to initialize and script signed controls that have not been marked "safe," a safety dialog box warning is displayed by Internet Explorer. To prevent this, the software developer/publisher can mark the ActiveX control "safe." The SIGNCODE program, which is part of the ActiveX software development kit (SDK), is used to sign the ActiveX code. Once the code is signed, even users who have set their Internet Explorer security settings to High are able to download, install, and register the marked ActiveX controls. [JOHN96] Programmer instructions for marking ActiveX controls can be found on Microsoft's Web site.

When a control is marked as "Safe," it is tantamount to an assertion by the software publisher that the ActiveX control will neither cause a security breach nor harm the end user's machine in any way. The flaw with this assumption is that software publishers can mark their controls as "Safe" even when they are not or have not been tested. Thus, Authenticode does not actually prevent malicious code; it merely establishes a line of accountability to the software publisher so they can be identified and, if need be, prosecuted for wrongful actions. An end user who chooses to set his Web browser security to "None" or ignores pop-up dialog security warnings when they occur has effectively defeated the ActiveX security mechanisms. In a corporate environment, end users accessing the Internet should be made aware of these security risks to avoid running anonymous, unsigned code without any safety precautions.

Despite the potential security risks in an Internet/extranet environment, ActiveX remains a popular development choice for intranet environments over which the organization has more control. This is particularly true in organizations with a large installed base of Windows applications and Intel platforms. ActiveX has very rich feature sets and functionality, can be programmed in any language, and its performance is highly optimized for a Windows environment.

JavaBeans, developed by Sun Microsystems, is a direct competitor of ActiveX. It is a framework for building applications from reusable Java components that run in a Java Virtual Machine (JVM). JavaBeans components are referred to as *Beans*. Like ActiveX, JavaBeans is designed to enable communication between object-oriented components within a framework container, such as a Web browser. While ActiveX controls can be developed in any language, they run best in a Windows environment. In contrast, JavaBeans are written in Java but run on any platform or operating system. This independence is summed up by the JavaBean mantra: "Write Once, Run Everywhere." It is achieved by compiling source code into platform-independent byte code. The byte code is interpreted in the runtime environment of a JVM.

According to Sun Microsystems, "JavaBeans acts as a bridge between proprietary component models and provides a seamless and powerful means for developers to build components that run in ActiveX container applications." [SUNM04] Most typically, like ActiveX, JavaBeans components are used to build applets and applications for user interface controls. IBM's VisualAge and Symantec's VisualCafe are two leading products for JavaBeans development.

JavaBeans is based on the Java security model, which uses both signed code and *sandbox* security measures. Like ActiveX, JavaBeans provides a means to digitally sign an applet and to verify that it has not been modified since it was signed. Java itself has many features that help guard against programmer errors or malicious code, such as protection from type-casting errors[7] and illegal memory access. Java applets have numerous security features and incorporate the security concept of a sandbox. A sandbox is a dedicated area in the browser that is used to run Java applets from unverified Web sites. The sandbox severely limits the functions the applet can perform, restricting or prohibiting file access, printing, program execution, and other activities on the end-user computer. These constraints are designed to prevent damage and unauthorized access to data and files, and unauthorized use of the end-user machine. Trusted applications are not subject to these restrictions and are allowed to perform any functions. JavaBeans' security and support for multiple platforms are often-cited reasons for its adoption as a component model in a client/server environment.

Enterprise JavaBeans (EJBs) extends the JavaBeans component model to provide distributed server components in a CORBA environment. JavaBeans and Java are not intended to support distributed components; and as such, they lack the transactional services and security interfaces that are implemented in EJBs. The EJB standard defines a container (a.k.a. an Object Transaction Monitor) for executing and accessing bean services, and is significantly more complex than JavaBeans to develop and implement.

In summary, in a distributed object network, client objects installed on a network communicate via an *Object Request Broker* (ORB) by invoking a method on a remote object. The ORB locates an instance of the object server class, invokes the requested method, and returns the results to the client object. The use of object servers and remote method invocations reduces the network traffic, much like the use of SQL remote procedure calls.

There are numerous vendor offerings for ORB middleware. Some are free and some are not. Some are based on proprietary standards and some are based on open standards. It is important to ensure that the features and functionality required by the application are well understood and compatible with the ORB middleware selected because there is so much variation in the ORB implementation.

Object frameworks are preassembled class libraries providing partially complete, specialized object-oriented application code. ActiveX, JavaBeans, CORBA, and COM+ are examples of frameworks used in client/server application development. Frameworks are designed for modularity, reusability, and extensibility so that complex applications can easily be developed and deployed from component parts.

5.5 Middleware Functions and Implementation

This section discusses the functions of client/server middleware and how they are implemented. Middleware anchors client/server application and network components together and coordinates their interaction. Middleware oversees the client request invoking a service, as well as the transmission and handling of the response over the network. In an n-tier client/server network, this requires that the middleware support pipes and server-side platforms that run the server components. However, middleware does *not* include the client-side interface or the server processing.

Middleware uses *pipes* to provide communications services between components and applications. Some of the major types of pipes used in n-tier networks include:

■ *Remote procedure calls (RPCs)*. In this communication model, a client issues a procedure call that is sent to and translated by a server that behaves as if it were local, when in fact it is a remote call. RPCs support multiple concurrent procedure calls through the use of threads or multiple server processes.

■ *Object Request Broker (ORB)*. This is an object-oriented communications model between clients and servers that uses an ORB to match communicating objects on the network and to route object requests between them.

■ *Message-Oriented Middleware (MOM)*. This communication model supports queued communications between applications. It allows applications to notify other applications of a triggering event. After receiving a message from an application, the MOM queues it for delivery. The actual delivery of the message is determined by the availability of the recipient application(s) and by preset timers or triggers.

Pipes support different types of messaging paradigms. Messaging between a client and server can be synchronous, in which case the client must wait for feedback on the status of the message receipt before further processing or message transmission. Alternatively, the messaging can be asynchronous. In this case, a client can send a request but need not wait for a response from the server, because the server can process it at an unspecified later time. This avoids tying up either the client or the server due to traffic or processing delays on the client, network, or server. Other types of messaging that the middleware can support include:

■ *Conversation:* an ongoing multi-transaction dialogue between a client and a server (e.g., TCP/IP sockets[8] and IBM'S CPI-C).[9]

■ *Request–response:* single message exchange between client and server components (e.g., RPCs and ORB method invocations).

■ *Queues:* a form of asynchronous communication between clients and servers in which messages are queued for the server (e.g., MOM and TP Monitor).

■ *Publish-and-subscribe:* a communication model that enables client and server components to subscribe to a topic of event with an event manager that receives message postings from other components. The event manager distributes the messages to subscribers. In advanced publish-and-subscribe environments, scripts and filters can be used to develop highly customized distribution lists. CORBA's event service and specialized MOM and TP Monitor services are examples that use this communication model.

■ *Broadcast:* a one-way message communication to all components on the network (e.g., specialized TP Monitor services).

Application servers that implement pipes and run server-side components are called *platforms*. Platforms perform a host of tasks to assist server components, such as load balancing, recovery actions, transaction processing, and other network management functions. TP Monitors, Object Transaction Monitors (OTMs), and Web application servers are examples of client/server platforms.

TP Monitors coordinate and synchronize complex transaction services. This involves process management, which is responsible for the initiation of server processes, assignment of work to servers, oversight of execution status, load balancing, and other functions. It also includes communication management and the handling of conversations, queuing, request–response, publish/subscribe, and broadcast messages. TP Monitors provide complete control of transaction processing from the start of a transaction on a client through the subsequent server processing and response. This is accomplished by enforcing ACID properties, which is an abbreviation for the following:

- *Atomicity.* The transaction is the fundamental, indivisible unit of work and must either succeed or fail in its entirety. Partial transaction processing is not allowed.
- *Consistency.* At the end of the transaction processing, the transaction must be completely successful or its actions must be cancelled and the system restored to its initial state.
- *Isolation.* A transaction's actions should not affect or be affected by other concurrently processed transactions. This implies that a multi-client system should appear to each client as a single-user system, such that transaction changes are serialized and only made visible to other clients and servers after they have been finalized. This ensures that concurrent processes will not corrupt each other's operations.
- *Durability.* Once a transaction has completed, its effects should become permanent, even if there are subsequent system failures.

To implement ACID procedures, TP Monitors employ a two-stage commit protocol to synchronize updates occurring on different client and server processes. The TP Monitor controls whether a transaction is *committed* (or accepted) or *refused* based on the report it receives from each participating client/server process. The commit process can be decomposed into phases, as shown in Figure 5.14. In the first phase, a transaction manager node sends *prepare-to-commit* commands to all subordinate nodes directly participating in the transaction processing. These subordinate nodes must, in turn, propagate *prepare-to-commit* commands to the processes they have begot as part of the transaction processing. The first

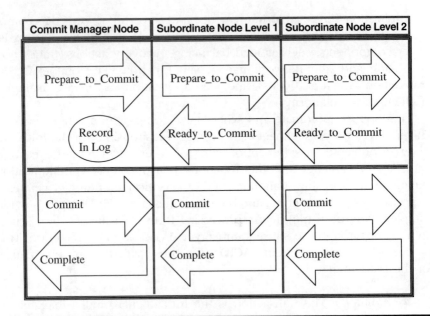

Commit Manager Node	Subordinate Node Level 1	Subordinate Node Level 2
Prepare_to_Commit	Prepare_to_Commit	Prepare_to_Commit
Record In Log	Ready_to_Commit	Ready_to_Commit
Commit	Commit	Commit
Complete	Complete	Complete

Figure 5.14 Overview of OSI two-phase commit process.

phase terminates when the commit manager receives *ready-to-commit* signals from *all* subordinate nodes involved in the transaction. This indicates that the transaction has successfully completed execution on the subordinate nodes and they are ready to authorize a final *commit* command. The transaction manager records the consensus in a secure, failsafe location so the information can be used, if need be, to initiate recovery actions if there are subsequent system failures. The second phase of the commit process begins once the transaction manager finalizes the decision to commit the transaction based on a unanimous "yes" vote from all participating nodes. It then issues the *commit* command to all its subordinate nodes, which they in turn forward to their subordinate nodes, throughout the entire network of nodes involved in the transaction process. The second phase of the commit terminates when all participating nodes have marked their part of the transaction as "committed." Once the controlling transaction manager receives commit confirmations from all subordinate nodes, the transaction is deemed complete and durable. The confirmation process terminates prematurely if any subordinate node issues a refuse reply, indicating that a processing failure has occurred. If this occurs, the transaction manager issues a command to all its other subordinates to roll back the work they have performed thus far on the transaction. The roll-back command is propagated from these subordinates to all other participating nodes until the system is restored to the state it was in prior to the start of the transaction. In this manner, a TP Monitor

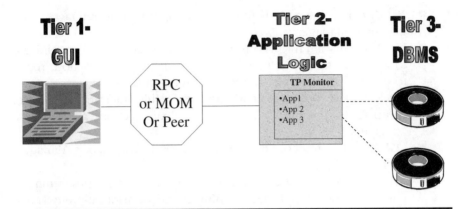

Figure 5.15   Example of TP-heavy client/server network.

guarantees that when transaction processing is complete, all participating parties are in agreement as to its success or failure.

TP Monitors are widely used in high-availability, high transaction volume environments. TP Monitors complement MOM extremely well, and they are often used together to support long-lived transactions, because TP Monitors can coordinate queue-based transactions and oversee the execution of server processes triggered by the queues. TP Monitors are essential to any mission-critical client/server application that must support large numbers of clients, and multiple servers or databases, by providing:

■ Firewall support
■ Extensive recovery procedures under ACID control
■ Load balancing and job prioritization
■ Unified infrastructure for heterogeneous server mix

When a TP Monitor is used to manage all aspects of the client/server communication processes, as just described, this is known as a *TP-Heavy* approach. An example of this is provided in Figure 5.15. When the TP Monitor is only used to manage the execution of stored procedures in a database environment, this is known as a *TP-Lite* approach.

The X/Open Distributed Transaction Processing (DTP) standards are widely implemented in TP Monitors. These standards are designed to support distributed transaction processing through the definition of interfaces that coordinate and synchronize communication exchanges between (1) applications, (2) resource managers (which provide access to shared resources, such as a DBMS, etc.), and (3) transaction monitors. Some of the key transaction processing interfaces defined in the X/Open standards include:

- *RM API.* This interface is used by an application to interact with a Resource Manager.
- *TX API.* This interface is used by an application to communicate to a Transaction Monitor when it is initiating a transaction, ready to commit it, or wants to abort the transaction.
- *XA API.* The Transaction Monitor uses this interface to control "prepare to commit," commit, or rollback actions of Resource Managers.
- *OSI-TP.* This protocol allows heterogeneous Transaction Managers to interoperate.
- *XATMI, TxRPC, and CPI-C.* These are transactional programming interfaces. XATMI is based on BEA Tuxedo's Application-to-Transaction Monitor Interface (XATMI), and is a message-oriented interface. TxRPC is based on the Distributed Computing Environment (DCE) remote procedure call interface. CPI-C is a peer-to-peer conversational interface based on IBM's CPI-C.

Application programs must implement both a transaction and a remote communications interface. The Transaction Monitor implements the rest to create a seamless distributed transaction processing environment.

IBM's CICS/MVS and IMS TP Monitors are examples of proprietary TP Monitors that have been used for years in legacy applications to ensure transactional integrity. Other newer TP Monitors are based on open standards, and include such products as Transarc's Encina and BEA's Tuxedo. Generally, TP Monitors from different vendors do not interoperate.

The Transaction Processing Performance Council (TPC) is a nonprofit corporation "founded to define transaction processing and database benchmarks and to disseminate objective, verifiable TPC performance data to the industry." [TPC04] This organization posts the results of various benchmarking tests, including cross-platform performance comparisons, to aid organizations in the evaluation and selection of TP Monitor products. The TPC Web site is located at http://www.tpc.org/.

TP Monitors built using ORBs are called Object Transaction Monitors. Examples of this technology include BizTalk Server 2004 (based on the Microsoft .NET Framework), IBM's IBM TXSeries™ for Multiplatforms (based on CORBA), and BEA's Tuxedo (based on CORBA). ORBs and TP Monitors leverage and complement each other's capabilities. Because ORBs and transaction middleware are now built into personal computer operating systems, Web browsers, and Java VMs, they can be used to assist TP Monitors with complex nested or long-lived transactions, and other communications tasks. For its part, the TP Monitor is ideally suited to manage the large numbers of runtime objects needed in a CORBA or COM client/server environment.

As the demand for powerful E-business Web sites has increased, Web application servers have played an increasingly important role in n-tier client/server architectures. The example shown in Figure 5.9 illustrates the use of Web servers and Web application servers in an n-tier infrastructure. A Web server hosts Web pages and applications for display by a Web browser. By installing the appropriate server software, almost any computer can be used as a Web server. Apache's free open source server, Microsoft's Internet Information Server (IIS), and Netscape's Enterprise server are examples of widely used Web server software.

When a client (i.e., a Web browser) requests access to a file, the Web server must determine the associated file extension, and whether or not there is a registered component available to handle it. If not, the Web server retrieves a default (or static) file for display in the client's Web browser. This type of operation is self-contained and stateless, in that all end users are presented with the same display or Web page.

If, however, there is a registered component associated with the file extension, the information is passed to a Web application server. The Web application server retrieves and executes the appropriate script. If the page being requested contains dynamic content (e.g., to be provided by database retrieval, servlet execution, etc.), the Web application server manages and controls the required session and dynamic page functions that must be performed. In object-oriented environments, a Web application (i.e., an OTM) server is needed to manage *stateful* dynamic content. System processing of stateful content is ordered and context specific. Each processing step of the transaction must be performed in a prescribed sequence, through a progression of predetermined states. For example, when making an online purchase, the user must enter a log-in ID and password prior to check-out and final billing. This type of strict control is necessary to ensure transactional integrity and enforcement of ACID properties. A brief synopsis of the differences between the handling of static and dynamic content is provided in Figure 5.16.

Web application servers play a critical role in managing interactions between clients and back-end servers. There are a variety of methods for inter-process communication between Web application servers, clients, and other servers, including: CGI (Common Gateway Interface), ASP (Active Server Pages), JSP (Java Server Pages),[10] RMI (Remote Method Invocations),[8] XML (Extended Markup Language), applets, servlets, ActiveX, CORBA, and EJB. For example, in a Java-based environment, Java applets and HTML might be used to control the display in an end-user Web browser, while servlets are used to implement the application processing logic and database processing on back-end servers.

Lower-end application servers tend to "support servlets and JSPs, but typically not EJBs;... [while] high-end [application servers]...typically sup-

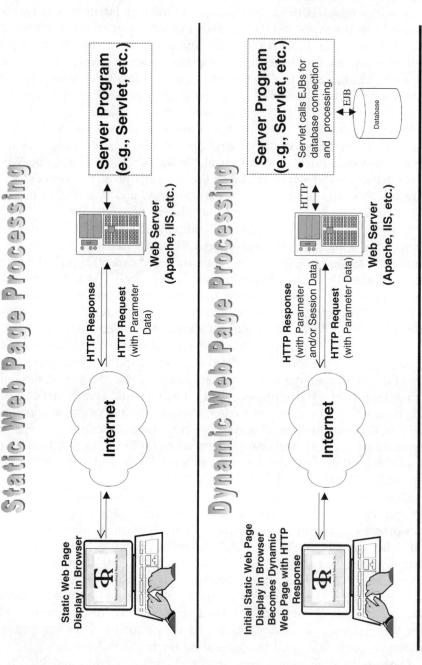

Figure 5.16 Static versus Dynamic Web Page Processing

port EJBs and Java messaging, and handle large transaction volumes." [COLL02] JBoss[12] is free, open-source application server software based on Java and the J2EE specification, which has a strong following at the low-end market. [JBOS04] Adoption of the JBoss server "rose from 13.9 percent in 2002 to 26.9 percent in 2003, for the highest percentage growth of any major application server, including BEA WebLogic and IBM Web-Sphere." Sun Microsystems is a strong second-tier market player with its Java Enterprise System. [EMIG04] IBM and BEA are examples of high-end Web application server products. Although these products are costly, particularly as compared to freeware, they come with extensive vendor support and are well-designed and fully configured to handle mission-critical applications and infrastructures.

In summary, middleware is a critical component of n-tier client/server networks. It is the vehicle by which the processing can be distributed across multiple remote locations, across multiple database and network environments in a manner that is transparent to the end user.

5.6 Middleware Standards

This section reviews and compares the leading middleware standards: CORBA, COM+, and XML/Web Services.

5.6.1 Common Object Request Broker Architecture (CORBA)

CORBA is an international middleware standard defined by the Object Management Group (OMG). The OMG is a nonprofit organization of over 800 hardware, software, and end-user companies that promotes "the theory and practice of object technology for the development of distributed computing systems." [OPEN04] The ISO (International Standards Organization) sanctions all OMG published standards. It should be emphasized that the OMG creates interface specifications for distributed object networking, and not the software or products that implement them. For detailed and current information on CORBA specifications, the interested reader should refer to the OMG Web site at http://www.omg.org.

CORBA consists of three major releases:

1. *CORBA 1.1.* This specification defines the IDL, DII, language bindings, and APIs for interfacing to an ORB.
2. *CORBA 2.0.* This specification defines the Internet Inter-ORB Protocol (IIOP), which provides a binary protocol for inter-ORB communication across multiple platforms.

3. *CORBA 3.0.* This specification is in progress and will provide significant additions to CORBA 1.1 and CORBA 2.0, including: [OMG04]
   - CORBA Components Model and an associated Implementation Definition Language (CIDL) This model specifies a component framework that is well integrated with Java and EJB
   - POA (Portable Object Adapter) to improve ORB portability across different vendors
   - CORBA scripting language to aid developers
   - "Pass by Value" features to facilitate information passing between objects
   - Multiple interface views for a single object
   - Improved Java language and Java/IDL mapping
   - Firewall specification for IIOP
   - Asynchronous messaging specification based on predefined Quality-of-Service (QoS) levels
   - Real-time and embedded CORBA ORB

CORBA is designed to help assemble mission-critical client/server networks from reusable components. It provides a means to implement transactional, secure, lockable, and persistent objects across a distributed infrastructure. As shown in Figure 5.17, the CORBA architecture is based on the following components:

- *Object Request Broker (ORB).* This is an object bus that manages requests and message exchanges between objects. In a CORBA environment, when an object requests the services of another, it is referred to as a client; and the object providing the service is designated as a server. In another context, it is possible that the roles of each object can be reversed. The ORB also implements the various *CORBAservices*, described next.
- *CORBAservices.* These define system-level object frameworks that extend the ORB. Frameworks are packaged with IDL interfaces and used to create, name, and activate objects. The OMG has defined standards for 15 object services, which are discussed later in this section.
- *CORBAfacilities.* These define horizontal and vertical (or Domain) application frameworks for objects in a particular business or scientific domain (e.g., Financial, Manufacturing, Transportation, Accounting, Life Sciences, etc.). In contrast to CORBAservices, which must be available on all platforms, the implementation of CORBAfacilities is optional.
- *Application Objects.* These are CORBA-compliant objects and applications that implement a well-defined IDL interface. Application objects are not part of the OMG standard — only the conventions for their implementation.

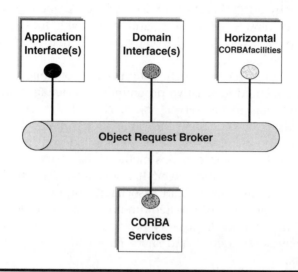

Figure 5.17   CORBA architecture components.

At the core of CORBA is the Interface Definition Language (IDL), which provides universal translation between communicating objects across a heterogeneous mix of programming languages, operating systems, and platforms. All CORBA objects must incorporate an IDL interface through which they communicate. The IDL (and its associated syntax and semantics), language bindings, and APIs for interfacing or binding distributed objects to the ORB are defined in the CORBA 1.1 specifications (which can be found at the OMG Web site). IDL is a subset of C++, with additional specialized extensions and keywords to support distributed processing. The IDL is used to specify an object's attributes, parent classes and associated inheritances, named methods, input/output parameters supported by the object interface, and results and exception types that can be returned by the object.

Once the object IDL is written, an IDL precompiler is used to translate the CORBA object definitions into a client stub and a server skeleton. The client stub and the server skeleton can be compiled into different programming languages. Thus, methods conforming to the IDL specifications can be written in and invoked from any language (e.g., C, C++, Smalltalk, COBOL, Java, etc.) that supports CORBA bindings.

The next step is to implement the server object in a native programming language, and to compile and link it to its IDL stub. This produces an executable server object that must then be registered in the *Implementation Repository* and, optionally, with a CORBA Naming Service. During registration, information is collected that is needed by the ORB to invoke and launch the server object. At runtime, the server objects are instantiated[13] on the server by an *Object Adapter*. The Object Adapter also records the runtime object references, classes, and types with the Implementation Repository so they will be available to an ORB that needs to locate or request the object services.

Similarly, after its IDL has been written and precompiled, a client object must be implemented in a native programming language with the appropriate method calls to the server object. Note, however, that the client does not have to be written in the same programming language as the server object. The client object is then compiled and linked to the IDL client stub. The client stub contains methods that marshal arguments sent to a remote instance of the server. It also contains methods needed to unmarshall the return results and associated output. The IDL interfaces are bound and stored in a distributed run-time database called the *Interface Repository*. The Interface Repository provides the means for client objects to dynamically access, store, and share information about their interface requirements.

As just described, the client and server IDL stubs support a static method invocation process in which the client and server interfaces are predetermined prior to compilation. CORBA also supports dynamic invocation interfaces (DIIs) to allow clients to discover server interfaces at runtime, and dynamic skeleton interfaces (DSIs) to provide a runtime binding mechanism for implementing new server interfaces. Although this facility makes it possible to add new object types at runtime, the dynamic invocation process reduces system performance by requiring extra processing overhead. These effects should be considered when deciding on a static or dynamic invocation approach. To invoke a dynamic method on an object, the client must follow these steps:

1. Obtain the method description from the Interface Repository.
2. Create an argument list.
3. Create the request.
4. Invoke the request.

The ORB acts as a matchmaker so that objects are aware of each other at runtime and can invoke each other's services. An ORB is implemented as a set of library objects linked to each client and to each server. The ORB employs an activation daemon that runs on the server host to ensure that client and server object connections are properly established. Once a connection is made, the client and server objects communicate directly with each other.

A client can only invoke an object's services through the methods specified in the object's interface. Through the ORB, a client object binds to a specific IDL stub associated with a particular object. If the client

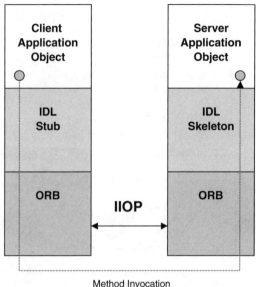

Method Invocation

Figure 5.18   CORBA ORB.

object should invoke a method on the server object through this stub, the ORB responds by locating the object and issuing calls to a published IDL skeleton that links the interface to the server object. The skeleton code converts the client calls to the format required by the object implementation, invokes and activates the server object, and returns the results and exceptions to the ORB for a reverse transmission back to the client object. The entire process from beginning to end is completely transparent to the client object. This is illustrated in Figure 5.18 and Figure 5.19.

To recap this discussion, the ORB components defined by CORBA include the following, as shown in Figure 5.20:

- *Client IDL Stubs.* These provide static interfaces to object services and are generated by an IDL compiler.
- *Dynamic Invocation Interface (DII).* This is used to discover methods invoked at runtime using standard CORBA APIs.
- *Interface Repository APIs.* This is used to obtain method signatures (or descriptions of registered component interfaces, methods, and parameters) dynamically at runtime. The interface repository is defined as part of the CORBA 2.0 specification.
- *ORB Interface.* This consists of APIs for performing local services (i.e., conversion of an object reference to a string and vice versa, etc.). This is the same for both client and server objects.

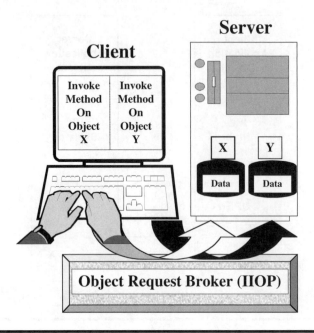

Figure 5.19   CORBA server object invocation.

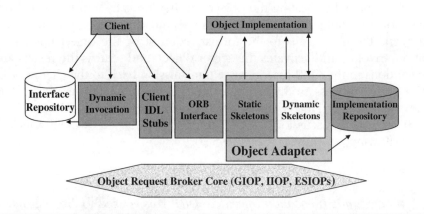

Figure 5.20   Structure of CORBA 2.0 ORB.

- *Server IDL Skeletons.* These provide static interfaces to each service provided by the server and are generated by an IDL compiler.
- *Dynamic Skeleton Interface (DSI).* This provides a runtime binding mechanism to allow servers to handle method calls from components that lack IDL stubs. The DSI is defined as part of the CORBA 2.0 specification.

- *Object Adapter (OA)*. This defines how an object is activated, and is responsible for creating new processes, creating new threads within existing processes, and reusing existing threads or processes. The OA accepts requests for service by server objects, and provides a runtime environment for instantiating server objects and assigning object IDs (or *object references*). It also registers the classes it supports and their runtime instances with the implementation repository. A server can use more than one object adapter.
- *Implementation Repository*. This provides a runtime repository of information about the classes a server supports, instantiated objects, and object IDs. It is also used to store additional information about the ORB.

CORBA 1.1 was originally conceived to provide portable object applications, without regard to the vendor implementation. In practice, this led to a lack of interoperability between different vendor offerings. The CORBA 2.0 specification was developed to address this deficiency and to provide support for distributed object calls. CORBA 2.0 specifies three major protocols:

1. *GIOP (General Inter-ORB Protocol)*. This protocol supports ORB-to-ORB communications across any connection-oriented transport protocol. GIOP defines interoperable object references (IORs), which associate profiles with object references that are passed between ORBs. The profiles describe how to communicate with the object when using different ORBs. It also defines seven message formats to support inter-ORB request and reply messaging.
2. *IIOP (Internet Inter-ORB Protocol)*. This protocol defines GIOP message processing over TCP/IP. IIOP makes it possible to use the Internet as a vehicle for ORB-to-ORB communications. To be certified as CORBA compliant, an ORB must implement IIOP or a half-bridge to it. A half-bridge processes IIOP messages on one side of the ORB, while passing other message formats through to the other side.
3. *ESIOP (Environment Specific Inter-ORB Protocol)*. This protocol allows CORBA and non-CORBA ORBs to communicate. This is useful for connecting legacy applications to a distributed CORBA environment. By encapsulating legacy code in a CORBA IDL wrapper, thereby creating a standard interface, the code can be integrated with other objects in a distributed object network. The CORBA 2.0 DCE CIOP (Distributed Computing Environment Common Inter-ORB Protocol) is one example of a CORBA ESIOP. DCE CIOP is designed to support multi-vendor, mission-critical, distributed

object-oriented applications. It leverages the services of DCE (using the DCE wire and RPC packet formats, etc.) while preserving the cross-platform neutrality of the ORB implementation. Thus, DCE CIOP is used to incorporate ORBs in a network environment where DCE is already installed. ESIOPs are specific to a particular network protocol.

CORBAservices augment and complement the functionality of the ORB. They are designed to help create customized, robust middleware solutions. The OMG has published standards for these object services, including [OMG04]:

- *Additional Structuring Mechanisms for the OTS.* This provides a low-level architecture for component-management middleware and other systems to support complex, advanced transaction processing.
- *Collection Service.* This provides CORBA interfaces for generic creation and handling of common object collections.
- *Concurrency Control Service.* This provides a vehicle for mediating "concurrent access to an object such that the consistency of the object is not compromised when accessed by concurrently executing computations."
- *Enhanced View of Time.* This specification provides a "generalized Clock interface to represent clocks with differing characteristics."
- *Event Service.* This defines supplier (of events) and consumer (of events) roles for objects that exchange event data using standard CORBA requests.
- *Externalization Service.* This defines a standard for "externalizing (recording the object's state in a stream of data) and internalizing objects."
- *Licensing Service.* This defines interfaces that support management of software licenses, allowing usage of objects to be "metered" so that their services can be billed.
- *Life Cycle Service.* This defines operations for creating, copying, removing, moving, and deleting objects.
- *Lightweight Services.* This specification defines a compatible subset of three existing CORBAservices (Event, Naming, and Time) to make these services suitable for use in resource-constrained systems.
- *Management of Event Domains.* This specification defines an "architecture and interfaces for managing event domains (sets of one or more event channels that are grouped together for the purposes of management, and/or for providing enhanced capabilities to the clients of those channels such as improved scalability.)"

- *Naming Service.* This defines the mechanism by which objects on the ORB locate other objects, through the use of names and network directories.
- *Notification Service.* This specification is an extension to the existing OMG Event Service. It supports transmission of "events in the form of a data structure; event subscription; discovery; QoS; and an optional event type repository."
- *Notification/JMS Interworking.* This specifies an architecture and interfaces for managing Notification Service in conjunction with the Java Message Service. This encompasses such aspects as "event message mapping; QoS mapping; event and message filtering; automatic federation between Notification Service channel concept and topic/queue concepts; and transaction support."
- *Persistence Service.* This provides interfaces for storing persistent information on heterogeneous types of storage servers (such as files, DBMS, etc.).
- *Properties Service.* This provides operations to dynamically associate named values and state information with objects that lack a static IDL interface.
- *Query Service.* This provides query operations for object collections, and is a superset of SQL.
- *Relationship Service.* This provides a way to create dynamic associations between objects and to enforce referential integrity constraints. It defines two kinds of objects: "relationships and roles."
- *Security Service.* This provides a complete framework for distributed security, including vehicles for authentication, access control lists, confidentiality, certification, accountability, and non-repudiation.
- *Telecoms Log Service.* This specification defines interfaces for logging events and subsequent querying of the log records.
- *Time Service.* This provides interfaces for synchronizing time in a distributed object environment by enabling a user "to obtain current time together with an error estimate associated with it."
- *Trading Object Service.* This provides a *"Yellow Pages"* for objects, facilitating "the offering and the discovery of instances of services of particular types."
- *Transaction Service.* This implements the two-phase commit protocol to support flat and nested transactions by providing interfaces that combine the transaction and the object paradigms.

The horizontal CORBAfacilities consist of the Printing Facility, the Secure Time and Internationalization and Secure Time Facility, and the Mobile Agent Facility. These facilities work in conjunction with CORBAservices and applications objects to support functionality across multiple business domains.

In response to overwhelming industry demand, the OMG established a Domain Technology Committee in 1996 to promote the development of vertical industry standards. Vertical or domain CORBAfacilities define standard interfaces for application objects used within specific industry segments. Thus far, the following industries have established OMG task forces for the development of vertical industry interfaces [OMG04]:

- Business Enterprise Integration (BEI) Domain Task Force
- Consultation, Command, Control, Communications, and Intelligence (C4I) Domain Task Force
- Finance Domain Task Force
- Life Sciences Research Domain Task Force
- Manufacturing Technology and Industrial Systems (ManTIS) Domain Task Force
- Software Based Communications Domain Task Force
- Systems Engineering Domain Special Interest Group

Business objects are self-contained, nonprocedural, interoperable CORBA components that implement an IDL interface and CORBAservices. The OMG is working to promote development of generic, plug-and-play CORBA business objects and BOF (Business Object Facility) IDL extensions to facilitate dynamic integration of components at runtime. A conceptual overview of CORBA business objects is provided in Figure 5.21.

Many of the extensions planned in CORBA 3.0 are designed to provide more robust support of large-scale distributed networks. The first generation of CORBA ORBs lacked the feature sets and functionality (such as concurrency, transactional integrity, fault tolerance, security, etc.) needed to support mission-critical applications. With new CORBAservices and CORBA 3.0 functionality, this is being addressed, although these specifications are futures and not yet reality.

CORBA has successfully achieved client-side code portability across multiple platforms and languages. However, the portability of server code is limited, particularly from different vendors. This issue is being addressed, in part, by the CORBA 3.0 POA (Portable Object Adapters) standards and by planned extensions and improvements to the IDL. Portable object adapters provide considerable new flexibility in mixing and matching ORBs supplied by different vendors. When a client request is made, the ORB forwards the request to a POA. The ORB uses an object key to determine the correct association between the POA and the requested server object. The POA then dispatches the request to a server object for execution. An application can have multiple POAs for the same server object. As part of another key initiative, the OMG has introduced a standard called Model Driven Architecture (MDA). This is a middleware standard

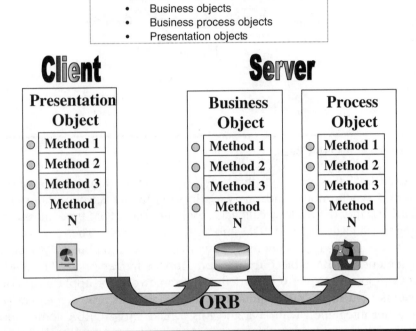

Figure 5.21 Types of CORBA business objects.

for integrating new systems with legacy systems with hooks into CORBA, XML, or Java.

Passing-by-Value will be supported in CORBA 3.0 as a new IDL construct (i.e., `valuetype`). This construct supports data and operations similar to those found in Java class definitions. When a `valuetype` is passed as an argument to a remote server, the object state is copied as part of the call. Thus, the operations invoked are local to that server and do not require transmission of request/reply messages over the network as they do in CORBA 2.0. This feature was needed to support remote Java interfaces.

Commercial, server-side ORBs do not scale well, particularly when it comes to supporting very large numbers of server objects. CORBA 2.0 uses a global repository to store object names. In large-scale applications, with millions of objects and object instances, this becomes unwieldy. Performance is also, in general, lower than that of approaches such as RMI. Server-side ORBS also lack the functionality provided by other more mature server products, such as TP Monitors; however, with the addition of asynchronous messaging, CORBA Beans, and other features in CORBA 3.0, this is being addressed.

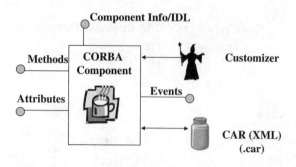

Figure 5.22   CORBA Bean.

CORBA 2.0 supported three types of invocations: synchronous, deferred synchronous, and one-way. In CORBA 3.0, asynchronous messaging is added to allow callbacks, polling, and MOM-type messaging using a statically defined IDL stub in a client object. These messaging types are needed, in conjunction with CORBA Component Model, to implement CORBA Object Transaction Monitor frameworks. The CORBA 3.0 Component Model (CCM) defines a framework for application components written in any programming language, and provides extensions for better integration with Java and EJB. Unlike EJB, CORBA components can be developed in multiple programming languages, so the object implementation is not tied to Java. CORBA bean containers (i.e., OTMs) can simultaneously host EJBs, as well as beans written in other languages. CORBA Beans share many features with EJBs, such as entity beans, session beans (for persistent and nonpersistent components), process beans, multiple interfaces, single inheritance, attributes, and customizers. Application programmers use customizers to adapt the appearance and behavior of the services provided by the CORBA bean to the client. Customizers give a client the ability to select services from a customized list presented through a user interface. A pictorial overview of a CORBA bean is shown in Figure 5.22. A CORBA bean is packaged in a zipped (.car) file — which is short for *component archive*. CORBA containers are built on top of the POA (portable object adaptor), and can incorporate CORBAservices such as security, events, notifications, MOMs, persistence, naming, and transactions. The CORBA Bean OTM framework is a powerful tool to manage object activation and deactivation, invocation requests, and enforcement of business logic within the component processing. This simplifies the object implementation because the framework handles execution flow and other system processing that would otherwise need to be explicitly programmed within the object structure.

CORBA excels in providing a flexible and neutral cross-platform, cross-language distributed network infrastructure. A plug-and-play IDL interface and the IIOP server-to-server backbone provide this. CORBA allows diverse legacy applications to coexist on the same network infrastructure, thereby permitting graceful migration to new technologies and applications. For example, because of its ability to operate on a diverse mix of platforms from a Web browser, Java is widely deployed in Internet applications. Java supports RMI over IIOP, providing a transparent mapping of RMI to CORBA over a standard wire protocol. Likewise, Enterprise JavaBeans (EJB) provides a secure, cross-platform infrastructure for deploying server-side components (i.e., beans) that communicate with each other and with the EJB framework. EJB supports a programming interface over IIOP; and as discussed earlier, CORBA supports IDL interfaces to EJBs. This means that CORBA clients can access EJBs implemented by CORBA EJB servers and can mix calls to CORBA and EJB objects within a transaction. In addition, "a transaction can span multiple EJB objects that are located on multiple CORBA-based EJB servers provided by different vendors." [SCALL00] CORBA downloads Java byte code from a Web server to a personal computer or Web computer, just like any Java class, and through IIOP, it ensures that the Java ORB can communicate with the ORBs of other vendors. In IT environments that must support COM, Java, COBOL, C, or other types of object programming paradigms, CORBA's platform and language neutrality make it a valuable mediator in expediting interoperability between Java and non-Java, CORBA-compliant servers for a wide range of complex applications.

5.6.2 Component Object Model (COM)/Distributed Component Model (DCOM)/COM+

Microsoft's COM family of standards constitutes the major competitor to CORBA. Microsoft has changed and added definitions to the COM standard over the years, so it has come to mean different things over time. To provide a basis for the discussion that follows, a summary of current COM standards is useful:

- *ActiveX.* This technology provides a vehicle for implementing COM components, known as *controls.* ActiveX components are independent, reusable software entities that can execute anywhere on a network running Windows (Windows 95/98/NT/2000) or Macintosh. ActiveX components are conceptually similar to a Java applet. They can be created and implemented using a variety of programming languages and development tools (including C++, Visual Basic, VBScript, etc.).

- *OLE.* This previously stood for *Object Linking and Embedding,* but now Microsoft's official position is that it does not have any translation. OLE is Microsoft's framework for supporting compound documents containing a variety of embedded content types (e.g., as text, sound, motion video, etc.). OLE is part of the COM/DCOM standards and, as such, an OLE document is also a COM component. OLE consists of hundreds of function calls and program interfaces collected in the Microsoft Foundation Class (MFC) Library.

- *Windows DNA.* This stands for *Windows Distributed InterNet Applications Architecture.* It is a COM+-based programming model and framework for building n-tier, high-performance, and distributed object applications. It defines a three-tier client/server model supported by (1) presentation layer tools (e.g., HTML, DHTML, and Microsoft script languages, components, Win32, etc.); (2) business layer tools [e.g., IIS, MTS, Microsoft Message Queue Server (MSMQ), Active Server Pages (ASP), COM Transaction Integrator (COMTI) for accessing IBM systems, etc.]; and (3) data access layer tools [based on Universal Data Access (UDA) technologies], such as OLE-DB.

- *DCOM.* This standard is used to deploy client and server COM components that interoperate and execute on separate machines. It serves the same basic functions as the CORBA IIOP; however, the implementation is quite different. In contrast to COM, which uses local remote procedure calls (LRPCs), DCOM uses DCE RPC, and its associated security features, interfaces, and packet formats. DCOM can be used across multiple network protocols, including TCP and HTTP.

- *COM+.* This provides an integrated runtime environment for supporting enterprise-strength, high-performance distributed applications. It combines and adds to the functionality of previous Microsoft technologies, including COM, DCOM, Microsoft Transaction Server (MTS), and Microsoft Message Queue Server (MSMQ). Some of the mission-critical functions provided by COM+ include (1) fine-grained security implemented at the method level, (2) load balancing, (3) object pooling for component reuse and JIT (just-in-time) object activation, both of which are needed for large-scale applications, (4) queued components, (5) publish and subscribe messaging and events, (6) transaction services, and (7) concurrency control. COM+ provides a runtime library to implement many default component features (i.e., IUknown, class factories, error information interfaces, security, etc.) to reduce the complexity of COM programming. [GORD00] "COM is a system-level component model that cannot rely on language-specific features to create objects. Thus, COM components are a little more complex.... COM+ attempts to make COM component development more like language-specific component development." [KIRT97]

■ *Longhorn.* This is Microsoft's code name for the next major release to replace Windows XP. Longhorn is a future product scheduled for release in 2006. Longhorn will utilize a new XML-based declarative programming language (i.e., XAML) that Microsoft is developing to make it easy to create client-side applications and COM components without programming.

At a high level, the COM family of standards and CORBA are very similar in the types of functions they perform. Both CORBA and COM are based on an object-oriented paradigm that defines objects as collections of methods and data. At the core of the COM standards is the COM object. All COM+, OLE, ActiveX, and DCOM objects are COM objects. The differences in Microsoft's naming of these COM components reflect different features and product packaging. However, they are all based on a standard Microsoft component model.

Like CORBA, COM+ utilizes an "object-bus" for interconnecting client and server objects across a distributed architecture. This is made possible by separating the object interface from its implementation through the use of an Interface Definition Language (IDL), which supports static and dynamic interfaces for method invocations. However, Microsoft's IDL is based on DCE and is not compatible with CORBA's IDL. To implement a COM component, one first starts by creating its class and interface definition using Microsoft's IDL. A MIDL precompiler creates client proxies and server stubs (analogous to CORBA's stub and skeleton), and converts the IDL into a *type library*, which is a binary representation of the IDL. MIDL provides language bindings for all its development environments (e.g., J++, C++, Visual Basic, etc.) and is language independent. After an interface is implemented, linked, and registered, it is ready for use.

The COM object is a collection of methods with an interface implementation that defines its class. A 128-bit Class ID (CLSID) uniquely identifies the component or class. In CORBA, one and only one interface defines an object's class. However, in COM, an object has multiple client interfaces. This is shown in Figure 5.23.

The COM interface is a low-level binary API containing an array of pointers. This array or table is called a *virtual table* (a.k.a. *vtable*). Each COM interface must have an associated vtable because COM invokes methods only through a pointer to the interface associated with that method. It is only through the vtable that a client program can interact with a COM server object. Figure 5.24 depicts a COM interface between a client and a server.

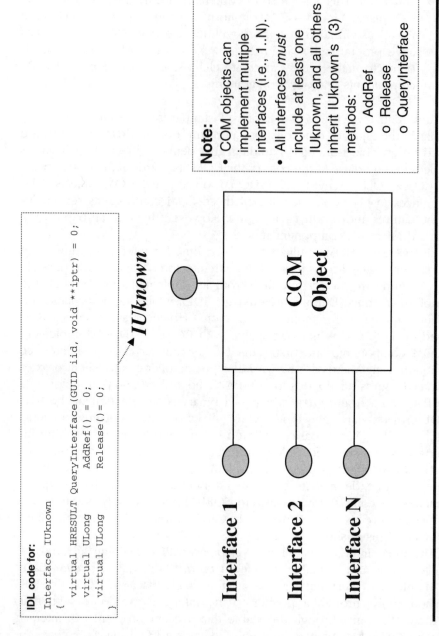

IDL code for:

```
Interface IUknown
{
    virtual HRESULT QueryInterface(GUID iid, void **iptr) = 0;
    virtual ULong    AddRef() = 0;
    virtual ULong    Release() = 0;
}
```

IUknown

COM Object

Interface 1

Interface 2

Interface N

Note:
- COM objects can implement multiple interfaces (i.e., 1..N).
- All interfaces *must* include at least one IUknown, and all others inherit IUknown's (3) methods:
  - o AddRef
  - o Release
  - o QueryInterface

Figure 5.23 Overview of COM object.

Note:

•The Virtual Table (vtable) contains an array of pointers to the method implementations and is shared by all class instances.

•*vtable* allows multiple client programs to dynamically link to object implementation at execution time.

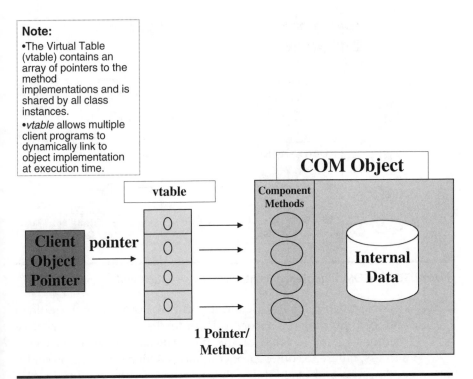

Figure 5.24   COM interface.

By convention, the symbolic name assigned to a COM interface must begin with "I" (e.g., IUknown). Each interface is assigned a unique runtime identifier, which is referred to as an *Interface Identifier (IID)*. The IID is produced by calling the DCOM API function CoCreateGrid, which executes an algorithm using the current data, time, network card ID, and a high-frequency counter. The result is a 128-bit globally unique identifier (*GUID*) that provides an unambiguous object reference.

All COM objects must implement at least one *IUnknown* interface, which is used to control client/server interactions and the life of the object. In addition to methods of their own, all other interfaces inherit these IUknown methods:

- *QueryInterface.* Before client/server interactions begin, the client is first given a pointer to access the COM object. To discover the pointers to the object's other interfaces, QueryInterface must be invoked. The client uses QueryInterface at runtime to determine whether a particular interface specified by an IID is supported by the object, and if so, to receive the pointer to the interface. This pointer is needed before the client can invoke the object services.

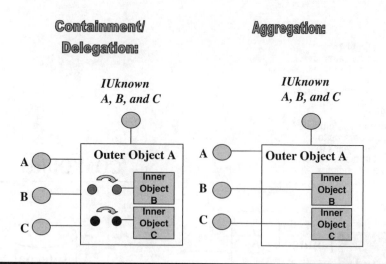

Figure 5.25   COM inheritance: containment/delegation and aggregation.

An object can refuse service to a client by not returning a pointer and by returning an error code instead. Most COM+ interface methods return a value of the type HResult (result), indicating success, failure, or other error conditions. QueryInterface promotes code reuse and extensibility through the use of *containment/delegation* and by *aggregation*. These concepts are illustrated in Figure 5.25. In containment/delegation, an outer object acts as a client to an inner, subordinate object. In this fashion, the outer object contains and delegates tasks to the inner object, thereby implementing itself. Aggregation is a specialized case of containment/delegation in which an outer object exposes interfaces belonging to an inner object as if they were implemented by the outer object. The inner object interacts directly with the client but delegates its IUknown functions to the outer object. This helps to decompose the implementation of the inner and outer objects. By implementing multiple object interfaces, new features can be added to a component without affecting the object's existing client interfaces. This is referred to as *versioning*. When versioning is used, a new interface is derived from an older one, but it is assigned a new IID. Each version of the interface supports the methods of its predecessor in addition to the new ones that are implemented. Through QueryInterface, clients can discover new interfaces and query them to determine the functions they support. This allows a client to invoke a function only if it is supported. In this way, the client can maintain compatibility with old and new interface versions. To aid in version control, it helps if the original and derived interfaces keep the same name with the addition of consecutive numbers (e.g., IClassFactory1, IClassFactory2, etc.) to keep them distinct.

■ *AddRef.* COM uses a reference counting scheme to control the life span of an object. An object must wait until it services all of its clients before it is allowed to self-destruct. Every object uses a 32-bit reference counter to keep track of the number of clients it is servicing. AddRef increments the reference counter every time a new component starts to use the object interface.

■ *Release.* Release works in conjunction with AddRef to manage an object's lifecycle. When a component is finished using an interface, it calls Release. Release responds by decrementing the reference counter by one. If the reference counter reaches zero (0), the object instance can be deleted to free up system memory. COM assumes that *all* client and server components properly implement the AddRef and Release methods to ensure proper object lifecycle management.

In addition to the IUknown interface, all server objects must implement the *IClassFactory* interface, which contains these two methods: [MICR04]

1. *CreateInstance.* This method creates an uninitialized instance of an object and returns a pointer to an object interface.
2. *LockServer.* This method increments the reference counter assigned to the class object to ensure that the server object remains alive.

The invocation of IClassFactory by a local client is illustrated in Figure 5.26. COM also defines the *IClassFactory2* interface, which is derived from IClassFactory, as a means to limit an object's creation to only those clients

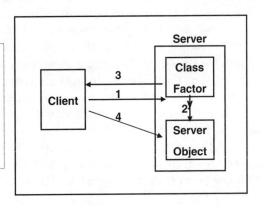

Note:
1. IClassFactory::CreateInstance is invoked by client.
2. IClassFactory creates object and obtains interface pointer.
3. Object interface pointer is returned to client.
4. Client invokes server object methods.

Figure 5.26   Local object creation with IClassFactory.

authorized to use them. IClassFactory2 supports the use of a license key that can be presented at runtime.

To recap the preceding discussion, a client initiates service by calling CoCreateInstance with the requested server object's CLSID and its associated IID. This creates an instance of the server object. When a client needs to create many objects of the same class, it can acquire a pointer to a class factory for that class. When the server object returns an interface pointer to the client, thereby granting it service, it also invokes AddRef to add to its internal reference counter. The client invokes the server object's interface methods via the interface pointer. If the client needs to invoke methods on another interface on the same object, it makes a QueryInterface call and passes the new IID to the current interface, and a pointer to the second interface is returned. When the client has completed using the services of an interface, it issues a Release call.

A server object can be packaged as either an in-process, local, or out-of-process server. The client accesses all server objects, regardless of location, via an interface pointer. An in-process server resides in a .DLL and is loaded directly into the client process when the client first accesses the server. A local server executes an .EXE in a separate process from the client but on the same machine and operating system. Clients use DCOM's Lightweight RPC (LRPC) for local server communication. Remote servers execute on remote machines, on possibly different operating systems. Clients use DCOM's DCE-RPC to communicate with remote servers. When the client and server are on different processes or machines, stubs and proxy objects are required. A client uses a proxy object to obtain the interface of a remote server. The proxy object provides a local representation of the server interface, while marshalling and passing necessary parameters to a server stub. The server stub, in turn, unmarshals incoming client requests and presents them to the designated object in the server. The stub object is created in the address space of the server and is used to manage the server object's actual interface pointer. The Microsoft IDL compiler is used to create these proxies and stubs. In this manner, the same client code for accessing in-process and out-of-process servers can be used (i.e., the server location is transparent to the client). COM employs a Service Control Manager (SCM) to locate and manage connections to remote servers. COM invokes the SCM when a request is made to create an object or to invoke IClassFactory. When a request for IClassFactory is made of the remote server, the local SCM contacts the remote SCM. The remote SCM, in turn, locates and launches the server, and establishes an RPC connection to the requested class factory. This is illustrated in Figure 5.27.

To respond to client requests for server objects, the server must implement the following generic functions: [ORFA99]

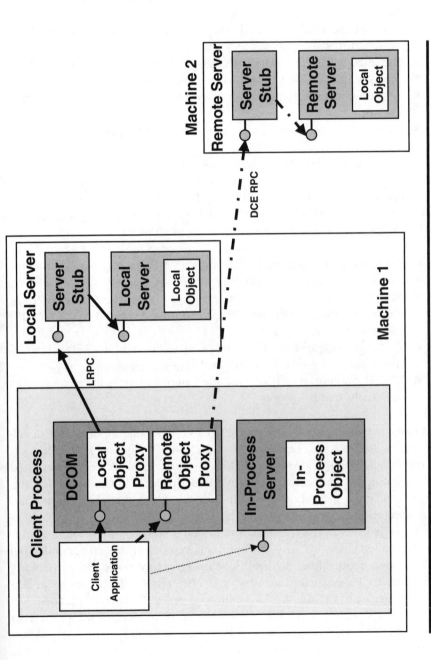

Figure 5.27 Local versus remote server processing.

- Implement IClassFactory interfaces for every CLSID; and if a class supports licensing, it must also implement an IClassFactory2 interface.
- Register the CLSID for each class supported with the NT Registry, which stores the pathname to the server .DLL/.EXE files.
- Initialize the COM library to access COM runtime services and APIs.
- Verify that the COM library is a compatible version.
- Implement a mechanism to self-destruct when there are no active clients.
- Uninitialize the COM library when it is no longer in use.

COM+ is an integral part of Windows and comes at no additional cost. It unifies COM, DCOM, and Microsoft Transaction Server (MTS) to enable development of large-scale, mission-critical distributed systems. COM+ was designed to automate many network management tasks that would otherwise require substantial programming to implement, including:

- Transaction management according to ACID properties
- Load balancing across multiple machines
- Security implemented on several levels (via Access Control Lists, Kerberos, SSL, certificates, IPSec, Security Service Provider (SSPI) extensions, etc.)
- Queued components for high availability and loosely coupled applications based on publish/subscribe messaging
- Object pooling, which keeps component instances alive in memory for ready use by clients

COM+ is well supported by a variety of vendors offering low-cost development tools such as Borland C++, Delphi, PowerSoft PowerBuilder, Microsoft Visual Basic, J++, and Visual C++. COM+ can be programmed in any language that can be compiled to produce binary code for the component implementation. The flip side of this is that the binary code implementation is platform specific.

COM+ is used in over 150 million systems, and has the largest worldwide installed base of any component system architecture. [MICR04] COM+'s free acquisition and low development tool costs are particularly appealing to IT organizations that are cost conscious or on a limited budget. Because COM+ works optimally on a Windows platform, in organizations with a large installed base of Windows desktops, it is a logical solution (although whether it is the best solution must be determined by careful requirements and detailed feature analysis). COM+ was designed to extend the development investment in COM by making it easier to leverage reuse of existing components in an enterprise-strength infrastructure. COM+ is a more mature technology than its competitor J2EE, and is widely implemented in COTS packaged component solutions. However, COM+ should be viewed as an interim solution, particularly with respect to new development. Microsoft has recently announced that it will continue support for COM and DCOM, but will be discontinuing its research and development on them. Instead, its efforts are focused on Indigo, a new service-oriented paradigm that will be implemented in Longhorn. According to Don Box, Microsoft software architect, "programs will use managed services based on the Extensible Markup Language to communicate with each other." [BOX04] In essence, Microsoft is redirecting its emphasis toward Web services and XML. According to Box, the problem with object-oriented technologies is that the programming requires a deep level of familiarity with detailed system handling. [BOX04] This makes it difficult for programmers to implement components and to reuse them on a variety of platforms. After studying the sections on CORBA and COM+ in this book, no doubt the reader is sympathetic to this argument.

Indigo is intended to provide "rich support for service-oriented design that is complementary to traditional object-oriented approaches. Indigo marries the best features of .NET Remoting, ASMX,[14] and .NET Enterprise Services into a unified programming and administration model. Indigo's deep support for standard protocols, including HTTP, XML, and SOAP, makes it easier to integrate applications and services without sacrificing security or reliability." [BOX04] Indigo will be based on XML, SOAP, and emerging Web Service protocols to provide a vehicle for supporting mission-critical functionality (i.e., security, reliability, and transactions) in SOAP-based applications. Indigo will be packaged with Longhorn and will also be available as a separate download for Windows XP and Windows Server™ 2003.

Microsoft has stated that "COM+ will continue to be supported in Indigo, and that existing COM+ components can be integrated within Indigo applications via the existing .NET/COM+ interoperability, or COM+ components can be migrated incrementally to .NET including the new Indigo features." Along with other caveats, Microsoft has advised developers to adhere to the following practices to ensure that COM+ applications can be transitioned in the future to Indigo: [CELA04]

1. Build services using the ASP.NET (ASMX) Web service model. (Enhance your ASMX service with WSE if you need the WSE feature set.)
2. Use object technology in a service's implementation.

According to [LONE04], "Moving developers away from the object-oriented world is a key element of Microsoft's battle against Java 2

Enterprise Edition (J2EE). J2EE is generally favored for complex back-end server software used in complex, mission-critical systems, such as running stock exchanges or high-volume Web sites. [LONE04] Longhorn promises to provide a developers platform where the emphasis is on how the application looks, as opposed to how it is put together.

5.6.3 Web Services

Web services are being promoted by a number of major vendors — including Microsoft, Intel, Sun Microsystems, and IBM — as the next-generation technology for business-to-business (B2B) and business-to-consumer (B2C) E-ecommerce. Web services allow programs written in different languages and executed on different platforms to interoperate. Although CORBA is intended to serve a similar mission, Web services operate over standard Internet protocols — such as TCP/IP, http, and XML — and are simpler to program and implement. The emergence of Web services reflects an industry shift from proprietary to open network protocols, with an emphasis on interoperability between disparate applications, systems, and networks.

From the perspective of IT management, large-scale client/server environments are expensive and complex to deploy and maintain, representing significant budget impacts and organizational risk. The goal of Web services is to reduce the costs and risks of integrating large client/server networks. However, Web services technology is very new and does not fully address security and mission-critical requirements. Therefore, most implementations are proof-of-concept, noncritical applications in an Intranet environment, where user access can be controlled. Sun Microsystems' J2EE and Microsoft's .Net frameworks are implementing Web services capabilities to support enterprise, large-scale networks.

According to [LAIR01], Web services can be defined as a "standard way to expose applications or resources on a network, built over HTTP and XML. A client connects to a URL and retrieves an answer: That's a Web service." Web services are based on a number of protocols, including XML (Extensible Markup Language); Simple Object Access Protocol (SOAP); Web Services Definition Language (WSDL); and Universal Description, Discovery, and Integration (UDDI). These protocols provide a foundation for message-based communication, which is the key to Web services. CORBA and DCOM/COM+ integrate applications, platforms, and networks using specialized programming interfaces and network transport mechanisms that rather tightly couple clients and servers. In contrast, Web services use a simple, very loosely coupled Web-based message exchange between clients and servers. Web service messaging relies on the following conventions:

■ Servers describe their services and provide instructions for accessing them using WSDL. WSDL provides a means to implement service contracts between servers and clients. According to the W3C, "WSDL is an XML format for describing network services as a set of endpoints operating on messages containing either document-oriented or procedure-oriented information. The operations and messages are described abstractly, and then bound to a concrete network protocol and message format to define an endpoint. Related concrete endpoints are combined into abstract endpoints (services). WSDL is extensible to allow description of endpoints and their messages regardless of what message formats or network protocols are used to communicate." [W3C01] WSDL files contain the following elements: (1) root element (i.e., <definitions>), which provides a complete description of all services names, operations, and associated methods exposed to the client; (2) message specifications (i.e., the method parameters sent to and returned from the server); (3) binding specification, which has "two attributes: style and transport. Style is an optional attribute that describes the nature of operations within this binding. The transport attribute specifies HTTP as the lower-level transport service that this binding will use;" and (4) a summary or implementation file, which is needed to publish the Web service at a UDDI registry. [SIDD01] The Universal Description Discovery, and Integration (UDDI) interface is an XML application that allows clients to peruse the service registries.

■ Clients access WSDL entries in a number of ways. Clients can locate service providers — by name, by business category, or by a number of other standard classification schemes (e.g., North American Industry Classification [NAICS], Standard Industrial Classification [SIG], etc.) — by accessing WSDL entries published in a service registry maintained by an independent service broker. The registry provides pointers to WSDL files that can be downloaded by the client. Alternatively, the client can obtain the WSDL files directly from the service provider, or through a search engine that searches the Web for WSDL entries, or by using WSIL (Web Service Invocation Language). WSIL is a Java API for invoking Web services that is independent of the transport protocol or service environment. "It allows stubless or completely dynamic invocation of Web services, based upon examination of the metadata about the service at runtime. It also allows updated implementations of a binding to be plugged into WSIF at runtime. It can also allow a new binding to be plugged in at runtime.... It allows the calling service to choose a binding deferred until runtime." WSIL can also invoke any service that is described in WSDL. [FREM02]

■ Having identified a service provider and the service it wishes to utilize, a client (which, in another context, might be a server) initiates a request for service from a server process via a remote method invocation. The client formats the request in a SOAP XML-based message and binds it to a URL and protocol, all according to the specifications provided in the WSDL file. Similar to Sun's RMI and CORBA's IIOP, SOAP handles the message exchange between the client and server. The server implementation remains hidden from the client.

The following discussion provides an overview of the major protocols used in Web services, including XML, SOAP, and UDDI.

XML is a license-free, platform- and application-independent standard developed by the World Wide Web Consortium (W3C). It establishes rules for storing, structuring, and formatting data and its presentation. According to the W3C, the design goals for XML are: [W3C04]

■ XML shall be straightforwardly usable over the Internet.
■ XML shall support a wide variety of applications.
■ XML shall be compatible with SGML.
■ It shall be easy to write programs that process XML documents.
■ The number of optional features in XML is to be kept to the absolute minimum, ideally zero.
■ XML documents should be human-legible and reasonably clear.
■ The XML design should be prepared quickly.
■ The design of XML shall be formal and concise.
■ XML documents shall be easy to create.
■ Terseness in XML markup is of minimal importance.

XML addresses inherent limitations of HTML. HTML documents intersperse both presentation instructions and the data to be displayed on a Web page. This intermingling makes it difficult to generalize HTML documents for use by other applications and programs. In contrast, XML documents contain only data elements and data tags. The tags describe the data contained in the XML document, facilitating programmatic treatment of the data by a wide range of applications.

All XML documents are stored in human-readable text format. Thus, any text editor (i.e., vi, Notepad, etc.) that supports ASCII/UNICODE can be used to open, create, and modify XML documents. A variety of specialized XML text editors are also available to facilitate XML document creation, including the following:

■ *Amaya*. This W3C sponsored text editor is written in C, for Windows, UNIX, and MacOSX operating systems. It is available at http://www.w3.org/Amaya/

- *MSXML*. This is an XML text editor built into Internet Explorer, releases 5.0 and above.
- *Mozilla*. This is an XML text editor built into Netscape's Browser, releases 6.0 and above.
- *XMLWriter*—This is a text editor based on MSXML, and is available at http://www.xmlwriter.net/.
- *XML SPY*. This text editor, which also provides support for XML schema bindings and code generation for multiple languages and platforms, is available at http://www.xmlspy.com.
- *Visual XML*. This text editor was written in Java and is available at http://www.pierlou.com/visxml/index.html.
- *BBSEdit* —This is an HTML and XML text editor for Macintosh, and can be downloaded at http://www.bbedit.com/.

After an XML document is created, it must be processed by an XML parser. This verifies that the XML document conforms to XML syntax and specifications. If an XML document passes the inspection of an XML parser, it is said to be a *well formed* document. If the XML document contains an optional reference to a Document Type Definition (DTD) file, a *validating XML parser* is needed to verify their consistency with respect to each other. A DTD file specifies the structure of the XML document. If the XML document complies with the referenced DTD, it is said to be a *valid* document. An XML parser that cannot process DTD references in an XML document is referred to as a *non-validating parser*. After an XML document is successfully parsed, it can be displayed using a Web browser.

XML documents are comprised of the following parts:

- *Prolog*. This consists of an XML declaration, and optional statements or comments.
  - The XML declaration is always the first entry in any XML document, and has the following general form:

    <?xml version="*version number*" encoding= "encoding type" standalone="yes|no" ?>

  where:
  - *version number* refers to the XML specification used by the document.
  - *encoding type* specifies the character codes used in the document.
  - *standalone* indicates the presence or absence of external file references (such as DTDs, CSS, schemas, namespaces, etc.).

- The prolog may also contain comments that are used for documentation and other purposes. The general form of a comment entry is:

  <!—insert comment text here -->

- The prolog may contain references to *namespaces*. Namespaces are referenced by a URI (Universal Resource Identifiers).[15] They are used to distinguish XML elements defined in different contexts and that have the same name. Because domain names are unique, they are often used to help specify XML namespaces.
- The prolog may contain references to Document Type Definitions (DTD) or schema(s). An XML parser uses these files to validate an XML document. They specify the meaning, content, values, and relationships between elements, attributes, and entities. There are a number of significant differences between schemas and DTDs, which are beyond the scope of this text. Briefly, DTD is an older, non-XML standard that uses Extended Backus-Naur Form (EBNF) syntax. Schemas are XML documents, providing support for namespaces and a greater number of data types. Both have overlapping and unique features, with different implementations.
- The prolog may contain references to style sheets (alternatively, style sheets can be referenced in the XML document body, described below.). Style sheets provide instructions on how to present the data in a Web browser. The two main style sheet languages are Cascading Style Sheets (CSS) and XSL (Extensible Style Sheet Language). The general form of this processing instruction is:

  <?xml-stylesheet type="style" href="sheet" ?>

  where:
  - *style* refers to the type of style sheet to be used. For example, if a cascading style sheet is to be used, this would be indicated by "text/css."
  - *sheet* is the specific name and location of the style sheet.
  - *xml-stylesheet* is a processing instruction.
- *Document body.* The document body contains data and associated tags, arranged in a hierarchical tree structure. XML element and attribute names can be of any length and can contain letters, digits, and special characters. They must begin with either a letter or an underscore, and are case sensitive.

- There is one and only one root element, and it is the first element of the XML document body. Other elements can be nested within the root element. The general form of an element is:

```
<element_name>Content</element_name>
```

For example: `<Author>Mary Mann</Author>`

- Empty elements can be added without any content to mark a section for the XML parser; for example:

```
<element_name/>
```

- Elements can be augmented by attributes, which provide descriptive information. Attributes consist of name and value pairs of the general form:

```
<element_name attribute="value"/>
```

- CDATA sections are used to include large blocks of text in an XML document that are not processed by the XML parser. The general form for including a CDATA section is:

```
<![CDATA[
Text Block
]]>
```

■ *Epilog.* This is an optional section containing additional comments and processing instructions. A number of standard epilog functions are available for performing specialized processing (e.g., to write a string to a text file, return an HTML comment with a copyright notice, return document tracing information, etc.).

Once an XML document has been created, it can be attached to a formatted Web page in a process known as *data binding*. When data binding is used, changes to the Web page and the XML document can be made independently. Informational Web pages (e.g., which might present weather reports, stock quotes, search results, etc.) often use data binding to dynamically display updated content, without making changes to the HTML code that generates the Web page. Data binding involves the use of an XML parser to produce data structures from an XML document. Generalized parsers are used to read, edit, add, and delete elements within an XML document. These parsers are used to programmatically manipulate

XML data. "Data binding also typically includes a unmarshalling (or loading) mechanism that can populate instances of the new data structures from data conforming to the underlying document structure and a marshalling mechanism that creates XML documents from an instance of the data structures." [THEB04]

DOM and SAX are the two most widely used XML parsers. The Document Object Model (DOM) API builds a hierarchical tree based on entities contained in an XML document. The tree is held in memory, simplifying program navigation and manipulation of the XML data structure. The Simple API for XML (SAX) parser processes an XML document in a single pass, generating *events* for each entity (i.e., a data element, data tag, etc.) it encounters. The events are used to return data for display or further processing. The use of these parsers requires specialized knowledge of their respective programming interfaces, and their APIs support relatively low-level programming functions. A number of vendors offer tools to further automate and improve the level of abstraction in processing XML documents to reduce the burden on a programmer. One such example is xmlspy® 2004, which includes a code generator that produces Java, C++, or Microsoft C# class files based on data elements defined in an XML Schema. According to [ALTO04], "An XML data binding allows you to programmatically work with XML documents from within your software application via a set of simple objects. It can be thought of as an abstraction layer between an XML document and your custom application (business logic), alleviating the need for developers to understand the technical intricacies of loading, editing, and saving an XML document using low-level XML parsing APIs." Microsoft supports data binding through the use of DSOs (Data Source Objects) and ActiveX Data Objects (ADO). A collection of records, or a data island, is attached to an HTML file, either directly within the file or, preferably, in an external XML file, as a DSO. ADO associates a method to the DSO. The method might perform validation, processing, or other data transformation actions.

A wide variety of vendors, standards organizations, and user communities are working together to develop a diverse and growing set of specialized XML framework standards to support horizontal and vertical industry applications. A few examples of this include:

- *ebXML (Electronic business using eXtensible Markup Language).* This is a set of protocols promoted by OASIS (Organization for the Advancement of Structured Information Standards) and the United Nations body for Trade Facilitation and Electronic Interfaces which is designed to support B2B (business-to-business) applications over the Internet. For more information, the interested reader is referred to the following Web site: http://www.ebxml.org.

- *CML (Chemical Markup Language)*. This defines a framework for storage and manipulation of molecular data and information using XML and Java. For more information, the interested reader is referred to the following Web site: http://www.xml-cml.org.
- *FIX (Financial Information Exchange protocol)*. This protocol is used for communicating securities information, such as stock quotes, market data, and trade orders. "FIX is widely popular, and, according to one study is used by 82% of brokers." [MALI03] For more information, the interested reader is referred to the following Web site: http://www.fixprotocol.org/cgi-bin/Welcome.cgi.
- *FpML (Financial products Markup Language)*. This is an XML-based standard for complex financial products. "FpML aims to standardize and streamline trading and exchange of information in the field of privately negotiated financial securities such as derivatives, swaps, and structured products. Currently, the standard covers interest rate swaps and Forward Rate Agreements (FRAs), equity, FX, and interest rate derivatives." For more information, the interested reader is referred to the following Web site: http://www.fpml.org/.
- *VoiceXML*. This framework is designed to support interactive voice communication using telephone, PC, or PDA devices. For more information, the interested reader is referred to the following Web site: http://www.voicexml.org.

Web services are based on the SOAP communications protocol. SOAP uses HTTP as a transport mechanism and XML as a standard message format. According to the W3C, SOAP is a one-way message paradigm that also supports request/response, multi-cast, and other complex client/server interactions. The core of the SOAP specification is the required XML message format, which consists of three elements (see Figure 5.28):

1. *Envelope*. This is mandatory and is the top element of the XML message document. The general form of the envelope has the following appearance: [W3CS04]

```
<?xml version="1.0?">
<soap:Envelope
xmlns:soap="http://www.w3.org/2001/12/
    soap-envelope"
soap:encodingStyle="http://www.w3.org/2001/12/
    soap-encoding">
...
Message information goes here
...
</soap:Envelope>
```

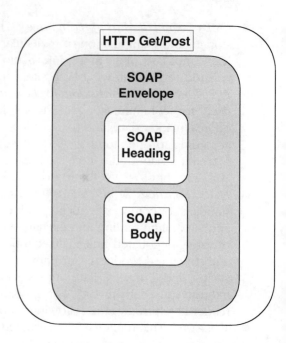

Note:

•An HTTP Get command requests a web page, and the HTTP Post command returns a data envelope across the Internet.

•Mandatory envelope encapsulates optional headers, and one or mandatory body elements, in this order.

•The headers and body are application specific and are not defined by SOAP standard.

Figure 5.28   SOAP message format.

2. *Header.* This is optional and is used to provide control information to a SOAP message recipient. It provides processing instructions for authentication, transaction management, message routes, intermediary node handling, etc. As SOAP messages are transmitted across a multi-hop route, intermediary nodes (designated with an "actor" attribute) can read, delete, or change header information. This capability is intended to provide a means for intermediary nodes to implement value added services.

3. *Body.* This is a mandatory section typically used for RPCs and error reporting. It contains the payload for the message recipient. By mapping application data to an XML format, SOAP allows applications to exchange data that they might not otherwise be able to directly.

One of the main concerns with respect to SOAP is its lack of security provisions. SOAP relies on the application programmer/network engineer to implement security because it is not addressed by the standard. Thus, there are a number of ways in which SOAP messaging can be compromised. One area of vulnerability is the SOAP message header. Through the actions of a hacker or by programming error, it is possible that intermediary nodes can intercept, read, modify, or delete the contents of a SOAP message en route. Another vulnerability, as reported in the CERT Vulnerability Note (VU #736923), is that the "Oracle Application Server 9iAS installs with SAP enabled by default, allowing remote, anonymous users to deploy and remove SOAP services." Thus, in theory, a client could inappropriately invoke or corrupt Oracle services that would otherwise be off-limits to them by taking advantage of this security loophole. SOAP also allows MIME attachments to a SOAP message through which worms and virus might be spread. [CLEW02] To guard against potential confidentiality, authentication, and data integrity vulnerabilities, the use of encryption (according to the XML-Encryption specification), digital signatures, VPNs, and other security measures should be considered when using Web services. The W3C is working on standards in these areas to augment XML.

The Universal Description, Discovery, and Integration (UDDI) protocol is sponsored by the OASIS standards consortium and major hardware and software vendors (including Intel, Microsoft, IBM, CISCO, and Oracle). UDDI leverages XML and Web services to provide a standard, interoperable platform for locating and using Web services over the Internet. It provides tools to help buyers and sellers locate each other, assess their respective offerings and mutual requirements, negotiate terms, and engage in various forms of B2B interaction. The UDDI protocol is not specific to any industry, and is designed to support worldwide commerce. [OASI04] Extensive documentation on UDDI can be found at Microsoft's and IBM's Web sites, as well as the OASIS UDDI Web site: http://www.uddi.org.

Although they provide the means to integrate large-scale, disparate applications and networks, the Web Services protocols (XML, SOAP, WSDL, and UDDI) do not fully address the requirements of mission-critical client/server applications. These applications require specialized standards, tools, and programming environments to support a host of network management and development needs. These include, but are not limited to:

■ Administrative services (e.g., auditing, logging, etc.)
■ Application integration
■ Business logic processing
■ Configuration management and control
■ Connection and communications services
■ Exception processing
■ Messaging services
■ Security (i.e., authentication, authorization, etc.)
■ Session management
■ Server-side processing (i.e., disk caching, load balancing, etc.)
■ Transaction management

Application frameworks are needed to provide a unified infrastructure for integrating, deploying, and managing the organization's hardware and software to address these concerns. Sun Microsystems' J2EE/Sun One (Open Net Environment) and Microsoft's .NET are the two major application framework technologies for building client/server applications. Both are designed to leverage Web services. Like CORBA and COM, they support many of the same functions but differ substantially in their philosophical perspective and implementation.

J2EE is based on the Java 2 SDK (Software Development Kit) and consists of the following major elements: [SUN04B]

- *J2EE specification.* This defines "the APIs to be provided with all J2EE platforms and includes full descriptions of the support levels expected for containers, clients, and components."
- *J2EE reference implementation.* This is "a complete implementation of the J2EE specification and is used by system vendors to compare with their own implementations."
- *Compatibility Test Suite.* This provides a suite of tests for all classes and methods required by J2EE, as well as interoperability tests.
- *Enterprise Java BluePrints.* This provides documentation and examples to illustrate "best practices for developing and deploying component-based enterprise applications in the J2EE platform."
- *Platform services.* This includes support for Enterprise JavaBeans, Java Servlets, JavaServer Pages, Java Connectivity Architecture (which is used to interface Java and non-Java applications), Java Message Service (to support asynchronous messaging with MOM middleware), Java Management extensions (to manage J2EE servers and applications), Java Naming and Directory Interface (to provide component location transparency), and Web services APIs. The Web services APIs are used to implement Web applications in Java, and support direct processing of XML documents and remote procedures. They include: [SUN04C]
  - Document oriented:
    - *Java API for XML Processing (JAXP).* This is used to process XML documents with XML parsers.
    - *Java Architecture for XML Binding (JAXB).* This is used to process XML documents using schema-derived JavaBeans component classes.
    - *SOAP with Attachments API for Java (SAAJ).* This is used to send SOAP messages over the Internet.
  - Procedure oriented:

- *Java API for XML-based RPC (JAX-RPC).* This is used to issue SOAP method calls over the Internet and to receive the results.
- *Java API for XML Registries (JAXR).* This is used to provide a standardized means of accessing business registries and sharing information in business-to-business communications.

It should be stressed that J2EE is a set of specifications, and not a product. Vendors license J2EE from Sun and implement it in their product offerings. Many vendors — such as IBM, BEA, Sybase, and Oracle — have licensed J2EE, and those that have passed the J2EE Compatibility Test Suite (CTS) are listed on the Sun Microsystems Web site.

The Sun ONE Platform is a comprehensive framework for creating and implementing applications based on J2EE and Web services. *"Services on demand"* is used by Sun Microsystems to refer collectively to personal computer applications, client/server applications, Web applications, and Web services interacting, anywhere, any time, and from any device. [SUN04D] Sun One is the delivery platform for Services on Demand. It encompasses "traditional Net-based services, such as security, authentication, and directory, along with more advanced capabilities, such as virtualized storage and composite services (those created by combining separate services.)" [SUN04E] Sun One provides an application development environment, with integrated tools built using NetBeans, Web application middleware, and hardware and software for application, portal, integration, and directory servers.

The J2EE philosophy is encapsulated by the motto: "Write Once, Run Anywhere." With Sun Microsystems' approach, all application development is in Java. However, many vendors can implement J2EE-compliant products that can be executed on a broad range of platforms. J2EE is an open standard (controlled solely by Sun Microsystems) and a mature technology, particularly in comparison with .NET. There is broad vendor support for J2EE products, including IBM's WebSphere middleware server, BEA's WebLogic Server, HP Web Services Platform, Oracle's 9*i* Web Services Framework, and others. Portability between J2EE commercial offerings has been limited, due to incompatibilities arising from varying degrees of conformance with the J2EE specifications and vendor value-added features. Therefore, J2EE may lead to vendor lock-in, albeit not Microsoft. The J2EE development environment is based on Java and EJBs, which can be complex to implement. J2EE software costs, relatively to .NET, are generally higher. However, when an organization must support multiple platforms, J2EE may be the only practical choice.

In contrast, Microsoft's .NET is based on the philosophy of supporting many programming languages, running on a Windows/Intel-based platform. Microsoft has made a strong commitment to the tight integration of its operating system and development environment to Web services. Accordingly, Web services are built into .NET (and in the future, Longhorn) in compliance with SOAP, WSDL, and UDDI standards. For organizations using predominantly Windows desktops, .NET is a compelling option. Microsoft's .NET encompasses the following elements: [MICR04B]

- *.NET Framework 1.1.* This is used for building "Web-based applications, smart client applications, and XML Web services — components that facilitate integration by sharing data and functionality over a network through standard, platform-independent protocols such as XML (Extensible Markup Language), SOAP, and HTTP."
- *Developer tools.* This includes Microsoft Visual Studio .NET, an integrated development environment (IDE).
- *.NET Framework.* This includes the Common Language Runtime (CLR), unified class libraries, and ASP.NET. CLR is the vehicle by which Microsoft supports numerous programming languages (e.g., Visual Basic.NET, Visual C#.NET, JScript.NET, and others). ASP.NET is a "set of technologies in the Microsoft .NET Framework for building Web applications and XML Web services. ASP.NET pages execute on the server and generate markup such as HTML, WML, or XML that is sent to a desktop or mobile browser." [MICR04C]
- *Enterprise servers.* This includes Microsoft Windows® Server 2003, Microsoft SQL Server™, and the Microsoft BizTalk® Server for Web services and Web-based applications.
- *Client software.* This includes Windows XP (for personal computers), Windows CE (for PDAs and other small footprint devices), and Microsoft Office XP.

5.7 Summary of Client/Server and Distributed Object Network Design Considerations

Distributed systems are intended to provide resource sharing, openness, concurrency, scalability, fault tolerance, and transparency. The extent to which these goals are satisfied is in large part determined by how well the client/server implementation is tailored to the organization's needs. For example, fat- and thin-client architectures support fundamentally different types of applications, and the wrong application of either could have severe performance, cost, and scalability impacts. There is an increasing industry trend toward the embrace of thin-client solutions that leverage

Table 5.1   Feature Comparison of CORBA, COM+, J2EE, and .NET

Feature	CORBA	COM+	J2EE	.NET
Support for many platforms	X		X	
Support for many languages	X	X		X
Product maturity	X	X		
Low cost		X		X
Excellent developer environment		X	X	X
Open standard	X		X	
Standard controlled by single vender		X	X	X

open standards and Internet protocols. Another major trend is toward distributed object computing and, most recently, service-oriented, Web-based networking.

CORBA, COM+, J2EE, and .NET have emerged as the major standards for the development of enterprise distributed object systems. ORBs for these technologies are ubiquitous and free, fostering continued growth in distributed object computing. Microsoft's DCOM ORB is built into Windows and NT, and CORBA ORBs are built into Java and a number of operating systems (including Linux, Solaris, HP-UX, and others). In addition, freeware CORBA ORBs are readily available for download from the Internet. A brief comparison of these technologies is presented in Table 5.1.

In the past, CORBA/J2EE and COM+ technologies were, for practical purposes, incompatible. This meant the organization had to standardize on one approach or another. Increasingly, standards and bridging products are available to provide interoperability between these approaches. For example, for COM–CORBA interoperability, there are products such as Orbix COMet, Iona's Static Bridge, ICL DAIS COM2CORBA, and Visual Edge's Object Bridge. [GERA99] These bridges act as middleware, translating method calls and parameter passing in the formats required by each respective technology. This allows COM clients to interact with CORBA servers, and vice versa. Because of the differences in interface and object implementation with each approach, these bridging solutions are nontrivial to implement, but they are nonetheless possible. The difficulties in bridging COM+ and CORBA reflect the relatively tight coupling of client and server interactions and implementation.

Web services, the newest client/server technology, loosens the coupling between clients and servers. Web services are based on XML and HTTP, and are capable of supporting a variety of transport mechanisms and client/server implementations. With Web services, it is possible to implement interfaces between COM+, CORBA, and legacy applications on the same network, thereby redefining application boundaries. Web services have the added advantage that they are easy to develop and implement. However, the standards for Web services are still evolving, and there are few commercial product offerings available. Fewer yet are industry implementations of significant scale. Two major concerns with Web services are security and performance. Web services do not have built-in security, and SOAP messages are sent in clear ASCII text. Use of encryption, SSL, or VPNs is essential to ensure the privacy and integrity of Web service transmissions. The use of XML text-based messaging also means that Web services protocols are inherently chattier and slower than their CORBA and DCOM counterparts, which transmit in binary format. This may result in the organizations having to upgrade or increase their Web servers to compensate for the additional bandwidth that must be supported.

MOM and OTMs are critical components of enterprise client/server networks. They are used to implement highly scalable networks, capable of supporting millions of objects, in a secure, manageable, and robust environment. These middleware components are increasingly important as distributed object computing takes hold across the Internet through Web services. Enterprise JavaBeans (EJBs) from Javasoft, CORBA Beans from OMG, and Microsoft's MTS are examples of OTMs with MOM capabilities.

Notes

1. Enterprise JavaBeans (EJB), as the name implies, is a Java-based, object-oriented architecture for implementing server components. It was developed by Sun Microsystems as an alternative to DCOM, and can be used with all major operating systems. EJB server programs are referred to as *servlets*. Servlets can be used, among other things, to provide Web-based interfaces for mainframe applications, and can be used as part of a migration strategy for extending the useful life of existing legacy code. Servlets are executable programs launched by a Web server instead of a Web browser. This contrasts with Java *applets,* which are downloadable Java executables used to present dynamic content in an end user's Web browser. Because servlets do not involve a download to an end-user machine, this reduces network traffic and the potential for infection by malicious code.
2. Total cost of ownership (TCO) encompasses all costs related to purchase, maintenance, upgrades, training, and support of computer/networking software and hardware. The original purchase price of hardware and software is often a relatively small portion of the TCO over the lifetime of the asset.

3. Other RPC standards exist, such as the Open Software Foundation's Distributed Computing Environment (DCE) and the IEEE ISO Remote Procedure Call Specification.

4. However, sometimes a three-tier network is considered a special type of n-tier network.

5. TOG is an "international vendor and technology-neutral consortium" that offers "proven certification methodology and conformance testing expertise, … [and] is an international facilitator in delivering the interoperability that single organizations require to help ensure independence." [OPEN04]

6. A cabinet is a single file that contains a number of compressed files, and is generally denoted by a .cab file suffix. Microsoft uses cabinet files when distributing its application programs (i.e., Excel, PowerPoint, etc.). When the programs are installed on an end-user machine, the compressed files in the cabinet are decompressed and copied to an appropriate folder.

7. Type casting allows a derived object to be viewed as an object of its base class, and is a manifestation of polymorphism. In essence, the exact object class is "cast away" and supplanted by the more generic parent class. The Java compiler checks for errors in the assignment of derived types to an object to ensure they are compatible.

8. Sockets are used to establish a bi-directional connection between a sender and receiver. There are 3 main socket types: 1) Stream sockets — which establish a connection that guarantees data will be received in the same order it is sent, 2) Datagram sockets — which provide a connectionless communication channel where sender and receiver applications assume responsibility for message ordering and integrity, 3) Raw sockets — which are used to implement custom low-level protocol functions. Stream and datagram sockets use IP port numbers to establish connections between different applications.

9. CPI-C — This stands for Common Programming Interface for Communications. It is an API for APPC (Advanced Program to Program Communication), and is an open standard based on SNA for distributed messaging between workstations, and midrange and mainframe servers.

10. Java Server Pages (JSPs) are HTML text files with embedded JSP code that is compiled into a servlet for execution on a Web server. JSPs provide dynamic scripting and handling of input to and output from Java servlets. JSP execution is triggered by the use of a .jsp file extension in a URL.

11. RMI (Remote Method Invocation) is a Java construct similar to Remote Procedure Calls (RPCs); however, it invokes methods on a remote object, as opposed to procedures, using the same syntax used for local calls. RMI allows an object and associated method data to be passed with procedure requests. RMI uses object serialization to *marshal and unmarshal* parameters and return values, respectively. RMI is based on the TCP/IP transport protocol and is a standard feature of the Sun Microsystems' Java Development Kit (JDK). RMI is implemented as (1) a client stub or proxy program (which defines the interfaces implemented on remote object, invokes service, serializes [i.e., marshals] outgoing parameter data, and deserializes

[i.e., unmarshals] return values from the client-side RRL [Remote Reference Layer] layer); (2) a server skeleton, (which accepts the client invocation from a server-side RRL, unmarshals arguments and invokes the methods on a server object, and then marshals return values to the server-side RRL layer); (3) a two-component, client and server Remote Reference Layer (which implements a specific remote reference protocol [e.g., IIOP] that is independent of the client stub and server skeleton, generates data streams to and from client stubs and server skeletons, and interfaces to the Transport layer); and (4) and the Transport Connection layer (which locates the remote server object, and establishes and maintains a connection between the client and server.)

12. Although it supports Sun Microsystems standards, JBOSS is not J2EE certified. This was a decision by JBOSS to keep the costs down. [EMIG04]

13. Instantiation of a class refers to the creation of a specific instance of the class as an executable file. Generally, in object-oriented parlance, class instantiation means the creation of an object. However, in Java, class instantiation is said to produce a specific class. In either case, the end result is the creation of an executable file. The difference is one of (confusing) nomenclature.

14. ASP.NET is Microsoft's development environment for Web applications and XML Web services. ASP.NET uses the file extension ".asmx" to indicate a XML Web service that can be addressed by a URL. For example, to access an XML Web service called HoroscopeServices.asmx, residing on a Web server named FortuneTeller, one would enter the following URL in the Web browser: http://FortuneTeller/HoroscopeServices.asmx.

15. URIs are similar to HTTP URLs, but they do not necessarily point to a specific location or page. URIs uniquely identify XML modules and XML objects.

References

[ALTO04]   Altova, XML Schema Driven Code Generator, 2004, http://www.altova.com/features_code.html.

[BOX04]   Box, D., *A Guide to Developing and Running Connected Systems with Indigo,* Microsoft Corporation, January 2004, http://msdn.microsoft.com/longhorn/default.aspx?pull=/msdnmag/issues/04/01/Indigo/default.aspx.

[CELA04]   Celarier, S., *Longhorn Developer FAQ,* Fern Creek, January 28, 2004, http://msdn.microsoft.com/longhorn/support/lhdevfaq/default.aspx.

[CLEW02]   Clewlow, C., SOAP and Security, QinetiQ LTd, 2002, http://www.qinetiq.com/services/information/white_paper0/soap_and.html.

[COLL02]   Frye, C., ADT's Programmers Report: Open-Source Servers Today, ADTmag.com, 6/3/2002, http://www.adtmag.com/article.asp?id=6416.

[EMIG04]   Emigh, J., JBoss' Open Source Plan, LinuxPlanet.com, March 18, 2004, http://linuxtoday.com/news_story.php3?ltsn=2004-03-18-024-26-NW-BZ-DV.

[FREM02]   Fremantle, P., *Applying the Web Services Invocation Framework,* IBM, June 2002, http://www-106.ibm.com/developerworks/webservices/library/ws-appwsif.html.

[GERA99]   Geraghty, R., Joyce, S., Moriarty, T., and Noone, G., *COM-CORBA Interoperability,* Prentice Hall, 1999, p. 46.

[GORD00]   Gordon, A., *The COM and COM+ Programming Primer,* Prentice Hall, copyright 2000.

[IONA03]   Iona Technology PLC, CORBA Means Business How IONA's CORBA Products Provide Business Value, December 2003, http://www.iona.com/whitepapers/OrbixE2ACORBATechWPV02-01.pdf.

[JAVA04]   Sun Microsystems, Inc., *Documentation JavaBeans Migration Assistant for ActiveX,* http://java.sun.com/products/javabeans/docs/jbmigratex.html JavaBeans, 2004.

[JBOS04]   JBOSS, 2004, http://jboss.org/overview.

[JOHN96]   Johns, P., *Signing and Marking ActiveX Controls,* Microsoft Corporation, October 15, 1996, http://msdn.microsoft.com/library/default.asp?url=/library/en-us/dnaxctrl/html/msdn_signmark.asp.

[KHOR02]   Khor, S., *Digital Code Signing Step-by-Step Guide (Applies to Microsoft® Office XP),* Microsoft Corporation, March 2002, http://msdn.microsoft.com/library/default.asp?url=/library/en-us/dnsmarttag/html/odc_ dcss.asp.

[KIRT97]   Kirtand, M., The COM+ Programming Model Makes it Easy to Write Components in Any Lanugate, December 1997, http://www.microsoft.com/msj/1297/complus2/complus2.aspx.

[LAIR01]   Laird, C., Open Sources Defining Web Services, SW Expert, June 2001.

[LONE04]   Loney, M., Microsoft Shines More Light on Longhorn, CNET News.com, January 27, 2004, http://news.com.com/2102-1046_3-5148148. html?tag=st_util_print.

[MALI03]   Malik, A., XML Standards for Financial Services, XML.com, March 26, 2003 http://www.xml.com/lpt/a/2003/03/26/financial.html.

[MICR04]   Microsoft Corporation, *Implementing IClassFactory,* 2004, http://msdn.microsoft.com/library/default.asp?url=/library/en-us/com/htm/comext_9lkp.asp.

[MICR04B]   Microsoft Corporation, *Technology Overview What Is Microsoft.NET?,* 2004, http://msdn.microsoft.com/netframework/technologyinfo/overview/.

[MICR04C]   Microsoft Corporation, *NET Framework Developer's Guide Exploring Existing XML Web Services Created Using ASP.Net,* 2004, http://msdn.microsoft.com/library/default.asp?url=/library/en-us/cpguide/html/cpconexploringexistingwebservices.asp.

[OMG04]   Object Management Group, 2004, http://www.omg.org.

[OASI04]   OASIS, *About UDDI,* 2004, http://www.uddi.org/about.html.

[OPEN04]   The Open Group, 2004, http://www.opengroup.org/overview/index.htm.

[ORFA99]   Orfali, R., Harkey, D., and Edwards, J., *Client/Server Survival Guide,* third edition, John Wiley & Sons, Inc., New York,1999, p. 511.

[SANS03]   SANS Institute, *The Twenty Most Critical Internet Security Vulnerabilities (Updated) ~ The Experts Consensus,* Version 4.0 October 8, 2003, Copyright 2001–2003, http://www.sans.org/top20/#threats.

[SCALL00]   Scallan, T., Gitelman, A., and Segue Software, *ejb Introduction to Enterprise JavaBeans,* March 8, 2000, http://www.segue.com

[SHOF97] Shoffner, M., JavaBeans vs. ActiveX: Strategic Analysis, *Javaworld. com,* February 97, http://www.javaworld.com/javaworld/jw-02-1997/ jw-02-activex-beans_p.html.

[SIDD01] Siddiqui, B., *Introduction to Web Services and WSDL,* IBM, November 2001, http://www-106.ibm.com/developerworks/library/ws-intwsdl/.

[SUNM04] Sun Microsystems, 2004, http://java.sun.com/products/javabeans/.

[SUN04B] Sun Microsystems, *Java Technology Java 2 Platform, Enterprise Edition (J2EE) Overview,* 2004, http://java.sun.com/j2ee/setstandard.html.

[SUN04C] Sun Microsystems, Overview of the Java APIs for XML, 2004, http://java.sun.com/webservices/docs/1.3/tutorial/doc/IntroWS4.html.

[SUN04D] Sun Microsystems, *Getting Started with Services on Demand and the SunOpen NetEnvironment Sun ONE,* 2001, http://wwws.sun.com/software/ sunone/wp-getstarted/wp-getstarted.pdf.

[SUN04E] Sun Microsystems, *Getting Started With Services on Demand and the Sun Open Net Environment (Sun ONE),* 2004, http://wwws.sun.com/software/ sunone/wp-getstarted/getstarted.html.

[THEB04] The Breeze Factor, LLC, *Breeze XML Binder 3.0 for Java™, Data Binding,* 2004, http://www.breezefactor.com/databinding.html.

[TPC04] The Transaction Processing Performance Council, 2004, http://www.tpc. org/default.asp.

[VAUH02] Vaughan-Nichols, S., Web services' fat protocols could wreck your network, TechRepublic, January 30, 2002, http://techrepublic.com.com/5102-6265-1040726.html.

[W3C01] Web Services Description Language (WSDL) 1.1, W3C, copyright 2001, http://www.w3.org/TR/wsdl.

[W3C04] W3C Consortium, copyright 2004, http://www.w3.org/TR/REC-xml/ #sec-origin-goals.

[W3SC04] W3Schools, *The SOAP Envelope Element,* Full Web Building Tutorials, http://www.w3schools.com/soap/soap_envelope.asp.

Chapter 6

Outsourcing

6.1 Outsourcing Overview

Outsourcing refers to the assignment of certain management, planning, implementation, or operational functions to an external third party. The number of providers offering outsourcing services is steadily growing, reflecting a corresponding increase in demand for these services. This demand is being fueled by the desire of organizations everywhere to contain costs through third-party arrangements, to minimize and share risks, and to obtain specialized expertise, products, and services. Outsourcers can be broadly categorized as follows:

- Device vendors and manufacturers (e.g., Cisco, Nortel, Lucent Technologies, and Siemens)
- Telecommunication providers (e.g., within the United States, AT&T and MCI; and internationally, British Telecom (BT), Equant, Cable and Wireless (C & W), Deutsche Telekom, etc).
- Consulting companies (e.g., Accenture, Deloitte Consulting, Booz Allen Hamilton, Pricewaterhouse Coopers [PwC], Ernst and Young, etc.)
- System integrators (e.g., IBM Global Services, EDS, Computer Sciences Corporation, etc.)
- Application development/software outsourcers (e.g., Microsoft, Siebel, Oracle, SAS, PeopleSoft, etc.)
- Specialized niche providers (e.g., Internet service providers [ISPs], application service providers [ASPs], etc.)

There are many motivations for outsourcing. Some of the most commonly cited reasons for outsourcing are to:

- *Achieve cost reduction or containment.* Organizations frequently turn to outsourcing as a way to lower their operating cost structure or to control it at a contractually fixed level.
- *Augment expertise and skills sets.* Outsourcing provides a means to bring expertise and world-class capabilities into the organization that might not be available internally.
- *Mitigate and manage risk(s).* Outsourcing offers a means to transfer risk to a third party. This may be particularly advantageous when the function being outsourced is difficult to manage and poses serious financial, legal, or social threats if it is not performed well.
- *Maintain focus on core competencies.* Organizations use outsourcing to avoid diversion of time, money, or resources in areas that could be better handled by a third party, thus allowing enhanced focus on more profitable or strategic endeavors.
- *Maintain state-of-the-art technology infrastructures.* Outsourcing is a way of gaining access to state-of-the-art technology, without making capital investments in rapidly depreciating assets. Outsourcing can provide the organization with flexibility in acquiring the use of new technologies without the associated difficulties of installing and managing the infrastructure internally.
- *Accrue positive cash flow benefits.* Through outsourcing, the organization can defer capital investments in technology that might be used more strategically elsewhere. In some outsourcing arrangements, staff, equipment, facilities, and other resources are transferred to a third-party outsourcer and are carried as assets on their books. The outsourcer then provides the services to the organization on an expense or lease basis. This can provide a cash infusion to the organization, along with other possible tax benefits.
- *Implement critical infrastructure support.* Demand for security, business continuity, and disaster recovery outsourcing is strong, particularly in regulated industries such as finance and health care. This type of outsourcing can encompass a very wide range of products and services. For example, it might involve contracts for secure, third-party, redundant power supplies, networking and telecommunications equipment, or extra data processing facilities. These products and services are needed to ward off and recover from major failures of mission-critical systems.

Some of the areas most likely to be outsourced include:

- *Internet-related outsourcing.* This includes Internet IP VPN connectivity, security, Web hosting, and related products and services.
- *ASPs (application service providers) software support.* ASPs deliver software and access to applications programs over the Internet using remote access protocols. According to ASPnews.com,[1] there are several types of ASPs serving different market niches: (1) enterprise ASPs (for delivery of high-end business applications); (2) local/regional ASPs (which provide application services for regional small to medium-sized businesses); (3) specialist ASPs (which provide applications for a specific need, such as Web site services for online banking or credit card processing); (4) vertical market ASPs (which serve specific industries, such as healthcare); and (5) volume business ASPs (which provide a volume supply or prepackaged applications services to small and medium-sized businesses).
- *Network management.* This includes services for router/switch management utilizing SNMP RMON I and II network management platforms.
- *E-commerce support.* This includes server support, load balancing, failsafe networking, database functions, security, and other related products and services.
- *Data storage.* This includes network attached storage (NAS), storage area networks (SANs), and electronic data vaulting solutions.
- *User support.* This includes help desk staffing, training activities, and other user support functions.
- *Business process outsourcing (BPO).* This refers to the outsourcing of specific, nonstrategic business functions to a third-party contractor. There are two general flavors of BPO. *Back-office outsourcing* involves internal business functions such as payroll or human resources. *Front-office outsourcing* includes technical support, marketing, and other customer-related services such as call centers.

BPO that is contracted internationally is referred to as *offshore outsourcing*. In the United States, offshore outsourcing has been a topic of considerable debate and concern over the possible loss of American jobs to workers overseas.[2] The idea of offshore outsourcing is to take advantage of labor savings accrued from wage arbitrage. For example, countries such as India and China offer a highly skilled, literate, English-speaking labor force at very competitive rates.

In addition to labor savings, other possible benefits of BPO include process improvement and automation. BPO firms are highly motivated to reduce costs and improve process efficiencies as a way of maximizing their profits and customer satisfaction, to their mutual benefit and that of their customers. Large BPOs are particularly well positioned to capitalize on new, specialized technologies (such as optical character recognition [OCR], etc.) and integrated IT solutions that may be difficult to justify on a single company basis, but which become economically viable when implemented for a large user community.

A number of big companies — such as General Electric and Siemens — are using offshore outsourcing for customer call centers, and other BPO functions, to capitalize on foreign labor economies. However, in general, offshore outsourcing is not yet the norm. According to Forrester Research, over 60 percent of the Fortune 1000 companies surveyed in the United States are doing little or nothing in regard to offshore outsourcing. These companies are labeled "bystanders" by Forrester Research. "Experimenters" comprise about a third of the Fortune 1000 population. These companies have some experience with offshore outsourcing, but to a limited degree, and not in areas that are strategic to their core business. "Committeds" and "Full Exploiters" comprise between 10 and 20 percent of the Fortune 1000. These companies have made very significant investments in offshore ventures, representing a sizeable portion of their IT budgets. In large part, the limited use of offshore outsourcing is related to the fact that it can be very difficult to manage overseas IT projects. Part of this difficulty may relate to legal restrictions and national security considerations that must be addressed when contracting work overseas. Furthermore, in addition to the difficulties inherent in IT projects, overseas ventures require diplomacy and an awareness of foreign customs and ways of doing business. Working with developing nations can be a considerable challenge if these factors are not properly addressed. Moreover, the process of moving jobs overseas can take three to four years to complete, and thus offshore outsourcing is not something that can be put into place quickly. [SURM03]

According to [KENN04], the major users of outsourcing services, in descending order, are the public government sector, the financial services industry, the healthcare industry, communications and media companies, and retailers. Typically, these organizations are involved in large-scale process reengineering, information technology infrastructure updates, security, and implementation of new and improved customer services. Some of these changes are required to meet federal and regulatory requirements, while others are needed to remain competitive. These buyers, as a whole, are particularly savvy users of technology, with access to shared user groups for information sharing. They are also very demanding, leveraging centralized purchasing for maximum volume discounts and competitive pricing, at the same time they are issuing requests for sophisticated, and often customized, products and services.

The value of recent BPO and IT outsourcing contracts is quite telling as to the scale of these initiatives. According to [CONS04], within the past year, the number of outsourcing deals valued at over $100 million grew almost 50 percent, to 244; and the number of outsourcing deals valued at over $1 billion grew almost 100 percent, to 29. IBM Global Services, Computer Sciences Corporation, Hewlett-Packard, and Cap Gemini Ernst & Young were among the biggest deal-makers. The largest purchasers of these services were public-sector organizations, "such as the U.K. National Health Service and the U.S. Department of Defense, which have brought in private sector specialists to upgrade their existing technology infrastructure." [CONS04] According to Datamonitor's IT Services Contract Tracker — a service that tracks every outsourcing and system integration contract worth over $1 million signed by major IT vendors — the largest outsourcing contract of the year was a $5.1 billion deal signed between Cap Gemini Ernst & Young and the U.K. Inland Revenue department.

Small to medium-sized businesses are demonstrating growing reliance upon ILECs, ISPs, ASPs, and cable operators to provide basic network services for voice, IP VPNs, security, Web hosting, messaging, and a variety of E-commerce applications. A full complement of vendor services, ranging from collocation and Web hosting to complete suites of managed services are appealing to this market segment. MCI and AT&T are two examples of major service providers selling a full range of solutions tailored to the needs of small and medium-sized businesses. Collocation is especially attractive to tight-squeezed, space-starved companies or those faced with expensive real estate. Collocation (a.k.a. co-location) refers to the practice of provisioning space for a company's telecommunications equipment on a service provider's premises. Generally, the customer owns the equipment but does not house it (however, in some vendor arrangements, equipment can be leased directly from the service provider and the customer does not need to own it). Basic collocation service includes space for housing equipment, diverse power sources and backup systems (including diesel/electric generators and UPS systems), a low-humidity, air-conditioned environment, fire protection, basic card swapping, and first-level "eyes and ears." This last (informal) term refers to visual inspection of major alarms and other maintenance and facilities oversight by the service provider, and notification of the company in the event of any problems. Co-location provides equipment storage in a private cage with a locked cabinet that can be entered "24/7" by the company's internal staff. It offers a fully secured location, with guards, fence parameters, and electronic gates, which are only accessible with ID badges or biometric scanners (such as fingerprint or retina scanners). Because collocated equipment resides next to the service provider facilities, it is also easy for organizations to connect to private line, Frame Relay, and other network services offered by their provider. By outsourcing managed services, even with limited funds and infrastructure, small enterprises can augment their capabilities and acquire state-of-the-art technology. Outsourcing may also allow a company to implement larger-scale projects than they might be able to undertake otherwise. By leveraging outsourcing to offer similar services with lower overhead and lower prices than their larger competitors, small to medium-sized companies are able to compete more effectively and to offer a more diverse range of services to their employees and customers.

The small and medium-sized business market is especially attractive to outsourcing vendors. In contrast to the large business market, this area is perceived as a growth opportunity. IT budget purse strings are very tight in large enterprises that typically want complex, tailored solutions at the lowest possible cost. Smaller companies do not have the bargaining power or sophistication of larger companies. Vendors are capitalizing on this by developing attractive bundled standard services and products on which they can realize higher profits and better economies of scale. From the companies' perspective, a bundled, one-stop-shopping solution is often much cheaper, and easier to implement and manage, with better features and functions than a solution implemented with internal resources or by selective outsourcing from multiple vendors. From the vendor perspective, working with a smaller infrastructure and bureaucracy shortens the approval cycle and gives sales staff direct access to decision makers who will buy their products and services. In a large company, an outsourcing project may be in direct conflict and competition with existing IT staff and budgets, and may have to go through a lengthy and elaborate approval process.

6.2 Steps in the Outsourcing Process

The following steps, which are summarized in Figure 6.1, describe the process of outsourcing network design, planning, implementation, and operational support:

1. Analyze existing environment and outsourcing requirements. This involves the following types of tasks:
   a. *Define goals and objectives.* Executive management should provide guidance on strategic initiatives and business goals to be achieved through outsourcing, and should encourage broad organizational support for subsequent efforts.
   b. *Document the existing environment.* A due diligence audit should be conducted to ascertain existing processes, procedures, and technology infrastructure that would be affected by outsourcing. Benchmarks on workloads, capacities, uptime performance, and

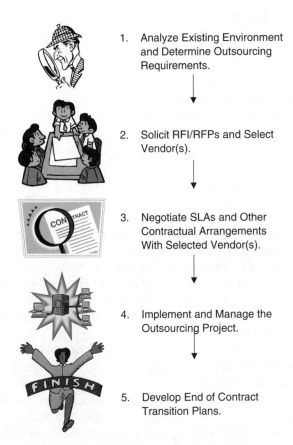

1. Analyze Existing Environment and Determine Outsourcing Requirements.

2. Solicit RFI/RFPs and Select Vendor(s).

3. Negotiate SLAs and Other Contractual Arrangements With Selected Vendor(s).

4. Implement and Manage the Outsourcing Project.

5. Develop End of Contract Transition Plans.

Figure 6.1   Summary of outsourcing process.

other metrics should be collected relative to the areas that are being considered for outsourcing.

c. *Ascertain baseline costs.* This includes a high-level financial analysis of the IT budget, staffing, and other relevant costs. It may also be helpful to conduct benchmarking studies against peer organizations and industry standards to determine how the organization's cost structure compares. This will provide insights into costs that outsourcing might help to reduce.

d. *Define outsourcing requirements.* This step determines what functions should be outsourced. This may involve a "gap analysis" to determine opportunities to improve the existing environment through judicious use of outsourcing. Preliminary service level agreements (SLAs) should also be defined at this juncture.

e. *Evaluate outsourcing alternatives.* This evaluation should consider various outsourcing scenarios, with associated ROI and risk analyses. This will help refine the scope of the outsourcing effort. It will also assist in the determination of whether outsourcing makes sense, and if so, which outsourcers should be considered for the job.

f. *Develop a plan outlining next steps.* This stage is necessary to obtain the necessary buy-in from management to proceed with consideration of an outsourcing solution. If management decides to continue the outsourcing initiative, the next step is usually a formal vendor proposal process.

2. Solicit vendor proposals and select vendors. This involves the following types of tasks:

a. *Prepare RFI (Request for Information) and distribute it to selected vendors and service providers.* Use the RFI to test vendor interest and initiative in supporting the outsourcing initiative.

b. *Collect and evaluate RFI responses.* The RFI responses provide a basis for selecting a preliminary slate of outsourcing candidates. A variety of metrics for evaluating the RFI are discussed later in this chapter.

c. *Schedule public/private question and answer sessions with selected vendors.* Typically, potential outsourcing service providers are invited to attend a public session with all the other prospective vendors. During this session, vendors can ask questions regarding the anticipated scope and requirements of the outsourcing effort. The session is also an opportunity for vendors to collect information on the existing environment and infrastructure, and other details that will be needed to respond to a Request for Proposal. Depending on the type of business, some follow-up sessions may be scheduled on a one-on-one basis with selected vendors to discuss outstanding questions and issues. In government outsourcing arrangements, care must be taken to avoid any hint of collusion with vendors. Government agencies can demonstrate their impartiality by providing all vendors with access to the same information at the same time, and by refusing individual vendor calls for information. In other types of organizations, such as a small, privately held enterprise, these considerations may not be an issue.

d. *Prepare RFP (Request for Proposal) and distribute it to selected vendors and service providers.* The RFP is useful in developing a mutual understanding with the vendor on the approach and methodology for proceeding with an outsourcing solution. The RFP usually includes a preliminary price estimate. For the selected vendor(s), the RFP process also provides the basis for determining the Statement of Work, service level agreements, and other contractual arrangements.

e. *Collect and evaluate RFP responses.* The RFP responses provide a basis for selecting the final outsourcing candidate(s). A variety of metrics for evaluating the RFP are discussed later in this chapter.

f. Select final outsourcing vendor(s).

3. Under the advice of legal counsel, negotiate SLAs and other contractual arrangements. This involves the following types of tasks:

a. *Document and finalize SLAs and associated performance requirements and metrics.* Metrics should be objective, tangible, and measurable and should relate to critical factors that will determine the success of the outsourcing project.

b. *Finalize the pricing arrangements.* Many different types of pricing arrangements are available. For example, some outsourcers charge on a fixed-fee basis, others charge on a per-usage basis, while others charge on a revenue-sharing or cost-savings basis. Pricing arrangements may also include penalty clauses if the service provider fails to satisfy the SLA and other contractual arrangements. Bonuses and other incentives may also be structured as part of the contract to reward vendors who exceed the project requirements. A mechanism should be defined for amending the terms of the outsourcing arrangement to accommodate changes in the scope of work to be performed, and to provide flexibility in the agreement should the need arise.

c. *Specify termination arrangements and limitations of liability.* A mutually agreed-upon schedule and process should be specified to allow either party to terminate the outsourcing arrangement, due to various factors that may arise. For example, this might include a clause that either party can terminate the agreement with written notice of at least a month, as long as certain outstanding expenses are reimbursed.

d. Specify the precise nature of the outsourcing work to be performed and other relevant details (such as staffing, hardware, and software to be supplied, etc.).

4. Implement and manage the outsourcing project. This involves the following:

a. *Establish a vehicle for project management and oversight.* This ensures that the outsourcing project is properly implemented and that SLAs and contractual obligations are being met by both the buyer and supplier of outsourcing products and services. The specific composition of the management team and its duties are highly dependent upon the nature of the organization and the type of products and services being outsourced. To the extent possible, it is advisable to follow common-sense, industry-standard practices for project management.

b. *Define a mechanism for resolving problems and disputes that may arise.* This should include management oversight and governance provisions to ensure that the outsourcing project is properly managed and executed.

5. Develop end-of-contract transition strategies and plans. This involves the following types of activities:

a. *Define a strategy for transitioning the organization when the outsourcing contract ends.* The reasons for terminating the contract will help determine the most appropriate course for the organization.

b. *Develop transition plans based on the exit strategy.* These plans depend on the organization's exit strategy, internal capabilities, and the final disposition of the outsourcing conclusion. If the outsourcing was contracted as a temporary service, then a plan should be developed for an eventual smooth transition of the outsourced activities, products, or services to the organization. However, if the outsourcing provides a vital service that still makes sense, it may be advisable to extend the outsourcing contract for an additional term. The important point is that it is a good idea to develop contingency plans for various types of conclusions to the outsourcing arrangements so as to minimize potential disruptions and ensure smooth transitions.

6.3 Major Considerations in Outsourcing

The decision to outsource must be prepared for very thoroughly and carefully. Once this decision is made, a choice must also be made as to which outsourcing provider(s) should be awarded a contract. The requirements analysis and RFI/RFP process are critical factors in making informed choices. When a company outsources its network design, planning, and implementation, it is giving away a strategic responsibility to an external company. For this reason, rarely does a large organization outsource these functions completely. If internal capabilities must be augmented to support these activities, then multiple consulting companies or telco/service providers may be contracted to prepare or validate network design alternatives. This is to avoid too much reliance on any one vendor, because vendor exclusivity may limit the organization's flexibility in future arrangements. In some cases, the service provider or consulting company providing outsourcing services may have its own internal, proprietary planning and management tools. This is understandable, as certain outsourcing functions may require specialized knowledge and expertise that is not easily transferred. These specialized tools reflect the outsourcer's intellectual capital and value-added service. However, ownership of proprietary internal data and detailed company information should remain with the company, not the outsourcer.

Contract management and continuous control over the outsourcing arrangement is absolutely necessary to ensure that the organization's interests are well served and that contractual obligations are met throughout the outsourcing engagement. Metrics should be associated with specific outsourcing activities or deliverables, and should be well defined and measurable. Once defined, metrics and SLAs should be built into the outsourcing contract. The metrics for evaluating the outsourcing initiative should be mutually agreed upon by both the vendor(s) and by the internal staff who will be responsible for the execution and management of the project. They should be clearly understood by all relevant parties.

Legal counsel should be engaged to ensure that the outsourcing arrangement does not expose the organization to undue risks and liabilities. One area of special concern relates to U.S. Government restrictions on the export of arms, equipment, software, and other information that might compromise national security. These restrictions may materially impact the outsourcing arrangements of large military or defense industries, or multinational companies. Organizations engaged in government or military work should pay particular attention to the arrangements of outsourcers with subcontractors to avoid being compromised and put at risk for penalties, fines, or other legal liabilities. This was a lesson that Raytheon Aircraft learned, after it engaged IBM to implement and manage its SAP system, and IBM then announced that it intended to use subcontractors from India to keep the costs down. Raytheon Aircraft's SAP system contained vital, sensitive information on the construction of jet aircraft — information that could not legally be accessed by foreign workers. Raytheon's senior management had to become involved to make sure that their system and network security would not be compromised by IBM's subcontracting arrangements. These concerns are not limited solely to the military and defense industries. In another example, a Pakistani clerical worker who had been processing the records of a U.S. hospital threatened to post confidential patient information on the Internet if she was not given an increase in salary. [OVER04] Large U.S. outsourcers — such as IBM, CSC, and EDS — routinely engage subcontractors from India's Tata Consultancy Services and from Wipro for BPO contracts. The benefits of offshore outsourcing should be considered relative to its risks in order to decide whether it is appropriate.

During the vendor evaluation and selection process, the organization should consider how adaptable the provider can be in providing access to new technologies or in dealing with unforeseen circumstances. Providers that are geared toward mass-market, commodity solutions are less likely to offer flexible terms than providers that specialize in custom, tailored solutions. Some outsourcing arrangements may need to be structured to provide flexibility in responding to changes in business conditions (precipitated by mergers, acquisitions, downsizing, security requirements, etc.). Thus, the outsourcing vendor responsiveness to this type of need may be a factor in the selection process.

The human dynamics of the interaction between internal and outsourcing staff are often an indication of how well flexibility can be provided for in the outsourcing arrangements. Vendor relationships that are cooperative, versus adversarial, promote good communications and facilitate compromise, if and when it is needed.

A well-documented process with defined accountabilities and responsibilities should be put into place at the project inception to ensure effective oversight and problem resolution throughout the life cycle of the outsourcing project. Many difficulties with outsourcing arrangements arise from an inability to manage vendor/client relationships and from human conflicts. For example, in some cases, a vendor may become a surrogate or rival to the IT organization that originally contracted its services. When vendors seek to pursue their own dealings with other internal corporate departments outside the original sponsoring organization, tensions and conflicts can arise. If the outsourcers are in a position to influence critical strategy and investment decisions, they may seek to serve their own interests over those of their clients. This may lead to the approval of risky, large projects dominated by the outsourcer. Other conflicts of interest may arise if a vendor services multiple clients that are in competition with each other. Continuity problems can also arise if outsourcing staff are reassigned to other clients during critical junctures in the engagement. These types of issues can be managed by establishing ground rules and strong management oversight right from the start.

Outsourcing frequently involves significant infrastructure change that must be carefully managed. Change management will help deal with these concerns and should start as early as possible in the outsourcing project to anticipate and address potential difficulties. Change management involves developing strategies for communicating and aligning the business objectives with the outsourcing initiatives. For example, one aspect of this might involve developing programs to retain or motivate internal staff being displaced by outside consultants.

In preparation for the RFI and RFP, a number of questions should be answered internally to address the rationale and requirements for outsourcing. These questions relate to the following:

- *What is the current cost structure?* This analysis should include the present costs of equipment, communications, people, and expenses directly related to the areas that are candidates for outsourcing. To the extent possible, both tangible and intangible soft costs should be included for an accurate ROI and baseline assessment of the existing environment.
- *What functions should be outsourced?* This should encompass a full review and analysis of existing processes, instruments, and human resources in order to decide which functions should be considered for outsourcing. Outsourcing is a good excuse to audit the present design and planning and to address areas that need improvement. After the preparations and planning for outsourcing are completed, the decision can be made to complete the work in-house, instead. This review by internal or external analysts may result in substantial savings in network design and planning expenses, in staff reduction, and in stabilizing network budgets.
- *What is the organizational dependency on network services and availability?* The organization's dependence on network availability is used to calculate the highest level of risk and downtime tolerance that should be specified in the service contract. It is not uncommon for certain vendors to fall short of desired performance goals, particularly in the early stages of an outsourcing engagement. To help prevent this, guaranteed service levels should be written into the outsourcing contract.
- *What level of privacy and security must the outsourcer provide with its service?* The security risks and liabilities to which the organization is subject will help determine the privacy and security requirements. To protect highly sensitive applications, the organization may require that the outsourcer not share its network design or network resources with other clients. Even more stringent security requirements may prohibit third-party vendors from gaining any access to the network and to its carried traffic.
- *What is the organization's core competency and focus?* Some organizations view the technology infrastructure as a key strategic component of their existence (for example, FedEx and the NYSE). Others may view the technology infrastructure as necessary, but not essential, to the furtherance of the organization's strategic goals. In organizations such as this, outsourcing may provide the means for the company to focus on what it does best, without having to build or buy a sophisticated network and IT infrastructure that will require skills sets outside its core competencies.
- *What level of capital investment is needed for internal implementation versus that which is required for a managed services solution?*

The availability of standardized or commodity offerings may facilitate this decision. If the organization must invest substantial amounts into technology infrastructure to achieve a solution that can be easily outsourced, this may tilt the decision in favor of a third-party solution. Otherwise, outsourcing is less of an organizational priority.

■ *What is the availability of skilled personnel?* This is a critical consideration that is frequently the deciding factor for outsourcing. Outsourcing provides a way to augment present and future staffing. However, staffing needs should be carefully quantified prior to the outsourcing decision so the appropriate levels can be established.

■ *What is the stability of the environment and its growth rate?* The stability of the organization's operating environment has potentially serious repercussions for the adequacy of the outsourcing arrangements. Where there is a substantial possibility of acquisition, merger, business unit sales, application portfolio changes, or other significant changes in business conditions, special provisions and contingencies should be made in the outsourcing contracts to account for them.

After answering these questions, the organization must decide which design, planning, modeling, implementation, and other related functions should be outsourced. In a broad sense, the following functions may be considered for outsourcing:

■ *Baseline evaluation of network configuration and performance.* This function involves taking a snapshot of the network and its performance. It might involve measuring key service and utilization indicators to determine optimal operational thresholds. It might also include gathering information on the network topology (including circuit, CPE, desktop, and software inventories, and collecting other device, network, and system documentation).

■ *Performance measurement.* For example, this might involve collecting performance statistics and measurements from various network segments using standard information sources, such as SNMP-MIBs and RMON-MIBs.

■ *Performance statistics reporting.* This involves collecting and maintaining performance information, and producing periodic and ad hoc reports to document the network status and usage.

■ *Design and planning of the network strategy.* This should relate to the overall organizational strategic objectives. Modeling tools are often used to perform "what-if" scenarios to help ensure the robustness and adequacy of proposed network architectures and

Table 6.1   Common Industry Practice for Outsourcing Network/IT Functions

Functions	Complete Outsourcing	Partial Outsourcing	Little or No Outsourcing
Network baselining	X		
Performance measurements		X	
Performance reports		X	
Design and planning			X
Implementation and maintenance	X		
Operations	X		

designs. This approach uses data on present and planned capacities, resource loads, utilization rates, and other factors to test system limits, performance under stress, and requirements for growth.

- *Implementation and maintenance of network/IT systems.* This includes physical implementation of the network, service activation, service assurance, and maintenance of network components.
- *Network/IT system operations.* This means that day-to-day operational responsibilities — for such functions as fault management, trouble ticketing and tracking, tests, service restoration, and repair — are turned over to the outsourcer.

Table 6.1 summarizes common industry practices with regard to the outsourcing of networking functions. Complete outsourcing is frequently seen for functions such as baseline evaluation, implementation, and operations. Partial outsourcing is typically seen for functions such as performance measurement and reporting. Network design and planning is an area that is rarely outsourced, due to its strategic organizational importance.

6.4   Request for Information

The Request for Information (RFI) is a tool to inform prospective vendors that an outsourcing effort is being considered and to give them an opportunity to respond with information that will help qualify them as a candidate for the job. Prior to issuing an RFI, the organization should consider which types of outsourcers are the most appropriate for satisfying the requirements at hand. Table 6.2 describes different types of outsourcers and the respective special considerations and benefits associated with

Table 6.2 Comparison of Service Provider Types

Vendor Type	Potential Benefits	Special Considerations
Mainframe vendor	Knowledge of data networks Reputation Knowledge of logical network design and management Vertical integration	Dependency on proprietary architecture Data orientation Typically, horizontal management integration Limited experience with physical network design
Telco suppliers	Knowledge of voice networks Reputation Knowledge of physical networks and their management Good knowledge of tariffs Voice and data integration options available	Dependency on proprietary architectures Little experience with logical network design
System integrators	Knowledge of both data and voice networks Knowledge of both logical and physical network design Implementation experiences	High cost Reputation Project driven
Consulting companies	Knowledge of both logical and physical network design Knowledge of both data and voice networks Aware of leading-edge technology	High cost Reputation Implementation experiences
Device vendor	Knowledge of own equipment Knowledge about design tools Implementation experiences	Little knowledge of other vendor's equipment Reputation Biased toward own products

them. An organization that is contemplating outsourcing should also ascertain the types of service about which it wishes to solicit vendor responses. Table 6.3 compares some of the trade-offs inherent in making choices between different types of service contacts. Armed with this knowledge, the organization can prepare a list of vendors to which it will send an RFI. As a general guideline, the organization should send the RFI to as many appropriate vendors as is feasible. This is to ensure that the

Table 6.3 Outsourcing Service Alternatives

Type of Service Contract	Benefits	Disadvantages
On-site design work	Rapid prototyping Dedicated personnel Continuous consultation	High cost No shared instrumentation Space requirements
Off-site design work	Shared instrumentation No space requirements Moderate cost More concentrated know-how	Slow communication Ad-hoc consultation No dedicated personnel Longer cycles for evaluating design alternatives
Virtual office	Moderate costs No space requirements More concentrated know-how	No dedicated personnel Periodic consultation Slower prototyping

organization receives enough vendor feedback to make informed choices on available solutions and to obtain competitive bids.

A generic outline of the RFI document is presented below, although the specific inclusion of any particular element should determined by the nature of the outsourcing project:

- *Cover letter.* The cover contains a formal request asking a vendor to submit an RFI. It also contains contact information and a brief overview of the proposed outsourcing effort.
- *RFI documentation.* The RFI is generally accompanied by descriptive information about the outsourcing project, and instructions and guidelines to aid vendors in preparing their responses. Some of the sections normally included in the RFI documentation include:
  - Executive summary of key goals, objectives, and benefits to be attained by the outsourcing project.
  - List of key contacts, and associated contact information, within the organization who are assigned to coordinate vendor interactions and responses.
  - Summary of key milestones and associated due dates. This should include a list of deadlines for vendor responses and also a schedule for vendor question-and answer-sessions and vendor selection decisions.
  - List of the RFI evaluation criteria and a description of the evaluation process that will be used for vendor selection.

- Special requirements for the dispensation and safeguarding of confidential and proprietary information. This is needed to protect the confidentiality of RFI details, from both a vendor and organizational perspective. The organization should specify which aspects of the RFI are to be kept confidential and contain company "proprietary" information. Likewise, the vendor should specify which aspects of its response contain vendor proprietary information and how the organization should protect this information upon receiving it. The provisions of this section should be prepared with the aid of legal counsel.

- Background information on the organization requesting the RFI. This might include information regarding the organization's line of business, the number and location of affected worksites, or other information that might impact the scope of the outsourcing engagement.

- Review of present systems. This might include information on software, hardware, LAN/WAN/MAN protocols, transaction volumes, and other relevant technical details on the current technical environment.

- Description of the outsourcing project. This should include a high-level description of the key deliverables and systems requirements being considered for outsourcing. Critical restrictions or constraints that the outsourcing solution must address can also be identified in this section.

- Request for information on vendor qualifications. This will be used to help determine which vendors will be invited to participate in the next step of the outsourcing evaluation, the RFP process. Vendors are typically asked to provide details and supporting documentation relating to:
  - Company size.
  - Revenue, profit, and financial stability. The vendor might be asked what portion of their revenue is derived from products and services similar to those being proposed in the RFI.
  - Annual reports.
  - References from other clients that have used its products and services relating to the vendor's integrity, execution, and flexibility.
  - Scope of vendor support across specified geographic areas.
  - Memberships in standard bodies and other professional organizations.
  - Product ownership.
  - Use of subcontractors, if any, and their role in the outsourcing effort.

- Credentials of executive management and technical staff, and proven track record in the area being outsourced. Technical staff should have the expertise and depth needed to satisfy the outsourcing requirements.
- Proven experiences in managing domestic and multinational endeavors.
- Whether customized instruments, tools, software, or other products under consideration for the outsourcing project will be used exclusively by the organization or shared with other clients.
- Ability to service company requirements as they scale and grow in the future.
- Request for information regarding the technical nature of the vendors solution. This will be used to help determine which vendor will be invited to participate in the RFP process. Vendors are typically asked to provide details and supporting documentation relating to:
  - Their understanding of the scope of service and requirements requested in the RFI.
  - The vendor's intentions with regard to the RFI (i.e., do they intend to bid and, if so, in what capacity?).
  - Summary of vendor approach to the outsourcing solution, including vendor-proposed alternatives to the RFI.
  - Hardware and software requirements, if applicable, that relate to the proposed outsourcing solution provided by the vendor.
  - Product details, including release date, expected product upgrades, product demonstration software, brochures, and other literature.
  - SLAs and product/price options and availability.
  - Availability of implementation assistance and system training and documentation.
  - Preliminary costs estimates, as applicable, for software, network, and hardware implementations, and for annual maintenance and support.

After the RFI has been prepared, it is helpful to prepare a log to track the status of the RFI responses. Figure 6.2 shows a sample log for this purpose. Ideally, this information should be collected in a spreadsheet or database. As the RFI responses are received, they should be date and time stamped, and a letter acknowledging their receipt should be sent to the vendor. Incoming RFIs should be checked to determine if they have

| RFI Number: _____ |
| Project Name: _____ |
| Project Description: _____ |
| Priority (H, M, L): _____ |
| Request Sent to: |
| Name: _____ |
| Company Name: _____ |
| Address: _____ |
| Phone: _____ |
| Date Sent: _____ |
| Date RFI Received: _____ |
| RFI Status (Received, Revision Needed, Not Received):_____ |
| Follow-up Actions: _____ |
| Comments: _____ |

Figure 6.2   RFI tracking log.

been submitted on time. If an RFI has not been submitted according to the established deadline, the organization will have to decide whether or not it will accept a late response and if it will follow up with a vendor to see why a response was not received.

After an RFI has been logged in, it should be reviewed for content. Ideally, RFIs should be evaluated by a project team assembled for this purpose. This will help ensure organizational consensus and fairness in the RFI evaluations. RFI responses should be evaluated with respect to their completeness and appropriateness. In conducting an appraisal of the RFI, the RFI evaluation team may wish to pay particular attention to the following:

- Company size
- Revenue and profit
- Reputation, by evaluating references from current clients
- Support for a certain geographic area
- Level of professionalism exhibited in RFI appearance and response
- Enthusiasm and desire of vendor to support the outsourcing initiative
- Timeliness of vendor responses

- Extent to which outsourcer will subcontract implementation solution to a third party
- References from current clients

Usually, the evaluation team first checks that these so-called "k.o." (or knock-out) or minimum-level criteria are satisfied. If so, the evaluation team can commence to review the responses with a more detailed eye on the vendor's technical capabilities. However, this is not intended to be an in-depth evaluation. The real purpose is to convince the evaluation team that the outsourcer could potentially provide the products or services required. As a next step, the vendor's cost estimates and payment terms may be evaluated for reasonableness. But, again, this is not an in-depth evaluation, as contract negotiations may take a substantially different form as the RFP process progresses. The expected result of the evaluation team's deliberations is a short list of outsourcers that do not violate any "k.o." criteria and that could do the job at a reasonable price. These are the vendors that will be invited to participate in the next step of the outsourcing process — the Request for Proposal.

6.5 Request for Proposal

The purpose of the Request for Proposal (RFP) is to solicit formal vendor proposals for the outsourcing project. The RFP is distinguished from a Request for Bid (RFB) in that the latter is used to solicit vendor proposals for a product or service that will be selected on the basis of price. In contrast, the RFP provides a competitive arena for vendor proposals, where a variety of alternative solutions and approaches can be presented. The organization should explicitly reserve the right to use nonproprietary ideas and concepts gathered from the vendor proposals to help come up with the best possible solution.

The RFP process is somewhat similar to the RFI process, with overlapping information collection. However, it is generally a more formal process, with stringent requirements for detail and supporting documentation in the vendor response. Vendors are usually invited to prepare an RFP response after being prequalified through an RFI process. This is not always true, as in some cases the organization may want to solicit a broad spectrum of vendor input and may choose to forego the RFI process. Instead, they may post notice of the RFP on a Web site or use mass advertising. Large government and public-sector organizations may favor this method, especially when the outsourcing project does not need to remain confidential or discreet. On the other hand, this approach might be totally inappropriate for a major financial institution that wishes to keep the details of a disaster recovery outsourcing project confidential and limited to a select few.

A generic RFP template for an outsourcing project is outlined below. The major elements of the RFP typically include a cover letter (which makes a formal request for a vendor proposal) and associated RFP documentation. The RFP documentation generally consists of the following sections:

- *Table of contents.* This provides an outline of the entire RFP package, including the documentation provided by the organization requesting the RFP, and the documentation required of vendors responding to the RFP.
- *Project overview.* This section provides information to prospective vendors on the nature of the outsourcing project. It usually consists of a:
  - Project statement, which provides an executive summary of key goals, objectives, and benefits to be attained by the outsourcing project.
  - Description of the intended scope of the outsourcing project, from both the organization's and the vendor's perspective. This may include background information on the organization's needs (e.g., relating to the number and location of affected worksites, etc.) and the extent to which the outsourcing solution will impact the organization's infrastructure. This may include a high-level description of the key deliverables and systems requirements being considered for outsourcing. Critical restrictions or constraints that the outsourcing solution must address can also be identified in this section.
  - Description of the organization's current environment. This might include a review of present systems, with information on software, hardware, LAN/WAN/MAN protocols, transaction volumes, and other relevant technical details pertaining to the outsourcing solution.
- *Instructions and guidelines.* This section provides information on such matters as the required format for the RFP response, scheduling of vendor question-and-answer sessions, deadlines for vendor responses, and the date(s) for announcement of final vendor selection decisions.
- *Evaluation criteria.* This section provides a list of the RFP evaluation criteria and a description of the evaluation process that will be used for vendor selection. It does not include the weighting of each evaluation factor, as this would give the vendor information that could be used to unfairly bias the presentation of their solution in the RFP.

- *Project requirements.* This is usually decomposed into the following sub-sections:
  - *Management requirements.* This section specifies requirements such as the required timing and delivery method for the outsourcing solution (which might be an on-site/off-site solution, etc.); the selection and assignment of vendor personnel (e.g., in some cases, it may be necessary to request that vendor staff not be reassigned during the course of the outsourcing engagement to ensure continuity and the appropriate skill mix); and project management and reporting responsibilities that the vendor must support.
  - *Technical requirements.* This section provides specifications relating to such matters as project deliverables and requirements for new hardware or software. If hardware is to be acquired, then detailed specifications should be included on the requirements for performance, reliability, operating system, compatibility with existing platforms and protocols, etc. If software is to be acquired, then detailed specifications should be included on requirements relating to functionality (including security, required platform, programming languages, database management systems, file formats, operating systems support, etc.), system documentation, and training to be provided to organization's staff, and so on. For networking infrastructure and support, required SLA levels should be specified.
  - *Vendor requirements.* This section defines the nature and form of the vendor responses. It typically includes these sub-sections:
    - *Qualifications and references.* This is needed to ascertain whether the vendor has the expertise and depth needed to satisfy the outsourcing requirements, and has a proven track record in similar endeavors. This should include information on the vendor's financial strength and viability, credentials of the outsourcer's management and assigned technical staff, scope of vendor support across specified geographic areas, memberships in standard bodies and other professional organizations, and the use of subcontractors, if any, and their role in the outsourcing effort.
    - *Overview* of vendor's project management approach, implementation methodology, and work and delivery schedules.
    - *Presentation of vendor-recommended solution.* Vendors should provide detailed product/service descriptions, with available SLAs and product/price options in their RFP response. This documentation might also address the availability of implementation assistance, system training, and documentation. In addition to written documentation provided in the

RFP, vendors may also be encouraged to present a proof-of-concept presentation or a live demonstration to assist in the evaluation of their capabilities in meeting the RFP requirements.

■ *Facilities requirements* and other support required by the vendor from the organization to implement the outsourcing project.

■ *Pricing of vendor solution(s).*

■ *Minimum terms and conditions.* This section lists minimum terms and conditions that the vendor must accept before it will be engaged on the outsourcing project. It may include provisions relating to:

– *Payment schedule, which may be linked to the delivery and acceptance of key project milestones.*

– *Anticipated length of contract duration or consulting engagement.*

– *Confidentiality.* This should specify the procedures by which confidential or proprietary information will be safeguarded. The organization should specify which aspects of the RFP are to be kept confidential and contain proprietary information. Likewise, the vendor should specify which aspects of its response contain proprietary information that must be given special consideration and treatment. The provisions of this section should be prepared with the aid of legal counsel.

– *Care of property.* This type of provision might be needed to ensure that the contractor takes responsibility for the care and custody of property furnished to it by the organization as a result of the outsourcing arrangement.

– *Indemnification against acts of God,* or other legal liabilities that might arise as a result of an agreement between the organization and the vendor.

– *Termination clauses.* This specifies the process by which either side may choose to terminate the contract.

– *Availability of funds.* This clause may be included in government and public-sector contracts that are contingent upon receiving ongoing budgetary approvals and funding.

– *Copyright infringement.* This typically involves a provision that the vendor is fully responsible and held liable if they violate any copyright laws in the course of the outsourcing engagement.

– *Restrictions on the assignment or subcontracting of the outsourcing contract to a third party.*

– *Restrictions on the vendor's public discussion of the contract or mention of it in commercial advertising.*

– *Warranties and terms, if applicable.*

1. Define major evaluation categories for reviewing RFPs. Within each major category, define sub-category evaluation criteria.

2. Collect RFP responses and measure each vendor's performance in every evaluation category.

3. Calculate and sort the vendor scores.

4. Select the top scoring outsourcing candidates for further evaluation and negotiation.

5. Select outsourcer(s) and sign agreements.

Figure 6.3   Summary of outsourcing evaluation process.

- *Sample contract.*
- *Appendices.* These are included as needed to provide additional supporting detail.

After the RFP has been distributed to vendors, it is advisable to prepare a log to track the status of the RFP responses. The log presented in Figure 6.2 (shown for the RFI) can be adapted for this purpose. Ideally, this information should be collected in a spreadsheet or database. As the RFP responses are received, they should be date and time stamped, and a letter acknowledging their receipt should be sent to the vendor.

6.6 Vendor Evaluation, Selection, and Checklists

After the RFPs have been received and logged, the real challenge — the evaluation of the RFP — begins. The first step in evaluating the RFP responses is to determine if the minimum requirements have been satisfied. If not, this may be grounds for rejecting the vendor proposal outright. Some of the more common reasons a proposal may be rejected relate to a failure on the vendor's part to:

- Submit the proposal in the required format.
- Adhere to the required timeline and deadlines for the proposal submission.
- Complete all the required sections and provisions mandated in the RFP. For example, this might include missing references to vendor costs or required clarifications.
- Meet the minimum business or technical requirements as put forth in the RFP.
- Provide formal written acceptance of the required RFP terms and conditions. For example, this might include lack of inclusion of a signed confidentiality or nondisclosure agreement in the vendor submission.

If the RFP is not acceptable, the organization must decide what follow-up actions, if any, should be taken with the vendor.

Assuming that the minimal standards for acceptance are satisfied, the evaluation team should commence a careful review of each vendor proposal with respect to the evaluation criteria defined at the onset of the RFP process. Briefly summarizing, these criteria generally encompass such considerations as:

- Vendor financial attributes (i.e., financial stability, reliance on relationships with third parties and subcontractors, length of time in business, etc.)
- Vendor personnel attributes (i.e., personnel qualifications, working relationships, cultural compatibility with the organization, etc.)
- Contractual attributes (i.e., clear and well-defined SLAs, vendor acceptance of terms and conditions, flexible terms, etc.)
- Pricing attributes (i.e., competitive payment terms and pricing, etc.)
- Technical attributes (i.e., effectiveness in satisfying technical requirements, which might include: security practices, business continuity/disaster recovery capabilities, data handling capabilities, use of authentication, authorization, access control, software validation, secure asset configuration, use of backups, provisions for monitoring, auditing, trouble tracking and ticketing, performance and system availability metrics, compatibility with existing systems and network platforms, etc.)

- Standards conformance attributes (i.e., compliance with industry standards and practices, conformity and compatibility with organization's standard processes and procedures, etc.)
- Quality attributes (i.e., vendor commitment to project, use of state-of-the-art technologies and approaches, etc.)

Implicit in the list above is the possibility that each decision factor may have a number of sub-category decision criteria associated with it. To help keep the proper perspective on each of the many factors that might influence the final decision on vendor selection, a two-stage evaluation process is recommended. First, an evaluation matrix should be prepared for each major decision criterion. An example is shown in Table 6.4 for the decision attributes associated with the vendor's *financial* credentials. As shown in this example, each sub-criterion is assigned a weight that reflects its overall importance with reference to the subsuming, more comprehensive decision criterion. A fair weighting scheme for each sub-criterion can be computed using the Saaty hierarchical decision method, group consensus, or some other method (see Appendix C for a discussion of these approaches). It is helpful to automate these calculations using a spreadsheet, particularly if there are lots of decision variables and vendor proposals to evaluate. At the time the RFP decision criteria are selected, it is also helpful to construct scoring guidelines to help provide a consistent basis for the evaluation team's review.

After scores are computed for each major decision category, a final score can be computed for each vendor's RFP submission. Table 6.5 shows how the results from Table 6.4, and other calculations for additional evaluation criteria (which are not shown) are used to assign an overall score to *Vendor X*. After every qualifying RFP has been scored, all the RFPs can be rank ordered. The highest-scoring RFPs should be given serious consideration as the top contenders for the contract award. During the scoring process, there may be several rounds of dialogue with the outsourcers to fine-tune the negotiations and to narrow the final selection down to two or three proposals. These discussions may involve vendor presentations, demonstrations, test procedures, and possible reconsideration of financial and other contractual terms. The RFP evaluation process is illustrated in Figure 6.3.

Prior to signing the outsourcing contract, the following items should be fully addressed and mutually agreed upon:

- *Points of contact.* Essentially, this establishes who is authorized to approve modifications to the contract. Somewhere in the agreement, provisions should be made to allow for contract modifications. They should allow either the outsourcer or the customer to

Table 6.4   Example: Scoring of Vendor X Financial Attributes

Vendor X Evaluation: Satisfaction of RFP Requirements Relating to: Vendor Financials	Decision Factor (Substitute financial metrics used in RFP evaluation)	Decision Factor Score (Maximum Score = 100; Minimum Score = 0) (a) (Note: This score is typically determined by organizational consensus)	Decision Factor Weighting (%) (b) (Note: Relative weights are determined by Saaty or consensus method)	Decision Factor Total Score (c) = (a) * (b)
	Financial Strength	$(a_1) = 75$	$(b_1) = 0.50$	$(c_1) = 37.5$
	Length of Time in Business	$(a_2) = 100$	$(b_2) = 0.30$	$(c_2) = 30.0$
	Vendor Asset Base	$(a_3) = 75$	$(b_3) = 0.20$	$(c_3) = 15.0$
Total		(Note: "100" = Significantly Exceeds Requirements, "0" = Fails to Meet Requirements)	100.00% (Note: the total of *all* decision weights must exactly equal 100%)	$\Sigma (c_i) = (37.5 + 30 + 15) = 82.5$

Table 6.5   Example: Overall Scoring of Vendor X RFP Submission

Vendor X Evaluation: Overall Satisfaction of all RFP Requirements	Decision Factor (Substitute RFP decision metrics)	Decision Factor Score (Maximum Score = 100; Minimum Score = 0) (a) (Note: This score is computed from $\Sigma (c_i)$)	Decision Factor Weighting (%) (b)	Decision Factor Total Score (c) = (a) * (b)
	Financials	$(\Sigma (c_j)) = 82.5$ (See Table 6.4 for example calculation)	$(b_1) = 0.10$	$(c_1) = 8.25$
	Personnel	$((\Sigma (c_j)) = 90.0$	$(b_2) = 0.10$	$(c_2) = 9.0$
	Contract	$\Sigma (c_j)) = 95.0$	$(b_3) = 0.05$	$(c_3) = 4.75$
	Price	$\Sigma (c_j)) = 80.0$	$(b_4) = 0.25$	$(c_4) = 20.0$
	Technical	$\Sigma (c_j)) = 92.0$	$(b_5) = 0.35$	$(c_5) = 32.2$
	Standards	$\Sigma (c_j)) = 0.0$	$(b_6) = 0.05$	$(c_6) = 0$
	Quality	$\Sigma (c_j)) = 70.0$	$(b_7) = 0.10$	$(c_7) = 7.0$
Total			100.00%	$\Sigma (c_i) = 81.2$

reopen negotiations, and to provide a vehicle for management to approve changes to the scope or priority of the work. Generally, changes should be made only after an analysis has demonstrated that the issue at hand is substantive and not an aberration. Trial agreements for new outsourcing services may prove satisfactory as a way to deal with the need for contract renegotiations.

▪ *Agreement duration.* The agreement should be written to limit it to a specific duration or period of time, such as a certain number of years. Alternatively, the outsourcing agreements might be scheduled for revision as planned hardware, networking nodes, facilities, or software upgrades are made. Above all, no one should be under the impression that the agreement is a commitment for eternity that remains in force despite changes in the business or networking environment.

▪ *Review.* Reviews are necessary to consider the impacts of a dynamically changing environment on the requirements and the structuring of the outsourcing arrangement. For the mutual benefit of all parties, these impacts should be openly discussed, and necessary changes should be incorporated into the existing contract as needed.

▪ *Service-level indicators.* Service-level indicators should be quantified in design and planning models, and the method of monitoring their compliance should be established. SLA conformance may be monitored using on-premise or off-premise monitoring.

▪ *User commitments.* The end user has certain responsibilities and commitments to ensure the ongoing success of the outsourcing arrangements. This includes providing outsourcers with accurate and timely information about their lines of business, strategic goals, critical success factors, networking environments, organizational changes, application portfolios, service expectations and indicators, intended use of network technology, and early notice to the vendor when there is an anticipated need to renegotiate the contract.

▪ *Reporting periods.* Performance reporting is used to identify potential problems and is a key tool for avoiding crisis management. Performance reports on key service indicators should be made regularly (i.e., weekly, monthly, etc.). Copies should be sent to the organizational entity in charge of the outsourcing project, and to appropriate information technology (IT) or information systems (IS) management staff. There should be mutual agreement on the format, content, and timing of the performance reports.

▪ *Costs and chargeback policy.* Contracting parties must agree on payment arrangements, chargeback reporting, alternative methods of bill verification, and expected inflation rates. Depending on the nature of the contract, the transfer of human resources may also need to be negotiated.

■ *Penalties for noncompliance.* Penalties for noncompliance generally include predefined credit and chargeback arrangements. However, in most cases, the economic damage caused by noncompliance is much more substantial than the penalty reimbursement.

■ *Employee transition.* Somewhere in the agreement, details relating to employee transition should be addressed. This section should include the names of employees to be transferred to the outsourcers, and the conditions of the takeover, such as salaries, job security, title, position, etc. Employee provisions for training and education should also be included in contractual arrangements for outsourcing, when appropriate.

■ *Billing and currency issues.* It is beneficial to tie the contract to a single currency. This is particularly important when entering into contractual arrangements with multinational companies.

■ *Periodic pricing reviews.* Subcontractors may lower or raise their rates over time. The contract should provide a means to track these changes and to make the necessary adjustments when needed.

6.7 Strategic Partnerships and Agreements

The emphasis of this chapter is on a traditional consulting or outsourcing contract for some or all of an organization's network/IT-related functions. The market for IT, networking, and outsourcing is extremely competitive. Large vendors have been increasingly involved in strategic arrangements with other vendors to reduce their cost structures, and to increase the diversity and quality of their products and services. There are many types of parties working together to provide outsourcing solutions, including manufacturers of monitoring tools, manufacturers of networking equipment, vendors of design and planning tools, outsourcers, consulting companies, and software and application development companies. These groups are creating very interesting partnerships with strategic synergies.

6.8 Keys to Successful Outsourcing

Outsourcing is an important component of IT planning and strategy. Market forces — such as offshore outsourcing — and the wide availability of product and service offerings have spurred tremendous vendor competition. As a result, the costs of outsourcing have steadily declined as service and quality have continued to improve. The buyer's market for outsourcing has been a major force in the trend toward more and more outsourcing of IT, networking, and other business functions.

Judicious outsourcing provides a means for the enterprise to focus on its core competencies, leaving less strategic areas to be supported by external professionals. Many aspects of network management and operations are widely outsourced. However, network design and capacity planning is one area that is not. There are particular reasons for this. Very frequently, telecommunications suppliers include modeling and simulation results to build a business case for the solutions offered in their proposals. The organization should view the results with a skeptical eye, because the telcos have a vested interest in promoting their approach, products, and services. Likewise, equipment vendors offer optimization services as part of their product offerings. However, typically, their internal modeling and simulation packages are fine-tuned for their equipment only and cannot be used to optimize other equipment on the network.

When difficulties arise in an outsourcing arrangement, it is mostly likely related to the following types of issues:

■ *Inadequate controls.* This may occur if there is a lack of strong project management. Independent companies or consulting firms can explode user budgets with high rates and costs. If the outsourcers also control critical governance functions, they may promote costly products and services that do not serve the organization's best interests. Cost control is a critical aspect of project management. For example, France Telecom recently "stunned" the management consulting industry by cutting its spending on consultants and other outside advisers by over a half a billion euros — or almost U.S. $570 million. The cuts were a result of a management review initiated by new executive leadership, and have allowed the company to return to profitability for the first time since 2000. In the words of the executive assigned to the financial turnaround of the company, Frank Dangeard, the consulting budget he inherited "just blew my mind" and "was completely out of control." With the reduced consulting expenditures, the company is now able to resume dividend payments. [JOHN04]

■ *Poor communications.* This often is a result of poor project management, and client/vendor relationship problems. "Soft" issues relating to organizational culture, lack of trust, mismatched expectations, and lack of common understanding of key goals and objectives can create serious problems that may undermine the outsourcing arrangement. This is why consideration of these matters is so critical during the RFP and vendor evaluation process.

▪ *Ineffective problem resolution.* This may be a result of and related to inadequate control and poor communications. It may also occur if the quality of the products and services provided by the outsourcer do not meet the organization's expectations or needs, and the matter is not dealt with quickly or effectively. Strong management and oversight of the outsourcing arrangement are needed to identify the potential for problems and to find proactive ways of avoiding or correcting them. In the absence of this, problems may go unnoticed for too long and may be difficult to resolve once the outsourcing relationship has soured. It is important to build an atmosphere of positive communication, where problems can be openly discussed without either side becoming defensive or combative.

▪ *Lack of due diligence in dealing with legal, security, and other risks and liabilities.* This consideration can encompass many things, and relates to the expression "the devil is in the details." For example, if the organization becomes too dependent on an outsourcer, its own viability may be compromised if the outsourcer suffers serious setbacks or goes out of business. In part, this can be avoided by working with multiple vendors and providers, but this may not be practical for all types of outsourcing arrangements. The organization should also be sure to retain and protect the ownership of its assets in the outsourcing arrangements. Another concern relates to the potential for hidden costs and underestimated complexities to arise. These may be difficult to assess at the project onset. The organization should factor the potential for this into their risk analysis and plan accordingly.

▪ *Inadequate technology transfer and resolution of intellectual property matters.* When a vendor contract expires, the organization no longer has access to the outsourcer's intellectual capital. This may present problems if the organization is ill-equipped to handle the functions that had previously been handled by the outsourcer. To avoid this, a technology transfer program should be planned from the inception of the outsourcing arrangement. The converse is also a consideration. It is possible for an organization to expose too much of its proprietary process and intellectual capital to an outsourcer during a consulting engagement. This should be guarded against through appropriate legal provisions and employee awareness from the onset of the vendor interactions.

Successful outsourcing experiences are the result of good management and oversight that places particular emphasis on:

▪ *Mutual goals and understanding.* All parties should fully understand the respective goals and objectives of the other and how they are to be achieved through the outsourcing arrangements.

- *Appropriate vendor selection.* The selection of a good outsourcer is the result of careful balancing of the RFP evaluation metrics. Price, flexibility, culture, and relationship considerations are paramount considerations in this process.

- *Proper contractual arrangements.* The outsourcing contract should specify SLAs, contract governance, and other contract details. These arrangements should strive for a proper balance between the organization's desire to control costs and the vendor's need to make a profit. If the vendor does not have an adequate profit margin built into the contract, the quality of the services and technology it offers may suffer.

- *Pertinent, measurable benchmarks and metrics.* These are needed to provide a means to monitor the progress and performance of the outsourcing arrangement.

- *Competitive outsourcing engagement.* It is generally not in the organization's best interests to limit its consideration or use to one vendor. It is advisable, where practical, to give portions of the outsourcing contract to multiple outsourcers. This helps preserve working relationships with other entities and provides a basis for flexibility in future negotiations.

- *Active, ongoing management participation.* Continuous management involvement is needed to reexamine the need for and execution of the outsourcing arrangements. This should encompass management review of the staffing, processes, technology, and other factors that determine whether an appropriate solution is still in place. The organization, and not the outsourcer, should control strategic decisions and contract governance.

6.9 Summary

In summary, outsourcing often provides a convenient way to augment staffing, skills, and facilities, while saving time and money. It is being used to support a full complement of network management functions relating to the support of LAN, WAN, help desk, applications development, and business process reengineering.

Notes

1. ASPnews.com is a Web site dedicated to "news and analysis for and about the Application service provider (ASP) and Web services industry." It can be found at http://www, ASPnews.com

2. The Organization for the Rights of American Workers (TORAW), labor unions, worker's rights advocacy groups, and other organizations have banded together to fight the trend toward outsourcing jobs to foreign countries. They are also encouraging individuals to boycott companies that use offshore outsourcing. Arguments put forth by TORAW against offshore outsourcing can be found at the following Web site: http://www.toraw.org. Arguments promoting offshore outsourcing can be found at the Outsourcing Institute's Web site: http://www.outsourcing.com

References

[CONS04] Consultant-News.com, Government sector drives 49% increase in global spending on major outsourcing projects in 2003, January, 22, 2004, http://www.consultan-news.com/Artivcle_Display.asp?ID=1251.

[JOHN04] Johnson, J., France Telecom stuns industry by slashing $570m from consulting bill, *Financial Times,* February 13, 2004.

[KENN04] Kennedy Information, Inc., The Global Consulting Market Place 2004–2006, Key Data, Trends, and Forecasts, 2004.

[OVER04] Overby, S., How to Safeguard Your Data in a Dangerous World, *CIO Magazine,* January 15, 2004, http://www.cio.com/ar chive/011504/outsourcing.html?printversion=yes.

[SURM03] Surmacz, J., Outsourcing Hype versus Reality, *CIO Magazine*, December 10, 2003, http://www2.cio.com/metrics/2003/metric639.html.

Chapter 7

Automated Design and Network Management Tools

7.1 Network Design Tools

The major steps involved in planning and designing a network include:

1. *Requirements analysis.* This involves collecting data on potential line types and costs, node types and costs, sources and destinations of traffic, and traffic flows between nodes.
2. *Topological design.* This involves using various design techniques, including heuristic design algorithms, to produce a network topology specifying link and node placements.
3. *Performance analysis.* This involves assessing the cost, reliability, and delay associated with the topological designs under consideration.

Network design tools help automate some or all of the above design activities. A good design tool can greatly assist the process of collecting the requirements data. For example, the major network design tools have built-in databases containing cost data for various node types, and tariff databases or cost generators for calculating circuit charges. As part of their service, the vendors of network design tools provide periodic updates to these databases so the information remains current. In addition to providing as much built-in data as possible, a good design tool should facilitate the entry and collection of organization-specific data needed to plan the network. This includes providing a means to collect data on the potential node locations and traffic flows. It might also include a traffic generator. The automated tool should relieve the designer, to the extent possible, of the burden of collecting and massaging a lot of requirements data. During the design phase, the network tool should provide a variety of tools, techniques, and algorithms so that many types of designs can be produced and evaluated. Automated tools are then needed to calculate the reliability, cost, and delay characteristics of each design candidate. Sensitivity analysis is also an important aspect of network design. Tools that allow the network designer to selectively modify certain aspects of the network and to recalculate the associated network performance characteristics are essential to a comprehensive network assessment. After completing the network design and analysis, reports and graphical displays documenting the design process are needed. In addition to supporting all these functions, a good network design tool should have an easy-to-use interactive, graphical user interface.

Some organizations rely almost exclusively on manual techniques to design a network. Although this approach certainly is flexible, it is usually wholly inadequate for a network of any substantial size. There are just too many things to consider, and too many calculations to make to develop an optimal network solution without the aid of automated tools. If there are n potential node placements in the network, there are $n * (n-1)/2$ potential lines and $2^{(n * (n-1)/2)}$ possible topologies. There are simply too many possibilities to enumerate or to test by hand! Organizations that rely on manual methods may do so because they are not aware that network design tools are available, or they may lack sufficient resources (i.e., staff or dollars) to acquire a network design tool. WAN design tools tend to be very expensive, and may cost tens of thousands of dollars. They may, therefore, be out of the price range of smaller organizations. Service providers and outside consultants frequently make use of automated design tools because, in a competitive bid situation, the organization offering the design with the best price and performance is likely to win the contract.

CISCO Network Modeling Tools (NMT) provides an example of a WAN design product for service providers. It is used to design IP, Frame Relay, and ATM wide area networks (WANs). NMT works in conjunction with Cisco's WAN Manager (CWM) product, which automatically collects configuration and other network data for use in the design analysis. NMT is used to perform sensitivity analysis to assess the impacts of changing traffic, reliability, and bandwidth requirements on the optimal network configuration and its associated costs. NMT is an online, menu-driven system that offers the following functionality to the network designer: [CISC02]

- Reports that document modeling results
- Graphic displays of the network configuration
- Updated tables containing network configuration data
- Data import and export facilities to provide flexibility in analyzing network data with other tools

NMT is just one of many network management tools offered by Cisco Systems. For a more complete listing of Cisco's network management products and services, the interested reader can refer to the Cisco Systems' Web site at: http://www.netsystech.com/en/US/products/sw/netmgtsw/index.html.

There are numerous network design tools available for WAN/LAN design and analysis. A few WAN design and planning tools are listed below; there is no attempt here to provide a comprehensive survey of all the network design tools on the market.

- *Network Design and Analysis Corporation (NDA) (Toronto, Canada).* NDA is a software company specializing in network design and pricing tools. It offers the tools listed below: [NETW04]
  - *Account Management System (AMS).* This is a Web-based application, based on Microsoft's DNA architecture, for pricing, quoting and ordering voice, data, and IP services.
  - *E-commerce.* This is a Web application to provide customer quotes and to generate customer orders using a built-in "pricing engine" that uses a combination of public, private, and contracted rates and business logic.
  - *ILEC-Pricer.* This is an online ILEC pricing tool that uses origination and destination NPA/NXX or CLLI codes to provide cost estimates.[1] The ILEC-Pricer is implemented as an ASP solution so only "a Web browser is needed to create a quote. Two of the four RBOCs, as well as many CLECs are using NDA's pricing engine because of its accuracy and content. Tariff updates are performed twice per month to provide the high level of accuracy required by carriers."
- *Netformx (Saratoga, California).* Netformx is a privately held company specializing in "guided selling and software development." It has partnerships with Cisco Systems, Nortel Networks, SBC Communications, GE Capital, IBM Global Services, and Verizon. Netformx's

DesignXpert Platinum suite consists of the following components: [NETF02]

- *DesignXpert®Platinum Edition*. This is a network design and device configuration tool.
- *Enterprise AutoDiscovery*. This is a network discovery, inventory, and documentation tool.
- *Multi-Vendor Knowledge Base*. This in an information repository containing over "42,000 network devices with a rules-based engine that assists the user in producing recommendations, configurations, orders, and solutions."
- *Knowledge Base Update Services*. This provides "real-time updates from manufacturers that cover changes in product, pricing, manufacturing constraints, specifications, software versions."

The selection and purchase of a network design tool should be predicated on the project objectives and requirements. This, in turn, provides the basis for developing a specification of the functional requirements and potential uses of the network design tool. For example, it may be necessary to have a tool that supports tariff databases. Some tools provide comprehensive tariff information for international and domestic networks. This functionality may be necessary if a large multinational network is being designed. If the organization is designing a large WAN that must be billed back to the users, then a tool may be needed to produce detailed cost and usage reports. In addition to considering the specific technical design requirements the tool should support, other factors that might be relevant include the availability of technical or consulting support, training, frequency and cost of tariff upgrades, the company financial status, and years in business.

In summary, the motivation for using a design tool is that it helps the network designer quickly develop a number of design options, under a variety of traffic and cost assumptions. The designs are then analyzed to discover which ones best support the requirements in the present and foreseeable future. Thus, all good design tools encourage exploration of multiple design options, with some allowance for intervention by the designer. The complexity of network design and analysis is such that it is not feasible to perform these tasks manually for medium- to large-scale networks.

7.2 Network Management Tools

Network design and planning are complementary to network management, because a well-designed network will be easier to manage. A network

• **Tasks:**	• **Implementation:**
– Gathering information on current configuration – Using data to change configuration – Storing data, maintaining inventory, & producing reports including vendor contact information, lease line circuit numbers, quantity of spares, etc.	– Manual versus automated tools – Simple tool: central storage of all network data – Advanced tool: Automatic gathering and storage of network data, automatic change of running configuration, etc.

Figure 7.1   Configuration management.

that is configured to collect network management data also provides a foundation for informed planning and new design decisions.

According to the ISO Network Management Forum, network management should encompass these functions: Security Management, Performance Management, Accounting Management, Configuration Management, and Fault Management. Figures 7.1 through 7.6 provide an overview of the network management tasks that automated tools support, as well as differences in the level of functionality between simple and advanced products. Chapter 8 discusses organizational impacts related to supporting these network management functions.

7.2.1 Network Management Protocols (CMISE/CMIP, SNMP, and RMON)

To perform network management tasks, data must be collected from every device in the network. This includes such information as the device name, device version, number of device interfaces, device parameters, device status, etc. To process and consolidate this information across multiple devices and platforms, network management protocols are needed. Network management protocols provide a uniform way of accessing standard metrics for network devices made by any manufacturer. The major network

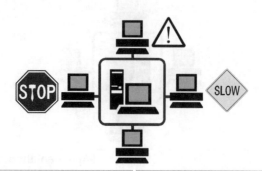

• **Tasks:**	• **Implementation:**
– Identify fault occurrence – Isolate cause – Correct, if possible	– Information gathering (event driven versus polling) – Simple tool: identifies problem(s), but not cause(s) – Advanced tool: Interprets network events, produces reports, & assists in fault correction & fault

Figure 7.2   Fault management.

• **Tasks:**	• **Implementation:**
– Gather data on network resource utilization – Set usage quotas – Bill users (may be based on a number of different approaches)	– Define appropriate metrics for system usage – Simple tool: monitors metrics exceeding quotas – Advanced tool: Automates billing, assists in forecasting resource requirements

Figure 7.3   Accounting management.

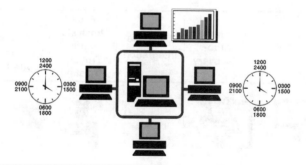

• **Tasks:**	• **Implementation:**
– Gather data on network resource utilization	– Define appropriate metrics for system performance
– Analyze data in both real time and off-line modes	– Simple tool: graphical display of network devices and links
– Perform simulation studies	– Advanced tool: Sets thresholds and error rates and then performs prescribed corrective actions. Collects historical data, which can in turn be used for simulation & other analysis

Figure 7.4 Performance management.

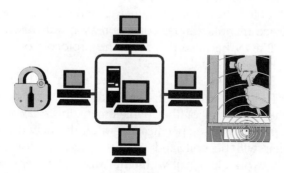

• **Tasks:**	• **Implementation:**
– Identify sensitive information & its location	– Locate access points (e.g., ftp, remote login, e-mail, etc.)
– Secure and maintain access points to information	– Secure access (encryption, packet filtering, etc.)
• **Note:**	– Identify potential or actual security breaches
– This is not the same as operating system security or physical security!	

Figure 7.5 Security management.

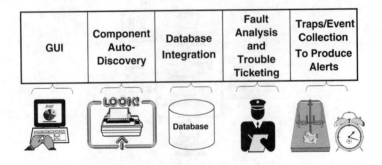

GUI	Component Auto-Discovery	Database Integration	Fault Analysis and Trouble Ticketing	Traps/Event Collection To Produce Alerts

Figure 7.6 Network management tool functionality.

management protocols include CMISE/CMIP and SNMP and RMON. These open standards have largely supplanted the proprietary, single-vendor network management protocols used in legacy mainframe environments.

CMISE/CMIP[2] provides a common network management protocol for all network devices conforming to the ISO Reference Model. It is designed to be a total network management solution. However, it requires a high degree of overhead and is difficult to implement. Therefore, its use is largely limited to telcos and service providers with very demanding, complex, and large-scale network management tasks and the highly skilled staff required to perform these tasks.

SNMP and RMON are complementary, and are very widely implemented by device manufacturers. For this reason, this discussion focuses on SNMP and RMON because these are the protocols of choice in most organizations, large or small.

The major SNMP standards are documented in a series of RFCs put out by the IETF.[3] At the core of SNMP (Simple Network Management Protocol) is the definition of the Management Information Base (MIB). The MIB provides a precise specification of all the network management information that will be collected over the network for each network device (a.k.a. network element). Within the MIB, SNMP represents network resources as objects. Each object relates to a specific variable associated with a particular aspect of a managed object (e.g., a printer object might consist of an interface object, status object, etc.), represented by either a scalar (i.e., a single value) or a table (containing multiple entries, as shown in the example given in Figure 7.7). The MIB defines how each variable is named, coded, and interpreted. The collection of all managed objects in the network is referred to collectively as the "MIB."[4]

The MIB conforms to the Structure of Management Information (SMI) for TCP/IP-based Internets. This SMI, in turn, is modeled after OSI's SMI. While the SMI is equivalent for both SNMP and OSI environments, the

Sample Routing Table

Destination	Next
2	3
3	4
3	2

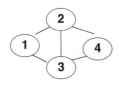

Note: Each table entry must be assigned an OID (Object ID) for retrieval purposes.

Figure 7.7   Sample routing table.

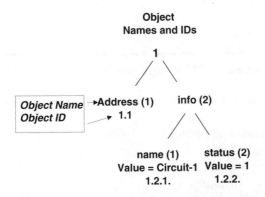

Figure 7.8   Example of object naming and OID.

actual objects defined in the MIB are different. SMI conformance is important, because it means that the MIB is capable of functioning in both current and future SNMP environments. In fact, the Internet SMI and the MIB are completely independent of any specific network management protocol, including SNMP.

MIBs are tree-like, hierarchical, structured data schemas defined according to the ISO Abstract Syntax Notation One. This syntax defines each MIB component as a node labeled by an object identifier and a short text description. Figure 7.8 illustrates a generic object and its associated id (OID) and name. Figure 7.9 presents an overview of the entire IETF MIB structure, with the Internet in the middle (and of which SNMP is only a small part).

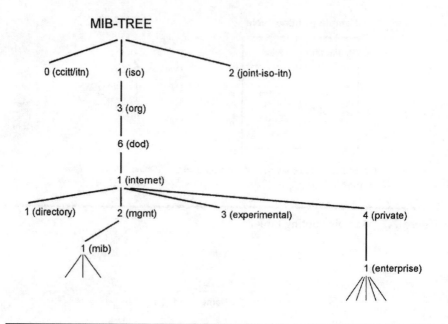

Figure 7.9   Overview of IETF management information base (MIB). (Source: IETF)

The SNMP MIB repository contains four types of attributes:

1. Management attributes
2. Private attributes
3. Experimental attributes
4. Directory attributes

MIB-I contains a limited list of objects dealing with IP internetworking routing variables. MIB-II extends this to provide support for a variety of media types, network devices, and SNMP statistics, and is not limited to TCP/IP.

The management attributes are the same for all agents supporting SNMP. Thus, SNMP managers may work with agents from various manufacturers. To offer more functionality, vendors may populate the private attributes with proprietary, value-added data.

Most real MIB implementations employ extensions, which provide additional information storage for various network management and operational tasks. Such extensions might include: [TERP96]

■ Enhanced security information relating to access of the various managed objects, the network management system, and the MIB (so it can be manipulated).

■ Extended data for user and user group administration. Users and user groups can themselves be managed MIB objects that must be created, maintained, and eventually deleted.

■ Configuration histories and profiles to provide data for reporting, backups, and alternate routing (as part of LAN fault and performance management functions).

■ Trouble tracking data to help expedite resolution of network problems. The augmented MIB is used to translate event reports into trouble tickets, assign work codes to staff, recognize and categorize, relate problems, escalate problems by severity, and close trouble tickets after eliminating the problem(s) and associated causes.

■ Extended set of performance indicators to support advanced performance management (relating to resource utilization, threshold violations, trends, bandwidth utilization between interconnected LANs, etc.).

RMON is a very well-known and popular example of an SNMP MIB extension.

Using a network management station, the utilization of the MIB and its extensions should be carefully monitored. This involves regular inspection of SNMP command frequencies, the number and types of traps being generated, information retrieval frequencies initiated by GetNextRequest (the significance of this SNMP command is described later), and the proportion of positive to negative poll responses.

In addition to the MIB, SNMP defines the following components:

■ Management agents
■ Management station
■ Network management protocol

7.2.1.1 Management Agents

Each agent possesses its own MIB view, which includes the Internet standard MIB and, typically, other extensions. However, the agent's MIB does not have to implement every group of defined variables in the formal IETF MIB specification. The management agent only contains the portion of the SNMP MIB monitoring and control data that is relevant to a particular device as implemented by the device manufacturer. This means, for example, that gateways need not support objects applicable only to hosts, and vice versa. This eliminates unnecessary overhead, facilitating SNMP implementation in smaller LAN components that have little excess capacity.

Agents respond to action and information requests or provide unsolicited information to the network management station. An agent performs two basic functions:

1. *MIB variable inspection.* The agent is responsible for examining the values of counters, thresholds, states, and other parameters.
2. *MIB variable alteration.* The agent is also responsible for resetting counters, thresholds, etc. For example, it is possible to reboot a node by setting a variable.

An agent MIB implementation can be hosted on several types of platforms, including: [TERP96]

- Object-oriented databases
- Relational databases
- Flat file databases
- Proprietary format databases
- Firmware

Thus, MIB information is distributed within agents. A typical configuration might include, at the agent level, a disk-based relational database, or a combination of PROM with static object attributes, and RAM with dynamically changing information.

7.2.1.2 The Management Station

The network management station consists of:

- Network interface and a monitor for viewing network information
- Network management applications
- Database of managed network entities

Managers execute network manager station (NMS) applications, and often provide a graphical user interface that depicts a network map of agents. Typically, the manager also archives MIB data for trend analysis. This is implemented in two different ways:

1. Each agent's MIB entries are copied into a dedicated MIB segment.
2. MIB entries are copied into a common area for immediate correlation and analysis.

The manager provides presentation and database services, which must be implemented carefully to ensure the appropriate features, functions, and performance are present to support network management. Some issues involved in this implementation are summarized below: [TERP96]

- Use of object-oriented databases
  - The advantages:
    - They are naturally well-suited to network management because the MIB itself is formally described as a set of abstract data types in an object-oriented hierarchy.
    - They have the ability to model interface behavior via stored methods.
  - The disadvantages :
    - This technique is not yet mature for product implementations.
    - There are no standards yet for query and manipulation languages.
    - The MIB object class hierarchy is broad and shallow in the inheritance tree.
    - Performance characteristics are not yet well documented.
- Use of relational databases
  - The advantages:
    - The technology is mature, stable, and is well supported.
    - There is a standard access language (SQL).
    - There are translators for translating ER (Entity-Relationship) models into relational schema.
    - There are many application choices and vendors.
  - The disadvantages:
    - They are not well-suited to storing OO (object-oriented) models.
    - Performance is highly dependent upon database tuning and the application.
- Use of other databases
  - Flat-file databases and other proprietary formats can be tailored very specifically to MIBs, and can be optimized for performance. However, this may lead a network management design and implementation that is more complex and time-consuming to develop, maintain, and change.

7.2.1.3 Network Management Protocol

The network management protocol defines the messaging format and conventions by which management stations and agents communicate. Objects (in the form of read-only Object IDs (OIDs), or read-write (OIDs), also known as Traps) are exchanged between an SNMP manager and

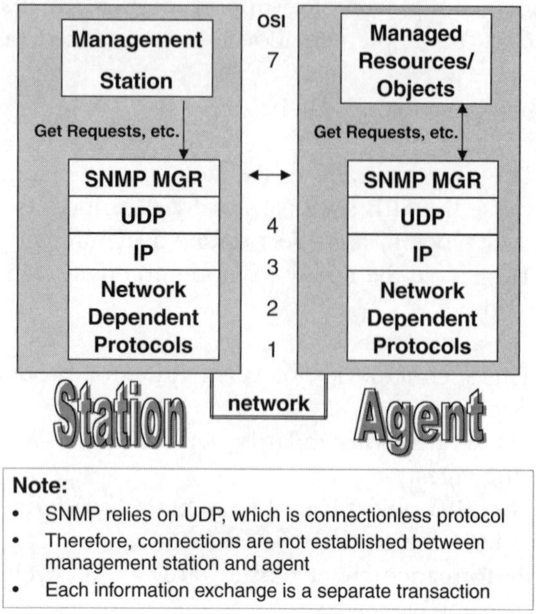

Figure 7.10 SNMP architecture.

agent through messages. SNMP uses the OSI level 4 Transport Layer User Datagram Protocol (UDP) for message transmission, as shown in Figure 7.10. The network management protocol defines the SNMP message types, which are sent as single packets, as illustrated in Figure 7.11 and summarized below:

- *GetResponse.* This is a poll initiated by a management station to request information from a management agent.
- *GetNextRequest.* This is a poll initiated by a management station to request the next item in the management agent's MIB table. This message must be sent repeatedly to obtain all the information contained in the agent's MIB, because items can only be retrieved one at a time.
- *GetNextResponse.* This is the agent's message reply to the management station's GetRequest.
- *SetRequest.* This is a command sent by a management station to an agent to set an item parameter. This command is used to perform device configuration management actions.
- *Trap.* This is an unsolicited message sent by an SNMP management agent. It is event driven because the agent does not need to be asked or polled by the management station to send the trap. Seven

SNMP Message:

SNMP Version #	Community Name	SNMP PDU

- **Community name:**
  - This is sent in plaintext and contains an unsecured password or IP address (which could be based on a domain name or some other name assigned to a group, department, etc.).
- **Five valid SNMP message types, or Protocol Data Units (PDUs):**
  - GetRequest
  - GetResponse
  - GetNextRequest
  - GetNextResponse
  - SetRequest
  - Trap
- **IETF RFC 1157 defines PDU conventions**

Figure 7.11   SNMP message format and types.

traps are defined in MIB-2. The first six (6) traps are defined by SNMP, and the last is a vendor-specific message type.

- System coldstart
- System warmstart
- Link down
- Link up
- Authentication failure
- Exterior Gateway Protocol (EGP) neighbor loss
- Enterprise specific (potentially unlimited in number)

SNMP does not support the concept of management station to management station communication. It only supports the manager and agent interaction. However, it does not set a limit on the number of management stations or agents that can coexist on the network.

SNMPv1, as its name implies, is a simple and easy protocol to implement. For this reason, it is widely supported by device manufacturers. However, a number of drawbacks with SNMPv1 led to the creation of a new protocol, SNMPv2. The SNMPv1 drawbacks included:

- It is officially standardized only for TCP/IP networks.
- It is inefficient for large table retrievals, due to the implementation of GetNextRequest.
- The password security provided by the cleartext community string in the SNMP message can be easily broken.

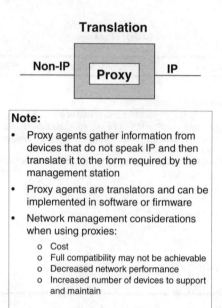

Figure 7.12   Proxy agent translation.

- This version supports only a limited number of MIB entries and message types, as described above.
- It cannot process bundled message requests, so each request must be sent separately, leading to increased network overhead and traffic.
- EGP MIB-2 cannot detect routing loops. The EGP (Exterior Gateway Protocol) is designed to detect a device's failure to acquire or reach its exterior gateway neighbors.

SNMPv2 was designed to address the deficiencies in SNMPv1, particularly with respect to security and functionality. Because it is a separate protocol, distinct from SNMPv1, SNMPv1 and SNMPv2 network management stations and agents are incompatible and cannot interact directly on the same network. If this is needed, proxy agents are used to translate between SNMPv1 and SNMPv2, as shown in Figure 7.12. SNMP proxy agent software permits an SNMP manager to monitor and control network elements that are otherwise not addressable using SNMP. For example, a network manager might wish to migrate to SNMP-based network management, but there are devices on the network that use a proprietary network management scheme. An SNMP proxy manages both device types in their native mode by acting as a protocol converter to translate the SNMP manager's commands into the proprietary scheme. This strategy

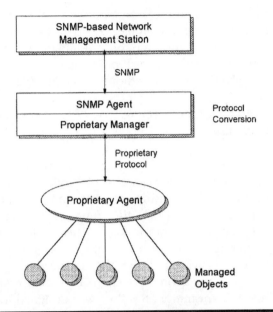

Figure 7.13   Proxy agent integration with SNMP.

facilitates migration from a proprietary environment to an open one. This is illustrated in Figure 7.13.

The SNMPv2 Structure of Management Information provides extensions to allow the capture of new data types and more object-related documentation. SNMPv2 also supports two new message formats or PDUs:

1. *GetBulkRequest.* This is a poll initiated by a management station to request and retrieve blocks of data from tables in the management agent's MIB table.
2. *InformRequest.* This message enables a manager to send trap information to another manager. SNMPv2 also defines a Manager-to-Manager MIB to support distributed management architecture.

SNMPv2 also added the following security capabilities:

■ MD5 processing capabilities were added to authenticate and time-stamp a message profile and its origin. MD5 is a message digest authentication algorithm developed by RSA, Inc.® MD5 computes a digest of the message that is used by the receiver to verify the contents of a message. The message can also be encrypted using DES, as part of another processing step. A message recipient must then perform these actions: (1) decrypt the message, (2) check the timestamp to verify the message is recent and is the latest message sent from the originator, (3) process the profile, and (4) verify the authentication. Using a MD5 message digest imposes at least a ten percent increase in message processing time, while using DES doubles the required message processing time. Thus, the added security of SNMPv2 is not without substantial cost and processing overhead.

■ A mechanism for timestamping loosely synchronized workstations was provided.

Like SNMPv1, the SNMPv2 security provisions lacked the strength needed for mission-critical networks. They both failed to adequately address the security of MIB data transmission, and access control. Access is based on a *shared* password (i.e., the community string) that could be altered or intercepted during transmission. This could lead to an unauthorized user gaining read-only or read-write access to network device(s). Thus, these protocols provide only limited control over message privacy and content, and authorization and control of device access, remote configuration, and administration. SNMPv3 was designed to address these serious security limitations by providing mechanisms for authentication, privacy, authorization, view-based access control (to limit what different user groups could see and do), and standards-based remote configuration. SNMPv3 does not attempt to deal with denial-of-service and traffic analysis attacks. [PERK04] While SNMPv1 and SNMPv3 are widely used within the industry, the market for SNMPv2 has not emerged because it lacks the simplicity of SNMPv1 and the security of SNMPv3.

RMON (Remote Networking Monitoring) refers to the IETF standards defined in RPCs 1757, 1513, and 2021. It defines an interface between an RMON agent and an RMON management application. Generally, RMON agents are used to monitor and troubleshoot LANs. Although RMON is not designed for WANs, some vendors offer WAN RMON products for analyzing T1/E1 and T2/E2 lines to a central site. RMON extends the SNMP MIB to define additional MIBs for network management. RMON has two versions, RMON 1 and RMON2, as shown in Figure 7.14. RMON 1 specifies several additional MIBs, as outlined below: [NETS04]

1. *Statistics*. This is used to collect real-time, current network statistics (e.g., percent broadcast traffic, percent multi-cast traffic, percent utilization, percent of errors, percent of collisions, etc.).
2. *History*. This refers to the collection of statistics over time.
3. *Alarm*. This MIB defines predetermined thresholds that may trigger an event when an alarm condition occurs.
4. *Host*. This MIB tracks individual host statistics, such as packets in/out, multi-cast/broadcast, bytes in/out, errors, etc.

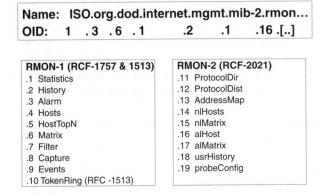

Figure 7.14   RMON nomenclature overview.

5. *hostTopN*. This MIB defines data collection on the "N" (statistically) most active hosts, tracking information such as packets in/out, multi-cast/broadcast, bytes in/out, errors, etc. Typically, N is set to 10.
6. *Matrix*. This MIB defines host-to-host conversation statistics, such as packet, byte, and error counts.
7. *Filters*. This MIB defines data relating to packet structure and content matching. It defines bit pattern matching at the data link, IP header, and UDP header level.
8. *Capture*. This MIB defines the collection of data that is uploaded to a management station for subsequent detailed packet analysis.
9. *Event*. This MIB defines data specifying reactions/actions to pre-determined conditions or thresholds.
10. *Tokenring*. This MIB defines Token-Ring related RMON extensions. This includes Token Ring Station (for Status and Statistics data); Token Ring Station (for Order data); Token Ring Station (for Configuration data); and Token Ring Source Routing (for Utilization data).

The first three MIBs — Host, HostTopN, and matrix — are directly involved with monitoring traffic flow and net flow switching. They provide the most granularity in traffic flow information. The filter, packet capture, and event MIBs are associated with the most processing and are intended for in-depth data analysis.

RMON 2, as the name implies, is the second version of RMON. It defines even richer data sets for network analysis by providing these additional MIBs:

11. *ProtocolDir.* This defines a master list of supported protocols for each probe. It is used by the network management system to determine which protocols (i.e., Internet Protocol, Novell, DECnet, Appletalk, Custom Protocol, etc.) the probe is able to decode.
12. *ProtocolDist.* This MIB defines segment protocol statistics, providing aggregate statistics on the traffic generated by each protocol on a per-segment basis.
13. *AddressMap.* This MIB is used to map network and MAC addresses to physical ports.
14. *nlHost.* This is used to define "host in/out" statistics based on a network layer address.
15. *nlMatrix.* This is used to define host-to-host network layer statistics capturing traffic data between host pairs at the network layer.
16. *alHost.* This is used to define "Host in/out" statistics based on application layer addresses.
17. *alMatrix.* This defines traffic statistics between host-to-host pairs based on an application layer address.
18. *usrHistory.* This is used for the same basic purpose as the RMON1 History group. It is used for data logging and collecting historical statistics that are user defined.
19. *probeConfig.* This is used to define standardized probe configuration parameters, including capabilities (supported RMON groups); software revision level; hardware revision level; current date and time settings; reset control (i.e., running, warmBoot, coldboot); download filename; IP address of TFTP download server; and download actions (e.g., to PROM or to RAM).

To summarize, RMON MIBs must be implemented by a device manufacturer so data can be collected by RMON (hardware/software) monitoring devices known as *probes.* A RMON software agent gathers the information for presentation on a network monitoring console with a graphical user interface, as shown in Figure 7.15. A variety of vendors support RMON network management. NetScout is an example of a vendor that implements the full RMON specification. For more information on NetScout's network management products, the reader is referred to their Web page at: http://www.netscout.com/products.

7.2.2 Network Management Implementation

Network management protocols are implemented in devices within the context of a network management architecture. In the past, network management was performed by a centralized global management director using proprietary protocols. A typical example of this is IBM's Tivoli

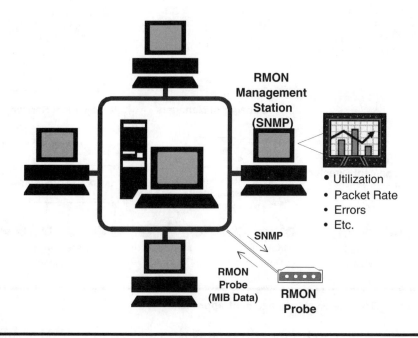

RMON
Management
Station
(SNMP)

• Utilization
• Packet Rate
• Errors
• Etc.

SNMP

RMON
Probe
(MIB Data)

RMON
Probe

Figure 7.15   Network monitoring using RMON.

NetView. Network devices (such as routers, hubs, switches, etc.) forwarded their network management data to the central monitoring point. A *Manager of Managers* architecture supports a hierarchy of subordinate management stations that ultimately report to a central source. Manager of Managers and network management platform architectures are often used with SNMP/RMON network management, as shown in Figure 7.16. For example, SNMPv3 supports a "Manager of Managers" approach, which allows a centralized management station to coordinate and collect information from secondary management stations. HP OpenView is one of the most comprehensive and widely used network management platforms, with over 135,000 installations. [HPOP04] It provides a complete end-to-end network management solution. As a platform, HP OpenView provides a vehicle for a variety of network management solution providers (e.g., CiscoWorks, Tivoli, NetMetrix, Oracle, etc.) to develop APIs supported in a single integrated environment. This also gives the network manager considerable flexibility in crafting a customized, tailored network management solution.

7.2.3 Web-Based Network Management

Web-based protocols, which are complementary to SNMP, are also being used for network management. This approach is likely to grow in popularity

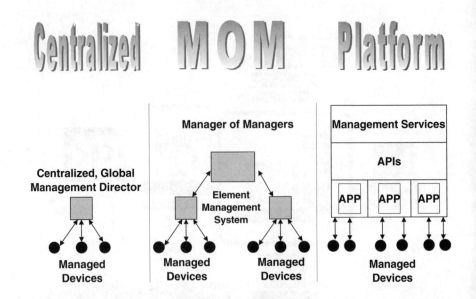

Figure 7.16   Network management approaches.

as TCP/IP, and Web protocols in general, continue to gain dominance within the IT landscape. At its simplest, this might involve the use of a Web browser and a managed entity running on a Web server. A more extensive network management approach might involve an enterprisewide deployment, as shown in Figure 7.17. Web-based network management requires a Web browser and a Web server using HTML (Hypertext Markup Language), URLs (uniform resource locators), and HTTP (Hypertext Transfer Protocol). To implement this approach, a Web-enabled router or hub — with a built-in Java (or CGI, etc.) Web server — is used to collect network data and to generate alarms. Reports are then distributed to clients for viewing via a Web browser. Chapter 5 provides a more precise description of this process. A number of products and services are available so organizations can implement this approach on their own or as third-party solutions from a variety of vendors and service providers (such as Asante, SNMP Research International, Tivoli, HP OpenView, etc.).

As Internet, extranet, and intranet computing becomes ubiquitous, so has the need for other types of powerful network management and analytical tools. Sane Solutions' NetTracker is an excellent example that addresses this niche. NetTracker provides detailed Web site traffic analysis that empowers end users (such as business managers, etc.) to optimize online channels. NetTracker collects log files from Web servers, E-commerce application servers, page-tagging servers, streaming media servers, proxy servers, and firewalls. NetTracker also supplies a well-packaged tool set for manipulating the Web site data. NetTracker is designed especially for

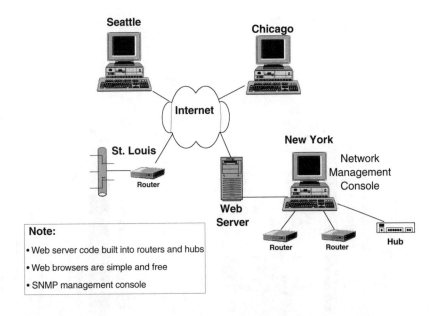

Seattle

Chicago

Internet

St. Louis

Router

New York

Network
Management
Console

Web
Server

Router   Router

Hub

Note:
• Web server code built into routers and hubs
• Web browsers are simple and free
• SNMP management console

Figure 7.17   Web-based network management.

Web site analytics, measurement, reporting, and integration with relational databases (e.g., Microsoft SQL, Oracle, and DB2) and Business Intelligence Solutions (such as Business Objects, Cognos, and MicroStrategy). Some of the highlights of NetTracker's major features include: [NETT04]

- Instant report generation and customized report wizards. As shown in Figure 7.18, reports provide information on:
    - Visitor segmentation
    - Content group analysis, which provides a summary of how recently and frequently visitors viewed each content group on the Web site (the content groups for a particular Web site are determined by the end user)
    - Scenarios, which track the conversion of Web site activity, such as a visitor response to an advertisement and a subsequent purchase
    - Trend reports
    - Path reports that investigate how visitors traverse a Web site
- Multi-tiered reporting, with drilldown capabilities, as shown in Figure 7.19
- Interactive calendars, which allow end users to easily select an activity/time period of interest, as shown in Figure 7.20

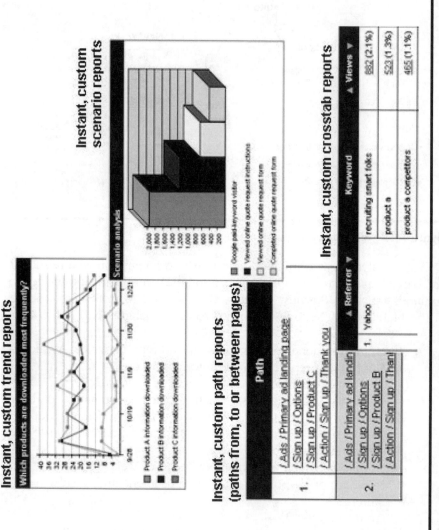

Figure 7.18 NetTracker custom report wizard and instant report generation. (*Source:* Sane Solutions.)

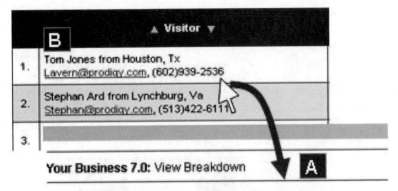

Figure 7.19 NetTracker multi-tiered reporting and drilldown capabilities. (*Source:* Sane Solutions.)

- Customized interfaces for different user groups, as shown in Figure 7.21
- Open database schema to allow data extraction and export to third-party databases and analytical tools, as shown in Figure 7.22

For more information on NetTracker, the interested reader is referred to its Web site at: http://www.sane.com.

7.2.4 Other Network Management Strategies

For the sake of completeness, two other alternative network management approaches are discussed. They generally are more complicated or are poorly supported by vendors. For this reason, they are not widely used.

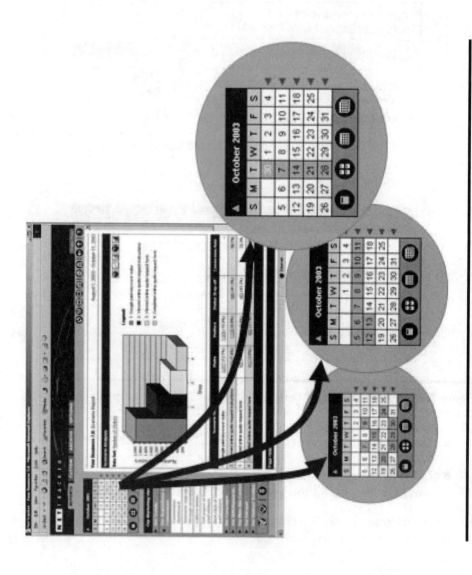

Figure 7.20 NetTracker interactive calendar capabilities. (*Source:* Sane Solutions.)

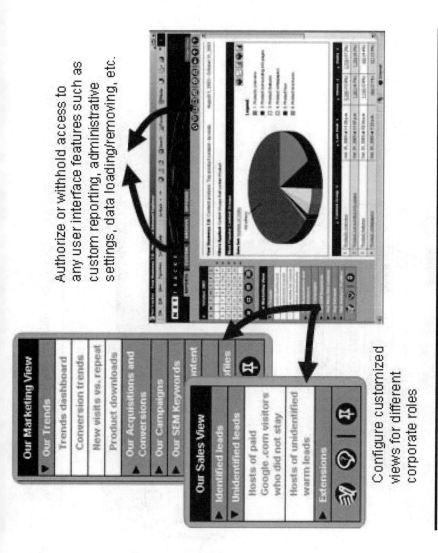

Figure 7.21 NetTracker customized interfaces. (*Source:* Sane Solutions.)

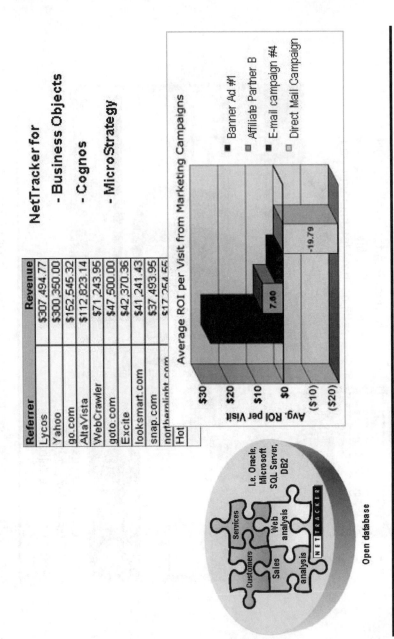

Figure 7.22 NetTracker open database scheme and analytics integration. (*Source:* Sane Solutions.)

CORBA (Common Object Request Broker) was designed to support enterprisewide, distributed client server processing, and can be used to implement network management tasks. Based on the functionality that it supports, CORBA is a potential alternative to the SNMP or CMIP standards. Although CORBA implementations are generally not available in vendor offerings, some organizations are considering writing their own network management software based on CORBA standards. One method of doing this would involve using *CORBAservices* (see Chapter 5 for a detailed discussion of CORBA) to implement OSI CMISE protocol functions. The software could then be compiled as an IDL (Interface Definition Language) so that it can be deployed on any platform or operating system as a network management entity. However, writing adapters that allow applications to interface with the CORBA ORB (Object Request Broker) can be difficult. This task is made more so when new services are added and the adapters must be modified accordingly. Because this approach is generally not commercially available from third parties, it would necessitate considerable commitment on the part of an organization that would wish to undertake it.

Web-based Enterprise Management (WBEM) by DTMF (Distributed Management Task Force) is yet another alternative network management approach. Although the WBEM standard was designed to provide comprehensive network management capabilities, it is complex and complicated to implement. It has not caught on in the industry, and its use is dominated by SNMP and other Web-based approaches.

7.3 Summary

This chapter discussed the need for automated tools to support complex network design, planning, and network management tasks. It also reviewed the major network management protocols, with particular emphasis on SNMP and RMON because these protocols are widely used and implemented in vendor products. The latest trend in network management is to leverage Web-based Internet standards for Web-based network management. With Web-based network management, it is possible to monitor an enterprise network from any personal computer and browser connected to it. Some useful Web sites for obtaining more information on network management protocols include:

- http://www.protocols.com
- www.omg.org
- http://www.ietf.org/html.charters/snmpv3-charter.html
- http://www.faqs.org/rfcs/

Notes

1. *NPA:* the first three digits of a North American telephone number (i.e., the "area code"). *NXX* refers to the next three digits of a North American telephone number, the central office code (where "N" is any number from two to nine and "X" is any number from zero to nine.). *CLLI* is an 11-digit alphanumeric code that refers to the common language location identifier that uniquely identifies each element (i.e., switches, intersections, etc.) of a signaling network. CLLI is a trademark of Telcordia Technologies, Inc. More information on CLLI codes and their licensing can be found at: http://www.commonlanguage.com/resources/commonlang/productshowroom/product/clli_tech/index.html.
2. CMISE (Common Management Information Service Element) is part of a family of protocols that define the exchange of management information between entities (manager/agent). The "Manager/Agent" model was copied in SNMP but implemented with less functionality. CMISE defines a user interface and corresponding services to be provided for each network component through the Common Management Information Service (CMIS) protocol. CMIP (Common Management Information Protocol) specifies the PDU (Protocol Data Unit) format and implementation of CMIS services through the Common Management Information Protocol (CMIP).
3. The IETF (Internet Engineering Task Force), a large, open, international body composed of researchers, vendors, network designers, and others, is concerned with the development, evolution, and "smooth" operation of the Internet. The IETF maintains an RFC Editor that publishes and archives all RFCs. For the most recent RFC information on IETF protocols and standards, the interested reader is referred to http://www.rfc-editor.org/overview.html.
4. The actual IETF MIB designation depends on the protocol, the protocol version, and the context. For example, the original SNMP MIB (a.k.a. MIB-I), defined in 1988 by RFC 1066, included variables needed to manage the TCP/IP protocol stack. It was updated by RFC 1213. The updated version, designated MIB-II, adds new variables. The SNMPv2 MIB is defined by RFC 3418, and so on.

References

[CISC02]   Cisco Systems, *Overview of the WAN Modeling Tools, Chapter 1,* Cisco WAN Design Tools User Guide, September 2002.
[HPOP04]   Hewlett-Packard, About HP Openview, 2004, http://www.openview.hp.com/about/indeix.html..
[NETF02]   Netformx, *Netformx's DesignXpert Platinum Suite,* http://www.netformx.com/pdfs/DesignXpertPlatinumSuiteFinal.pdf.
[NETS04]   NetScout, http://www.netscout.com/products/.
[NETT04]   Sane Solutions, NetTracker Web Analytics Solutions, 2004, http://www.sane.com.

[NETW04] Network Design and Analysis, Corporation, 2004, http://www.ndacorp. com/amsintro.html.

[PERK04] Perkins, D., A Consolidated Overview of Version 3 of the Simple Network Management Protocol (SNMPv3), Network Working Group — Internet, February 15, 2004, ftp://ftp.rfc-editor.org/in-notes/internet-drafts/draft-perkins-snmpv3-overview-00.txt.

[TERP96] Terplan, K., *Effective Management of Local Area Networks, 2nd Edition*, McGraw-Hill, New York, 1996.

Chapter 8

Technical Considerations in Network Design and Planning

8.1 Overview of the Network Design Process

Network design is a painstaking, iterative process. The first step of this process is to define the requirements the network must satisfy. This involves collecting information on anticipated traffic loads, traffic types (e.g., data, video, etc.), and sources and destinations of traffic. This information is used, in turn, to estimate the network capacity needed. These requirements are used as input to the second step, the design process. In this step, various design techniques and algorithms are used to produce a network topology. The design process involves specifying link and node placements, traffic routing paths, and equipment sizing. After a candidate network solution is developed, it must be analyzed to determine its cost, reliability, and delay characteristics. This third step is called performance analysis. When these three steps are completed, the first design iteration is finished. Then the entire process is repeated, either with revised input data (e.g., using revised traffic estimates, etc.) or by using a new design approach. This process is summarized in Figure 8.1.

The basic idea of this iterative process is to produce a variety of networks from which to choose. Unfortunately, for most realistic design problems, it is not possible from a mathematical perspective to know what the optimal network should look like. To compensate for this inability

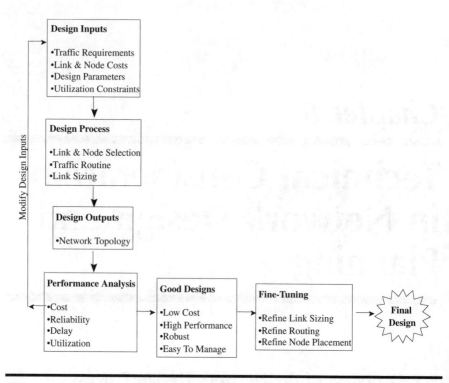

Figure 8.1   Overview of network design process.

to derive an analytically perfect solution, the network designer must use a judicious form of trial and error to determine his or her best options. After surveying a variety of designs, the designer can select the one that appears to provide the best performance at the lowest cost.

Because network design involves exploring as many alternatives as possible, automated heuristic design tools are often used to produce quick, approximate solutions. Once the overall topology and major design aspects have been decided, it may be appropriate to use additional, more exact solution techniques to refine the details of the network design. This fine-tuning represents the final stage of the network design process.

This chapter takes the reader through each of the major steps involved in network design: requirements analysis, topological design, and performance analysis.

8.2 Data Collection

The network requirements must be known before the network can be designed. However, it is not easy to collect all the information needed to design a network. Often, this is one of the most time-consuming aspects of the design process. Data must be collected to determine the sources and destinations of traffic, the volume and character of the traffic flows, and the types and associated costs of the line facilities to transport the traffic. It is rare that these statistics are readily available in the succinct summary form needed by network design algorithms and procedures.

Most typically, a considerable volume of data is collected from a variety of sources. For existing networks, it may be possible to collect information on:

- Session type(s)
- Length of session(s): average, minimum, and maximum time
- Source and destination of data transmissions
- Number of packets, characters sent, etc.
- Application type
- Time of session
- Traffic direction and routing

To relate these findings to the user requirements, interviews and data collection activities may be needed to determine the following:

- Required response time
- Current and planned services and applications to be supported (i.e., e-mail, database transactions and updates, file transfer, etc.)
- Anticipated future services and applications needs
- Communications protocols, including network management protocols, to be supported
- Network management functions to be supported
- Network reliability requirements
- Implementation and maintenance budget

This data must then be related, culled, and summarized if it is to be useful. Ultimately, the traffic data must be consolidated to a single estimate of the source–destination traffic for each node. Between each source and destination of traffic, a line speed must be selected that will have sufficient capacity[1] to transport the traffic requirement.

The line speed and the line endpoints are used, in turn, to determine the line cost. Usually, only one line speed is used when designing a network (or a significant sub-network portion), because multiple line speeds complicate the network manageability and may introduce incompatibilities in transporting the data. Table 8.1 illustrates the data in the summarized form needed by most network design procedures.

Thus far the issue of node[2] costs has not been addressed. The reason for this is that these costs seldom play a major role in the design of the

Table 8.1   Sample Traffic and Cost Data Needed for Network Design

Traffic Source	Traffic Destination	Estimated Traffic (Bytes)	Usable Line Capacity (Bytes)	Estimated Line Cost ($ Monthly)
(1)	(2)	(3)	(4)	(5)
City A	City B	80,000	1,000,000 (T1)	1,000.00
City A	City C	770,000	1,000,000 (T1)	3,500.00
City B	City N	500,000	1,000,000 (T1)	6,006.00
City B	City C	30,500	1,000,000 (T1)	5,135.00

network topology. *After* a network has been designed, the node costs are factored into the total network cost (see Section 8.6.4).

When collecting traffic and cost data, it is helpful to maintain a perspective on the level of design needed. The level of design — that is, be it high-level or finely detailed — helps to determine the amount of data that should be collected, and when more detailed data is required and when it is not. It may be necessary to develop multiple views of the traffic requirements and candidate node sets, in order to perform sensitivity analysis. It is easier to develop strategies for dealing with missing or inconsistent data when the design objective is clear.

To the extent that it is practical, the traffic and cost data collected should be current and representative. In general, it is easier to collect static data than to collect dynamic data. Static data is data that remains fairly constant over time, while dynamic data may be highly variable over time. If the magnitude of the dynamic traffic is large, then it makes sense to concentrate more effort on trying to estimate it more accurately, because it may have a substantial impact on the network performance. Wherever possible, automated tools and methods should be used to collect the data. However, this may not be feasible, and sometimes the data must be collected manually. However, a manual process increases the likelihood of errors and limits the amount of data that can be analyzed.

8.2.1 Automated Data Generation

There may be situations where no data is available on existing traffic and cost patterns. This may be the case when an entirely new network is being implemented. When actual data is unavailable, an option is to use traffic and cost generators. Traffic and cost generators can also be used to augment actual data (particularly when it is missing or inconsistent) or to produce data for benchmark studies.

For a traffic and cost matrix similar to Table 8.1, for a network with n sources and destinations, the number of potential table entries is:[3]

$$\binom{n}{2}$$

As stated in [CAHN98]:

> There is only one thing certain about a table with 5000 or 10,000 entries. **If you create such a table by hand it will contain thousands and thousands of errors and take weeks of work.**

Thus, there is strong motivation for using automated tools to generate or augment the traffic and cost data needed to design a network.

8.2.2 Traffic Generators

A traffic generator, as its name implies, is used to automatically produce a traffic matrix (i.e., the first three columns of Table 8.1) based on a predetermined model of the traffic flow. Many design tools have traffic generators built into them. Alternatively, stand-alone software routines are available that can be used to produce traffic matrices. A traffic generator can easily produce matrices representative of increasing or decreasing traffic volumes. Each of these matrices can be used to design a network, and the results studied to analyze how well a design will handle increasing traffic loads over time. The traffic generator might also be used to produce traffic matrices based on a uniform or random traffic model. While these are not realistic traffic distributions for most networks, they may, nonetheless, provide useful results for benchmarking studies. Other, more realistic traffic models can be used to produce traffic matrices that more accurately conform to observed or expected traffic flows. For example, it may be appropriate to use a model that adjusts traffic flows as a function of node population, geographic distance between other sites, and anticipated link utilization levels. [KERS89] Other models might be based on the type of traffic (i.e., e-mail, World Wide Web traffic, and client/server traffic) that is to be carried by the network. The interested reader is referred to [CAHN98], which lists a number of exemplar software routines for producing traffic matrices to conform to a variety of model conditions and assumptions. The major decision when using a traffic generator is the selection of the traffic flow model that will be used to produce the traffic matrix.

8.2.3 Cost Generators and Tariff Data

Cost generators are similar in concept to traffic generators. Cost generators are used to produce a cost matrix (i.e., columns (1), (2), and (5) in Table 8.1). In an actual design situation, it may be necessary to produce several cost matrices representing the various line and connectivity options. In this case, each circuit or line option should be represented in a separate traffic/cost matrix.

A major challenge in creating cost matrices is that the tariff[4] structures (which determine the costs for lines between two points) are complex, and contain many anomalies and inconsistencies. For example, tariffs are not based solely on a simple distance calculation. Thus, it is possible that a longer circuit may actually cost less than a shorter circuit. Two circuits of the same length may have different costs, depending on where they begin and terminate. In the United States, circuits within a LATA[5] may cost less than circuits that begin and end across LATA boundaries. Depending on the geographic scope of the network, domestic (United States) and international tariffs may apply. Given the complexities of the tariff structures, the "ideal" way of keeping track of them is to have a single look-up table containing all the published tariff rates between all points. Although comprehensive, accurate tariff information is available, it is very expensive, costing thousands of dollars. Thus, the "ideal" solution for gathering tariff data may be too costly to be practical.

Access to a comprehensive tariff database does not guarantee that the desired cost data is obtainable for all points. For example, tariff information is only available for direct services that are actually provided. If no direct service is available between two points, then obviously a tariff will not be published for these points. Furthermore, even if accurate tariff information is available, it may still be too complex to use easily.

When the tariff data is either too costly or complex to use directly, alternative methods must be employed to estimate the line costs. It should be noted that while these alternative methods may be easier and cheaper than the purchase of a commercial tariff tool, they are not as accurate. The need for accuracy must be weighed against the trade-off of using a simpler scheme to estimate line costs.

One simple cost model is based on a linear distance function. In this model, line costs between two nodes i and j are estimated by a function containing a fixed cost component, $\mathcal{F}$, and a variable distance based component, $\mathcal{U}$. [KERS89] It should be noted that the $\mathcal{F}$ and $\mathcal{U}$ cost components will vary by link type. The linear cost function can be summarized as:

$$\text{Cost}_{ij} = \mathcal{F} + \mathcal{U} \, (\text{dist}_{ij}) \tag{8.1}$$

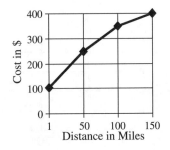

Figure 8.2   Example of piece-wise linear cost function.

where:

$Cost_{ij}$	= cost for line between two nodes i and j
$\mathcal{F}$	= fixed cost component
$\mathcal{U}$	= variable cost based on distance
$dist_{ij}$	= distance between nodes i and j

When the locations of nodes i and j are expressed as V and H[6] coordinates (i.e., (V_i, H_i), and (V_j, H_j), respectively), the distance, $dist_{ij}$, between the nodes is easily calculated using a standard distance formula: [KERS89]

$$dist_{ij} = \sqrt{\left(V_i - V_j\right)^2 \Big/ 10 + \left(\left(H_i - H_j\right)^2 \Big/ 10\right)} \qquad (8.2)$$

This simple model can also be used to simplify a complex tariff structure. Linear regression can be used to transform selected points from the tariff table into a linear cost relationship. The fixed cost component, $\mathcal{F}$, and the variable cost component, $\mathcal{U}$, can be derived by taking the partial derivatives of the cost function with respect to $\mathcal{F}$, and with respect to $\mathcal{U}$, respectively. [CAHN98, p.151] This simplified model can perform well in special cases where the tariff structure is highly linear.

A somewhat more realistic estimate of cost may be possible using a piece-wise linear function. The piece-wise linear cost function is very similar to the linear cost function, except that the $\mathcal{F}$ (fixed) and $\mathcal{U}$ (variable) cost components vary according to distance. An example of a piece-wise linear function is presented below and in Figure 8.2.

$Cost_{ij} = \$100 + \$3/(dist_{ij})$ (for $dist_{ij}$ between 0 and 50 miles)
$Cost_{ij} = \$100 + \$2/(dist_{ij})$ (for $dist_{ij}$ between 50 and 100 miles)
$Cost_{ij} = \$100 + \$1/(dist_{ij})$ (for $dist_{ij}$ between 100 and 150 miles) (8.3)

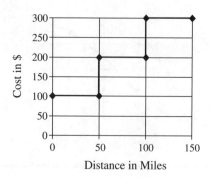

Figure 8.3 Example of step-wise linear cost function.

where:
Cost$_{ij}$ = cost for line between two nodes i and j
dist$_{ij}$ = distance between nodes i and j

Many service providers use this model to price private lines. Note that there are no additional usage fees in the model presented here. With this type of cost model, there is an economic incentive to fill the line with as much traffic as possible for as much time as possible.

A step-wise linear function is illustrated in Figure 8.3. A hypothetical step-wise linear function is given below. Note that in this function, the fixed costs are only constant within a given range, and there is no longer a variable cost component.

Cost$_{ijk}$ = \$100 (for dist$_{ijk}$ between 0 and 50 miles)
Cost$_{ijk}$ = \$200 (for dist$_{ijk}$ between 50 and 100 miles)
Cost$_{ijk}$ = \$300 (for dist$_{ijk}$ between 100 and 150 miles) (8.4)

where:
Cost$_{ijk}$ = cost for line between two nodes i and j up to point k
dist$_{ijk}$ = distance between nodes i and j

When international circuits must be priced, the cost models may need to be extended to provide more realistic estimates. For example, if a line is installed across international boundaries, adjustments may be needed in the cost model to account for differences in the tariff structures of each respective country. In addition, lines installed across international boundaries are usually priced as the sum of two half-circuits. A communication supplier in one country supplies half of the line, while a communication supplier in the other country supplies the other half of the line.

8.3 Technical Requirements Specification

8.3.1 Node Selection, Placement, and Sizing

This section discusses matters relating to the selection, placement, and sizing of network nodes. This discussion begins with a review of the devices commonly used as network nodes:

- *Bridge.* This device is used to interconnect networks or sub-network components that share the same protocols. Bridges are fairly simple and do not examine data as it passes through them.
- *Concentrator.* These devices allow multiple devices to dynamically share a smaller number of lines than would otherwise be possible. Concentrators are sometimes called MAUs (Multiple Access Units). Concentrators only allow one device at a time to use the communications channel, although many devices can connect to the concentrator. This is a significant difference from multiplexers.
- *Digital Service Units (DSUs) and Channel Service Units (CSUs).* These devices are used to connect digital devices to digital circuits.
- *Gateway.* Gateways are used to interconnect otherwise incompatible devices using protocol conversion techniques.
- *Hub.* A device that is used to interconnect workstations or devices.
- *Modem.* A MODulator/DEModulator device is used to connect digital devices to analog circuits.
- *Multiplexer.* These devices are also known as MUXes. They allow multiple devices to transmit signals over a shared channel, thereby reducing line and hardware costs. In general, multiplexers do not require the devices on the line to operate using the same protocols. There are two types of multiplexers. One type uses Frequency Division Multiplexing (FDM). This type of multiplexing divides the channel frequency band into two or more narrower bands, each acting as a separate channel. The second type of multiplexer uses Time Division Multiplexing (TDM). This type of multiplexing divides the line into separate channels by assigning each channel to a repeating time slot.
- *Intelligent multiplexer, or statistical time division multiplexer (STDM).* These devices use statistical sampling techniques to allocate time slots for data transmission based on need, thereby improving the line efficiency and capacity. The design of STDMs is based on the fact that most devices transmit data for only a relatively small percentage of the time they are actually in use. STDMs aggregate (both synchronous and asynchronous) traffic from multiple lower-speed devices onto a single higher-speed line.

■ *Router.* This is a protocol-specific device that transmits data from sender to receiver over the best route, where "best" may mean the cheapest, fastest, or least congested route.

■ *Switch.* These devices are used to route transmissions to specific destinations. Two of the most common types of switches are circuit switches and packet switches.

The selection of a specific device depends on many factors. These factors may include the node cost and the requirements that have been established for protocol compatibility and network functionality. This is a context-specific decision that must be made on a case-by-case basis.

As previously discussed, the selection of a particular *type* of node device has little impact on the design of the network topology. However, the placement of nodes within the network does impact the network topology. Typically, nodes are placed near major sources and destinations of traffic. However, this is not always true, as sometimes node placements are based on organizational or functional requirements that do not strictly relate to traffic flow. For example, a node can be placed at the site of a corporate headquarters, which may or may not be a major source of traffic. If the node locations must be taken as given and cannot be changed, then the decisions on node placement are straightforward.

However, in other cases, the network designer is asked to suggest optimal node placements. One node placement algorithm — the Center of Mass algorithm — suggests candidate locations based on traffic and cost considerations. A potential shortcoming of the Center of Mass (COM) algorithm is that it may suggest node placements in areas that are not feasible or practical. Alternatively, the ADD and DROP algorithms can be used to select an optimal subset of node locations, based on a predefined set of candidate nodes. Thus, potential sites for node placements must be known in advance when using these latter two algorithms. These algorithms are discussed in Chapter 2.

Once the node placements and the network topology have been established, the traffic flows through the node can be estimated. This is needed to size the node. Measuring the capacity of a node is generally more difficult than measuring the capacity of a link. Depending on the device, the node capacity may depend upon the processor speed, the amount of memory available, the protocols used, and software implementation. It may also reflect constraints on the type, amount, and mix of traffic that can be processed by the device. These factors should be considered when estimating the rated versus the actual usable node capacity.

Queuing models can be used to test whether or not the actual node capacity is sufficient to handle the predicted traffic flow. Queuing analysis allows the network designer to estimate the processing times and delays expected for traffic flowing through the node, for various node capacities and utilization rates. Section 8.6.2 provides an introduction to the queuing models needed to perform this analysis. The node should be sized so that it is adequate to support current and future traffic flows. If the node capacity is too low, or the traffic flows are too high, the node utilization and traffic processing times will increase correspondingly. Queuing analysis can be used to determine an acceptable level of node utilization that will avoid excessive performance degradation.

In sizing the node's throughput requirement, it may also be necessary to estimate the number of entry points or ports needed on the node. A straightforward way of producing a preliminary estimate is given below:

$$\text{Number of Ports} = \frac{\left(\text{Total Traffic Through Ports in bps}\right)}{\left(\text{Usable Port Capacity in bps}\right)}$$

This estimate is likely to be too low because it does not allow for excess capacity to handle unanticipated traffic peaks. Queuing models similar to those used for node sizing can be use to adjust the number of ports upward to a more appropriate figure. A queuing model allows one to examine the cost of additional ports versus the improvements in throughput. Queuing analysis is a very useful tool for port sizing.

8.3.2 Network Topology

The network topology defines the logical and physical configuration of the network components. The basic types of network topologies are listed below. Figure 8.4 provides an illustrative example of each network type.

- *Star.* In this topology, all links are connected through a central node.
- *Ring.* In this topology, all the links are connected in a logical or physical circle.
- *Tree.* In this topology, there is a single path between all nodes.
- *Mesh.* In this topology, nodes can connect directly to each other, although they do not necessarily all have to interconnect.

The selection of an appropriate network topology depends on a number of factors — including protocol and technology requirements. Chapter 2 discusses in detail the issues involved in selecting a network topology.

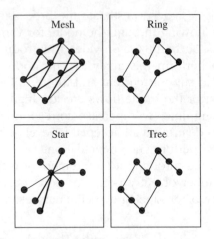

Figure 8.4   Sample network topologies.

Table 8.2   Taxonomy of Network Routing

Routing Characteristics
One route between nodes versus multiple routes between nodes
Fixed routing versus dynamic routing
Minimum hop routing versus minimum distance routing versus arbitrary routing
Bifurcated routing versus non-bifurcated routing

8.3.3 Routing Strategies

A number of schemes are used to determine the routing of traffic in a network. These routing schemes can be described in the terms listed in Table 8.2. [CAHN98, p. 249]

The meanings of "one route between nodes," "multiple routes between nodes," and "arbitrary routing" are self-explanatory. Fixed routing means the traffic routing is predetermined and invariant, irrespective of conditions in the network. In contrast, dynamic routing may change, depending on network conditions (i.e., traffic congestion, node or link failures, etc.). Minimum hop routing attempts to send the traffic through the least possible number of intermediate nodes. Minimum distance routing is used to send traffic over the shortest possible path. Bifurcated routing splits traffic into two or more streams that may be carried over different paths through the network. Nonbifurcated routing requires that all the traffic associated with a given transmission be sent over the same path.

The actual routing scheme used depends on the characteristics of the devices and the technology employed in the network. This is demonstrated in the examples below: [CAHN98, p. 159]

■ Router networks frequently use a fixed minimum routing or minimum hop routing scheme.
■ SNA uses a static, arbitrary, multiple, bifurcated routing scheme.
■ Multiplexer-based networks generally use minimum distance or minimum hop routing schemes.

8.3.4 Architectures

Network architectures define the rules and structures for the network operations and functions. Before the introduction of network architectures, application programs interacted directly with the communications devices. This meant that programs were written explicitly to work with specific devices and network technologies. If the network was changed, the application programs were modified accordingly. The purpose of a network architecture is to shield application programs from specific network details and device operations. As long as the application programs adhere to the standards defined by the architecture, the architecture can handle specific device and network implementation details.

Most networks are organized as a series of layers, each built upon its predecessor. The number of layers used and the function of each network layer vary by protocol and by vendor. The purpose of each layer is to off-load work from successively higher layers. This divide and conquer strategy is designed to reduce the complexity of managing the various network functions. Layer n on one machine operates with layer n on another machine. The rules and conventions used in this interaction are collectively known as the layer n protocol. Peer processes represent entities operating at the same layer on different machines. Interacting peer processes pass data and control information to the layer immediately below, until the lowest level is reached. Between each adjacent pair of layers is an interface that defines the operations and services the lower layer offers to the upper one. The network architecture defines these layers and protocols. Among other things, the network architecture provides a means for: establishing and terminating device connections, specifying traffic direction (i.e., simplex, full-duplex, half-duplex), error control handling, and methods of controlling congestion and traffic flow.

There are both open and proprietary network architectures. IBM's proprietary network architecture — Systems Network Architecture (SNA) — is an example of one of the earliest network architectures. It is based on a layered architecture. Although SNA was originally designed for centralized, mainframe communications, it has been continually updated over the years and supports distributed and peer-to-peer communications. Xerox's XNS (Xerox's Network Services), Digital's DecNet, Novell's NetWare, and Banyan's VINES are examples of proprietary LAN network architectures.

One of the primary issues in using a proprietary architecture is that it tends to lock the user into a single vendor vision or solution. This may also make it more difficult to incorporate other third-party products and services into the network. In response to strong market pressures and technological evolution, proprietary architectures have gradually evolved toward more open standards.

The International Systems Interconnection (ISO) protocol was developed by the International Organization for Standardization to provide an open network architecture. The OSI model is based on seven layers. The lowest level is the physical layer, which specifies how bits are physically transmitted over a communications channel. The next layer is the data-link layer. This layer creates and converts data into frames so that transmissions between adjacent network nodes can be synchronized. The third layer is the network layer. This layer determines how packets received from the transport layer are routed from source to destination within the network. The fourth layer is the transport layer. This layer is responsible for providing end-to-end control and information exchange. It accepts data from the session layer and passes it to the network layer. The next higher layer is the session layer. The session layer allows users on different machines to establish sessions between each other. The presentation layer performs syntax and semantics checks on the data transmissions, and structures the data in required display and control formats. It may also perform cryptographic functions. The highest layer, the application layer, provides an interface to the end user. It also employs a variety of protocols (e.g., a file transfer protocol). Appendix B elaborates on this discussion and presents a comparison of the OSI architecture to TCP/IP.

Network architecture considerations come into play when the network is being implemented because the network architecture has profound impacts on the types of devices, systems, and services the network can support and on how the network can be interconnected with other systems and networks. The network architecture also has a significant impact on what and how new products and services can be integrated into the network, because any additions must be compatible with the architecture in use. Some of the key decisions involved in selecting a network architecture include:

■ *Open or proprietary architecture.* In making this decision, it is helpful to keep in mind that the full promise of open architectures has yet to be achieved. Although there is steady progress toward open architectures, they are not fully implemented in the marketplace. An expedient compromise might be to select a network or network components that encompass a subset of OSI functionality.

■ *Selection of network management protocol.* Network management protocols come in both proprietary and open varieties. The requirements for network management may dictate the selection of one over the other. For example, it may be necessary to manage a diverse array of third-party devices. This might influence the decision to adopt a network management protocol that can successfully integrate device management across multiple platforms.

■ *Specification of application requirements.* Depending on the requirements at hand, a particular network architecture may be selected because it facilitates important applications that must be supported by the network.

■ *Special device requirements.* A requirement for specific devices that are supported by a network architecture may influence its selection.

■ *Selection of communication services.* All network architectures support traditional digital (e.g., T1, fractional T1, and T3 lines) and analog lines. However, the use of various other network services may dictate specific network architecture requirements. For example, the network architecture must explicitly support satellite data links, if satellite services are to be used. As another example, networks offering Frame Relay, SMDS, or ATM services also require specialized network architectures.

■ *Future plans and expected growth.* Plans for future network migrations may influence the selection of a network architecture, particularly if the network architecture under consideration is moving in a direction consistent with the evolution of the organization's needs.

8.3.5 Network Management

The importance of being able to manage the network after it is implemented is gaining increasing recognition in the marketplace. This is reflected in the emergence of both proprietary and open (e.g., SNMP[7] and CMIP[8]) network management protocols. These protocols provide a means to collect information and to perform network management functions relating to:

■ *Configuration management.* This involves collecting information on the current network configuration, and managing changes to the network configuration.

- *Performance management.* This involves gathering data on the utilization of network resources, analyzing this data, and acting on the insights provided by the data to maintain optimal system performance.
- *Fault management.* This involves identifying system faults as they occur, isolating the cause of the fault(s), and correcting the fault(s), if possible.
- *Security management.* This involves identifying locations of sensitive data, and securing the network access points as appropriate to limit the potential for unauthorized intrusions.
- *Accounting management.* This involves gathering data on resource utilization. It may also involve setting usage quotas and generating billing and usage reports.

The selection of a network management protocol can have significant impacts on the network costs and on the selection of the network devices and systems. For example, IBM's proprietary network management system — NetView — is expensive and requires an IBM operating system; however, it provides comprehensive network management functionality for both SNA and non-SNA networks. In selecting a network management approach, the benefits must be weighed against costs, compatibility issues, and the network requirements.

In general, network management is easier when the network is simple, homogeneous, and reliable. Designing a network with this in mind means that complexity should be avoided unless it serves a good purpose. Network complexity should reflect a requirement for services and functions that cannot be provided by simpler solutions. All other things being equal, it is better to have a network comprised of similar, compatible components and services. A network that is robust, reliable, and engineered to support growth will be easier to maintain than a network with limited capacity. Thus, a network with good manageability characteristics should be given preference over designs that are more difficult to manage, particularly when these benefits can be achieved without incurring significantly higher costs. Network manageability can also be enhanced by careful vendor selection. In this context, vendors that guarantee the quality and continuity of network products and services are preferable to those that do not.

Network management encompasses all the processes needed to keep the network up and running at agreed-upon service levels. Network management involves the use of various management instruments to optimize the network operation at a reasonable cost. Network management is most effective when a single department or organization controls it. The major players and functions in the network management process are:

- Clients
- Client contact point(s)
- Operations support
- Fault tracking
- Performance monitoring
- Change control
- Planning and design
- Billing and finance

Clients represent internal or external customers or any other users of management services. Clients may report problems, request changes, order equipment or facilities, or ask for information through an assigned contact point. Ideally, this should be a single point of contact. The principal activities of the contact point include:

- Receiving problem reports
- Handling calls
- Handling and processing inquiries
- Receiving change requests
- Handling orders
- Making service requests
- Opening and referring trouble tickets
- Closing trouble tickets

The contact point forwards trouble tickets (i.e., problem reports) to operations support. In turn, operations support may respond with the following types of activities:

- Problem determination by handling trouble tickets
- Problem diagnosis
- Corrective actions
- Repair and replacement of software or equipment
- Referrals to third parties
- Backup and reconfiguration activities
- Recovery processes
- Logging and documenting events and actions

It is possible that various troubleshooting activities by clients or operations support may result in change control requests. Problem reports and change requests should be managed by a designated group (usually in operations support) assigned to fault monitoring. The principal functions of fault monitoring include:

■ Manual tracking of reported or monitored faults
■ Tracking progress on status of problem resolution and escalating the level of intervention, if necessary
■ Information distribution to appropriate parties
■ Referral to other groups for resolution and action

Fault monitoring is a key aspect of correcting service- and quality-related problems. Fault monitoring often results in requests for various system changes. These requests are typically handled by a change control group. Change control deals with:

■ Managing, processing, and tracking service orders
■ Routing service orders
■ Supervising the handling of changes

After the change requests have been processed and validated, they should be reviewed and acted upon by the group designated to perform planning and design. Planning and design performs the following tasks:

■ Needs analysis
■ Projecting application load
■ Sizing resources
■ Authorizing and tracking changes
■ Raising purchase orders
■ Producing implementation plans
■ Establishing company standards
■ Quality assurance

The recommendations made by planning and design are generally then passed on to finance and billing and to implementation and maintenance. Implementation and maintenance make changes and process work orders approved by planning and design and by change control. In addition, this area is in charge of:

■ Implementing change requests and work orders
■ Maintaining network resources
■ Performing periodic inspections
■ Maintaining database(s) to tracks various network components and their configuration
■ Performing network provisioning

Network status and performance information should be continuously monitored. Ideally, fault monitoring should be proactive in detecting problems, and in opening and referring trouble tickets to the appropriate departments for resolution. Performance monitoring deals with:

- Monitoring the system and network performance
- Monitoring service level agreements and how well they have been satisfied
- Monitoring third-party and vendor performance
- Performing optimization, modeling, and network tuning activities
- Reporting usage statistics and trends to management and to users
- Reporting service quality status to Finance and Billing

Security management is also a vital part of network management. It is responsible for ensuring secure communication and protecting the network operations. It supports the following functions:

- Threat analysis
- Administration (access control, partitioning, authentication)
- Detection (evaluating services and solutions)
- Recovery (evaluating services and solutions)
- Protecting the network and network management systems

Systems administration is responsible for administering such functions as:

- Software version control
- Software distribution
- Systems management (upgrades, disk space management, job control)
- Administering the user-definable tables (user profiles, router tables, security servers)
- Local and remote configuring resources
- Names and address management
- Applications management

Finance and billing is the focal point for receiving status reports regarding service level violations, network plans, designs, and changes, and invoices from third parties. Finance and billing is responsible for:

- Asset management
- Costing services
- Billing clients
- Usage and outage collection
- Calculating rebates to clients

- Bill verification
- Software license control

The instruments available to support each of the network management functions are highly varied in sophistication, scope, and ease of use. The tools and organizational processes needed to support the network management functions are dependent upon the business context in which the network is being operated.

8.3.6 Security

Network security requirements are not explicitly considered during the execution of topological network design algorithms. Nonetheless, security considerations may have a considerable impact on the choice of network devices and services. For example, an often-cited reason for a private network, as opposed to a public network, is the need for control and security.

There are many ways to compromise a network's security, either inadvertently or deliberately. Therefore, to be effective, network security must be comprehensive and should operate on several levels. Threats can occur from both internal and external sources, and can be broadly grouped into the following categories:

- *Unauthorized access to information*. This type of threat includes wiretapping, and people correctly guessing a password to gain access to a system they are not authorized to use.
- *Masquerading*. This type of threat occurs when someone gains access to the network by pretending to be someone else. An example of this type of threat is a Trojan horse. An example of a Trojan horse is a software routine that appears to be benign and legitimate, but is not. A Trojan horse masquerading as a log-on procedure can prompt a network user to supply the password required to gain entry to the system. The network user may never even know that he has given away his password to the Trojan horse!
- *Unauthorized access to physical network components*. This type of threat might occur if someone were to cut through a communications link while making building repairs or if a bomb were to explode where it could disrupt the network.
- *Repudiation*. This threat occurs when someone denies having used the network facilities in an inappropriate or improper manner. An example of this is someone sending harassing e-mail to another person, while denying it.

- *Unauthorized denial of service.* This threat occurs when a user prevents other users from gaining the access to the network to which they are entitled. This might occur if the network is inundated with traffic flooding the network, thus blocking entry to the system. This type of threat can be caused by intentional or unintentional acts.

One level of security is offered by *protocol security,* and thus it is important to assess the level of vulnerability posed by the presence or lack of good protocol security in the network. For example, SNMP and other network management protocols that have been designed with security in mind can be used to identify and protect the network against unauthorized use. IP[9] networks, on the other hand, are potentially vulnerable to source address spoofing. Spoofing is a form of masquerading where packets appear to come from a source that they did not. IP networks are also susceptible to packet flooding caused by an open connection. This creates system overloads that may lead to a denial of service on the network. A good defense against this type of attack is to configure routers and firewalls in the network to filter out incoming packets that are not from approved sources. In future versions of IP, new security provisions will undoubtedly become available. For example, in IP Version 6, IP Authentication Headers and IP Encapsulating Security Payload are provided. Other security protocols to protect Internet traffic include Secure Socket Layer (SSL) and Secure Hypertext Transport Protocol (SHTTP).

Operational security provides a second level of network security. Operational security involves disabling network services that are not necessary or appropriate for various types of users. In this context, remote log-in and file transfer protocols may be disabled or controlled so that viruses and unauthorized personnel are prevented from gaining entry to the network. Operational security also involves such good practices as changing passwords regularly, constant use of updated anti-virus programs, ongoing monitoring of anonymous and guest system access, and enabling and reviewing security logs and alerts.

Network security can also be implemented at the physical level. This approach attempts to safeguard access to the network by securing network components and limiting access to authorized personnel only.

Network security can be implemented at the data level. This involves the use of encryption technology to protect the confidentiality of data transmissions. Use of encryption technology implies that both the sender and the receiver must employ compatible procedures to encrypt and decrypt data. This, in turn, has implications on the management and implementation of the network services.

There are two major forms of encryption: single-key and public/private key. An overview of single key cryptography is provided in Figure 8.5.

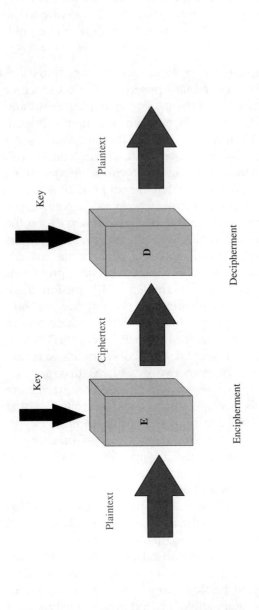

Legend:

- Plaintext - Data Before Encryption
- Ciphertext - Data After Encryption
- E - Encryption Function
- D - Decryption Function
- Key - Parameter Used In Cipher To Ensure Secrecy
- Random Seed - Randomly Selected Number Used To Generate Public And Secret Keys

Figure 8.5 Secret key example.

DES is a widely used single key encryption scheme. With DES, the data to be encrypted is subjected to an initial permutation. The data is then broken down into 64-bit data fields that are in turn split. The two resulting 32-bit fields are then further permuted in a number of iterations. Like all secret key encryption schemes, DES uses the same key to perform both encryption and decryption. The DES key, by convention, is 56 bits long.

Seminal work by Diffie and Hellman, and by Rivest, Shamir, and Adleman (RSV), led to the development of public/private key cryptography. In this scheme, data is encrypted with a public key that can be known by many, and is decrypted by a private key known only to one. The beauty of public key encryption is that it is computationally infeasible to derive the decipherment algorithm from the encipherment algorithm. Therefore, dissemination of the encryption key does not compromise the confidentiality of the decryption process. Because the encryption key can be made public, anyone wishing to send a secure message can do so. This is in contrast to secret key schemes that require both the sender and receiver to know and safeguard the key. Public key encryption is illustrated in Figure 8.6.

One application of public key cryptography is the generation of digital signatures. A digital signature assures the receiver that the message is authentic; that is, the receiver knows the true identity of the sender and that the contents of the message cannot be modified without leaving a trace. A digital signature is very useful for safeguarding contractual and business-related transmissions because it provides a means for third-party arbitration and validation of the digital signature. Public and private keys belong to the sender, who creates keys based on an initial random number selection (or random seed). The message recipient applies the encipherment function using the sender's public key. If the result is plaintext, then the message is considered valid. Digital signatures are illustrated in Figure 8.7.

In summary, comprehensive network security involves active use of protocol, operational, and encryption measures. Good management oversight and employee training complement these security measures.

8.4 Representation of Networks Using Graph Theory

There are numerous rigorous mathematical techniques for solving network design problems based on graph theory. Graph theory provides a convenient and useful notation for representing networks. This, in turn, makes is easier to computerize the implementation of network design algorithms.

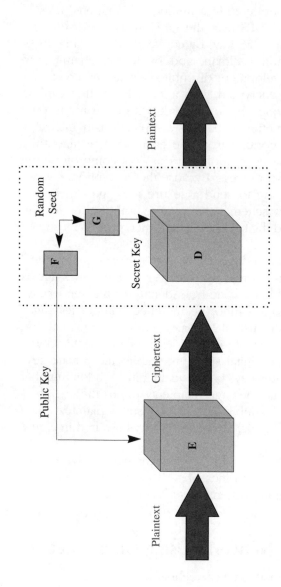

Legend:
- Plaintext - Data Before Encryption
- Ciphertext - Data After Encryption
- E - Encryption Function
- D - Decryption Function
- Key - Parameter Used In Cipher To Ensure Secrecy
- Random Seed - Randomly Selected Number Used To Generate Public & Secret Keys

Figure 8.6   Public key cryptography.

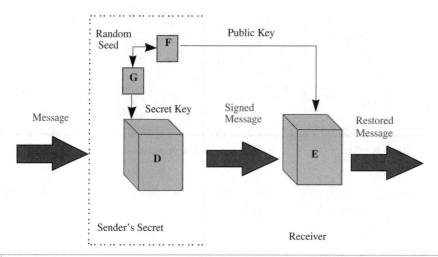

Legend:
- Plaintext - Data Before Encryption
- Ciphertext - Data After Encryption
- E - Encryption Function
- D - Decryption Function
- Key - Parameter Used In Cipher To Ensure Secrecy
- Random Seed - Randomly Selected Number Used To Generate Public   And Secret Keys (i.e., F and G)

Figure 8.7   **Digital signature.**

8.4.1 Definitions and Nomenclature

When introducing network design algorithms in later sections, the following definitions and nomenclature relating to graph theory are necessary.

8.4.1.1 Definition: Graph

A graph G is defined by its vertex set (nodes) V and its edges (links) E.

8.4.1.2 Definition: Link

A link is a *bi-directional* edge in which the ordering of the nodes attached to the link does not matter. A link can be used to represent network traffic flowing in either direction. Full-duplex lines in a communications

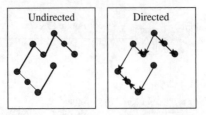

Figure 8.8 Example of an undirected and a directed graph.

network support traffic in both directions simultaneously and are often represented as links in a graph.

8.4.1.3 Definition: Undirected Graph

An undirected graph contains only bi-directional links. See Figure 8.8 for an illustration.

8.4.1.4 Definition: Arc

An arc is a link with a specified direction between two nodes. Half-duplex lines in a communications network handle traffic in only one direction at a time and can be represented as arcs in a graph.

8.4.1.5 Definition: Directed Graph

A directed graph is a graph containing arcs. See Figure 8.8 for an illustration and comparison with an undirected graph.

8.4.1.6 Definition: Self-Loop

A self-loop is a link that begins and ends with the same node. See Figure 8.9 for an illustration of a self-loop.

8.4.1.7 Definition: Parallel Link

Two links are considered parallel if they start and terminate on the same nodes. See Figure 8.9 for an illustration of parallel links and a comparison with a self-loop.

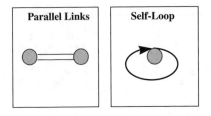

Figure 8.9 **Parallel links versus self-loop.**

8.4.1.8 Definition: Simple Graph

A simple graph is a graph without parallel links or self-loops. Most network design algorithms assume that the network is represented as a simple graph.

8.4.1.9 Definition: Adjacency

Two nodes i and j are adjacent if there exists a link (i, j) between them. Adjacent nodes are also called *neighbors*.

8.4.1.10 Definition: Degree

The degree of a node is the number of links incident on the node or the number of neighbors the node has.

8.4.1.11 Definition: Incident Link

A link is said to be incident on a node if the node is one of the link's endpoints.

8.4.1.12 Definition: Path

A path is a sequence of links that begins at an initial node, s, and ends at a specified node, t. A path is sometimes designated as (s, t).

8.4.1.13 Definition: Cycle

A cycle exists if the starting node, s, in a path (s, t) is the same as the terminating node, t.

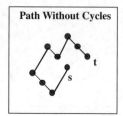 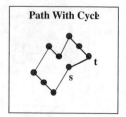

Figure 8.10 Example of path without cycles and path with cycles.

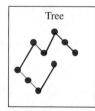 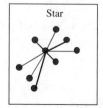

Figure 8.11 Tree graph versus star graph.

8.4.1.14 Definition: Simple Cycle

A simple cycle exists if the starting node *s* is the same as the terminating node *t* and all intermediate nodes between *s* and *t* appear only once. See Figure 8.10 for an example of a graph with a cycle and a graph with no cycles.

8.4.1.15 Definition: Connected Graph

A graph is considered connected if at least one path exists between every pair of nodes.

8.4.1.16 Definition: Strongly Connected Graph

A directed graph with a directed path from every node to every other node is considered a strongly connected graph.

8.4.1.17 Definition: Tree

A tree is a graph that does not contain cycles. Any tree with *n* nodes will contain (n–1) edges. See Figure 8.11 for an example of a tree graph.

8.4.1.18 Definition: Minimal Spanning Tree

A minimal spanning tree is a connected graph that links all nodes with the least possible total cost or length and does not contain cycles. (Assume that a weight is associated with each link in the graph. This weight might represent the length or cost of the link.)

8.4.1.19 Definition: Star

A graph is considered a star if only one node has a degree greater than 1. See Figure 8.11 for an example of a star graph and a comparison with a tree graph.

8.5 Introduction to Algorithms

Many network design problems are solved using special techniques called "algorithms." Given the importance of algorithms in solving network design problems, it is worthwhile at this juncture to formally define and review important properties of algorithms.

8.5.1 Definition of Algorithms

An algorithm is a well-defined procedure for solving a problem in a finite number of steps. An algorithm is based on a model that characterizes the essential features of the problem. The algorithm specifies a methodology to solve the problem using the model representation.

Algorithms are characterized by a number of properties. These properties are necessary to ensure that the algorithm correctly solves the problem for which it is intended, or correctly identifies when a solution is impossible to find, in a finite number of steps. These properties include:

- *Specified inputs.* The inputs to an algorithm must be from a prespecified set.
- *Specified outputs.* For each set of input values, the algorithm must produce outputs from a prespecified set. The output values produced by the algorithm comprise the solution to the problem.
- *Finiteness.* An algorithm must produce a desired output after a finite number of steps.
- *Effectiveness.* It must be possible to perform each step of the algorithm exactly as specified.
- *Generality.* The algorithm should be applicable to all problems of the desired form. It should not be limited to a particular set of input values or special cases.

Table 8.3   List of Potential Link Costs

From Node	To Node	Link Cost
A	B	1
A	C	6
A	D	+∞
B	C	5
B	D	+∞
C	D	2

An example of a "greedy" algorithm is now presented. The algorithm is considered greedy because it selects the best choice immediately available at each step, without regard to the long-term consequences of each selection in totality and in relation to each other.

We use the greedy algorithm to find the cheapest set of links to connect all of a given set of terminals. The graphical representation in Table 8.3 is used to model this network design problem. Using this representation, all terminal devices are modeled as nodes (a, b, c, and d), and all communications lines are modeled as links. Associated with each possible link (i, j) — where i is the starting node and j is the terminating node — is a weight, representing the cost of the link if it is used in the network. A cost of "+∞" is used to indicate when a link is prohibitive in cost or is not available.

8.5.2 Greedy Algorithm (Also Known as Kruskal's Algorithm)

1. Sort all possible links in ascending order and put in a link list.
2. Check to see if all the nodes are connected.
   - If all the nodes are connected, then terminate the algorithm, with the message "Solution Complete."
   - If all the nodes are not connected, continue to the next step.
3. Select the link at the top of the list.
   - If no links are on the list, then terminate the algorithm. Check to see if all nodes are connected; and if not, then terminate the algorithm with the message "Solution Cannot Be Found."
4. Check to see if the link selected creates a cycle in the network.
   - If the link creates a cycle, remove it from the list. Return to Step 2.
   - If the link does not create a cycle, add it to the network, and remove link from link list. Return to Step 2.

Table 8.4   Sorted Link List

From Node	To Node	Link Cost
A	B	1
C	D	2
B	C	5
A	C	6
A	D	$+\infty$
B	D	$+\infty$

Figure 8.12   First link selected by greedy algorithm.

Now solve the sample problem using the algorithm specified above:

1. Sort all possible links in ascending order and put in a link list (see Table 8.4).
2. Check to see if all the nodes are connected. Because none of the nodes are connected, proceed to the next step.
3. Select the link at the top of the list. *This is link AB.*
4. Check to see if the link selected creates a cycle in the network. It does not, so add link AB to the solution and remove it from the link list. One obtains the partial solution shown in Figure 8.12 and then proceeds with the algorithm.
5. Check to see if all the nodes are connected. They are not, so proceed to the next step of the algorithm.
6. Select the link at the top of the list. *This is link CD.*
7. Check to see if the link selected creates a cycle in the network. It does not, so add link CD to the solution and remove it from the link list. One obtains the partial solution shown in Figure 8.13 and then proceeds with the algorithm.
8. Check to see if all the nodes are connected. They are not, so proceed to the next step.
9. Select the link at the top of the list. *This is link BC.*
10. Check to see if the link selected creates a cycle in the network. It does not, so add link BC to the solution and remove it from the

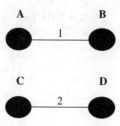

Figure 8.13 Second link selected by greedy algorithm.

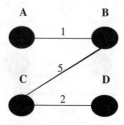

Figure 8.14 Third and final link selected by greedy algorithm.

link list. One obtains the partial solution shown in Figure 8.14 and then proceeds with the algorithm.

11. Check to see if all the nodes are connected. All the nodes are now connected, so terminate the algorithm with the message "Solution Complete." Thus, Figure 8.14 represents the final network solution.

Use the checklist below to verify that the greedy algorithm exhibits all the necessary properties defined above.

■ *Specified inputs.* The inputs to the algorithm are the prespecified nodes, potential links, and potential link costs.
■ *Specified outputs.* The algorithm produces outputs (link selections) from the prespecified link set. The outputs produced by the algorithm comprise the solution to the problem.
■ *Finiteness.* The algorithm produces a desired output after a finite number of steps. The algorithm stops when all the nodes are connected, or after all the candidate links have been examined, whichever comes first.
■ *Effectiveness.* It is possible to perform each step of the algorithm exactly as specified.

- *Generality.* The algorithm is applicable to all problems of the desired form. It is not limited to a particular set of input values or special cases.

The greedy algorithm just described provides an optimal network solution when there are no restrictions on the amount of traffic that can be placed on the links. In this form, the algorithm is called an "unconstrained optimization technique." However, it is not realistic to assume that links can carry an indefinite amount of traffic. When steps to check the line capacity restrictions are added to the algorithm, the algorithm becomes a constrained optimization technique. When the algorithm is constrained, and the restrictions on the algorithm are active (e.g., this would occur if the traffic limit is reached at some point, and therefore a link selected for inclusion in the design must be rejected), there is no longer a guarantee that the algorithm will produce an optimal result.

Many studies have been conducted on the constrained form of Kruskal's greedy algorithm. These studies show that although the greedy algorithm does not necessarily produce an optimal, best-case result, in general it produces very good results that are close to optimal. [KERS93]

In summary, an algorithm is considered "good" if it always provides a correct answer to the problem or indicates when a correct answer cannot be found. A good algorithm is also efficient. The next section discusses what it means to be an efficient algorithm.

8.5.3 Introduction to Computational Complexity Analysis

The efficiency of an algorithm can be measured in several ways. One estimate of efficiency is based on the amount of computer time needed to solve the problem using the algorithm. This is also known as the *time complexity* of the algorithm. A second estimate of efficiency is the amount of computer memory needed to implement the algorithm. This is also referred to as the *space complexity* of the algorithm. Space complexity is very closely tied to the particular data structures used to implement the algorithm.

In general, the actual running time of an algorithm implemented in software will largely depend upon how well the algorithm was coded, the computer used to run the algorithm, and the type of data used by the program. However, in complexity analysis one seeks to evaluate an algorithm's performance *independent of its actual implementation.* To do this, one must consider factors that remain constant irrespective of the algorithm's implementation.

Because one wants a measure of complexity that does not depend on processing speed, space complexity is ignored because it is so closely tied with implementation details. Instead, one can use time complexity as a measure of an algorithm's efficiency. One can measure time complexity in terms of the number of *operations* required by the algorithm instead of the actual CPU time required by the algorithm. Expressing time complexity in these units allows one to compare the efficiency of algorithms that are very different.

For example, let N be the number of inputs to an algorithm. If Algorithm A requires a number of operations proportional to N^2, and Algorithm B requires an number of operations proportional to N, one can see that Algorithm B is more efficient. If N is 4, then Algorithm B will require approximately four operations, while Algorithm A will require 16. As N becomes larger, the difference in efficiency between Algorithms A and B becomes more apparent.

If Algorithm B requires time proportional to f(N), this also implies that given any reasonable computer implementation of the algorithm, there is some constant of proportionality C such that Algorithm B requires no more than (C * f(N)) operations to solve a problem of size N. Algorithm B is said to be of order f (N) — which is denoted O (f (N)) — and f (N) is the algorithm's growth rate function. Because this notation uses the capital letter O to denote *order*, it is called the Big O notation.

Some examples and explanations for the Big O notation are given below:

- If an algorithm is O(1), this means it requires a constant time that is independent of the problem's input size N.
- If an algorithm is O(N), this means it requires time that is directly proportional to the problem's input size N.
- If an algorithm is $O(N^2)$, this means it requires time that is directly proportional to the problem's input size N^2.

Table 8.5 summarizes commonly used terminology describing computational complexity. The terms listed below are sorted from low to high complexity. In general, network design algorithms are considered efficient if they are of $O(N^2)$ complexity or less. The greedy algorithm examined in the previous section can be shown to be O(N log N), where N is the number of edges examined, and is considered very computationally efficient. [KERS93] Most of the effort expended in executing the greedy algorithm presented in the previous section goes toward creating the sorted link list created in the first step.

Brute force and exhaustive search algorithms are considered strategies of last resort. Using these methods, all the potential solution candidates are examined one by one, even after the best one has been found. This is because these methods do not recognize the optimal solution until the

Table 8.5   Computational Complexity Terminology

Complexity	Terminology
O(1)	Constant complexity
O(log n)	Logarithmic complexity
O(n)	Linear complexity
O(n log n)	n log n complexity
O(n^b)	Polynomial complexity
O(b^n), where b > 1	Exponential complexity
O(n!)	Factorial complexity

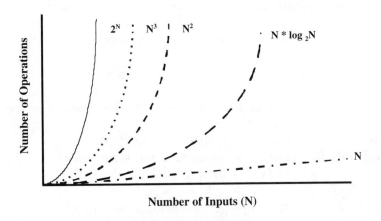

Figure 8.15   Growth rate function comparison.

end after all the candidates have been compared. Brute force and exhaustive search methods are usually O(b^N) or worse. The worst computational complexity is factorial complexity, which is generally associated with n-p (i.e., non-polynomial time) complete problems. Public key cryptography and Traveling Salesman problems are two examples of n-p complete problems. In general, when the input size is large, n-p complete problems are exceedingly difficult to solve and require very large amounts of computing time. Figure 8.15 shows the effect of increasing computational complexity on the number of operations required by the algorithm to solve the problem.

A Big "O" estimate of an algorithm's time complexity expresses how the time required to solve the program changes *as the input grows in size.* Big "O" estimates do not directly translate into actual computer time because the Big "O" method uses a simplified reference function to estimate complexity. The simplified reference function *omits* constants and other terms that may affect actual computer time. Thus, the Big "O" method provides a *lower* bound on computer time. This is illustrated in the examples that follow.

Example 1: If algorithm is $O(N^3 + N^2 + 3N)$, one can ignore the low-order terms in the growth rate function. Therefore, the algorithm's complexity can be expressed by a simplified growth function: $O(N^3)$.

Example 2: If algorithm is $O(5N^3 + 2N^2 + 3N)$, one can ignore multiplicative constants in the high-order terms of the growth rate function. Therefore, the algorithm's complexity can be expressed by a simplified growth function: $O(N^3)$.

Example 3: If algorithm is $O(N^3) + O(N^3)$, one can combine the growth rate functions. Therefore, the algorithm's complexity can be expressed by a simplified growth function: $O(2 * N^3) = O(N^3)$.

A formal generalization of these examples is given below.

If:	one is given f(x) and g(x), functions from the set of real numbers
Then:	one can say that $f(x) = O(g(x))$
If and only if:	there are constants C and k such that $\mid f(x) \mid \le C \mid (g(x) \mid$ whenever $x > k$

An example to illustrate this generalization is given by:

Show:	$f(x) = O(x^2 + 2x + 1)$ is $O(x^2)$
Solution:	Because $0 \le (x^2 + 2x + 1) \le (x^2 + 2x^2 + x^2) = 4x^2$, whenever $x > 1 = k$, and $C = 4$
Then it follows:	$f(x) = O(x^2)$

Complexity analysis can be used to examine worst-case, best-case, and average scenarios. Worst-case analysis tries to determine the largest number of operations needed to guarantee a solution will be found. Best-case analysis seeks to determine the smallest number of operations needed to find a solution. Average-case analysis is used to determine the average number of operations used to solve the problem, assuming all inputs are of a given size.

Note that time complexity is not the only valid criterion to evaluate algorithms. For example, other important criteria might include the style and ease of the algorithm's implementation. The appeal of time complexity, as described here, is the fact that it is independent of specific implementation details and provides a very robust way of characterizing algorithms so they can be compared.

Also note that different orders of complexity do not always matter much in an actual situation. For example, a more complex algorithm might run faster than a less complex algorithm with a high constant of proportionality, especially if the number of inputs to the problem is small. Time complexity, as presented here, is particularly important when solving big problems. When there are very few inputs, an algorithm of any complexity will usually run quickly.

In summary, computational complexity is important because it provides a measure of the computer time required to solve a problem using a given algorithm. If it takes too long to solve a problem, the algorithm may not be useful or practical. In the context of network design, one needs algorithms that can provide reasonable solutions in a reasonable amount of time. Consider the fact that if there are n potential node locations in a network, then there are $2^{(n * (n-1)/2)}$ potential topologies. Thus, brute force and computationally complex design techniques are simply not suitable for network problems of any substantial size.

8.5.4 Network Design Techniques

Three major types of techniques are used in network design: heuristic, exact, and simulation techniques. Heuristic algorithms are the preferred method for solving many network design problems because they provide a close to optimal solution in a reasonable amount of time. For this reason, the bulk of our attention in this book concentrates on heuristic solution techniques or algorithms.

Linear programming is a powerful, exact solution technique that is also used in network design. Linear programming methods are based on the simplex method. The simplex method guarantees that an optimal solution will be found in a finite amount of time. Otherwise, the simplex algorithm will show that an optimal solution does not exist or is not feasible, given the constraints that have been specified. Linear programming models require that both the objective function and the constraint functions be linear. Linear programming also requires that the decision variables be continuous, as opposed to discrete.

In the context of network design, one might want to find a low-cost network (i.e., one wants to minimize a cost function based on links costs or tariffs), subject to constraints on where links can be placed and constraints on the amount of traffic that the links can carry. Although linear functions may provide useful approximations for the costs and constraints to be modeled in a network design problem, in many cases

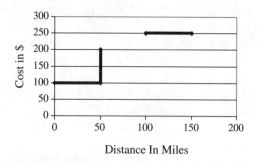

Figure 8.16   Graph with discontinuities.

they do not. The tariffs that determine the line costs are usually nonlinear and may exhibit numerous discontinuities. A discontinuity exists when there is a sharp price jump from one point to the next, or when certain price/line combinations are not available, as illustrated in Figure 8.16. Linear programming can be used successfully when the cost and constraint functions can be approximated by a linear or piece-wise linear function that is accurate within the range of interest. This implies that the linear programming approach is best applied, in the context of network design, when the neighborhood of the solution can be approximated *a priori.*

When the decision variables for designing the network are constrained to discrete integer values, the linear programming problem becomes an integer programming problem. Integer programming problems, in general, are much more difficult to solve than linear programming problems, except in selected special cases. In the case of network design, it might be desirable to limit the constraints to zero (0) or one (1) values to reflect whether or not a link is being used. However, these restrictions complicate the problem a great deal. In this context, "complicated" means that the computational complexity of the integer programming technique increases to the point where it may be impractical to use.

Simulation is a third commonly used technique in network design. It is often used when the design problem cannot be expressed in an analytical form that can be solved easily or exactly. A simulation tool is used to build a simplified representation of the network. Experiments are then conducted to see how the proposed network will perform as the traffic and other inputs are modified. In general, simulation approaches are very time-consuming and computationally intensive. However, they are helpful in examining the system performance under a variety of conditions to test the robustness of the design. There are many software packages available that are designed exclusively for simulation and modeling studies.

In summary, as network size increases, so does the computational complexity of the network design process. This occurs to such an extent that for most problems of practical interest, exact optimal solution techniques are impractical, except in special cases. Heuristic design techniques providing approximate solutions are usually used instead. Although in many cases, good solutions can be obtained with heuristic techniques, they do not necessarily guarantee that the best solution will be found. After an approximate solution is found using a heuristic method, it may be helpful to use an exact solution technique or simulation to fine-tune the design.

8.6 Performance Analysis

Once a network has been designed, its performance characteristics can be analyzed. The most common measures of network performance are cost, delay, and reliability. This section presents an overview of the methods used to calculate these performance measures. However, other measures — for example, link, memory, and node utilization — of network performance are possible and may also be helpful in evaluating the quality of the network design. For more information on other network performance measures and their calculation, the reader is referred to [CAHN98] and [KERS93].

8.6.1 Queuing Essentials

This section summarizes some of the major results from queuing theory that are applicable to the analysis of network delay and reliability. It concentrates on results that are easily grasped using basic techniques of algebra and probability theory. Readers who want more information on this subject are referred to [KERS93], [GROSS74], and [WAGN75]. In addition to providing detailed theoretical derivations and proofs, these references also discuss other more complex queuing models that are beyond the scope of this book.

Queuing theory was originally developed to help design and analyze telephone networks. Because the rationale for a network is to provide a means to share resources, there is an implicit assumption that the number of resources available in the network will be less than the total number of potential system users. Therefore, it is possible that some users may have to wait until others relinquish their control over a telecommunications facility or device. Queuing analysis is used to estimate the waiting times and delays that system users will experience, given a particular network configuration.

Queuing theory is very broadly applicable to the analysis of systems characterized by a stochastic[10] input process, a stochastic service mechanism, and a queue discipline. The key descriptive features of the input process are the size of the input population, the source of the input population, inter-arrival times between inputs, and how the inputs are grouped, if at all. The features of interest for the service mechanism are the number of servers, the number of inputs that can be served simultaneously, and the average service time. The queue discipline describes the behavior of the inputs once they are within a queue. Given information on these characteristics, queuing theory can be used to estimate service times, queue lengths, delays (while in the queue and while being serviced), and the required number of service mechanisms.

Some queuing notation is needed for the discussion that follows, including some essential definitions and nomenclature.

8.6.1.1 Standard Queuing Notation:

The following notation specifies the assumptions made in the queuing model:

Arrival Process/Service Process/Number of Parallel Servers/Optional Additions

Optional additions to this notation are:

Limit on Number in the System/Number in the Source Population/Type of Queue Discipline

The arrival and service processes are defined by probability distributions that describe the expected time between arrivals and the expected service time. The number of servers or channels operating in parallel must be a positive integer. Table 8.6 summarizes abbreviations and assumptions commonly used in queuing models.

Table 8.7 provides some examples to illustrate these abbreviations.

A Markovian distribution is synonymous with an exponential distribution. This distribution has the interesting property that the probability a system input that has already waited T time units must wait another X time units before being served is the same as the probability that an input just arriving to the system will wait X time units. Thus, the system is "memoryless," in that the time an arrival has already spent in the system does not in any way influence the time the arrival will remain in the system.

Other widely used queuing notation is summarized in Table 8.8.

Table 8.6 Summary of Standard Queueing Nomenclature

Standard Abbreviation	Meaning
M	Markovian or exponential distribution
E_k	Erlangian or Gamma distribution with k identical phases
D	Deterministic distribution
G	General distribution
c	Number of servers or channels in parallel
FCFS	First come, first served
LCFS	Last come, first served
RSS	Random selection for service
PR	Priority service

Table 8.7 Queuing Abbreviations

Queuing Model Abbreviation	Meaning
M/M/c	Markovian input process and Markovian service distribution with c parallel servers
M/M/1/n/m	Markovian input process and Markovian service distribution with one server, a maximum system capacity of n, and a total potential universe of m customers
M/C/3/m/m	Markovian input process and constant service distribution with three servers and a maximum system capacity of n and a total potential universe of m customers

Using the notation in Table 8.8, one can introduce Little's law. This is a very powerful relationship that holds for many queuing systems. Little's law says that the average number waiting in the queuing system is equal to the average arrival rate of the inputs to the system multiplied by the average time spent in the queue. Mathematically, this is expressed as: [GROSS74, p. 60]

$$L_q = \lambda * W_q \tag{8.5}$$

Another important queuing relationship derived from Little's law says that the average number of inputs to the system is equal to the average

Table 8.8 Queuing Notation

Notation	Meaning
$1/\lambda$	Mean inter-arrival time between system inputs
$1/\mu$	Mean service time for server(s)
λ	Mean arrival rate for system inputs
μ	Mean service rate for server(s)
ρ	Traffic intensity = $(\lambda/(c * \mu))$, where c = number of service channels = utilization factor measuring maximum rate at which work entering system can be processed
L	Expected number in the system, at steady state, including those in service
L_q	Expected number in the queue, at steady state, excluding those in service
W	Expected time spent in the system, including service time, at steady state
W_q	Expected time spent in the queue, excluding service time, at steady state
N(t)	Number of units in the system at time t
$N_q(t)$	Number of units in the queue at time t

arrival rate of inputs to the system multiplied by the average time spent in the system. Mathematically, this is expressed as: [GROSS74, p. 60]

$$L = \lambda * W \qquad (8.6)$$

The intuitive explanation for these relationships goes along the following lines. An arrival, A, entering the system will wait an average of W_q time units before entering service. Upon being served, the arrival can count the number of new arrivals behind it. On average, this number will be L_q. The average time between each of the new arrivals is $1/\lambda$ units of time, because by definition this is the inter-arrival rate. Therefore, the total time it took for the L_q arrivals to line up behind A must equal A's waiting time W_q. A similar logical analysis holds for the calculation of L in Equation (8.6).

A number of steady-state models have been developed to describe queuing systems that are applicable to network analysis. A steady-state model is used to describe the long run behavior of a queuing system after it has stabilized. It also represents the average frequency with which the system will occupy a given state[11] over a long period of time.

8.6.2 Analysis of Loss and Delay in Networks

The following discussion provides examples of queues relating to network analysis, and illustrates how the queuing models can be used to calculate delays and service times. Before using any queuing model, it is important to first identify the type of queue being modeled (e.g., is it an M/M/2 queue or some other type of queue?). Next, the assumptions of the queue should be examined to see if they are reasonable for the situation being modeled. A fundamental assumption inherent in the models presented here is that the system will eventually reach a steady state. This is not a reasonable assumption for a system that is always in flux and changing. One way that this assumption can be checked is to examine the utilization factor ρ (see definition in Table 8.8). As ρ approaches unity (1), a queuing system will become congested with infinitely long queues, and will not reach a steady state. The steady-state queuing models presented below also assume that the inter-arrival times and the service times can be modeled by exponential probability distributions.

Sometimes not enough is known about the arrival and service mechanism to specify a probability distribution, or perhaps what is known about these processes is too complex to be analyzed. In this case, one may have to make do with the first and second moments of the probability distributions. This corresponds to the mean and variance, respectively. These measures allow one to calculate the squared coefficient of variation of the distribution. The coefficient of variation, C^2, is defined as:

$$C^2 = V(X) / \left(\overline{X}\right)^2$$

where:
$V(X)$ = variance of probability distribution
$\overline{X}$ = mean of probability distribution

When the arrival rate is deterministic and constant, C^2 is equal to zero. When the arrival rate is exponentially distributed, C^2 is equal to one. When C^2 is small, the arrival rate tends to be evenly spaced. As C^2 approaches one, the distribution approaches a Poisson distribution, with exponential inter-arrival time, and the process is said to be random. When C^2 exceeds one, the probability distribution becomes bursty, with large intermittent peaks.

Except where explicitly stated otherwise, all the steady-state queuing models present herein assume an infinite population awaiting service. In reality, this is seldom the case. However, these formulas still provide a good approximation when the population exceeds 250 and the number of servers is small. When the population is less than 50, a finite population model should be used. [MINO96]

The steady-state models introduced here also assume that traffic patterns are consistent and do not vary according to the time of day and source of traffic. This, too, is an unrealistic assumption for most telecommunication systems. Despite the fact that this assumption is rarely satisfied in practice, the steady state models still tend to give very good results.

The steady-state models also assume that all the inputs are independent of each other. In a network, it is entirely likely that problems in one area of the network will contribute to other failures in the network. This might occur, for example, when too much traffic from one source creates a system overload that causes major portions of the network to overload and fail as well. Despite the fact that this assumption is also rarely satisfied in practice, the models still provide useful results in the context of network design.

One of the compelling reasons for using steady-state queuing models is that, despite their inherent inaccuracies and simplifications of reality, they often yield robust, good results. The models are also useful because of their simplicity and closed form solution. If one tries to interject more realism in the model, the result is often an intractable formula that cannot be solved (at least as easily). The requirements for realism in a model must always be weighed against the resulting effort that will be required to solve a more complex model.

8.6.2.1 M/M/1 Model

This type of queue might be used to represent the flow of jobs to a single print server, or the flow of traffic on a single T1 link. Using standard queuing notation, the major steady-state relationships for M/M/1 queues are given below. For these relationships to hold, λ/μ must be less than 1. When the (λ/μ) ratio equals or exceeds one (1), the queue will grow without bound and there will be no system steady state.

Probability that the system will be empty = $1 - (\lambda/\mu)$ (8.7)

Probability that there will n inputs in the system = $\rho^n (1 - \rho)$ (8.8)

Expected number of inputs in the system = $L = \rho/(1 - \rho)$ (8.9)

Expected number of inputs in the queue = $L_q = \rho^2/(1 - \rho)$ (8.10)

Expected total time in system = $W = 1/(\mu - \lambda)$ (8.11)

Expected delay time in queue = $W_q = \rho/((1 - \rho)*\mu)$ (8.12)

8.6.2.1.1 M/M/1 Example of Database Access Delay

A very large number of users request information from a database management system. The average inter-arrival time of each request is 500 milliseconds. The database look-up program requires an exponential service time averaging 200 milliseconds. How long will each request have to wait on average before being processed?

Answer:

As given in the problem, λ = 1/0.50 seconds = 2 seconds; μ = 1/0.20 seconds = 5 seconds; and λ/μ = ρ = 2/5

Therefore, the expected delay waiting in queue is:

$$W_q = \rho / ((1 - \rho) * \mu) = 0.4/((1 - 0.4) * 5) = 0.1333 \text{ seconds}$$

8.6.2.1.2 M/M/1 Example of Delay on Communications Link

A very large number of users share a communications link. The users generate, on average, one transmission per minute. The message lengths are exponentially distributed, with an average of 10,000 characters. The communications link has a capacity of 9600 bps. One wants to know:

1. What is the average service time to transmit a message?
2. What is the average line utilization?
3. What is the probability that there are no messages in the system?

Answer:

(1) The average service time is the message length divided by the channel speed:

$1/\mu$ = (10,000 characters * 8 bits per character)/9600 bits per second
= 8.3 seconds

(2) The average line utilization is λ/μ = ρ = (0.0167)/(0.12) or the line is utilized at an average rate of 13.9 percent.

λ = 1 message per minute * 60 seconds per minute = 0.0167 messages per second

μ = 1/8.3 = 0.12 messages per second

(3) The probability that the system will be empty is $1 - (\lambda/\mu) = 1 - (0.139) = 0.861$, or 86.1 percent of the time the line is empty.

8.6.2.2 M/M/1/k Model

This type of queue is used to represent a system that has a maximum total capacity of k positions, including those in queue and those in service. This type of queue might be used to model a network node with a limited number of ports or buffer size. Using standard queuing notation, the major steady-state relationships for a M/M/1/k queue are summarized as:

Probability that the system will be empty $= (1 - \rho)/(1 - \rho^{k+1})$ (8.13)

where $\rho \neq 1$

Probability that there will n inputs in the system $= p_k$ (8.14)
$= \rho^n (1 - \rho))/(1 - \rho^{k+1})$

for $n = 0, 1, \ldots, k$

Expected number of inputs in the system $=$

$$= L = \frac{\rho - (k+1)\rho^{k+1} + k\rho^{k+2}}{(1-\rho)(1-\rho^{k+1})} \qquad (8.15)$$

Expected number of inputs in the queue $= L_q = L - \rho(1 - p_k)$ (8.16)

Expected total time in system $= W = L/(\lambda*(1-p_k))$ (8.17)

Expected delay time in queue $= W_q = L_q/(\lambda*(1-p_k))$ (8.18)

8.6.2.2.1 M/M/1/5 Example of Jobs Lost in Front-End Processor due to Limited Buffering Size

Assume the buffers in a front-end processor (FEP) can handle, at most, five (5) input streams (i.e., $k = 5$). When the FEP is busy, up to a maximum of four (4) jobs are buffered in queue. The average number of jobs arriving per minute is five (5), while the average number of jobs the FEP processes per minute is six (6). How many jobs, on average, are lost, or turned away, due to inadequate buffering capacity?

Answer:

A job will be turned away when it arrives at the system and there are already four jobs (one in service and three in the queue) ahead of it. Thus, to find the number of jobs that are turned away, calculate the probability that a job will arrive when there are already four jobs in the queue and one in service, and multiply this by the arrival rate.

λ = 5 per minute ; μ = 6 per minute; λ/μ = ρ = 5/6 = .833
Probability that there will 5 inputs in the system = $p_5 = \rho^5(1 - \rho))/(1 - \rho^6)$
$$= (0.833)^5/1 - (0.833)^6)$$
$$= 0.10$$

Therefore, $\lambda^* p_5$ = (5 per minute) * (0.1) = 0.5 jobs per minute are turned away.

8.6.2.3 M/M/c Model

This type of queue might be used to represent the flow of traffic to c dial-up ports, or the flow of calls to a PBX with c lines, or traffic through a communications link with c multiple trunks. Using standard queuing notation, the major steady-state relationships for M/M/c queues are listed below. For these relationships to hold, $\lambda/(c * \mu)$ must be less than 1. When this ratio equals or exceeds 1, the queue will grow without bound and there will be no system steady state. Although the equations are more complicated than the ones introduced thus far, they are nonetheless easily solved using a calculator or computer.

The probability that the system will be empty = P_0 =

$$\left[\sum_{n=0}^{c-1} (\lambda/\mu)^n / n! + \frac{(\lambda/\mu)^c}{c!\left(1 - \frac{\lambda}{c\mu}\right)} \right]^{-1} \tag{8.19}$$

The probability that the system will have n in the system = P_n =

$$\begin{cases} \left(\dfrac{(\lambda/\mu)^n}{n!} \right) * P_0 & \text{for } 0 \le n \le c \\[4mm] \left(\dfrac{(\lambda/\mu)^n}{c!c^{n-c}} \right) * P_0 & \text{for } n > c \end{cases} \tag{8.20}$$

Expected number of inputs in the system = $L = \rho + L_q$ (8.21)

Expected number of inputs in the queue = $L_q = \dfrac{(\lambda / \mu)^c \rho\, P_0}{c!(1 - \rho)^2}$ (8.22)

Expected total time in system = $W = 1/\mu + W_q$ (8.23)

Expected delay time in queue = $W_q = L_q / \lambda$ (8.24)

8.6.2.3.1 M/M/c Example of Print Server Processing Times

Two print servers are available to handle incoming print jobs. Arriving print jobs form a single queue from which they are served on a first-come, first-served (FCFS) basis. The accounting records show that the print servers handle an average of 24 print jobs per hour. The mean service time per print job has been measured at two (2) minutes. It seems reasonable, based on the accounting data, to assume that the inter-arrival times and service times are exponentially distributed. The network manager is thinking of removing and relocating one of the print servers to another location. However, the system users want a third print server added to speed up the processing of their jobs. What service levels should be expected with one, two, or three print servers?

Answer:

The appropriate model to use for the current configuration is (M/M/2) with λ = 24 per hour, and μ = 30 per hour. The probability that both servers are expected to be idle is equal to the probability that there are no print jobs in the system. From Equation 8.19, this is calculated as:

$$P_0 = \left[\sum_{n=0}^{1} (24/30)^n / n! + \frac{(24/30)2}{2!\left(1 - \dfrac{24}{60}\right)} \right]^{-1} = \left[1 + 0.8 + 0.5333 \right]^{-1} = 0.43$$

One server will be idle when there is only one print job in the system. The fraction of time this will occur is:

$$P_1 = \left(\frac{24}{30} \right) * P_0 = 0.344$$

Both servers will be busy whenever two or more print jobs are in the system. This is computed as:

$$P(\text{Busy}) = 1 - P_0 - P_1 = 0.226$$

The expected number of jobs in the queue is given in Equation (8.22):

$$L_q = \frac{\left(\dfrac{24}{30}\right)^2 \left(\dfrac{24}{60}\right)(0.433)}{2\left(1 - \dfrac{24}{60}\right)} = 0.153$$

The expected number of jobs in the system is calculated according to Equation (8.21):

$$L = \rho + L_q = 0.153 = 0.8 = 0.953$$

The corresponding W and W_q waiting times, as computed from Equation (8.23) and Equation (8.24) are:

$$W = 1/\mu + W_q = 0.953/24 = 0.397 \text{ hours} * 60 \text{ minutes per hour}$$
$$= 2.3825 \text{ minutes}$$
$$W_q = L_q/\mu = 0.153/24 = 0.006375 \text{ hours} * 60 \text{ minutes per hour}$$
$$= 0.3825 \text{ minutes}$$

Similar calculations can be performed for a single print server and for three print servers. The results of these calculations are summarized in Table 8.9.

Table 8.9 Comparison of M/M/1, M/M/2, and M/M/3 Queues

	1 Print Server	2 Print Servers	3 Print Servers
P_0	0.2	0.43	0.44
W_q	8 minutes	0.3825 minutes	0.1043 minutes
W	10 minutes	2.3825 minutes	2.1043 minutes

8.6.2.3.2 M/M/1 versus M/M/c Example and Comparison of Expected Service Times

Given the same situation as presented in the previous example, what would happen to the expected waiting times if a single, upgraded print server were to replace the two print servers currently used? The network manager is thinking of installing a print server that would have the capacity to process 60 jobs per hour.

Answer:

The new option being considered equates to an M/M/1 queue. The new print server has an improved service rate of $\mu = 60$ jobs per hour. The calculations for the expected waiting times are shown below.

$P_0 = 1 - (\lambda/\mu) = 1 - (24/60) = 0.6$

$W = 1/(\lambda-\mu) = 1/(60 - 24) = 0.0277$ hours * 60 minutes/hour = 1.662 minutes

$W_q = \rho/((1 - \rho)*\mu) = 0.4/((1 - 0.4) * 60) = 0.1111$ hours * 60 minutes/hour = 0.666 minutes

These calculations demonstrate a classic result that it is always better to have a single, more powerful server whose service rate equals the sum of *c* servers, than it is to have *c* servers with *c* queues. Another important implication of this is that when a network is properly configured from a queuing perspective, it may be possible to provide better service at no additional cost.

8.6.2.4 M/G/1 Model

In the models presented thus far, it is assumed that all the message lengths are randomly and exponentially distributed. In the case of packet-switched networks, this is not a valid assumption. Now consider the effects that packet-switched data has on the delay in the network. Consider an M/G/1 queue in which arrivals are independent, a single server is present, and the arrival rate is general. Using the Pallaczek-Khintchine formula, it can be shown that the average waiting time in an M/G/1 system is: [GROSS74, p. 226]

$$W_q = \frac{\lambda E[1/\mu^2]}{2(1-\rho)} \tag{8.25}$$

Where: E[1/μ²] = second moment of the service distribution, the second moment is defined as:

$$E\left[1/\mu^2\right] = \sum_{j+1}^{M} P_j \mu_j^2 \qquad (8.26)$$

where:
Pj   = probability of a message being type j
1/μ = service time for message of type j

The variance V of the service distribution is given by:

$$V = E\left[\left(1/u\right)^2\right] - \left(E\left[1/\mu\right]\right)^2 \qquad (8.27)$$

8.6.2.4.1 M/G/1 Example of Packet-Switched Transmission Delay

There are two networks. Both networks use 56-Kbps lines that are, on average, 50 percent utilized. Both networks transmit 1000-bit messages, on average. The first network transmits exponentially distributed message lengths. The second network transmits packets of constant message length. Compare the waiting times for message processing in the two network configurations.

Answer:

First consider the case when the message length is constant. From the data, the average arrival rate can be computed as:

$$\lambda = \frac{\left(56000\right)\left(0.5\right)}{1000} = 28\,\text{messages/second}$$

The mean service time is computed from the data as:

$$\left[\left(1/\mu\right)^2\right] = 0.00648\,\frac{\text{sec}^2}{\text{msg}}$$

The waiting time can now be computed from Equation (8.25) as:

$$W_q = \frac{\lambda E[1/\mu^2]}{2(1-\rho)} = \frac{(28\,m/s)0.00648}{2(1-0.5)} = 0.18144 \text{ sec/msg}$$

Now consider the case when the message length is exponentially distributed. This corresponds to an M/M/1 queue. The waiting time calculation for this queue is given by Equation (8.12) and is calculated as:

$$\mu = \frac{1,000 \text{ bits}}{56,000 \text{ bits / sec}} = 0.018 \text{ seconds}$$

$$W_q = \frac{\rho}{\mu(1-\rho)} = \frac{0.5}{0.018(1-0.5)} = 55.5 \text{ sec/msg}$$

Thus, the delay when the message lengths vary according to an exponential distribution is considerably longer than when the message lengths are constant. Note that in both of these cases, the average message length is the same. This is an important result that demonstrates why, all other things being equal, packet-switched networks are more efficient than networks that transmit messages of varying length.

8.6.2.5 M/M/c/k Model

For an M/M/c/k model, one can assume that c is greater than k. This model corresponds to the situation where the system only has room for those in service. It has no waiting room. When all the servers are busy, the arrivals will be turned away and denied service. Certain types of telephone systems that cannot buffer calls exhibit this behavior. The steady-state equations for this queue are given below.

The probability that the system will be empty = P_0 =

$$\left[\sum_{n=0}^{c-1}(\lambda/\mu)^n/n! + \frac{(\lambda/\mu)^c}{c!}\left(\frac{1-\rho^{k-c+1}}{1-\rho}\right)\right]^{-1} \tag{8.28}$$

where:

$$\lambda_n = \begin{cases} \lambda & 0 \le n < k \\ 0 & n \ge k \end{cases} \qquad \mu_n = \begin{cases} n\mu & 1 \le n \le c \\ c\mu & c < n \le k \end{cases}$$

$$\rho = \lambda\big/c\mu$$

The probability that the system will contain n inputs = P_n =

$$
\begin{cases}
\dfrac{(\lambda/\mu)^n}{c!\,c^{n-c}}\,P_0 & \text{for } c < n \le k \\[3em]
\dfrac{(\lambda/\mu)^n}{n!}\,P_0 & \text{for } 0 \le n \le c
\end{cases}
\tag{8.29}
$$

The effective arrival rate is less than the service rate under steady-state conditions, and is calculated as:

$$
\lambda_e = \lambda/(1 - P_k).
\tag{8.30}
$$

The expected queue length, L_q, is calculated from the use of sum calculus and its definition as:

$$
= \frac{\left(\dfrac{\lambda}{\mu}\right)^c (\rho \times p_0)}{c!\,(1-\rho)^2}\left\{1 - \left[(k-c)\right]\left(1-\rho\right)+1\right]\rho^{k-c}\right\}
\tag{8.31}
$$

where:

$$
\rho = \frac{\lambda}{c\mu}
$$

Because the carried load is the same as the mean number of busy servers, one can calculate the expected number in the system as:

$$
L = L_q + \tfrac{\lambda}{\mu}\left(1 - p_k\right)
\tag{8.32}
$$

From Little's law, it follows that the waiting times are:

$$
W_q = L_q / \lambda\left(1 - P_k\right)
\tag{8.33}
$$

and

$$
W = L_q + \frac{1}{\mu}
\tag{8.34}
$$

We now consider a special case of the M/M/c/k model: the M/M/c/c queue. For this queue, the effective arrival and service rates are:

$$\lambda_n = \begin{cases} \lambda, & 0 \le n < c \\ 0 & n \ge c \end{cases} \tag{8.35}$$

$$\mu_n = \begin{cases} n\mu, & 1 \le n < c \\ 0 & \text{elsewhere} \end{cases} \tag{8.36}$$

The probability that the system will be empty = P_0 =

$$\left[\sum_{n=0}^{c} (\lambda/\mu)^n / n! \right]^{-1} \tag{8.37}$$

The probability that the system will have n in the system = P_n =

$$\frac{e^{-c\rho} (c\rho)^n / n!}{\sum_{j=0}^{c} e^{-c\rho} (c\rho)^j / j!} \tag{8.38}$$

where: $\rho = \lambda/c\mu$

8.6.2.6 Erlang's Loss Formula

In the M/M/c/c model, the system is saturated when all the channels are busy (i.e., for P_c). By multiplying the numerator and the denominator of Equation (8.38) by $e^{-c\rho}$, one can obtain a truncated Poisson distribution with parameter values ($c * \rho = \lambda/\mu$). These values can be obtained from tables of the Poisson distribution, or from an Erlang B table, thereby simplifying the calculations for P_c below. This formula is perhaps better known as Erlang's loss formula. In the context of a telephone system, it describes the probability that an incoming call will receive a busy signal and will be turned away. This formula was used to design telephone systems that satisfy predefined levels of acceptable service.

$$P_c = \frac{e^{-cp}\left(cp\right)^c / c!}{\sum\limits_{j+0}^{c} e^{-cp}\left(cp\right)^j / j!} \qquad (8.39)$$

8.6.2.6.1 M/M/c/c/Example of Blocked Phone Lines

An office has four shared phone lines. Currently, half the time a call is attempted, all the phone lines are busy. The average call lasts two minutes. What is the probability that a call will be attempted when all four lines are busy?

Answer:

In this problem, one is given the fact that $P_4 = 0.5$. Therefore, using Equation (8.39), and an Erlang loss or Poisson table, the value for P_4 is obtained when $4 \, \rho = \lambda/\mu = 6.5$. Therefore, the implied arrival rate is:

$$\lambda = 6.5 * \mu = 3.25 \text{ calls per minute}$$

8.6.2.7 Total Network Delay

Thus far, queuing models have been used to estimate delay on a single network component. This section illustrates how these calculations can be extended to estimate the overall network delay or response time. Note that, in general, multiple calculations are needed to compute the overall network delay, as demonstrated below. These calculations can be involved and tedious — especially for a large network — and in actual practice it is best to use software routines to automate these calculations. Many network design tools offer built-in routines to estimate delay using techniques similar to those described here.

By way of example, one can construct a delay model for a packet-switched network. In packet-switched networks, propagation delay, link delay, and node delay are the major components of network delay. This can be summarized as:

Delay Total = Total average link delay + Total average node delay
+ Total average propagation delay   (8.40)

Assuming an M/M/1 service model to approximate traffic routing on the network, total average link delay on the network can be estimated by:

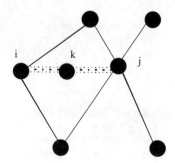

:·:·:·:·:· The shortest path from (i,j) is through the shortest path
from (i,k) and the shortest path from (k,j).

Figure 8.17   Nested shortest paths.

$$D_{link} = \frac{1}{\displaystyle\sum_{i=1}^{I}\sum_{j=1}^{J}R_{ij}} \sum_{l=1}^{L}(D_l * F_l) \tag{8.41}$$

where:

R_{ij} = traffic requirement from node i to node j

F_l = flow on link l

D_l = delay on link l = $\dfrac{(P / S_{ij})}{1-(F_l / C_{ij})}$

P = packet length

S_{ij} = link speed from node i to node j

C_{ij} = link utilization from node i to node j

Thus, the total average network delay is the sum of the expected delay on all the links. The unknown variable in the above equation for D_{link} is link flow. A shortest path algorithm can be used at this stage to assign link flows to solve for this variable.

A shortest path algorithm computes the shortest path between two given points, and is based on the insight that the shortest paths are contained within each other. This is illustrated in Figure 8.17. Thus, as described in [KERS93, p. 157]:

"If a node, k, is part of the shortest path from i to j, then the shortest i, j-path must be the shortest i, k-path followed by the shortest j, k-path. Thus we can find shortest paths using the following recursion:

$$d_{ij} = \min_k (d_{ik} - d_{kj})$$

where d_{xy} is the length of the shortest path from x to y."

Dijkstra's algorithm and Bellman's algorithm, two well-known examples of shortest path algorithms, specify ways to initiate the recursion equation given above so that a solution can be found.

Node delay is a function of node technology. Assume, for the purposes of illustration, that this delay is a constant 120 milliseconds. Thus, total average node delay is estimated by:

$$D_{node} = C_{node} * A \tag{8.42}$$

where:
D_{node} = total average node delay
C_{node} = 120 milliseconds/node
A   = average number of nodes per shortest routing path

Propagation delay is the time it takes electromagnetic waves to traverse the transmission link. Propagation delay is proportional to the physical distance between the node originating a transmission and the node receiving the transmission. If the tariff database is used to compute link costs, it is also likely that V (vertical) and H (horizontal) distance coordinates for each node are available so that the mileage between nodes can be estimated. This mileage multiplied by the physical limit of electronic speed — 8 microseconds per mile — can be used to obtain an estimate of the total average propagation delay, as indicated below. In general, propagation delays are small compared to the queuing and transmission delays calculated in the previous section.

$$D_{prop} = \frac{C_{prop} * \sum_{l=1}^{L} F_l * M_l}{\sum_{i=1}^{I} \sum_{j=1}^{J} R_{ij}} \tag{8.43}$$

where:
D_{prop} = total average propagation delay
R_{ij} = traffic requirement from node i to node j
C_{prop} = 8 microseconds/mile

F_l = flow on link l
M_l = length in miles of link l

8.6.2.8 Other Types of Queues

There are other types of queues that have not been discussed thus far but which are useful in network analysis. These queues include:

- *Queues with state-dependent services.* In this model, the service rate is not constant and may change according to the number in the system, either by slowing up or by increasing in speed.
- *Queues with multiple traffic types.* These types of queues are common in network applications. To handle the arriving traffic, one may wish to employ *a priority* service regime, to give faster service to some kinds of arrivals relative to the others.
- *Queues with impatience.* This type of queue is designed to model the situation in which an arrival joins the queue but then becomes impatient and leaves before it receives service. This is called *reneging.* Another type of impatience, called *balking,* occurs when the arrival leaves upon finding that a queue exists. A third type of impatience is associated with *jockeying,* in which an arrival will switch from one server to the next to try to gain an advantage in being served.

These models interject more realism in the network analysis, at the expense of simplicity. The interested reader is referred to [KERS93] and [KLIE75] for more information in this area.

8.6.2.9 Summary

Queuing is broadly applicable to many network problems. This section presented examples of queuing analysis for the following types of network configurations:

- Delay in a T1-based WAN
- Transmission speed in a packet-switched network
- Printer-server on a LAN with limited buffering
- Expected call blockage in a PBX
- Total network delay in an APPN packet-switched network

The queueing models presented thus far are based on steady-state assumptions. These assumptions were made to keep the models simple and easy to solve. It is important to maintain a perspective on what degree of accuracy is needed in estimating the network delays. If further refinements to the delay estimates will require substantially more time and effort, then this may well temper the decision to attempt to introduce more realism into the models. The models introduced here provide a high-level approximation of reality, and thus there may well be discrepancies between the anticipated (i.e., calculated) system performance and actual observed performance.

8.6.3 Analysis of Network Reliability

There are numerous ways that network reliability can be estimated. This section surveys three commonly used methods for analyzing reliability: (1) component failure analysis, (2) graphical tree analysis, and (3) k-connectivity analysis. The interested reader is referred to [KERS93] for a comprehensive treatment of this subject.

One measure of network reliability is the fraction of time the network and all of its respective components are operational. This can be expressed as:

$$\text{Reliability} = 1 - (\text{MTTR})/(\text{MTBF}) \tag{8.44}$$

where MTTR is the mean time to repair failure, and MTBF is the mean time before failure. For example, if the above equation is used to compute a reliability of 0.98, this means that the network and its components are operational 98 percent of the time.

Because the network is comprised of multiple components, each component contributes to the possibility that something will fail. One commonly used model of component failure assumes that failures will occur according to the exponential probability distribution. An exponential random variable with parameter λ is a continuous random variable whose probability density function is given for some $\lambda > 0$ by:

$$f(x) = \left\{\lambda e^{-\lambda x} \quad \text{if } x \geq 0 \quad and \quad f(x) = \left\{0 \quad \text{if } x < 0 \right. \right. \tag{8.45}$$

The cumulative distribution function F of an exponential variable is given by:

$$F(a) = 1 - e^{-\lambda a} \quad \text{for } a \geq 0 \tag{8.46}$$

These definitions are now used in an illustrative sample problem. Assume that the failure of a network component can be modeled by an exponential distribution with parameter $\lambda = 0.001$ and X = time units in days. Thus, for this component:

$$f(x) = 0.001e^{-0.001X}$$

and

$$F(x) = 1 - e^{-0.001X}$$

Using the cumulative probability density function F (x) above, one can compute the probabilities of failure over various time periods:

Probability that the network component will fail within 100 days $\;= 1 - e^{-1} = 0.1$
Probability that the network component will fail within 1,000 days $= 1 - e^{-1} = 0.63$
Probability that the network component will fail within 10,000 days $= 1 - e^{-10} = 0.99$

This model describes the probability of failure for a single network component. A more generalized model of network failure is given in the simple model below. This model says that the network is connected only when all the nodes and links in the network are working. It also assumes that all the nodes have the same probability of failure p, and all the links have the same probability of failure p'. A final assumption made by this model is that the network is a tree containing n nodes. This model of the probability that the network will fail can be written as: [CAHN98]

$$\text{Probability (failure)} = \left(1 - p\right)^{n} \times \left(1 - p'\right)^{n-1} \qquad (8.47)$$

When the networks involved are more complex than a simple tree, this formula no longer holds. When the network is a tree, there is only one path between nodes. However, in a more complex network, there may be more than one path, even many paths, between nodes that would allow the network to continue to function if one path were disconnected. With the previous approach, all combinations of paths between nodes should be examined to determine the probabilities of link failures associated with a network failure. For a network of any substantial size, this gives rise to many combinations; that is, there is a combinatorial explosion in the size of the solution space. This type of problem is, in general, very computationally intensive.

The following discussion focuses on alternative strategies for estimating the reliability of complex networks containing cycles and multiple point-to-point connections. Except for the smallest of networks, the calculations are sufficiently involved as to require a computer.

Graph reduction is one technique used to simplify the reliability analysis of complex networks. The idea of graph reduction is to replace all or

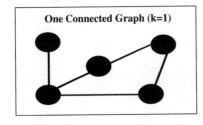
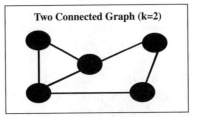

Figure 8.18 Example of one-connected and two-connected graphs.

part of the network with an equivalent, yet simpler graphical tree representation. For example, one type of reduction is parallel reduction. If there are parallel edges between two nodes and the two edges are operational with probabilities p and p', respectively, then it is possible to replace the two edges with a single edge whose probability is equal to the probability that either or both of the edges are operational. Other transformations are possible (e.g., to reduce a series of edges, etc.) and are likely necessary to sufficiently transform a complex network so that its reliability can be calculated. For more information on graph reduction techniques, the reader can refer to [KERS93].

"K-connectivity" analysis is also useful in assessing network reliability. It strives to characterize the survivability of a network in the event of a single-component failure, a two-component failure, or a k-component failure. If a network contains k separate, disjoint paths between every pair of nodes in the network, then the network is said to be k-connected. Disjoint paths have no elements in common. Paths are edge-disjoint if they have no edges in common. Likewise, paths are node-disjoint if they have no nodes in common. An example of a one- and a two-connected graph is provided in Figure 8.18.

It is possible to test for k-connectivity, either node or edge, by solving a sequence of maximum flow problems. This is a direct result of work by Kleitman, who showed that: [KLEI69]

> Given a graph G with a set of vertices and edges (V, E), G is said to be k-connected if for any node $v \in V$ there are k node-disjoint paths from v to each other node and the graph G' formed by removing v and all its incident edges from G is (k–1) connected.

Thus, it is possible to determine the level of k-connectivity in a network by performing the following iterative procedure described in [KERS93]. The computational complexity of this process is $O(k^2N^2)$.

...it is only necessary to find k paths from any node, say v_1, to all others, k–1 paths from another node, say v_2 to all others in the graph with v_1 removed, and k–2 paths from v_3 to all others in the graph with v_1 and v_2 removed, etc. [KERS93]

Thus far, the discussion has focused on link failures. Clearly, if there is a node failure, the network will not be fully operational. However, a network failure caused by a node can not be corrected by the network topology, because the topology deals strictly with the interconnections between nodes. To compensate for possible link failures, one can design a topology that provides alternative routing paths. In the case of a node failure, if the node is out of service, the only way to restore the network is to put the node (or some replacement) back in service. In practice, back-up or redundant node capacity is designed into the network to compensate for this potential vulnerability.

Let us demonstrate how k-connectivity can be used to assess the impact of node failures. One begins by transforming the network representation to an easier one to analyze. Suppose one is given the undirected graph in Figure 8.19. To transform the network, begin by selecting a target node. Then transform all the incoming and outgoing links from that node into directed links, as shown in Step 2 in Figure 8.19. Finally, split the target node into two nodes: i and i′. We connect nodes i and i′ with a new link. All incoming links stay with node i and all the outgoing links stay with node i′. This is shown in Step 3 of Figure 8.19. Once the nodes are represented as links, the k-connectivity algorithm presented above can be used to determine the level of k-node connectivity in the network.

This section concludes with some guiding principles. Single points of failure should be avoided in the network. To prevent a single line failure from disabling the network, the network should be designed to provide 2-k edge-disjoint connectivity or better. This will provide an alternative route to transmit traffic if one link should fail. However, multiple k-connectivity does not come cheaply. In general, a multi-connected network is substantially more expensive than a similar network of lower connectivity. A common target is to strive for 2-connectivity, and to compensate for weakness in the topology by using more reliable network components. However, k-connectivity alone does not guarantee that the network will be reliable. Consider the network illustrated in Figure 8.20. In this network, there are two paths for routing traffic between any two pairs of nodes. However, should the center node fail, the entire network is disconnected. A single node whose removal will disconnect the network is called an articulation point. One solution to this problem is to avoid a design where any one link or node would have this impact. It is apparent that the failure of some nodes and links may have more impact on the network

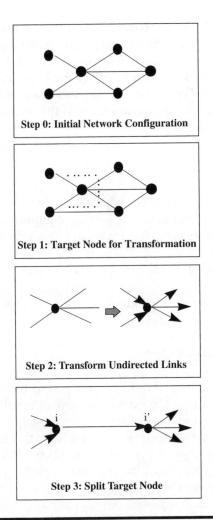

Figure 8.19   k-Connectivity reliability analysis.

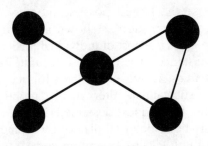

Figure 8.20   Example of two-connected graph with single node point of failure.

than the failure of others. To the extent that is possible, one wants to design networks with excess capacity strategically located in the network. While it is desirable to have excess capacity in the network — for performance reasons so that traffic of varying intensity can be easily carried without excessive delays — one would also want to add network capacity where it will make the most difference on the overall network reliability. [KERS93]

8.6.4 Network Costs

Many of the costs associated with the network are obtainable from manufacturers and organizations leasing equipment and providing services. Important costs associated with the network include:

- Tariffs costs for all network links (these are location-specific charges)
- Monthly charges for other network expenses (including device costs, software costs, etc.)
- Installation charges
- Usage-sensitivity charges

All network expenses should be reduced to the same units and time scale. For example, one time costs — such as installation costs and one time purchases — should be converted to monthly costs by amortization. For example, if a network device costs $9000, this lump sum should be converted to a monthly charge. Thus, a $9000 device with an expected useful life of three years has a straight-line amortized monthly cost of $250.

Likewise, usage-sensitive charges should be converted to consistent time and dollar units. Usage charges, as the name implies, vary according to the actual system usage. When calculating the network costs, a decision must be made as to whether or not an average cost or a worst-case cost calculation is needed. In the former case, the average monthly usage fee, and in the latter case the largest possible monthly usage fee should be used in the final cost calculation as shown in Equation 8.49.

A tariff database gives the cost of individual links in the network, based on their type and distance. It is the most accurate source of information on monthly line charges. However, the fees for accurate, up-to-date tariff information can be substantial. Alternatively, monthly link costs can be estimated using the techniques described in Section 8.2.3. Once the individual link costs have been tabulated, by whatever means, they are summed to obtain the total link operating cost:

$$\text{Total monthly line costs} = \sum_{n=1}^{N} \sum_{i=1}^{I} \sum_{j=1}^{J} O_{nij} * M_{nij} \quad (8.48)$$

where:

O_{nij} = cost of link type n, from node i to node j

M_{nij} = number of type n links between nodes i and j

A similar calculation can be performed for the node costs. The total monthly network cost is computed as the sum of all monthly charges, as indicated below:

Total Monthly Costs = Monthly Line Costs + Monthly Amortized Costs + Monthly Usage Costs + Other Monthly Service Costs $\quad (8.49)$

Notes

1. Capacity: the capacity of a line refers to the amount of traffic it can carry. Traffic is usually expressed in "bits per second," which is abbreviated as bps. The actual carrying capacity of a line depends on technology, because the technology determines the amount and nature of "overhead" traffic, which must be carried on the line.
2. Node: in the context of the network topology, a connection point between lines. It is a very general term for a terminal, a processor, a multiplexer, etc.
3. The notation refers to the number of combinations of n sources and destination nodes taken two at a time, with each set containing two different nodes and no set containing exactly the same two nodes.
4. Tariff: a published rate for a specific communications service, equipment, or facility that is legally binding on the communications carrier or supplier.
5. LATA: A Local Access and Transport Area (LATA) defines geographic regions within the United States within which the Bell Operating Companies (BOCs) can offer services. Different LATAs have different tariff rates.
6. V and H: the Vertical and Horizontal coordinate system was developed by AT&T to provide a convenient method of computing the distance between two points using a standard distance formula.
7. SNMP (Simple Network Management Protocol): protocol defined to work with TCP/IP and establish standards for collecting information and for performing security, performance, fault, accounting, and configuration functions associated with network management.
8. CMIP (Common Management Information Protocol): protocol designed, like SNMP, to support network management functions. However, it is more comprehensive in scope and is designed to work with all systems conforming to OSI standards. It also requires considerably more overhead to implement than does SNMP.

9. IP (Internet Protocol): controls the network layer protocol of the TCP/IP protocol suite.
10. Stochastic process: a process with events that can be described by a probability distribution function. This is in contrast to a deterministic process whose behavior is certain and completely known.
11. State: the state of a queuing system refers to the total number in the system, both in queue and in service. The notation for describing the number of units in the system (including those in queue and those in service) is $N(t)$. Similarly, the notation for describing the number of units in the queue at time t is: $N_q(t)$.

References

[CAHN98] Cahn, R., *The Art and Science of Network Design*, Morgan Kaufmann, 1998.

[GIFF78] Giffen, W., *Queuing Basic Theory and Applications*, Grid Series in Industrial Engineering, Columbus, OH, 1978.

[GROSS74] Gross, D. and Harris, C., *Fundamentals of Queuing Theory*, John Wiley & Sons, New York, 1974.

[KERS89] Kershenbaum, A., Interview with T. Rubinson on April 27, 1989.

[KERS93] Kershenbaum, A., *Telecommunications Network Design Algorithms*, McGraw-Hill, New York, 1993.

[KLEI69] Kleitman, D., Methods of investigating connectivity of large graphs, *IEEE Transactions on Circuit Theory (Corresp.)*, CT-16:232–233, 1969.

[KLIE75] Kleinrock, L., *Queuing Systems*, Volumes 1 and 2, Wiley-Interscience, New York, 1975.

[MINO96] Minoli, D., Queuing fundamentals for telecommunications, *Datapro*, McGraw-Hill Companies, Inc., New York, June 1996.

[PILI97] Piliouras, B., Interview with T. Rubinson on August 6, 1997.

[ROSS80] Ross, S., *Introduction to Probability Models*, second edition, Academic Press, New York, 1980.

[RUBI92] Rubinson, T., A Fuzzy Multiple Attribute Design and Decision Procedure for Long Term Network Planning, Ph.D. dissertation, June 1992.

[SHOO91] Shooman, A. and Kershenbaum, A., Exact graph-reduction algorithms for network reliability analysis, *IEEE Proceedings from Globecomm '91*, August 19, 1991.

[WAGN75] Wagner, H., *Principles of Operations Research*, 2nd edition, Prentice Hall, Inc., Englewood Cliffs, NJ, 1975.

Chapter 9

Business Continuity and Disaster Recovery Planning and Design

9.1 Management Considerations

The purpose of business continuity (BC) efforts is to sustain a company's overall functioning at a predetermined essential level despite catastrophic events that might occur. It also involves putting in place procedures to re-establish full functioning as quickly and smoothly as possible after a major business interruption. The purpose of disaster recovery (DR) is to restore failed infrastructure components so the organization can resume normal operations. Thus, BC and DR management work hand-in-hand to ensure that appropriate policies, practices, and infrastructure consistent with organizational objectives are in place to avert or to recover from a calamity.

The terms "business continuity" and "disaster recovery," as used in industry, have broad connotations and organizational impacts beyond the scope of this book, which is focused on networking and IT systems and technology. A few of the many possible types of business interruptions that might arise include terrorist activities, natural disasters, fire, nuclear accidents, equipment malfunction, software failures, power outages, and stock market glitches. These disasters may result in network and system failures, loss of critical data, or possibly wholesale destruction of the data center or workplace. The extent to which these types of events precipitate a crisis depends in large part on the nature of the enterprise and how it is structured. Broadly speaking, a disaster can be defined as an event that cripples the enterprise's ability to deliver essential functions and services and that threatens its very existence. BC/DR planning involves all-encompassing consideration of the people, facilities, software, hardware, network services, technologies, processes, and procedures required to ensure continuity and restoration of operations in the event of a major disruption or crisis. Thus, a comprehensive BC strategy might include personnel considerations, such as how a labor strike might impact the enterprise. Typically, BC/DR planning involves preparing a *"business resumption plan*, which specifies a means of maintaining essential services at the crisis location; a *business recovery plan*, which specifies a means of recovering business functions at an alternate location; and a *contingency plan*, which specifies a means of dealing with external events that can seriously impact the organization." [SEAR03] An organization that needs to ensure continuous operations[1] during a major crisis must develop a data continuance environment designed to address a wide range of potential failures and associated recovery actions.

Since September 11, 2001, legal and regulatory requirements — arising from the Gramm-Leach-Bliley Act, HIPAA, etc. — have become increasingly strict in mandating BC/DR planning in such vital areas as financial services, health care, insurance, and government. Even enterprises not directly affected by September 11 have a heightened awareness of vulnerabilities arising from reliance upon telecommunications technology and an increased concern with BC/DR planning. [STRO03] The emergence of E-commerce and "24/7" business operations is also fueling the need for BC/DR planning because the costs of downtime can be enormous, as shown in Table 9.1. Industry cognizance of these costs is reflected by growing corporate investment in BC. In large enterprises, BC/DR planning has emerged just behind security as a major corporate priority.

9.1.1 Define the Business Objectives Served by BC/DR Planning

Chapter 1 presented guidelines for defining business objectives, identifying risks and project requirements, and developing implementation strategies and plans. These guidelines are readily applicable to business continuity and disaster recovery planning. Although business objectives will vary with each type of organization and their mode of operation, several essential objectives should be served in the context of BC/DR:

Table 9.1   Average Costs of Downtime for Selected High-Volume Transaction Processing Applications

Business Application	Average Cost $/Hour of Downtime	Industry
Brokerage operations	$6.5 million	Financial
Credit card/sales authorization	$2.6 million	Financial
Pay-per-view television	$1.1 million	Media
Home shopping (TV)	$113,000	Retail
Home catalog sales	$90,000	Retail
Airline reservations	$89,500	Transportation

Source: The Fibre Channel Industry Association, http://www.fibrechannel.org.

- Keep the business alive.
- Establish appropriate balance between costs of implementing and not implementing DR/BC.
- Maintain open channels of communication with customers and employees.
- Comply with fiduciary and legal responsibilities, to include health and safety, insurance, regulatory and legal requirements, and other audit and control requirements.

Industry surveys, including those conducted by Gartner Research Group, indicate that in general, companies are poorly prepared to handle disasters. Although companies may survive a disaster in the short term, a significant number go bankrupt within five years. [WITT01] [OLAV03] Smaller companies, operating on smaller margins and without the benefit of a robust infrastructure are even more likely to go under after experiencing a major business interruption. Businesses should carefully weigh the cost of business continuity and disaster recovery planning and implementation against the potential for losses. The approach a company will adopt will be highly dependent upon the risks to which they are subject and the extent to which they wish to insulate themselves from those risks.

Given what may be at stake, it behooves senior management to be actively involved in seeing that an adequate BC/DR plan is in place. Management should be fully aware of the potential risks of a disaster to the organization when establishing the scope and budget of the BC/DR efforts. This includes cognizance of the organizational and personal liabilities that may arise from a lack of due diligence, and potential business and revenue impacts arising from various types of BC/DR emergencies. In securing senior management support for BC/DR initiatives, it is often helpful to compare the organizational readiness relative to industry standards and to that of competitors. Senior management serves a key role in the sponsorship of BC activities and in demonstrating the organization's commitment to the planning process. Perhaps most importantly, senior management should provide the oversight to see that a team is in place with accountability and responsibility for BC/DR initiatives. Senior management should also assist in ensuring that business objectives and requirements are properly aligned with the IT BC/DR processes and procedures put in place. This is needed to ensure that the plan is properly directed and executed in an emergency. BC/DR programs are most effective when they are coordinated at the corporate level. This helps to ensure a unified approach across multiple business units and disparate IT infrastructures.

9.1.2 Determine Potential Risks, Dependencies, Costs, and Benefits: Risk Assessment and Business Impact Analysis

After a BC/DR team has been sponsored and assembled, one of its first tasks is to identify the risks against which the enterprise must safeguard. Some of the types of vulnerabilities and risks that should be considered include:

- Disaster risk factors
- Financial risk factors (to the market as a whole and to the enterprise)
- Intangible impact risk factors (including negative perceptions, and the loss of customers and revenue)
- Loss of facilities risk factors
- Personnel and labor risk factors
- Productivity risk factors
- Regulatory and liability risk factors
- Security risk factors
- Societal impact risk factors
- Technology-related risk factors

Using a process similar to that outlined in Chapter 1, these risks should be quantified and prioritized according to relevant decision criteria. The prioritization should reflect the probability of risk occurrence and its possible impact on the organization. Some risks may pose greater harm, depending on their timing and duration. These effects should also be factored into the prioritization process. The results of this risk analysis should be documented in a Risk Assessment (RA) Report. The Risk Assessment Report should also evaluate existing controls for mitigating risk exposures, and provide suggestions for reducing the impact of failures, should they occur. Various strategies should be considered to reduce risks in a cost-effective manner, and may involve judicious balancing of insurance, outsourcing, and infrastructure investment. In some cases, the organization may decide to put aside monetary reserves to deal with potential emergencies at the time of their occurrence.

The next task in developing a comprehensive BC/DR plan is to identify mission-critical applications and functions. It is generally neither realistic nor economical to implement a BC plan that guarantees normal operations in any and all crisis situations. Therefore, the organization must determine which functions are essential so that the available budget and resources can be allocated appropriately. This is done in a Business Impact Analysis (BIA), which documents vital business functions, processes, and applications, and their interdependencies. The types of applications that might need to be supported by a BC plan include network communications, e-mail, payroll, customer services, shipping and receiving, and others. There are likely to be a host of business functions that must be safeguarded and provided for during an outage, so this study can be quite extensive in a large enterprise. Gathering this information from all the business units is challenging and requires substantial organizational commitment, particularly when there are substantive differences in how the business units operate. To conduct an inventory of IT infrastructure and applications for the BIA, some sort of asset/configuration management must be in place. Other components of the BIA should include:

- Calculation of the financial impacts of each business function and how these impacts are affected by the timing and duration of failures. This calculation should consider the impacts of peak workload disruptions.
- Assessment of operational impacts arising from a loss of business function.
- Classification of business functions by criticality (i.e., Absolutely Essential Impact, Important Impact, Significant Impact, and Low Impact) and recovery order. Prioritizing the business functions that need to be maintained helps the organization to develop a staged recovery approach.
- Identification of possible points of failure, with an emphasis on single points of failure and those representing the greatest risk of vulnerability. As an aid to this process, it may also be helpful to develop risk scenarios with the associated outcomes and impacts on the enterprise. For example, national, regional, and single-site

disasters might be modeled to determine how this would affect the company's preparedness for an emergency.

■ Specification of recovery point and recovery time objectives for each application:

- *Recovery Point Objective (RPO).* This metric describes the number of lost or uncommitted transactions that can be tolerated in the event of a disaster. It relates to how quickly data must be restored. For example, if an organization has established an RPO of two hours for a given function, procedures must be in place to restore the relevant data to the state they were in two hours ago or less. To accomplish this, backups are needed *at least* every two hours. Any data created or modified at a time less than the RPO (i.e., two hours) that is not copied and duplicated will be lost or will need to be restored by some sort of recovery process. If the application RPO is set to zero (0), then a mechanism — such as synchronous remote copy — must be in place to duplicate transactions as they are processed.

- *Recovery Time Objective (RTO).* This metric describes the elapsed time — in minutes or hours — between a failure and the time at which business operations and transaction processing can resume.

RTOs and RPOs are of particular concern to the financial and healthcare industries because they are regulated. Fines may be levied if recovery times exceed federal or regulatory requirements. For example, the RTO for core clearing and settlement organizations is federally mandated to be two hours. This reflects the fact that in the financial services industry, time is of the essence in performing market functions. Furthermore, interdependencies between firms increase the impact of a failure in any single firm, putting the market as a whole at risk. Substantial infrastructure investment and additional costs must be shouldered to achieve these low RTO/RPOs. However, within these regulated industries, some other lesscritical or non-market-related processes may be allowed more downtime. Knowing which processes can be delayed helps minimize BC/DR costs and simplify emergency planning. In [SUNG03], three recovery tiers are suggested for financial institutions:

Tier 1: applications requiring zero (0) to two (2) hours of RTO, and an RPO of zero (0) hours. This includes applications such as "funds transfer, online banking, trading, and ATM applications."

Tier 2: applications requiring less than twenty-four (24) hours of RTO, and an RPO of zero (0) to two (2) hours. This includes applications such as "e-mail functions and call center software."

Tier 3: applications tolerating greater than twenty-four (24) hours of RTO, and an RPO of twenty-four (24) to thirty-two (32) hours. This includes applications such as "payroll, general ledger, and development or R&D efforts."

Despite an organization's best efforts, in an actual recovery situation, there always remains the possibility of a delay between the expected and actual RTO/RPOs. This delay is referred to as a *business continuity gap*. In the case of regulated industries, there are very strong incentives to ensure that business continuity gaps are as close to nonexistent as possible. *Business resiliency* — that is, the degree to which an organization can adapt and recover from failures, outages, or changes in the environment — is a key requirement in this type of continuous data continuance environment. It involves developing alternative hardware, software, labor, processes, and procedures in the event that swift changes are needed in normal operations. The requirement for business resiliency has far-reaching implications on the overall design and implementation of a BC/DR networking infrastructure. Resilient networks are flexible and tolerate (at least some) alteration, whether necessitated by disaster or changing business conditions. Resilient network infrastructures are the result of ongoing proactive, disciplined, and comprehensive business and IT planning and coordination.

Accurate and thorough RA and BIA are essential in identifying the business resiliency factors that must be accounted for in a BC/DR plan. Business functions, application priorities, and end-user requirements must be carefully worked out and documented. Recovery procedures must be tested regularly and meticulously to verify that staff, technology, facilities, and other resources are up to the task. An emergency is not the time to discover that key functions have not been provided for and that insufficient resources have been allocated to BC/DR. However, because circumstances and systems are always changing, it is difficult to maintain the proper level of accuracy in the RA and BIA. Automation of configuration management, asset management, and other accounting functions can help, but the organization must also institute disciplined controls, policies, and practices to collect the information needed and to keep it current. Thus, the RA and BIA should be viewed as part of a dynamic, and not a static, infrastructure planning process.

9.1.3 Design and Implement BC/DR Plan

After completing the RA and BIA, the next step is to design an actionable BC/DR solution specifically tailored to the organization's business objectives and requirements. Although this book focuses on networking aspects, an enterprise BC/DR should contain provisions (including policies, processes, procedures, and documentation) relating to the following:

- *BC/DR team personnel:* to ensure that a BC/DR team is in place with well-delineated roles and responsibilities. This will include team members who are active before (i.e., they are involved in planning) and after (i.e., they are involved in execution) an emergency situation arises.
- *Customers:* to provide customers with vehicles for interaction, communication, completing important transactions, etc.
- *Employees:* to provide employees with vehicles for interaction, communication, work assistance, etc. A few of the many options that should be considered to provide employees with access to voice or data communications include remote access (e.g., Internet dial-up accounts to let employees access internal corporate systems from alternative facilities or from home), backup storage access, authentication/authorization security measures, VPNs, and wireless and mobile cellular technologies.
- *Facilities:* to put in place processes and procedures to rebuild worksites or to provide alternative backup facilities and locations, etc.
  - If the primary worksite is destroyed, a secondary worksite may be needed. If the secondary worksite is geographically removed from the primary site, the employees from the primary site may not be available to man the secondary site. A secondary workforce may be needed to support the BC/DR efforts. Thus, arrangements for alternative worksites may have substantial personnel implications (in terms of salaries, overtime, training, etc.) that will need to be addressed in the BC/DR plan.
- *Local, state, and federal authorities:* to establish contact and coordination procedures for interacting with appropriate authorities for the various types of disasters that might occur.
- *IT/network systems:* to establish appropriate safeguards, processes, procedures, and backup facilities for data, applications, networks, platforms, and IT system infrastructure components to ensure smooth and timely BC/DR.
  - Reliable networks, highly available networks, and continuous networks warrant different types of resources and recovery procedures. The requirements for each type of network should be reflected in the BC/DR plan. RTO/RPO objectives drive the need for asynchronous data replication, synchronous mirroring, SANs, and other technologies to provide backup of critical data and applications. For example, continuous network requirements

may necessitate redundant network services (such as mesh networks, SONET, etc.) and synchronous data mirroring, while a noncritical network may only need tape backup services.

- IT systems and network administrators should ensure appropriate network management, fault management, performance, configuration management, version control, asset management, and security systems are in place to support BC/DR efforts, both before and after a business interruption occurs. Data and application requirements may necessitate special provisions to ensure system integrity and security are not compromised during a disaster scenario.

■ *Vendors and third-parties* (which might include local and long-distance telcos, outsourcers, managed services personnel, insurers, suppliers, etc.): to provide vehicles for initiating and coordinating BC/DR activities with internal and external personnel. Well-delineated roles and responsibilities relating to BC/DR should be defined for third parties and vendors, including hand-off and migration procedures for a smooth transition of activities and services in an emergency.

To summarize, the key steps involved in developing and implementing the BC/DR plan are to:

■ Define roles and responsibilities of the BC/DR team, including senior management participation, and the use of consultants and outsourcers.
■ Prepare BC/DR project plans with detailed task and resource assignments. A tracking mechanism should be implemented to document schedules and progress against deliverables, completion of milestones, and the budget status of BC/DR tasks and activities.
■ Conduct an RA and a BIA and document the results.
■ Develop a recovery schedule that gives priority to mission-critical applications and services based on the results of the RA and BIA.
■ Investigate BC/DR solution alternatives and strategies. This most likely will involve working with vendors and third parties to augment the organization's resources in a disaster.
■ Identify insurance requirements and implement risk-reduction strategies.
■ Create comprehensive organizational policies and procedures supporting the BC/DR efforts.
■ Develop and implement IT/networking processes, procedures, and practices. As discussed previously, this involves addressing a host of issues relating to the entire data continuance environment. With

respect to network services, three major strategies for BC/DR are to outsource, duplicate facilities at one or more sites, and build redundancy into the network infrastructure. These strategies are discussed in more detail in the remainder of this chapter.

- Develop and implement security processes, procedures, and guidelines to preserve the integrity of corporate data and information. This should address both electronic and paper data sources that may be affected in a disaster.
- Finalize preparation, documentation, and dissemination of a formal business resumption plan, a business recovery plan, and a contingency plan to personnel assigned to BC/DR efforts. These plans encapsulate and incorporate previous efforts and represent officially sanctioned and approved BC/DR policies, procedures, and practices.
- Develop and disseminate employee training and awareness materials relating to BC/DR. This may also involve a broad-scale corporate skills assessment initiative to determine employee competencies and where they need to be augmented to support BC/DR efforts.
- Develop and implement processes and procedures to test the effectiveness of BC/DR planning. This should include rehearsals and disaster drills. This is necessary to ensure that proper measures are in place to respond to an emergency, and that employees are aware of and familiar with them.
- Perform periodic auditing and verification to identify potential problems in the BC/DR response and to gauge the overall effectiveness in the BC/DR program. Ideally, this should be done by an independent third party. This evaluation might assess the degree to which management is involved in the entire BC/DR process, and the presence of proper documentation relating to policies, procedures, network design, configuration management, application software, etc. A detailed audit and verification process is needed to ferret out potential business continuity gaps that need to be corrected before it is too late.

As can be inferred from the previous discussion, the BC/DR process involves creating extensive documentation. In a real emergency, this documentation will be crucial to the overall success of the DR response. Thus, special attention should focus on keeping the BC/DR plans current and at the appropriate level of detail. Missing documentation, unclear internal and external communication channels, outdated plans, and mismatched production and recovery network configurations will severely compromise BC/DR efforts.

Cost is one of the most widely cited reasons why organizations fail to implement a comprehensive and well-documented BC/DR program. However, short-cuts are not advisable. BC/DR is an expensive endeavor and requires significant organizational discipline and commitment to succeed. In a difficult economic climate, it is hard for IT groups to muster the resources for an adequate BC/DR data continuance environment. Consequently, it is more imperative than ever to work smart to maximize the pay-off of IT initiatives. Throughout this book, there is discussion on ways to minimize network complexity (which include consolidation, standardization, scalability, use of IP and client/server-based architectures, etc.) and to maximize network security. Good business and sound network design practices expedite normal operations and complement BC/DR efforts.

9.1.4 Keep Informed about BC/DR Standards and Recommendations

There are numerous organizations and standards bodies involved in developing recommendations and tools (including templates, examples of BC/DR plans, etc.) for BC/DR. For example, in the United States, two important organizations include the National Fireman's Protection Agency (NFPA) — which has produced the USA NFPA 1600 Standard on Disaster/Emergency Management and Business Continuity Programs — and the Federal Emergency Management Agency (FEMA) Disaster Planning for Business and Industry. Other countries have developed standards tailored to meet their needs — for example, the Canadian Standards Association CAN/CSA-7731-M95, Emergency Planning for Industry, and the British Standard BS 7799 Standards in Information Security Management. For additional sources of information on BC/DR planning, the interested reader is referred to the following:

■ Federal sources:
  - Federal Emergency Management Agency (FEMA). This independent federal agency has offices across the United States, with approximately 4000 "standby disaster assistance employees who are available to help out after disasters … FEMA works in partnership with other organizations that are part of the nation's emergency management system. These partners include state and local emergency management agencies, 27 federal agencies and the American Red Cross." The FEMA Web site is located at http://www.fema.gov.

- National Fire Protection Association (NFPA). This international organization's mission is to develop recommendations, codes, standards, research, and training on emergency management, business continuity, and disaster recovery. The MFPA Web site is located at http://www.davislogic.com/NFPA1600.htm. Detailed information on the NFPA 1600 — Standard for Disaster/Emergency Management and Business Continuity Programs can be found at http://www.nfpa.org/Codes/NFPA_Codes_and_Standards/List_of_NFPA_documents/NFPA_1600.asp.
- The following document provides a good overview of the management aspects of BC/DR planning: "Identification of the Core Competencies Required of Executive Level Business Crisis and Continuity Managers," by Greg L. Shaw and John R. Harrald, *Journal of Homeland Security and Emergency Management*: Vol. 1: No. 1, Article 1, 2004, http://www.bepress.com/jhsem/vol1/iss1/1.

■ Industry sources:
- Diaster-Resource.com. This Web site provides an "online DISASTER RESOURCE GUIDE… to help you find information, vendors, organizations and many resources to help you prepare for (mitigate) or recover from any type of natural or other type of disaster." It also provides contract information for federal and state agencies dealing with BC/DR. (This can be found at http://www.disaster-resource.com/content_page/govmnt.shtml.) The Web site is located at http://www.disaster-resource.com/.
- The Business Continuity Institute (BCI). This international organization, represented by 19 countries, is chartered "to promote the highest standards of professional competence and commercial ethics in the provision and maintenance of business continuity management services." The organization sponsors numerous conferences, training and workshops, books, and publications on BC. The BCI Web site is located at http://www.thebci.org.
- DRI International (DRII). This international organization works in concert with BCI to provide training and certification programs in BC/DR. It produces an annual Professional Practices document (which can be downloaded free at http://www.drii.org/displaycommon.cfm?an=2). This document is used as a preparation guide for DRII certification examinations, and covers the subject areas listed below:
  ■ Risk evaluation and control
  ■ Business impact analysis
  ■ Developing business continuity strategies

- Emergency response and operations
- Developing and implementing business continuity plans
- Awareness programs and training
- Maintaining and exercising the business continuity plans
- Crisis communications
- Coordination with external agencies

  – Securities Industry Association (SIA) Business Continuity Planning Committee. This organization was formed in response to the attacks on the World Trade Center on September 11, 2001, and is represented by "80 firms, industry utilities, exchanges, and other organizations...." The group is divided into seven subcommittees that address major business continuity issues affecting the securities industry." White papers, reports, and recommendations relating to BC/DR in the financial industry can be accessed at http://www.sia.com/business_continuity/html/reports.html.

  – Business Continuity Planners Association (BCPA). This is a nonprofit association of "business professionals responsible for, or participating in, business recovery, crisis management, emergency management, contingency planning, disaster preparedness planning, or a related professional vocation … The mission of BCPA is to provide a professional and educational environment for the exchange of experience, dissemination of information, [and] professional growth..." The BCPA Web site is located at http://www.bcpa.org/.

  – International Association of Emergency Managers (IAEM). This is a "nonprofit educational organization dedicated to promoting the goals of saving lives and protecting property during emergencies and disasters." The IAEM Web site is located at http://www.iaem.com/index.shtml.

9.2 Technical Considerations

According to a recent research report from Gartner Dataquest, on "average 40 percent of downtime is caused by application failures such as performance issues or 'bugs,' 40 percent by operator error, and approximately 20 percent by system or environmental failures. About 60 percent of these system or environmental segment failures are caused by hardware problems. Overall, less than 5 percent of application downtime results from disasters." [EQUA03] These results indicate that most outages are not caused by random disasters, but by preventable or predictable events. This means that careful planning and best practices can and should address many of the events the organization is likely to encounter on a routine basis. Developing BC/DR strategies to deal with unlikely, devastating events is far more challenging. This is due to the inherent difficulty of predicting the future, particularly as is relates to improbable events, and the high associated cost to provide adequate measures of protection (that may or may not ever be called into play) against far-reaching catastrophes. The sections that follow discuss technical considerations for fashioning networking and IT solutions for BC/DR. Three R's are a recurring theme for fashioning a BC/DR strategy: resiliency, redundancy, and replacement. Resiliency refers to providing alternative ways of performing the same service within the same infrastructure. Redundancy refers to having reserve equipment (i.e., circuits, computers, servers, routers, hot spares, etc.) available to support normal operations should primary network components fail. Replacement refers to having duplicate facilities available as an alternative base of operation should the primary worksite become disabled.

9.2.1 Facilities Solutions

This section discusses basic strategies used to achieve redundancy and replacement of data center and worksite facilities.

For larger enterprises, particularly those operating across a geographically dispersed area, it makes sense to consider operating multiple, networked data centers. Although this is more complex and expensive than operating a single data center, it protects the enterprise against a single point of failure. However, to be effective in supporting BC/DR, the networking between data centers must be robust and diverse. It should support such features as automatic traffic rerouting, self-correcting capabilities in the event of a link or other network failure, and alternative means of connectivity that can substitute for each other in the event of an emergency. For example, a private line network might be supported by a VPN backup that could be used in the event of a disaster (this is discussed in more detail in Section 9.2.2).

Cisco and a number of other vendors offer intelligent networking equipment to maximize the utilization of multiple data centers and to facilitate a switch-over should one site fail. Intelligent switch- or router-based devices[2] can be used to perform a multitude of tasks, including: load balancing, enhanced server security, dynamic site selection, traffic redirection, and a variety of network management functions. Load balancing is useful in distributing large volumes of traffic among a number of data centers or servers. This can be accomplished by defining virtual server clusters and by using load balancing tools to maintain even traffic distribution between them. Some of the more common load balancing

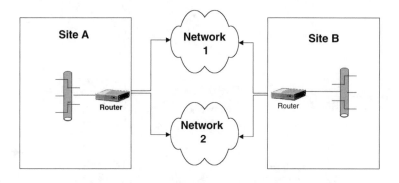

Figure 9.1   Dual network configuration for disaster recovery.

algorithms include round robin, adaptive, least connections, and fastest response.[3] Many load balancing products have built-in denial-of-service protection and other security features. E-commerce and online sites may use dynamic site selection features to direct Internet or intranet traffic to servers based on loading factors or other predefined metrics to help maintain continuous operations.

Dual networking is a common BC/DR strategy, particularly in environments that require high availability and continuous data replication. This is illustrated in Figure 9.1. In this scheme, a secondary network is used to back up a primary network. If a failure is detected in the primary network, traffic is automatically rerouted to the secondary network. Network diversity between the two sites is achieved by employing intelligent Frame Relay, ATM, or private IP-based solutions employing SNMP and RMON type 1 and 2 network management monitoring. Service providers promote this approach by offering diversity and discounted pricing for secondary PVCs, ports, and local access. As shown in Figure 9.1, if there is a failure in the (e.g., Frame Relay) communications between sites A and B, a transparent switch can be made from the Network 1 PVC(s) to the Network 2 PVC(s). The network start and termination points are the same for Networks 1 and 2, thus simplifying the diversity solution.

To ensure a greater degree of reliability in the communications, some enterprises incorporate an additional Site C location into their dual-network infrastructure. This is illustrated in Figure 9.2. If communications are lost between locations A and B, a disaster recovery scenario can be initiated to reestablish communications between A and C, and from C and B. Note that this approach is more expensive and complex in that it requires backup data center connectivity between sites A and C, and between sites B and C. The highest level of business continuance (with the lowest RTO/RPO) is ensured by complete data center mirroring, in which both the production and backup data center are fully synchronized. This

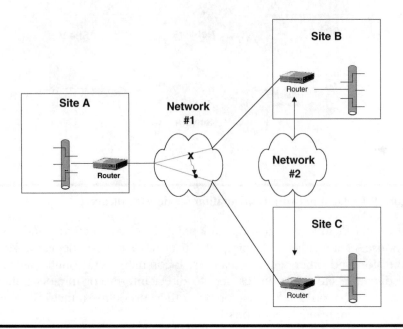

Figure 9.2 Data mirroring configuration for disaster recovery.

solution is usually implemented using DWDM MAN and data storage replication technologies.

Even if all the above facilities and networking provisions are in place to ensure robust, diverse connectivity between sites, this does not guarantee that the data centers will be able to support BC/DR functions. Each data center should be sized to support the *total* peak load for all the data centers to which it serves as backup. Otherwise, the data center will lack sufficient capacity to support the workload it might have to take on in the event of a catastrophe. Furthermore, the platforms, application software, system components, and even certain low-tech conditions must be compatible across all data centers. For example, if a data center must produce billing statements on a specific form and type of preprinted paper that it would not ordinarily stock, it should nonetheless have sufficient supplies available to do the job as required in an emergency situation. BC/DR requires considerable attention to details of this nature.

Hot sites are another alternative for BC/DR. A hot site is an office or data center that has been fully configured, wired, and furnished to support the enterprise's business operations, and remains in standby mode in the event that it is needed. If an enterprise's data processing center becomes inoperable, its data processing and other operations are moved to the hot site. Hot sites are intended as temporary solutions, typically for a month or two, or less. Leading hot site vendors include IBM Business Continuity and Disaster Services, SunGard Recovery Services, and Hewlett-Packard Business Recovery Services. Hot sites can be expensive;[4] and in practice, it can be difficult to manage a smooth transfer of operation to the hot site in a time of emergency. For these reasons, many large companies prefer to maintain multiple data centers under their own control, augmenting their data center with synchronous data mirroring[5] and other capabilities to support high or continuous availability applications. However, as the requirement for high-availability systems becomes more prevalent and regulated, there is growing demand for hot sites. This is particularly true in the financial services industry, which accounts for over 65 percent of all hot site recoveries. [SCHR03]

To make their offerings more attractive, commercial hot site vendors are offering new services such as PC/LAN electronic data vaulting[6] and mobile/portable office space. This latter option is designed for smaller business operations and is much cheaper than a corresponding fixed hot site solution. Mobile hot sites are set up in trailers, while portable hot sites are constructed when and where they are needed.

In *US Hot Site Market Analysis & Forecast* [SCHR03], Tari Schreider presents a number of interesting statistics on hot site usage. An average of 40 companies per year use commercial hot sites to recover from disasters typically resulting from, in order of prevalence, loss of power, hardware problems, and fire. Almost 44 percent of these recoveries were precipitated by regional events that simultaneously affected multiple hot site subscribers. However, all subscribers were successfully serviced by their hot site provider and none were denied access to their hot site because of excessive demand. This report also points out that roughly half of the companies that use hot sites do so for less than 72 hours.

A *cold site* is yet another type of BC/DR solution. Conditioned office space is rented or leased from a commercial third-party vendor. However, the enterprise (i.e., subscriber) must provide, install, and maintain all the equipment needed to support its DR operations. A cold site is less expensive[7] than a hot site, but the enterprise must also put in considerably more time and effort to get it ready for use. To aid this approach, some vendors are offering shipping options and other arrangements so subscribers can quickly install leased computer and networking equipment in a BC/DR situation.

9.2.2 Network Solutions

Good network design, by its very nature, provides built-in protection to minimize the impacts of link and network equipment failures. This section summarizes major ways that networking diversity is achieved so that these considerations can be factored into a BC/DR plan:

- *Carrier diversity.* For critical WAN networks, carrier diversity may be desirable as an extra measure of resiliency and redundancy. WAN networks are, by nature, long distance (with the possible exception of regional WANs). The primary providers of long-distance service are the IXCs, such as AT&T, MCI, Sprint, and Qwest. If carrier diversity is needed, then contracts with multiple vendors will need to be negotiated to provide these services.
- *Local loop diversity.* Generally, local loop diversity is not a given and must be specially arranged with the local provider (e.g., Verizon, SBC Communications, BellSouth, etc.). Local loop diversity provides alternative traffic routing to the CO (or central office). Smaller organizations must present a forceful case to their local carrier to make arrangements for loop diversity. Arrangements for local loop diversity are somewhat easier for larger enterprises to negotiate because of their larger buying power.
- *Layer 2 WAN access technologies.* OSI layer 2 access technologies provide physical addressing and establish a point-to-point link for data transport across a physical network. The use of more than one access protocol complicates the physical and management aspects of the networking infrastructure. However, this complication may be offset by the benefits of having more than one access protocol available to support communications in an emergency. Many well-established, attractively priced options for incorporating multiple access technologies into the network infrastructure are available. Their use in supporting BC/DR is standard industry practice. For example:
  - POTS (plain old telephone service) can be backed up with spare lines or secondary POTS lines provided by other carriers. Cable modem and cellular/wireless communications can also be used to replace POTS (and in some cases they are replacing POTS altogether as a primary service). Legacy X.25 is sometimes used, particularly in remote parts of the world, as a backup to dial-up service.
  - Frame Relay networks are increasingly being backed up with public IP-based VPNs with security or with a private IP-based network. (This practice is in widespread use as a migration strategy to MPLS; however, it also supports BC/DR.) Alternatively, ISDN can be used to substitute for Frame Relay. Yet another option is to make provisions for additional backup PVC circuits in the network.
  - DSL can substitute for cable modem or analog dial-up services.
  - VSAT (Very Small Aperture Terminal) supports high-bandwidth voice, data, and video communications over a private

infrastructure. VSAT satellite communications can operate in remote areas that would otherwise be difficult to reach with land lines. VSAT can be tailored to meet individualized home and business requirements for security, reliability, and capacity, and it is relatively quick and easy to set up and operate. As such, it is well-suited as a backup solution for such entities as retail stores and broadcasting companies.

■ *Network layer 3 IP design.* OSI layer 3 technologies are responsible for the routing of data from point A to point B across a WAN, using such protocols as OSPF, BGP, and Cisco's IPGRP/EIGRP. Layer 3 protocols function similarly to a postman. When encountering a blocked road along his mail route, a postman needs to know alternative routes to get to the same destination. Distance vector and link state protocols are two major layer 3 approaches for routing diversity. Using a cost matrix, distance vector protocols (such as RIP, BGP, and IGRP) look for the shortest (i.e., cheapest) way to reroute traffic. Link state protocols (such as OSPF) consider the link speed of alternative routes to avoid using a shorter route that is slower. This is akin to the choice of traversing a single-lane road over a shorter distance or a four-lane highway over a longer distance. EIGRP is a distance vector protocol with link state attributes. Private networks using OSPF are giving way to public BGP networks as enterprises come to rely more upon public IP technology for communications. BGP works in conjunction with TCP to provide a means to exchange routing information between gateway hosts on the Internet. BGP4, the latest version of BGP, is designed to compensate for serious limitations of IPv4. In the early 1990s, IPv4's deficiencies — including exhaustion of class B addresses[8] and explosion in the size of routing tables held in memory by routers — threatened the very survival of Internet growth. To provide efficient IP routing and to help overcome IPv4 address limitations, BGP4 uses *Classless Interdomain Routing* (CIDR)[9] to implement *route summarization* and *address aggregation*. To understand how this is accomplished, some background on IP addresses is included here. As originally implemented on the Internet, the first byte of the IP address[10] was used to specify an address class, which in turn implicitly determined the destination node. By convention, Class A subnet masks were designated by 255.0.0.0, Class B subnet masks by 255.255.0.0, and Class C subnet masks by 255.255.255.0. Thus, by examining the first byte of an IP address, a router could determine if traffic was routed to a Class A, Class B, or Class C destination. In contrast, CIDR explicitly includes subnet information with the routing information, thus eliminating the need to infer an address class. This frees subnet designations from being tied to a particular class, allowing them to assume an arbitrary length and designation.[11] For example, under BGP4, an IP address of 255.162.226.0/24 refers generically to all addresses that match on the first 24 bits. The remaining 8 bits of the IP address can assume any value. Routing tables with designated address/mask pairs are used by routers to determine how to direct traffic to the next network hop until the destination is reached. Using CIDR, a single address prefix can designate a block of IP addresses, reducing the size and number of entries in routing tables, simplifying Internet connection, and reducing the load on routers. This is referred to as *route summarization*. Route summarization is analogous to POTS phone numbers, in which the first three (3) digits dialed are sent to a switch that determines whether the call should be forwarded to a local or an outside region. This layer 3 aspect of BGP4 provides economies of scale over layer 2 protocols, in that the first few digits of the address can be used to quickly ascertain where data is to be routed, allowing extraneous local network regions to be bypassed. CIDR specifies three hierarchical routing levels, with provisions for additional levels if need be: (1) site level (through aggregation of all site addresses into a single routing prefix), (2) network service provider level, and (3) continent boundary level. CIDR also provides mechanisms for network administrators to secure, tune, and configure network routing. For example, BGP4 allows the configuration of destination addresses that are not used publicly. Because BGP and BGP4 have been successful in overcoming weaknesses in IPv4, for the time being, there is limited economic incentive for vendors or end users to initiate major conversion efforts to IPv6, particularly because IPv6 is not backward compatible with IPv4. When ISPs took hold in large numbers in the late 1990s through the early 2000s, BGP's advantages in supporting the route summarization and hierarchical infrastructures needed for advanced communications made it a clear choice, and it has become a *de facto* standard.

■ *Network services resiliency.* This encompasses a wide variety of products, services, and tools. The selection of these is driven by the enterprise philosophy and core competencies. It can be likened to differences between an audiophile and a college student in their approach to buying stereo equipment. An audiophile might opt for performance over cost, and purchase finely tuned separate stereo components. A college student, on the other hand, might opt for cost and space savings over performance, and purchase an integrated stereo system. Similarly, the network designer might choose to configure key components — such as a router — to support a variety of network management tasks. However, because the router's main function is to route traffic, eventually this may be comprised if processing demands are too great as a result of doing too many other tasks. This extra workload can be compensated for by augmenting the network with devices that perform specialized functions. Although this approach provides the best opportunity to tune and optimize the performance of each network component, it also generally results in higher costs and complexity. Similar considerations apply when deciding upon outsourcing and other third-party arrangements. A few of the choices for adding network resiliency include:

- *Outsourcing support.* This is available from a variety of vendors — including Novell, Cisco, IBM, Hewlett-Packard, AT&T, and MCI, to name a few — for a gamut of products and services. This includes, but is by no means limited to, Web hosting, network monitoring, incident management, redundant equipment, data storage, and network backup services.

- *DNS/DHCP products.* These are needed to help network administrators manage and automate network maintenance functions in a TCP/IP environment. For example, DNS/DHCP tools are available to perform a variety of network management services, such as (1) maintenance of DNS[12] servers with up-to-date addressing and configuration information, and error checking at the time of data entry; (2) resolution of addressing conflicts; (3) backup and recovery services for DNS servers; (4) maintenance of uninterrupted DHCP[13] services (this is important because if the DHCP service is interrupted, network devices will lose their IP addresses and connection); and (5) assistance in creating logical, as opposed to physical, network groups for improved maintainability and resiliency.

- *Additional network management and security components, as needed.* An example of this is the use of Secure Socket Layer (SSL) to isolate a LAN from the WAN and to secure Web-based communications. SSL provides site and user authentication, and encryption to ensure the privacy and integrity between computer connections between a Web server and a browser. The highly CPU-intensive nature of RSA public key cryptography used in the SSL handshake between server and user can have a detrimental impact on the performance of high-traffic, secure Web servers. In response, vendors developed SSL accelerators and offloaders to handle the RSA public key exchange. In addition to the encryption functions performed by SSL accelerators, SSL offloaders provide bulk data encryption (when receiving SSL traffic) and decryption (when transmitting SLL traffic) capabilities to offload these tasks from Web servers and content switches.[14] To ensure operational continuity in high-availability networks, hot, standby SSL servers, accelerators, and offloaders can be configured. In the event a primary SSL component fails, the failover components can seamlessly assume operations. SSL was designed to secure Web-based communications between a browser and server; however, it does not deal directly with the encryption of e-mail and IM chats, and other personal computer communications, which may be sent in cleartext. Typically, PC applications are secured end-to-end across a network using IP VPNs with IPSec[15] (and DES3 encryption and/or PKI[16] encryption). However, with this approach, IP VPN users typically need a separate client on their resident PC devices to initiate a secure session. This requires an initial download of client software and additional periodic software updates via push technology. With the emergence of Voice-Over-IP using SIP, SSL may provide an easier means to secure Internet communications. Because SSL is integrated into every browser and Internet Web server, the end user does not have to download SSL client software onto his or her machine. Once all the enterprise communication (i.e., in the form of pagers, telephones, etc.) becomes Web based, SSL can facilitate secure, truly mobile communications. If this occurs, the increased reliance upon SSL will have to be factored into the BC/DR plans. This might result in the use of additional content switching devices, SSL offloaders and accelerators, and failover standby equipment.

9.2.3 Data Storage Solutions

This section presents some of the most commonly used data storage alternatives for BC/DR. When evaluating the suitability of these alternatives, it is important to keep in mind the nature of the business to be protected. For example, a distributed, high-volume client/server environment might require near-real-time backup capabilities, while a small business office might only require overnight backup capabilities. Requirements relating to the backup data storage location, timing, and scope requirements must be determined before a technology solution is designed and implemented. This involves asking where backup data should be created and stored. Should backup data reside locally for ease and convenience, or should it be stored remotely as an added security precaution against a catastrophic facilities failure? It also involves asking how frequently backups must be made and the speed with which the data must be restored should a failure occur. The answers to these questions must be documented in the form of RTO/RPOs for each application. If an application is not very critical or time sensitive, then perhaps a batch-oriented recovery process will suffice. If an application is more time sensitive — with an RTO of eight hours or less — then *shadowing* may be needed. In shadowing, as updates are made to a production data/file source, they are also applied to a duplicate (i.e., shadow) copy of the data/file. This is an asynchronous process, in that updates are made to the copy after they have been made to the original. Thus, the production and the shadow versions of the data/file may be slightly out of synchronization at any given point in time. The replication and recovery process should not exceed the level set by the RTO/RPO. If an application is extremely time sensitive and cannot afford a single lost transaction, then synchronous replication or *mirroring* may be needed. When mirroring is used, all updates to the production data/file are synchronously applied to a duplicate copy at a remote site. This approach provides the fastest recovery with the least possible loss of data. However, due to physical limitations of the technology,[17] mirroring is only feasible if the distance between the production data center and the backup sites is 100 to 200 kilometers or less. This has significant implications on the financial industry because it severely limits the distance over which synchronous mirroring can be supported. Since September 11, 2001, the financial industry has sought to establish backup sites that are sufficiently far from primary processing sites so as to be immune from the effects of a similar terrorist attack. Given the constraints of technology, some financial firms are employing synchronous backup at a secondary data center located within the 200-kilometer limit, and beyond this they are using asynchronous shadowing at yet another backup data center. A final question to ask is: what level of backup is needed to satisfy RTO/RPO goals? Should the replication be performed at the device storage level, at the application level, at the host level, or at the data center level? The answers to all these questions will guide the analyst in selecting the appropriate data storage technologies, as discussed below.

9.2.3.1 Tape Backups

Tape backups have long been a mainstay in the corporate BC/DR repertoire. This is one of the cheapest and easiest solutions to implement, in large part because the technology is so old and well-established. Tape backups are particularly well-suited to data archival and restoration of corrupted data. Data integrity is an ongoing enterprise challenge because there are so many ways it can be compromised in the course of daily operations and in a disaster scenario. Archived tape backups provide a way to cleanly reload and roll back data to a predetermined point in time. Applications that have a low transaction volume, are not mission critical, and are not time sensitive are particularly good candidates for tape backup as a BC/DR solution.

If tape backups will be used for BC/DR, then organizational policies and procedures must be formalized with respect to the specific applications and files that will be backed up, the type of backups to be made (differential, incremental, or full), scheduling and frequency of the backups, and storage and handling of the tapes. Ideally, an automated, centralized catalog and library system should be in place to document the tape contents, and other relevant system parameters and information to expedite handling tape backups in a recovery situation. Robots can also be used to automate tape handling and management.

One of the more common backup configurations involves the use of dedicated tape drives attached to a personal computer, host, controller, or server via a parallel SCSI[18] connection. Although this approach does not promote resource sharing, it provides convenience and maximum performance.

Tape drives can also be configured on a LAN, with or without NAS (Networked Attached Storage). In this case, both the tape drive(s) and the NAS will contribute to and share the LAN bandwidth. When tape drives are used on a LAN, they can seriously impact network performance, particularly if large amounts of data must be copied. If this is the case, it may become necessary to schedule the backups so as not to interfere with normal LAN traffic.

Another type of tape backup configuration employs an application server to back up data directly onto tape, bypassing the LAN altogether. When a shared-storage option is used with this configuration, an application server must manage the backup process and the use of shared resources (for example, tape drives shared among other application servers must be reserved, etc.). Alternatively, a dedicated storage option can be used, in which the application server is assigned a dedicated tape drive for backup purposes. While both of these direct backup methods minimize the traffic load on the LAN, they also substantially increase the work load on the application servers. These effects must be considered when deciding on the best tape backup strategy.

To relieve congestion, more and more organizations are using SANs to offload designated file transfer and backup operations entirely from the LAN. Tape drives play an integral role as inexpensive storage media and can coexist with other storage devices (such as RAID) configured as part of a SAN. This is discussed in more detail in the sections that follow.

Depending on the risks and requirements of the organization, the tapes can be stored off-site for safekeeping. Off-site tape storage can take many forms, including a bank safety deposit box, a remote corporate location, or climate-controlled commercial storage (from a provider such as Iron Mountain). Electronic Vaulting (EV) is a relatively new service being used to improve the accessibility of data for BC/DR. EV is a third-party service for archiving and storing data at a secure, remote location on dedicated direct-access storage device(s). EV provides quick retrieval of files and data, as well as automatic backup (including full, incremental, and differential). To use this service, electronic vaulting software must be installed at the subscriber site on personal computers, LANs, or mainframe computers. The electronic vaulting software provides a user-friendly GUI (Graphical User Interface) for performing various backup, recovery, and archival tasks. A communications link (i.e., via Internet, Frame Relay, ISDN, etc.) must exist between the subscriber and the EV vendor. This communications link should be sized to adequately support the data transfer rates needed for BC/DR. Typically, a backup operation is first performed on selected files and data onto tape or some other electronic media. File compression and encryption are generally used in this process. This backup, in turn, is loaded onto DASD into the electronic vault. From this point forward, backup, archive, and retrieval operations are performed according to predetermined schedules or as needed. The major considerations involved in deciding to use this service are the costs (relating to EV fees, communications costs, and DASD/RAID costs) and timing (relating to the inherent latency associated with data upload and download from the remote EV).

There are significant reliability and time drawbacks to using tape backups in a recovery process. First, tape backup is an inherently slow process. An enterprise that stores lots of data on tape must be prepared for a long recovery time. "As the amount of data stored increases, the bandwidth limitations (about 17 Mbytes per second) of tape means that it is only well-suited to applications where acceptable recovery times can be counted in days rather than minutes or hours." [CISC03B] Second, tapes are subject to failure, such that a complete loss of data may result. The tape may be (logically) corrupted at the time the backup is created, due to system error; or it may be (physically) damaged, without any outward sign of failure until such time as it is needed to restore data. Tape backup serves a valuable purpose in the BC/DR strategy of many corporations; however, it is best thought of as one piece of a larger backup solution, and not a total solution.

9.2.3.2 Network Attached Storage (NAS)

Network Attached Storage (NAS) provides disk storage to augment that of workstations and servers on a LAN,[19] and is used to relieve other servers

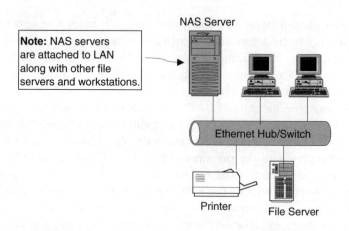

Figure 9.3   Example of LAN with NAS.

(e.g., application, database, etc.) on the network from the burden of file access and transport. This is shown in Figure 9.3. This arrangement reduces or eliminates the need for multiple direct access storage (DAS)[20] devices, and allows end users working with a heterogeneous mix of computers and software to access the same files and data.

An NAS device has its own assigned IP address, and consists of hard disk storage, software, and an NAS server. RAID (Redundant Array of Independent Disks)[21] storage is often used, although single-disk and other types of storage are also possible. Typically, Web browser-enabled software is used to configure and map file locations, and to perform other NAS network management functions. The server is dedicated exclusively to managing NAS file access and file sharing activities on the network. This limited functionality is referred to as a thin server approach. The NAS operating system is usually stripped to the essentials and is optimized to run on specific hardware and software, thus making it less prone to system failures and security breaches. This also makes it easier and faster to diagnose and fix a problem, should one arise. NAS devices are available for use with a wide variety of operating systems, including: off-the-shelf desktop or server operating systems (e.g., Windows, UNIX, and Linux); embedded operating systems (e.g., Windows CE); and real-time operating systems (e.g., VxWorks and QNX). To provide heterogeneous server access, most NAS servers support CIFS[22] and NFS[23] and multiple network protocols (which typically include Microsoft's Internetwork Packet Exchange and NetBEUI, Novell's Netware Internetwork Packet Exchange, Sun Microsystems' Network File System, and TCP/IP).

NAS is playing an increasingly important role in BC/DR because of its network efficiencies and failsafe features. For example, NAS devices can be added and removed as needed, without shutting down other servers on the network. NAS is designed to facilitate file handling for demanding applications, such as read-intensive applications (e.g., streaming video, MP3 audio, e-mail, and Web services) that must quickly respond to many simultaneous requests for server access. The NAS thin server architecture allows the NAS to act as an intermediary between the application server and the data while maintaining its independence from the application. NAS systems are also used to perform remote asynchronous data replication in background mode for applications that must remain online. There are no distance limitations with this type of replication. To further minimize the potential disruption of a system outage, high-availability NAS devices can be configured with an automatic switch to a secondary NAS server that remains on standby until needed. The process of automatically switching to a mirrored server (or data center or any other network component) is referred to as *failover*. When successfully executed, system failovers are transparent to end users.

A variety of security measures can be used in conjunction with NAS to facilitate disaster recovery. For example, RADIUS[24] authentication and authorization can be used to control dial-up user access to NAS resources. In a typical configuration, a remote user dials into an access server — via an ISDN line, remote IP dial-up, or some other form of dial-up service. The access server, in turn, sends an authentication request to a separately configured RADIUS server. The RADIUS server performs user authentication and authorizes network access based on prestored permission and configuration information. Centralization of the authentication process in this manner is inherently more secure and scalable than a distributed process (with multiple points of entry and maintainability). Each NAS device is configured with a list of allowed users and their associated attributes, which the RADIUS server must authenticate and authorize. To use RADIUS, the remote user must be running PPP (over an ISDN or dial-up interface), and have PAP,[25] CHAP,[26] or some form of caller ID enabled. The standard process for RADIUS authentication follows these major steps:

- The end user dials into and establishes a PPP connection with a network access server.
- The end user and access server engage in an authentication process, usually via CHAP or EAP,[27] which involves the exchange of authentication information.
- The access server sends an authentication request — including information about itself, the access port, and an encrypted password — to the RADIUS server. The password is encrypted using a secret key shared with the RADIUS server.
- The RADIUS server checks the request against the user information on file, and sends either an acceptance or a rejection of the authentication request to the access server.

■ Filters (such as the use of router or firewall-based access control lists) and other security measures can also be put in place to further enhance security and control of the network access.

Compared with traditional application servers, file servers, and DAS, NAS technology offers a number of advantages, including:

■ *Reduced management costs.* NAS allows data storage to be consolidated from multiple direct access servers, which is easier to manage and troubleshoot.
■ *Improved system scalability.* NAS allows additional storage to be added easily and quickly as needed. Networked storage allows enhanced resiliency and utilization across multiple servers.
■ *Improved security.* NAS supports the establishment of user permissions, attributes, and security policies. It is designed to work in conjunction with other network security features — such as RADIUS, etc. — to maintain network integrity.
■ *Improved disaster recovery.* NAS supports automatic backup and failover features to support high-availability systems.

Common-sense design principles and an awareness of NAS limitations should be borne in mind to ensure the realization of the advantages described above. The NAS competes with and shares the LAN bandwidth to which it is attached. Thus, heavy NAS activity may result in significant degradation of the entire LAN performance. To avoid this, where possible, it may make sense to judiciously segment the LAN to provide a more even distribution of traffic and to avoid congestion. Another consideration that should be borne in mind is that NAS servers are typically file oriented and not transaction or database oriented. Thus, NAS is not well-suited to handling a heavy mix of bulk file transfers and online transaction data. When traffic loads become sufficiently large or are dominated by transaction processing, it makes sense to consider a SAN solution as an alternative. SANs are specifically designed to support the storage requirements of very high-availability, low-latency, and transaction-oriented applications.

NAS products can be roughly divided into low-end appliances and high-end appliances. The largest market share is for the low-end/low-cost NAS solution, which is used mostly by small and medium-sized businesses. The lower-end NAS products generally use standard hardware with a streamlined, customized version of the operating system. Microsoft's NAS products are among the most popular of this type and run on a multitude of machines from such vendors as Hewlett-Packard, IBM, Dell, EMC, Iomega and Quantum, and others. Windows-based NAS devices are fully configured and ready to go right out of the box, requiring very little setup time.

According to Gartner Dataquest, Microsoft-powered NAS unit shipments grew 94 percent between 2001 and 2002, and represent a market share in excess of 41 percent. [HORW03] [FRAU03] The dominance of Windows as a *de facto* standard contributes to the success of Microsoft's NAS products. Network administrators already familiar with Windows find it easy to install and maintain Windows-based NAS devices. Windows NAS devices also support Windows file server and third-party software for virus protection, business continuance, network management, and a host of other tools. This is augmented by powerful and easy-to-use file management software that Microsoft provides with its NAS products. Using a Web browser, network administrators can manage data replication and failover tasks remotely. In addition to ease of use and rich feature set functionality, Windows-based NAS devices are priced very competitively. Low-end Windows-based NAS devices start at around $2000, while comparable proprietary offerings cost twice as much or even more. Original equipment manufacturers (OEMs) — such as Hewlett Packard, EMC, and CLARiiON — are using Microsoft's NAS instead of developing their own platform, and these licensing arrangements are also helping to keep Microsoft's prices down. [HORW03]

Large enterprises with demanding applications may require a high-end NAS approach. A variety of vendors offer specialized NAS appliances tailored to support large-scale enterprise solutions, including Network Appliance (FAS 900 Series Enterprise Servers), EMC (Celerra), and Hewlett-Packard E7000. Of these vendors, Network Appliance is the market leader. High-end NAS devices offer significant hardware enhancements, as reflected in the "use of RAID, the number of processors, support of iSCSI and Fibre Channel connectivity and the amount of RAM." [HORW03]

The market in this area is very competitive for vendors as IT organizations look for ways to keep down costs. In contrast to the growth Microsoft experienced in the low-end NAS market, proprietary high-end NAS shipments declined by 36 percent in 2002, from 79 percent to 53 percent, and the trend is continuing. [HORW03] This appears to be due in large part to the expense of high-end proprietary NAS platforms, which can cost in the tens of thousands and upwards of hundreds of thousands of dollars. [FRAU03B] Microsoft also continues to be a factor in driving down prices. Microsoft recently introduced Windows Storage Server 2003 to compete with higher-end proprietary NAS products. Windows Storage Server 2003 is designed to handle up to 256 terabytes, with no theoretical limit on the number of file systems it can process. It offers many features to support large-scale BC/DR data storage requirements, including "file serving, server consolidation, backup/restore and replication, and integration with storage area networks." [MICR03] Network Appliance is marketing its NAS devices as a better solution for integrating Windows, UNIX, and

Linux systems. However, "Microsoft's licensing deals may allow manufacturers to undercut NAS devices from Network Appliance on price while adding an improved operating system."[FRAU03B]

9.2.3.3 Storage Area Networks (SANs)

SANs provide scalable, interconnected storage pools that can extend from a workstation or server across a LAN, MAN, or WAN. A SAN is a network dedicated exclusively to data and file transport, and is separate and distinct from a LAN. SANs consist of the components described in the subsections below.

9.2.3.3.1 Storage Platforms

Typical types of SAN storage platforms include:

- RAID
- Cached disk arrays[28]
- Tape drive systems (including automated tape libraries, stackloaders,[29] and RAIT[30])
- JBOD: an acronym for "Just a Bunch of Disks"

9.2.3.3.2 Server Platforms

Typical SAN server components include:

- *Host bus adapters (HBAs).* These are installed in the PCI slot of the SAN server, and with the appropriate cabling provide the connection to the SAN interface.[31] The HBA serves the same purpose on the SAN as a *Network Interface Card* (NIC) serves on the LAN. It should be noted that the HBA used must be matched to the type of SAN interface in use. For example, Fibre Channel (FC) HBAs are used only with FC SANs, etc. Other specialized NICs are available to adapt servers and other devices for intercommunication using alternative SAN protocols. For example, iSCSI NICs can be used in servers that lack native iSCSI capabilities to enable the interconnection of iSCSI SAN devices using standard Ethernet switches. Several types of iSCSI NICs are offered in the market place (e.g., Fast Ethernet NIC, Gigabit Ethernet NIC, and Gigabit Ethernet NIC with TCP offloading).
- *Switches, routers, hubs, and bridges.* These devices support the same types of actions performed by similar devices used in LANs, MANs, and WANs. However, they must be compatible with the SAN interface in use.

- *SAN software.* Server operating systems generally do not allow multiple servers to share the same disk storage. To overcome this limitation, SAN software is used to assign and allocate storage across multiple servers, and to perform other security and network functions. The three main ways of doing this are:
  - *Switch zoning.* This is the oldest SAN partitioning method. It is used to create SAN segments that are similar to a VLAN in a LAN environment. Switch zoning assigns portion(s) of the shared storage to a specific client or server, based on the port to which they are connected on the (SAN) hub or switch. SAN zoning is fairly easy to configure using management software provided with the SAN switch or hub.
  - *Logical unit number (LUN) masking.* This is similar to switch zoning, but it allows storage to be segmented according to a device ID, as opposed to a port assignment. For example, LUN masking can be used to partition a single RAID into multiple LUNs, each of which is assigned to a specific server. LUNs can be defined across individual disks, part(s) of multiple disks, groups of disks, or tape libraries. However, this partitioning method does not allow servers to share files across different LUNs.
  - *Storage virtualization.* This approach is used when multiple servers must simultaneously access a heterogeneous mix of storage devices (i.e., IDE, RAID, direct attached SCSI, etc.). In this scheme, the SAN is split into logical server and storage zones. A specialized server/controller server manages requests for access to storage and arbitrates contention between competing requests. Storage virtualization masks the details of the SAN infrastructure to end users, much like "...the way we access electrical power. We flip a switch, and the light comes on. We don't care about the mechanics of the switch, junction boxes, breaker panels, power poles, substations, or the power grid. We don't want to have to think about the electrical power infrastructure. We simply want power on-demand. Call it power virtualization." [BENN02] Arbitration or virtualization software allows computers running different operating systems (e.g., Macintosh OS, Sun Solaris, IBM, AIX, and Windows, etc.) to access the same files and disk partitions transparently like on a LAN.

9.2.3.3.3 Interconnecting Fabric or Interface for Data Transport

Just like any network, SAN interfaces must support schemes for device addressing, security, routing, data handling, and other network functions.

Differences in these schemes are reflected in the variety of SAN interface standards and approaches available in the marketplace. Generally, they are not interoperable. The major types of SAN interfaces include:

- *ESCON (Enterprise Systems CONnection).* This interface is used to interconnect IBM S/390 computers, workstations, and SAN storage devices using optical fiber. ESCON fiber supports full-duplex, point-to-point connections. The major components include an ESCON channel, control unit, and Director. The ESCON Channel is a host-based channel connection that directs traffic flow between host storage and other devices. The ESCON control unit manages communications between input/output devices (such as a cluster controller, tape drive, or DASD) and an ESCON Channel or Director. The ESCON Director is a dynamic switch that serves as a communications hub for ESCON channels allowing simultaneous duplex data transfer between any two ports.
- *Small Computer System Interface (SCSI).* This ANSI standard was originally developed in the 1980s to define a parallel communication interface between hardware devices — such as disk drives, tape drives, CD-ROM drives, and printers — and transport of data on personal computers and workstations. There are several versions of SCSI[32] in the marketplace. Older versions support only a few devices per bus over short operating distances. SCSI-3 was defined to support Fibre Channel, the Serial Bus Protocol, and the Serial Storage Protocol (SSP). SCSI-3 also provides a command set to optimize use of new technologies incorporated into a variety of devices (such as tape drives, RAID, graphics, media changers, CD-ROM jukeboxes, object-based storage, etc.).
- *Fibre Channel.* This ANSI standard evolved from SCSI to support direct communications between many servers and storage devices over optical fiber, coaxial cable, or twisted-pair telephone wire. Fibre Channel provides block-level data transmission, with quality-of-service features, guaranteed packet delivery, and flow control mechanisms. Although Fibre Channel is supposed to be backward compatible with SCSI, in practice there may be compatibility issues. Fibre Channel is much faster, with the majority of devices operating at 1 or 2 Gbps. [INTE02] Recently, the International Committee for Information Technology Standards (INCITS)[33] approved a 10-Gbps Fibre Channel ANSI standard, so FC will continue to support high-bandwidth applications of the future. FC operates over longer distances than SCSI and is supplanting it in the SAN marketplace. Fibre Channel supports three topological configurations:

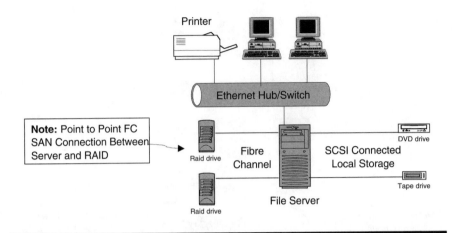

Figure 9.4   Example of Fibre Channel SAN with point to point configuration coexisting with local SCSI devices.

- *Point-to-point.* In this configuration, devices are connected by dedicated, point-to-point, bi-directional, full-duplex links. This is illustrated in Figure 9.4.

- *Arbitrated Loop (FC-AL).* This configuration is similar to Token Ring. It is the oldest version of FC and also one of the cheapest. FC-AL is a closed-loop configuration with point-to-point connections *between* ports and uni-directional traffic flow. It supports up to 127 nodes active at any given time. FC-AL supports stand-alone private loops or public arbitrated loops (which are attached to fabric switches to create a larger SAN). FC hubs can be attached within a single loop — as shown in Figure 9.5 — or they can be used to attach multiple loops, or they can be attached to an FC fabric switch. In an FC-AL configuration, a single port must wait to be given permission (i.e., it must arbitrate) to transmit over the loop. Once access to the loop is granted, the port is allowed to establish communication with another port and to use the full link bandwidth. Other ports remain in a monitoring mode during this time. Once transmission is complete, the connection is closed, and the loop is released to another port. A port cannot arbitrate again until all other ports have been given a chance to transmit. FC-AL is usually implemented in dual–loop configurations for fault tolerance.

- *Switched fabric Fibre Channel.* The term "fabric" refers to the hardware ("Fibre" switches, hubs, and loops) that connects

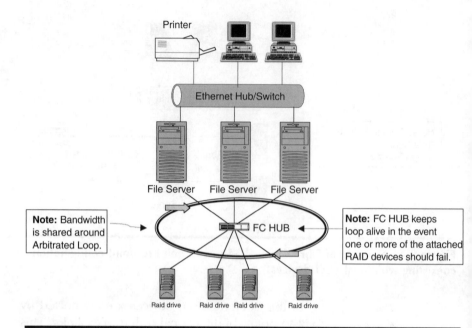

Figure 9.5 **Example of Fibre Channel SAN with arbitrated loop configuration.**

workstations and servers to the storage devices on the SAN. Switched FC is designed to support both direct channel and switched point-to-point communications. It is by far the most commonly implemented SAN topology. An example of this topology is illustrated in Figure 9.6. This ANSI standard defines a block data transport, and a serial communications interface capable of supporting more than 15.5 million devices. With this many addresses, device polling is not a feasible option for detecting the presence or absence of devices. Instead, FC uses a Name Service. Devices are automatically detected as they are added to the fabric, after which they must log in with a fabric switch as part of an initialization process. The FC fabric switch maintains a Simple Name Server (SNS). The name server consists of an object database of attached devices and their associated permanent device ID,[34] class-of-service, fabric address, upper layer protocol support, and other information. When a device logs in to a switch on a specific port,[35] the switch correlates the device ID with the port address, and updates the internal name server. A State Change Notification (SCN) is also used to notify FC servers as storage resources are added or changed on the network. This dynamic addressing and tracking capability removes the possibility of human error and facilitates additions,

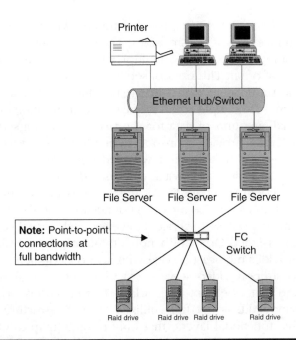

Figure 9.6   Example of Fibre Channel SAN with switched fabric configuration.

moves, and changes in the SAN. FC servers can query the SNS to request a list of all SCSI-3 storage devices. The FC server can use this list to poll and establish sessions with appropriate storage devices. In cases where it is not appropriate for a server to establish communication with selected storage devices (e.g., a Windows NT server is not compatible with a UNIX storage device), zoning or other partitioning methods can be used to restrict the discovery of potential storage targets. A destination ID is embedded in the frame header of packet transmissions between communicating devices, thus making simultaneous transmission between multiple node pairs on the SAN possible. The FC fabric switch is similar to a telephone network, in that every node can establish a direct, temporary connection using the full network bandwidth. In addition, FC also supports connectionless transmission using time division multiplexing. FC uses several classes of service to determine which type of connection to use: (1) Class 1 Service provides a dedicated connection between two devices using the entire link bandwidth. Frames are received in the order that they are transmitted and an acknowledgment is sent to the transmitter upon receipt. This class of service is good for high-bandwidth applications.

(2) Class 2 Service provides frame switched, acknowledged, connectionless transmissions. The bandwidth is shared by transmitting devices using multiplexing. Frames may be delivered out of order. This class of service is good for connections where the set-up time would be greater than the latency of transmission. An example of this is a database application that exchanges small amounts of data for short periods with a server. (3) Class 3 Service provides transport similar to Class 2, except that frame delivery is neither acknowledged nor guaranteed. In the event of congestion, the switch can discard Class 3 frames. This class is recommended for FC-AL architectures and real time broadcast. (4) Class 4 Service provides connection-oriented virtual circuits, similar to Class 1 service. However, each virtual connection has an assigned QoS and bandwidth allocation. (5) Class 6 Service provides multicast service with acknowledgment. It is designed to support video servers. (6) Class Intermix allows Classes 2 and 3 to be sent over Class 1 connections, if the Class 1 bandwidth is not fully utilized. The FC structure is based on five functional layers (that only roughly map to OSI). In brief, they are: (1) FC-0 is the lowest level. It specifies the physical components of the SAN, including the media, transmitters, receivers, and connectors that can be used with Fibre Channel. It also specifies the Open Fibre Control System (OFC), which ensures that the optical power level of switch lasers operate within established safety standards. (2) FC-1 is used for byte encoding and decoding, and error control. (3) FC- Al was added after the original FC specification was developed, so it does not have a number. It is designed to support FC-AL topologies. (4) FC-2 implements transmission features defined by the different FC classes of service. It is responsible for various link services (i.e., log in/log off, establishing links, removing links, etc.), data packetization, reassembly, framing, and sequencing. (5) FC-3 is currently under development and was only recently approved. It is concerned with advanced services relating to distribution and negotiation of timeout values, broadcast/multicast, and security. (6) FC- 4 defines the application interfaces supported by Fibre Channel. Some of the protocols supported at this level include SCSI, IPI (Intelligent Peripheral Interface), HIPPI (High Performance Parallel Interface), IP, AAL5 (ATM adaptation layer), and IEEE 802.2. These upper-level protocols are usually implemented in software, while the other FC layers are generally implemented in hardware. It should be noted that FC vendors do not necessarily implement the FC layers[36] completely

or separately in their products. Switched fabric SANs can have a cascaded or non-cascaded architecture. In the former, fabric switches are interconnected to form a large mesh so that a device connected to one switch can access any other node connected to a switch within the fabric. In the latter, switches are not interconnected. Thus, this non-cascaded configuration is more vulnerable to failure than the mesh configuration. Fibre Channel is a mature technology that provides high performance and reliability with complete error checking. It also supports high bandwidth transmission at gigabit speeds. Factors limiting its deployment generally relate to its cost and scalability. Switched FC is one of the most expensive SAN solutions to implement, and in practice is limited to a geographical range of about 200 to 500 meters. [INT02] FC also suffers when the SAN grows to the upper limits of its design. As switches are added or zoning changes are made to the FC fabric, the SNS data must be updated so FC servers can locate FC storage devices. Because each switch maintains its own copy of the SNS database, the exchange of update information to other switches across the fabric can seriously impact network performance. For more information on Fibre Channel, the interested reader is referred to http://fibrechannel.org and to http://www.t11.org/index.htm.

■ *IP-based SANs.* This approach supports SAN data storage, retrieval, and transmission across a LAN, MAN, or WAN, extending the effective reach of SCSI and FC over an IP infrastructure. SAN connectivity over a WAN is increasingly used for remote data replication and fault tolerance. Service providers, carriers, and vendors offer a wide variety of IP SAN storage and WAN connectivity solutions specifically tailored to BC/DR requirements. The three main enabling protocols for IP SANs are Internet SCSI (iSCSI), Fibre Channel-over-IP (FCIP), and Internet Fibre Channel Protocol (iFCP). Each protocol supports the SCSC-3 serial interface and standard command set, and transmission over TCP/IP at gigabit rates. These protocols are discussed below:

– *iSCSI.* This IETF protocol is used to create SANs using standard Ethernet hardware. An example of an iSCSI SAN is illustrated in Figure 9.7. iSCSI offers interoperability between SCSI, Ethernet, and TCP/IP at transmission speeds up to 10 Gbps. Through gateways, it can be used to connect to other types of SCSI interfaces, such as FC. Servers in an iSCSI network are referred to as "iSCSI Initiators" and storage devices are referred to as "iSCSI Targets." "Initiator" software is installed on SAN servers

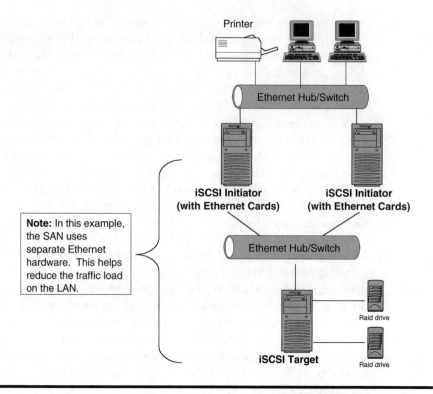

Figure 9.7   Example of iSCSI SAN with dedicated Ethernet configuration.

to enable communication with iSCSI Targets. Correspondingly, "Target" software is installed on servers (such as disk controllers for shared disk systems, iSCSI routers, or other target storage devices). An iSCSI Initiator can automatically discover iSCSI Targets and their associated IP address by issuing a query to an Internet Storage Name Service (iSNS)[37] server. This returns a list of IP addresses to which the Initiator is permitted access. After an Initiator has identified Target devices, they must properly log in before Target devices will accept iSCSI PDUs (Protocol Data Units) from them. In this way, iSCSI supports mutual authentication between Initiators and Targets, and appropriate authorization for access to SAN resources. If additional security is needed, other measures may be needed — such as IPSec to ensure the confidentiality and integrity of the transmission. When a data storage request is finally issued by an iSCSI Initiator, the server operating system generates the necessary SCSI commands and data. These commands and data are, in turn, encapsulated or encrypted, and converted into IP packets[38] with appropriate headers and trailers. The SCSI commands and data are sent over an IP network as part of the IP packet payload. On most LANs, Ethernet is used to further encapsulate IP for physical transport over the network. When the IP packets are received at the destination — the iSCSI Target — they are decrypted if necessary, and disassembled, with the SCSI commands extracted from the data. The SCSI commands and data are then sent to the Target server's operating system for further processing. Because iSCSI is bi-directional, a reverse process can be initiated by the iSCSI Target. This process is transparent and occurs without user intervention. iSCSI enables the interconnection of iSCSI SANs; however, it does *not* work with Fibre Channel SANs. If existing Fibre Channel SANs must be interconnected, FCIP or iFCP solutions will be needed. iSCSI SANs can be built using commodity Ethernet hardware, so they are much cheaper than comparable Fibre Channel SANs. However, iSCSI is not as reliable because it is based on the best effort delivery philosophy of TCP/IP. It also is a new technology and it is not nearly as mature as Fibre Channel. According to [INT02], iSCSI SANs are most suitable for organizations such as: ISPs, SSPs (storage service providers), those requiring remote data replication and recovery, those requiring real-time access to geographically dispersed data, and those with limited IT budgets and resources that need a SAN infrastructure.

— *Fibre Channel-over-IP (FCIP)*. This IETF protocol was designed to enable FC transmission across an IP network, thereby overcoming Fibre Channel's distance limitations. This protocol complements FC by providing a means to interconnect remote FC SANs via FCIP gateways. This is illustrated in Figure 9.8. FCIP uses TCP/IP to establish connectivity and to provide data transport between remote FC SANs, while preserving the underlying FC fabric and protocols. This is accomplished through *tunneling*. The advantage of tunneling is that it allows the transmission of private data over a public network. A common example of tunneling is the use of VPNs and the PPTP protocol[39] to send data securely over the Internet. Tunneling works by encapsulating or embedding Fibre Channel frames (or packets) within TCP/IP packets that are transported over the IP network. FCIP supports IPSec and the IKE[40] protocol to secure and safeguard communications. For example, as shown in the diagram, a FC server at Site A might want to send data to a storage device at Site B for replication and disaster recovery purposes. Data from the FC server is sent to an FC switch at Site A, which in turn,

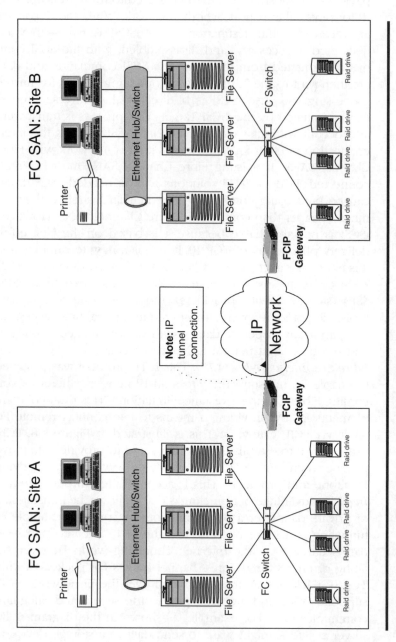

Figure 9.8   Example of FCIP SAN interconnection.

sends the data to an FCIP gateway. The FCIP gateway creates a tunnel over the Internet (or IP network), thereby connecting the two FC SANs. FCIP gateways are usually connected to a FC switch port at the respective entry point of each local SAN. The FCIP gateway at the sending site encapsulates FC frames in TCP/IP and forwards the packets through the tunnel. At the receiving site B, the FCIP gateway strips away the TCP/IP wrapper and forwards the native FC frames to a local FC switch. The local FC switch, in turn, forwards the data to the appropriate FC device for processing. FCIP is transparent to both the FC fabric and the IP network. The only devices with assigned IP addresses and involvement in protocol mapping are the FCIP gateways. IP discovery is limited to the FCIP gateways. FCIP uses the Service Locator Protocol (SLP) to identify FCIP gateways in the IP network, because iSNS — used by iSCSI and iFCP for this purpose — is not supported by FCIP gateways. The FC discovery process remains intact for the FC fabric. FCIP is recommended when two or more Fibre Channel SANs must be connected in an existing infrastructure, because iSCSI cannot be used for this purpose. Generally, FCIP is better suited to point-to-point SAN connections than to multi-point connections. Companies that have invested heavily in FC and need to leverage this investment in their disaster recovery solutions — which might include serverless backup, synchronous and asynchronous mirroring — find FCIP an attractive option. As TCP/IP becomes increasingly important as a corporate standard, IP solutions of this type become more appealing to the IT staff responsible for installing and maintaining the network infrastructure.

- *Internet FC Protocol (iFCP)*. This IETF protocol is used to connect individual Fibre Channel (FC) devices to an IP-based SAN using Gigabit Ethernet switching and routing devices. With iFCP, storage devices can be located anywhere within the IP network and no longer are limited to being part of a FC SAN. iFCP is a gateway-to-gateway protocol designed to complement FC, FCIP, and iSCSI. It can be used as part of a migration strategy for integrating FC and iSCSI SANs, and can replace or coexist with FCIP. Figure 9.9 shows an illustrative example of this coexistence in a SAN. In an iFCP SAN, the FC devices (e.g., switches, disk arrays, HBAs, etc.) must connect to a local iFCP gateway via an F-Port. On its N-Port (i.e. FC) side, the FC device appears to be part of a standard FC fabric. On its F-Port (i.e., iFCP) side, it appears to be part of an IP network. An iFCP gateway encapsulates incoming FC frames into IP and converts

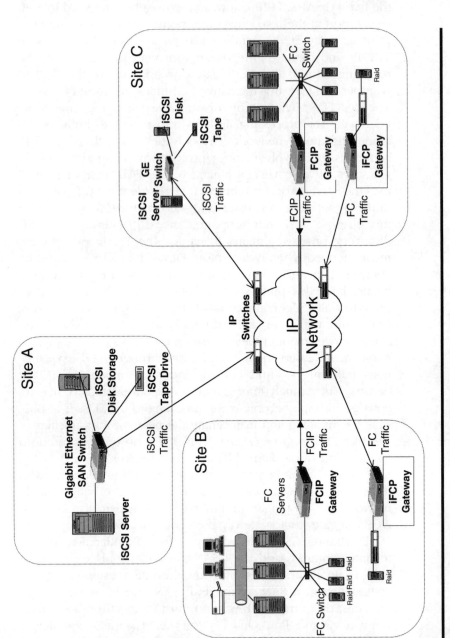

Figure 9.9   Example of multi-protocol SAN with IP backbone.

the FC transport session into a TCP/IP session. This involves mapping lower (FC-2) transport layer services to TCP/IP (i.e., Ethernet), while retaining the upper (FC-4) application layer's services.[41] A receiving iFCP gateway converts the IP transmission back to a FC session as required for device-to-device, device-to-SAN, or SAN-to-SAN communication. When a FC Initiator sends an SNS query looking for Target storage devices, the request is directed to an iSNS server. The iSNS server performs intelligent name look-up and device discovery for the iFCP gateway, in a role similar to that performed by the name server in a FC SAN. The iSNS server also returns N-Port and IP addressing and zoning configuration information used to establish iFCP sessions between FC devices and iFCP gateways. Figure 9.10 diagrams a generic iFCP configuration with various port attachments and an iSNS server. An iSNS server can be located anywhere in the IP network where it is accessible by iFCP gateways, and can be configured for centralized or decentralized access. In the former case, a single iSNS server can be used to manage the entire iFCP domain. In the latter case, multiple local iSNS servers can be used, each with its own store of relevant information and IP addresses for its respective domain. Recapping this discussion, iFCP switches perform several major functions: (1) target device discovery for FC Initiators, (2) translation of the FC device address into a corresponding remote gateway IP address and an associated FC N-Port address beyond the gateway (this function is performed in conjunction with the iSNS server), and (3) switching and routing between IP gateway regions, with firewall and fault isolation capabilities. According to [SNIA04B], the main benefits iFCP offers are: (1) high-performance IP and Ethernet SAN infrastructure, (2) support for Fibre Channel end devices or existing Fibre Channel SANs, (3) familiar Ethernet and IP management tools, (4) multipoint routing for multiple remote data center requirements, (5) data integrity with TCP/IP session control over the WAN, (6) support for "SAN-aware" applications, and (7) integration with data replication applications. iFCP is particularly well suited for large enterprises that must support demanding SAN applications and that have already invested heavily in FC and IP infrastructure. iFCP allows them to leverage their investment and expertise, and is a natural migration path for connecting remote FC SANs for DR/BC. In contrast with a pure FC SAN solution, iFCP is much more scalable, in terms of geographic distance and the number of devices that can be easily supported. "Since Gigabit

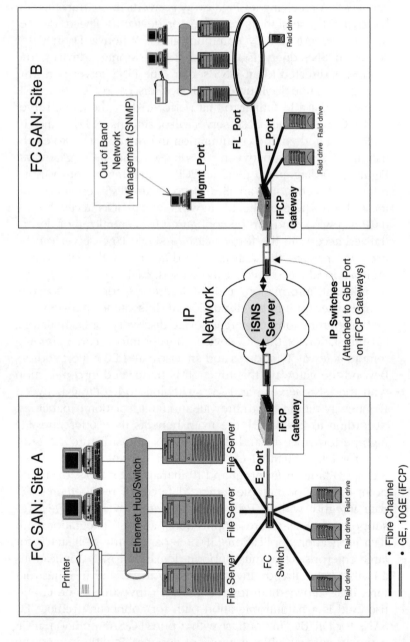

Figure 9.10 iFCP gateways with various types of port attachments.

Ethernet switch vendors supply high port density and high availability switches at about half the per-port cost of Fibre Channel switches, it is possible to cost-effectively extend the SAN infrastructure as requirements increase." [SNIA04B] With iFCP, the organization can continue to employ FC devices, while replacing FC switches with lower-cost IP switches as it makes sense to do so. With iFCP, cost economies can be realized, along with the performance benefits of FC. FC products are mature and widely interoperable, and in particular, FC storage arrays have better reliability and throughput than comparable IP offerings (although this gap is expected to close in the next few years). It makes sense for an organization to maintain the best of both worlds when they are already in place. Relative to FCIP, iFCP is superior in providing networked connectivity in a large-scale enterprise SAN deployment. Recall that FCIP creates point-to-point connections between two SANs so they appear as one. This results in the FCIP switches having to handle all inter-fabric traffic. This load is increased if a failure occurs in one SAN, because a state change will be detected on the other side that will lead to the propagation of State Change Notifications (SCNs) throughout the interconnected SAN. This, in turn, can result in significant increases in network traffic and congestion. If lower-speed WAN lines are used to connect the SAN, this will also exacerbate potential traffic flow and performance issues. In general, by its very nature, iFCP offers more scalability than an FCIP tunneling approach. iFCP and iSCSI are deemed the most fiercely competitive of the IP SAN technologies. The most commonly expressed reservations about iSCSI vis-à-vis iFCP relate to its lower performance and lack of market maturity. These concerns will become less important in the future now that iSCSI has been approved as a standard and interoperable iSCSI products are being offered in the marketplace. However, for the time being, iFCP offers the best features of FC performance and reliability, with the manageability, scalability, and cost economies of IP. A major industry concern with iFCP as an enterprise solution is that only one vendor — McData Systems (formerly Nishan Systems) — offers complete iFCP solutions. In contrast, most SAN vendors have implemented FCIP for WANs. Analysts also note that it is telling that McData has iSCSI ports on its iFCP switches, reflecting both the importance and competition between the two. Enterprises building a SAN from scratch are much less likely to employ iFCP over iSCSI because of its inherently higher cost (due to FC components) and complexity (due to the requirement that two infrastructures — FC and IP — must be maintained). For the sake of completeness, it should be mentioned that there are proprietary approaches similar to iFCP on the market. Two examples of similar IP gateway approaches include Cisco's proprietary Virtual SAN and LightSand's Autonomous Regions with Domain Address Translation (AR/DAT). [SPAN03] For more information on the iFCP standard, the reader should consult the following:

- http://www.ietf.org — the Web site of the IETF, which is responsible for the iFCP specification.
- www.snia.org — the Web site of the Storage Network Industry Association, a nonprofit industry trade association, with the stated mission of advancing "the adoption of storage networks as complete and trusted solutions."

- *InfiniBand Architecture (IBA)*. This specification is sponsored by a trade association of major vendors (including Intel, Dell, Compaq, IBM, and many others). It defines a point-to-point, dedicated, channel-based, switched architecture for communications at gigabit speeds (2.5 Gbps, 10 Gbps, or 30 Gbps). The name is derived from the concatenation and shortening of "infinite bandwidth." The specification was first released in 2000, and defines a new type of computer bus that is particularly well suited for SAN servers and storage devices. It is compatible with and can coexist with Fibre Channel, iSCSI, and IP through the use of InfiniBand bridges, switches, and routers. This is illustrated in Figure 9.11. InfiniBand supports communication between devices that may reside in the same machine or across widely distributed locations within a SAN fabric. Primary components of an InfiniBand network include:

- *IBA subnet(s)*. A subnet consists of a subnet manager, end nodes, switches, and links. A subnet is the smallest IBA network unit, and can be interconnected to other subnets via switches and routers to form a larger network. The subnet manager resides on an end node or switch and is responsible for configuring and managing each local subnet.
- *End nodes*. These are source and destination devices, and include storage devices, servers, and channel adapters. IBA does not distinguish between various types of end nodes, except by *form factor*. A form factor defines standard end node sizes — standard (20 × 100 × 220 mm), wide (twice the width of standard), tall (twice the height of standard), and tall wide (twice the height and width of standard).

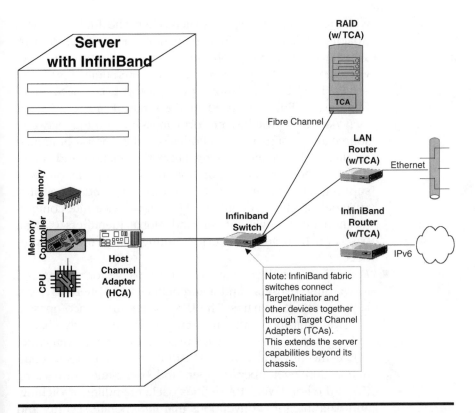

Figure 9.11   Native InfiniBand example.

- *Channel adapters.* There are two types of IBA channel adapters. The Host Channel Adapter (HCA) is an interface residing within a server that controls access to the server's CPU, memory, and the IBA fabric. The HCA communicates with a target channel adapter (TCA) or a switch. The HCA is usually an IBA adapter card consisting of a PCI interface and an I/O controller. This is a vendor-specific solution that allows existing server platforms to be quickly modified to support IBA. In a native implementation, the HCA can be integrated with the system motherboard. This approach allows complete or partial replacement of the server's PCI bus with IBA (note: this is the approach shown in Figure 9.11). However, vendor support for this approach is more limited. In this configuration, *partitioning* is needed to distinguish and properly control access to private and publicly shared devices on the IBA network. With partitioning, multiple clients can share access to target devices on an IBA network while remaining unaware of each other. The Target Channel Adapter (TCA) works in conjunction with a HCA or IBA switch to enable geographically distributed input/output devices (e.g., RAID, tape drives, etc.) to communicate with the server. The TCA must support the protocol (i.e., SCSI, Fibre Channel, or Ethernet) used by the target device.

■ *Switches*. IBA switches perform message routing between source and destination end nodes based on preprogrammed routing tables created during network initiation and modification. The switch configuration, number of ports, protocol support, and other switch features are vendor specific. Switches use local identifiers (LIDs) for device intercommunication on a single subnet, and global identifiers (GIDs), based on IPv6 addressing, for routing across multiple subnets.

■ *Routers*. IBA-enabled routers are responsible for routing and forwarding packets, and interconnecting subnets to form larger network domains. The IBA specification encompasses physical, link, network, and transport layers. Briefly characterizing some of the key features of the IBA architecture: (1) At the physical level, IBA supports bi-directional communication over fiber, copper, and backplane connectors. (2) At the link layer, IBA defines packet handling, switching, and data integrity conventions that incorporate security and QoS features. For example, IBA supports the logical construct of *virtual* and *management lanes*. Up to 15 virtual and one management lane can be defined for each physical link to provide priority and QoS handling for designated communications. During transmission, IBA distinguishes between data packets, which contain transaction information, and management packets, which are used to control addressing, subnet directing, and other link management functions. Cyclic redundancy checking (CRC) and data encryption are standard on all communications and are also performed at this layer. (3) At the network layer, IBA uses IPv6 addressing for the packet source and destination. This promotes compatibility with TCP/IP networking standards, and vastly increases the potential device addressing space. (4) At the transport layer, IBA handles packet delivery, partitioning, channel multiplexing, and transport services. The transport layer provides mechanisms for both reliable and unreliable communication that can be implemented in software, hardware, or firmware by vendors. IBA supports *clustering* — the interconnection of two or more devices to form a single logical device unit. Clustering requires very high-bandwidth, highly reliable, and very low latency transport. To ensure this level of performance, IBA implements the Virtual Interface Architecture (VIA)[42] specification. Much of this implementation is in firmware to optimize performance and speed as much as possible. Remote Direct Memory Access (RDMA) is another emerging, enabling technology complementary to VIA and IBA that is currently under development by the RDMA Consortium and the IETF. RDMA enables a local system to have direct read and write access to the memory of a remote system, and vice versa. RDMA-over-IP is a related protocol that, as the name implies, is designed to support RDMA operations over TCP/IP. Vendors have incorporated RDMA-over-TCP/IP into a number of product offerings, most notably Ethernet NICs, which offer RDMA support (with and without TCP/IP offloading). These products provide an inexpensive interface for high-performance server, database, or clustered environments using IBA over a TCP/IP transport layer. TCP offloading — which relieves a host server's CPU from the burden of TCP/IP protocol processing — serves a distinctly different function than RDMA. Recall that TCP/IP was designed to support a diverse WAN/LAN application environment, and operates over an unreliable Ethernet link layer. However, TCP's flexibility comes at the expense of significant CPU utilization, network delays, and memory requirements due to the protocol processing, memory-to-memory data copying, user-to-kernel transitions, and other overhead functions that must be performed. IBA operates over a reliable link layer that offers considerable improvements in latency and performance over TCP/IP. In high-performance network environments, IBA, TCP/IP offloading, and RDMA are used together to minimize CPU and memory demands on servers and to reduce delays to the lowest practical limits. Originally, InfiniBand was conceived as a replacement for the Peripheral Component Interconnect (PCI) bus used in most personal computers and servers. A prototypical PCI bus configuration is shown in Figure 9.12. As CPU and bandwidth capabilities and application demands have increased, they stretched and outstripped the capabilities of current PCI bus technology. Some of the major limitations of PCI — which IBA was designed to address — include (1) *Limited transmission*

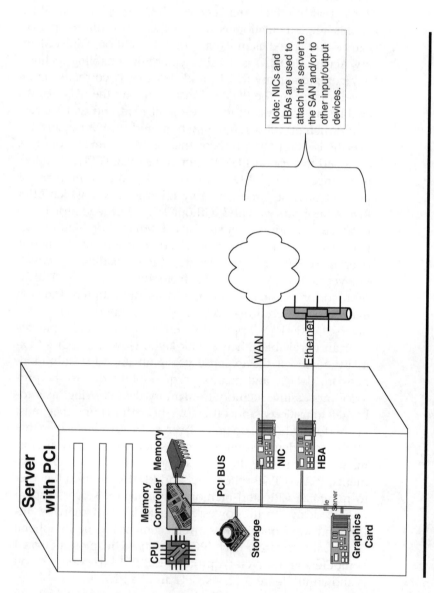

Figure 9.12 PCI bus architecture.

speed. The maximum transmission speed for current PCI and PCI-X technology is 528 Mbps and 1064 Mbps, respectively. (2) *Single path to the CPU and memory that must be shared by all devices attached to the PCI bus.* This creates resources contention and introduces processing delays. The effects of this interdependency are further exacerbated when devices on the PCI bus execute load instructions that involve reading and retrieving data (a common and normal occurrence in most applications), and the CPU must wait for the data to become available before it can begin processing. (3) *Limited scalability.* For example, PCI-X supports a maximum of four cards at 66 MHz; this number drops to one at speeds of 133 MHz. (4) *Limited fault tolerance.* When a PCI card fails, the server must typically be taken offline, because the failure of any one component on the bus can cause the entire bus to fail. (5) *Limited server capacity.* With PCI technology, the server components (including NICs, HBA, etc.) are contained in a chassis that must have sufficient space for the devices needed. Analyst opinions are mixed regarding if and when IBA will become a commonplace replacement for PCI. [SHAN03, SHAN04, CONN04, INFI04, JACO02] Intel and Microsoft, two early IBA proponents, have withdrawn their support for native IBA implementations, leaving this to other smaller vendors. Microsoft explained its plans to drop support for native IBA in its Windows.Net Server product line by stating that it believes the market will continue to leverage their investment in existing Ethernet infrastructure. As summarized in [CONN04], InfiniBand technology is not inexpensive, it is very new, and it is not yet widely implemented, either in vendor offerings or customer data centers. However, other sources— including the InfiniBand trade association — are more upbeat. For example, IDC — a technology and telecommunications market research and consulting firm — predicts that by 2005, half of all servers will be IBA-enabled. [INFI04B] Although it is unlikely that IBA will replace the PCI bus in the near future, it shows considerable promise as a vehicle for building server clusters for SAN storage (see Figure 9.13 for an example), BC/DR, and supercomputing applications. An IBA feature that is very useful for SANs is the ability to offload server workloads using *third-party I/O.* Communicating servers using third-party I/O only require host server intervention to initiate an operation. Thereafter, servers interact directly without host involvement,

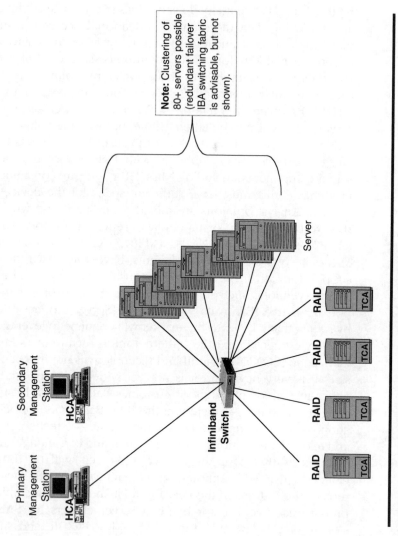

Figure 9.13   Example of InfiniBand server cluster.

thereby reducing the host CPU load. IBA's inherent redundancy and fault tolerance is well suited to SAN and BC/DR applications. IBA provides multiple paths between sources and destinations and self-healing capabilities in the event of a link or transmission failure. IBA has built-in capabilities enabling transparent failover between redundantly configured, hot pluggable I/O devices. The device modularization provided by IBA's switched fabric architecture provides scalability and facilitates fault isolation and resolution. Because input/output devices need not exist in a specific location, they can be conveniently placed for repair and maintenance, with minimal or no disruption to other network components. In contrast to PCI, with InfiniBand, the server footprint can be shrunk to the size of a singe HCA adapter card, because the server boundaries can extend beyond the housing chassis to a potentially vast network of interconnected input/output devices. This is accomplished through networking and switching, and the decoupling of input/output devices from the server CPU and memory. InfiniBand *server blades* are an example of this. An IBA server blade is a very small footprint chassis that typically houses a CPU, memory, power supply, and limited storage — usually just sufficient to hold the operating system — on a passive backplane with little or no circuitry. A server blade is generally a dedicated server and does not need to share resources with other processes. If need be, its memory or storage can be augmented by direct attachment to other devices (such as RAID, etc.). For these reasons, a server blade can be stripped to the barebones minimum. This is advantageous because a blade is easier to repair, upgrade, and house because of its reduced function and size. Server blades are being used in high-performance SAN environments to cluster many servers together to provide data mirroring and failover capabilities. Low-end servers are also being clustered as blades to create parallel computing capabilities that rival higher-cost, high-performance server solutions. We have briefly outlined IBA and its key features here. The actual IBA specification is quite extensive and, of necessity, many details have been omitted. For a comprehensive treatment, the interested reader is referred to the complete specification, which can be found at http://www. infinibandta.org.

According to the market research firm iSuppli Corporation, demand for enterprise SAN solutions is growing at about 30 percent per year and is expected to continue to remain strong. [GARD04] Virtually all large enterprises have installed SANs, and small to medium-sized enterprises are following suit as more affordable SAN solutions (especially iSCSI) become available. Nearly all the leading hardware vendors offer SAN solutions, and there is considerable competition in every storage niche. The SAN market is dominated by switch vendors (such as Brocade, McData, Vixel, and Qlogic) and storage/server vendors (such as EMC, HP, IBM, LSI Storage, Sun Microsystems, and Dell).

Common and compelling reasons cited for the growth in the SAN market relate to its strengths as outlined below:

- *Scalability.* Storage capacity can be increased as required to support E-commerce, online transaction processing, and other BC/DR applications. SAN technology has been designed to accomplish this with little disruption to existing systems and networking infrastructure.
- *Storage consolidation.* Storage resources can be pooled and shared across a large number of users, systems, networks, and locations. By consolidating multiple low-end storage systems into fewer centralized systems, economies of scale can be achieved, as well as improved resource utilization and simplified device management.
- *High data/system availability.* SANs can be used to provide network resiliency and redundancy, device clustering, failover, mirroring, and data replication capabilities in support of enterprise BC/DR efforts.
- *High performance.* SANs are designed to relieve traffic from LANs, so the performance of each respective network can be optimized. Fibre Channel, InfiniBand, and other SAN technologies are being developed to support very high bandwidth applications with a high degree of reliability.

As organizations deploy SANs, they also must grapple with the following major considerations:

- *Compatibility.* There are interoperability problems between vendors, across SAN standards, and within the same SAN standard. There are a multitude of vendor solutions for a diverse array of devices/components, including network adaptor cards, switches, storage arrays, tape libraries, operating system and application software, network management systems, and many others. Most of these are based on proprietary architectures that do not interoperate well or at all with other platforms. This is compounded by a plethora of emerging and inconsistent standards and industry approaches. This reflects the immaturity of the industry as a whole and a desire by vendors for market differentiation.

■ *Limited management software and tool support.* Full-enterprise, excellent commercial SAN management software is not yet widely available and it is expensive and difficult to develop internally. This is, in part, due to the relative infancy of the industry and the fact that many key SAN standards (including SAN network management) have yet to be solidified.

■ *Cost.* SAN devices/components are substantially higher cost than corresponding basic SCSI/EIDE storage devices/components. SANs do not make sense for isolated, scratch storage. SANs become cost effective when economies of scale can be realized. Even in these cases, SAN solutions represent a very significant organizational investment.

SANs make the most sense for organizations that need high-speed, high-availability, and high-reliability data storage solutions. Even so, SAN performance is heavily impacted by all of the following: (1) storage devices (e.g., tape, RAID, etc.) supported on the network; (2) server configuration (including the system bus, e.g., PCI, PCI-X, or InfiniBand), the memory system, CPU speed, the file system(s), and attached host adapters (e.g., SCSI, FC, or iSCSI); (3) system overhead on the network (e.g., number of clients' programs running, application software, operating system, networking protocols, compression and encryption, etc.); and (4) the SAN architecture (e.g., shared, networked, loop, or fabric) and configuration (e.g., number and type of network devices such as bridges, switches, and routers, bandwidth availability and utilization, etc.).

Fibre Channel is the most widely used SAN technology, and is designed to provide excellent performance and reliability for high-bandwidth and high-volume operations. Newer IP-based SAN solutions offer flexibility and lower cost over larger geographic distances. In general, SANs are expensive, and a high availability SAN is at least twice the cost of a simple SAN. This is due to the cost incurred for redundant components and costly dual switch fabrics. Ultimately, the choice of a SAN solution must be guided by the organization's needs and requirements in the context of the existing and planned hardware, software, systems, data, and networking infrastructure.

9.2.3.4 Data Storage Practices

This section reviews industry-standard BC/DR practices employing tape, NAS, and SAN technologies. These technologies are complementary and

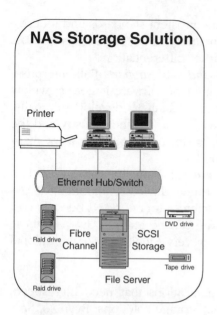
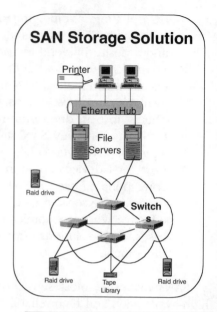

Figure 9.14   NAS versus SAN storage.

overlapping, and many organizations use some combination of the three. Tape storage and retrieval is perhaps the most widely used as there are many hardware and software products, and institutionalized practices to support it. NAS solutions leverage existing legacy LAN infrastructure in performing storage management. Conceptually, a NAS is a sophisticated file server, generally accessed through an Ethernet connection. For smaller-scale organizations and data storage needs, this approach offers simplicity. It also provides better performance, bandwidth, and reliability than tape. NAS solutions do not scale well, and at some point, for the most demanding BC/DR requirements, a SAN approach may be needed. This is illustrated in Figure 9.14. SANs were conceived to remedy the limitations of Ethernet and IP in supporting large-scale data storage, replication, and backup requirements. SANs create a network for data storage tasks that is separate from the LAN, so as to lessen the impact on normal end-user applications. With SAN technology, critical backup and restore procedures can be scheduled, automated, and performed at any time as dictated by DC/DR requirements. SANs do not compete with the SCSI and direct storage market, and do not make sense for direct desktop connections.

SANs are uniquely positioned to provide storage consolidation, scalability, centralized management, improved data access and security, while relieving the LAN and application servers from these data storage tasks. To maximize resource utilization and virtualization, there is a growing trend toward combined NAS and SAN storage configurations. One approach is to employ a NAS gateway — also known as NAS head — to link NAS and SAN networks. NAS gateways use a stripped down, embedded operating system with limited disk, and lack a monitor or keyboard. This trend is also reflected in the emergence of NAS and SAN products that combine features of the other. For example, some SANs now support Ethernet, while some NASs support Fibre Channel. As progress continues toward a more compatible coexistence of technology, end users will have greater choice and flexibility in crafting network storage solutions that best suit their needs.

Data storage and replication are among the most vital BC/DR functions. Replication involves duplication of databases, files, volumes, networks, and data centers. If a primary storage data source is disabled or inaccessible, a secondary one can be used to restore or resume functioning. The interconnectivity of computer and networking and data center resources, by its very nature, provides resiliency and redundancy to support BC/DR. Replication solutions take many forms, including:

■ *Tape backup and storage at offsite location(s).* It is a common practice for enterprises to create tape backups at a primary site and then to transfer these tapes to a safe off-site location for storage. Tape vaulting is used to facilitate this process. Because tape backup and restoration are inherently slow, this method is not suitable as a stand-alone BC/DR strategy for mission-critical, high-bandwidth enterprise systems.

■ *LAN-free backup.* This refers to the use of a SAN to offload a LAN from data transfer and file management tasks. A specialized file server is used to control activities such as backups and restores to tape or RAID storage devices. This approach is shown in Figure 9.15.

■ *Serverless backup.* Like LAN-free backup, this method also employs a SAN to offload file transfer and data movement from the LAN. In addition, it also relieves the SAN file server of backup tasks. Disk-to-tape or disk-to-disk backup is performed without the help of an application or SAN file server, and without consuming SAN bandwidth. Instead, backup tasks are performed by a specialized SAN bridge or router. In contrast to LAN-free backup, this approach uses resources dedicated solely to backup operations. Thus, backup tasks can be scheduled at any prescribed time, and are not constrained to wait for a backup window to open up on the SAN file

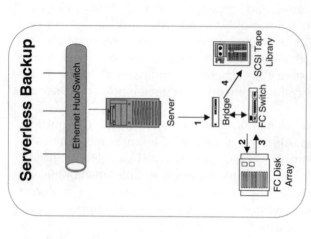

Serverless Backup

1. SAN server initiates backup w/ bridge via extended copy request.
2. Bridge initiates data transfer from disk array via FC switch.
3. Disk arrays sends data through switch, which forwards it to bridge.
4. Data is copied onto tape
   Note: SAN file server initiates backup process and then bridge manages it.

LAN-Free Backup

1. SAN file server initiates backup with FC drive.
2. Data is transferred to server and held in memory.
3. Server sends data to SCSI tape library via SCSI bridge.
4. Data is copied onto tape
   Note: SAN file server manages entire backup

Figure 9.15   LAN-free versus serverless backup processing.

server. The backup process typically involves disk imaging, in which a snapshot of the data is created with pointers to the data. An intelligent agent — generally built into the vendor-supplied bridge/router software — creates a list of logical backup entities that must be processed and issues a standard SCSI extended copy command to begin the copy process. This frees the SAN file server to work on other file management tasks and reduces the traffic load on the SAN network.

■ *Synchronous mirroring.* This backup approach provides the fastest recovery time with the least possible amount of data loss. It is used when the application RTO objectives are very low and are measured in minutes. Synchronous mirroring maintains a duplicate data copy that is continuously updated as changes are made to a primary data source (which might be a file, database, or disk volume). Synchronous mirroring is frequently used to support the backup, restoration, archiving, migration, and distribution of data for business continuance and disaster recovery purposes. Synchronous mirroring requires specialized storage systems and software, and high-speed communications links to achieve the very low latency needed to keep the primary and secondary data sources in lock-step. Typically, fiber optic MANs are used to connect remote mirroring sites to provide the fastest possible connection. Even so, synchronous mirroring imposes a significant burden on application software and introduces processing delays. The speed of light defines the lower bound on transmission speed and the associated maximum distance over which mirroring can occur. This distance is about 100 to 200 kilometers. Beyond this distance, session timeouts and intolerable network delays occur, which make synchronous mirroring infeasible. There are two basic approaches to mirroring: *storage-centric* and *server-centric*. Of the two, storage-centric mirroring is significantly easier to implement and manage, and is used more often. It can be used with any file system or combination of file systems. In a storage-centric approach, the disk storage device assumes complete responsibility for the backup operations, independently from and transparently to the host application processor. When data is written to the primary disk, the storage device, through the use of specialized software, initiates a mirroring sequence to copy the updates. This is illustrated in Figure 9.16. Copies can be stored on a different portion of a primary disk or to another disk. When a server-centric approach is used, as the name implies, the backup operations are controlled by a processor or server to which the primary and backup disks are linked. This method assures a higher level of data integrity, which must be

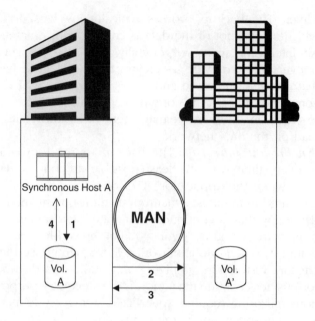

1. Host A writes transaction data to volume (A) storage device.
2. Volume A (software) sends copy of (A) transaction data over MAN to Volume A'. Volume A' (software) writes data to Volume (A') storage device.
3. Volume A' (software) sends acknowledgement of transaction completion to Volume A (software).
4. The current write operation must complete at Volume A and Volume A' before a new write operation is allowed from Host A.

Figure 9.16   Remote synchronous mirroring: storage-centric approach.

weighed against its higher cost and complexity. In a dual write server configuration, both the primary and the secondary disks are linked to the same processor. Software residing on this processor issues write commands simultaneously to each disk as updates are made. This is shown in Figure 9.17. In a relay server-centric configuration, when the primary host processor issues a write command to the primary volume, the command is also forwarded to the secondary storage server. The secondary storage server has software that manages and controls the write activities on the secondary disk. It also communicates with the primary host to acknowledge the successful completion of input/output activities.

■ *Asynchronous mirroring.* This approach can be used to extend the distance between synchronous backup sites to facilitate BC/DR activities. For example, a financial firm might use synchronous mirroring between a New York City data center and a remote New

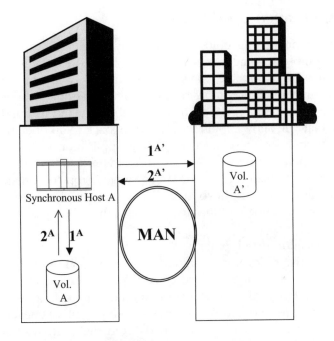

1. Host A writes data to Volume A storage device. Simultaneously, Host A writes transaction data to Volume (A').
2. Volume (A) acknowledges completion of write operations to Host A. Volume (A') does the same. After acknowledgement is received from Volume A and Volume A' at Host A, a new write operation can resume.

Figure 9.17   Remote synchronous mirroring: server-centric dual write approach.

Jersey data center, and asynchronous mirroring between the New Jersey data center and a Philadelphia data center, to extend the area over which a recovery could be made in the event of a catastrophic event. Asynchronous mirroring is also used for high-volume, bulk data transfers or migrations between disk volumes or remote sites, and is much better suited to this task than synchronous mirroring. *Semi-synchronous mirroring* is a variation on asynchronous mirroring. In this mode, data is first written to a primary disk. When the write operation is complete, the secondary volume is updated. Only when the primary and secondary systems have both been synchronized will write activity be allowed to resume on the primary disk. This is very close to the synchronous mirroring process described above, except that with this approach there is the possibility that the primary and secondary disk will be out of synchronization by one transaction.

9.2.4 Application Level Solutions

As computer processing chips, operating systems, and application programs inexorably become more complicated, it is more difficult to conceive of and test all the possible ways a failure might occur, in tandem and together. Failures will inevitably occur despite all efforts to prevent them. Part of the BC/DR strategy is to recognize this fact and to build countermeasures that will allow rapid recovery or rollback to an operational state. Adhering to industry best practices during application development and production operation goes a long way towards reducing the potential for application performance errors or "bugs" which stop the organization in its tracks. Although it is beyond the scope of this book to provide a comprehensive treatment of this area, some of these best practices include:

■ Application development:
  – Software developers should adhere to good programming practices, including the use of self-documenting code, code reusability (employing object oriented design techniques, etc.), and incremental module/program development whenever possible.
  – Software should be designed, from the start, with scalability, efficiency, security, and maintainability in mind.
■ Testing and verification:
  – The organization should employ peer reviews and independent test procedures prior to releasing new software for production use.
  – Configuration management and version control software should be used to track changes in software and to allow rollbacks if needed. If different versions of the software employ different parameters or runtime procedures, this should be thoroughly documented.
  – A formal change control process should be instituted, and standardized test procedures and test databases should be developed and utilized prior to releasing any new software for production use.
■ Production operation and maintenance:
  – Production runs should be monitored to verify that the expected system outputs are being produced. Error reports and help requests should trigger action and follow-up by production and operations staff.
  – The organization should employ periodic reviews and independent test procedures to verify that production software is working as intended.
  – Production runs should be logged, cataloged, and documented as appropriate.

9.2.5 Outsourcing Solutions

Almost every aspect of BC/DR can be outsourced, in part or in full. Although the list below is by no means exhaustive, some of the more common areas companies are outsourcing as part of their BC/DR initiatives include:

- *Network diversity, redundancy, and failsafe configuration.* Examples include:
  - Network architecture design for BC/DR, which is usually provided free of charge by telecommunications carriers and telcos
  - Resilient layer 2 and layer 3 networking services, such as remote LAN dial and other remote dial options; redundant voice services (ISDN); conferencing options (i.e., video, audio, and Internet conferencing); paging and cellular services, and others
  - Dual or redundant networking with traffic rerouting and redirection capabilities
  - Redundant routers, switches, and other network components to protect against equipment failure
  - Circuit redundancy to protect against circuit or hub failure
- *Network boundary protection.* Examples include:
  - Managed services for firewalls
  - Intrusion detection systems.
  - VPNs
  - Content filtering systems
  - Virus protection
- *Data center continuity and data storage services.* Examples include:
  - Data center managed services
  - Operating system and application backups
  - Data mirroring backup (i.e., SAN services)
  - Data replication backup (i.e., NAS services)
  - Electronic vaulting and off-site storage
  - Managed hosting services, such as Web hosting, application hosting, and document hosting
- *Network management outsourcing.* Examples include:
  - Remote or on-site operations management, performance management, configuration management, security management and monitoring, incident management, and fault management of network components and infrastructure
- *Site backup.* Examples include:
  - Redundant data centers and work locations for business operations
  - Hot sites

- Cold sites
- Alternative power and utility sources
■ *BC/DR planning, implementation, and auditing services.* Examples include:
  - Vulnerability assessment
  - Penetration and readiness testing
  - BC/DC documentation preparation and review

The reasons cited most often for outsourcing are to reduce costs, to have access to experts specializing in BC/DR, and to share or transfer the enterprise risk to another third party. With this in mind, the following observations are offered to aid in deciding what products and services should be outsourced:

■ Outsourcing may be necessary if the organization does not have the requisite skill sets or resources to plan, design, implement, maintain, or audit the BC/DR plan. Each aspect of the PB/DR process requires different competencies, which the organization may wish to augment by using outside services and products.

■ It is generally best to maintain control over core business competencies in-house. In-house expertise may be difficult to transfer to a consultant or third party; and furthermore, doing so might compromise the organization's control over its intellectual property (which encompasses its knowledge base, products, policies, procedures, and practices). Data security and privacy are important considerations and must be safeguarded during the engagement of third parties. When these concerns are paramount, they may mitigate against the use of third parties altogether. Nonstrategic business functions — such as mail room processing, etc. — may be more readily outsourced without compromising the organization's core business.

■ Cost savings are likely to be greatest when economies of scale can be achieved by outsourcing to a third party. This is particularly true for products and services that have become heavily commoditized. For example, telco services are offered at very competitive rates that the organization is unlikely to match when using a private solution.

■ Outsourcing may provide an objective perspective that may be difficult to achieve within the organization. For example, business process reengineering to support BC/DR may be politically charged, such that it is best sponsored by upper management and executed with the assistance of outside resources.

■ Implementing a thorough and ongoing RFI (Request for Information), RFP (Request for Proposal), and contract negotiation process with vendors and providers will help assure that services and products are purchased at competitive rates and support organizational requirements. In Chapter 6, the RFI, RFP, and outsourcing process is discussed in greater detail.

9.3 Summary

As companies face increased pressure to maintain *"24/7"* operations without fail, BC/DR has become an organizational imperative and a significant new cost of doing business. Each organization will have many specific business requirements that must be supported by BC/DR efforts. However, several "A-ssential" requirements the BC/DR plan address should address are universal:

■ *Accessibility.* Employees, customers, vendors, and suppliers should have a vehicle to maintain essential communications.
■ *Adaptability.* The BC/DR plans and infrastructure should be scalable and capable of adapting to changing business, operating, and technology conditions.
■ *Alacrity.* The BC/DR response should be as rapid and as smooth as possible.
■ *Availability.* Key systems, data, and other business functions should remain operational, as required.

The checklist below summarizes the major steps in formulating a BC/DR plan:

■ Analyze risks and requirements:
  – Establish management sponsorship of BC/DR initiative.
  – Create BC/DR project team(s) with assigned accountabilities.
  – Perform risk assessment and business impact analysis.
  – Develop strategies and recommendations for mitigating risks
■ Develop the BC/DR plan:
  – Obtain funding approval and commitment from senior management.
  – Review existing plans and provisions for BC/DR, and develop recommendations and strategies to improve and augment them, where necessary.
  – Assess and define processes, procedures, policies, facilities, staffing, technology, and other resources needed to meet RPO and RTO objectives and recovery order.

- Develop BC/DR implementation plan that addresses personnel, equipment, facilities, and other backup capabilities needed to properly address recovery objectives.
- Develop testing methodology.
■ Implement and maintain the BC/DR plan:
- Maintain, test, and validate resiliency and backup services on an ongoing basis. Revise where needed to correct flaws in planning or to address changing business conditions.
- Develop and disseminate appropriate employee training and information to promote awareness of company BC/DR efforts.
- Maintain up-to-date documentation on all aspects of BC/DR planning.
- Perform ongoing management and operational evaluation of BC/DR needs and requirements.

Notes

1. IT systems (including application software, hardware, and other network components) characterized by *continuous operations* support 24/7 end-user accessibility. Continuous operations have uptime requirements in excess of 99.99 percent, corresponding to less than 53 minutes per year of lost time. No planned outages are allowed in this type of system, even for maintenance or other changes. A system with *continuous availability* (i.e., it is available over 99.9999 percent of the time, corresponding to less than 32 seconds per year of lost time) provides an even more stringent level of end-user service — in that neither planned nor unplanned outages are allowed. To ensure this level of operation, fault avoidance, fault tolerance, and fast recovery and restart mechanisms must be built into all system components.

2. Switch-based products are generally used with Ethernet hubs. These are sometimes alternatively described as session-level switches, content-intelligent switches, or layer 4 to 7 switches. If throughput requirements exceed 1000 Mbps, a router-based solution may be more appropriate.

3. The round-robin approach sends traffic requests to the next server in a list, irrespective of the server load. Adaptive algorithms use a variety of metrics, including historical data, to minimize the chance of overloading a particular server. Software agents running on the server can be used to report traffic loads to the load balancer so it can make informed decisions on traffic routing and server assignments. The least connections algorithm sends traffic to the server with the least number of clients. The fastest response algorithm sends traffic to the server that responds first.

4. Hot site "subscriptions are based on the number of MIPs, DASD, and peripherals required to recover a 'like' computer configuration. Subscriptions average 52 months in length and costs range from $250 to $120,000 a month." [SCHR03]

5. Synchronous data mirroring is used to back up mission-critical data. A primary data source and its "mirror" are synchronized so they can be swapped at any time without any downtime for the application server. Data mirroring is discussed in more detail later in this chapter.

6. Electronic data vaulting is discussed in more detail later in this section. Data vaulting allows data to be sent directly from a PC over a LAN to a designated hot site for safekeeping.

7. According to [SCHR03], cold site "subscriptions range from 12 to 60 months and costs generally range from $500 to $2000 per month."

8. By the year 1992, over half of the 16,384 Class B addresses had been allocated.

9. For complete details on CIDR, the interested reader is referred to http://www.faqs.org/rfcs/rfc1519.html.

10. Under IPv4, IP addresses are 32 bits, usually listed in decimal notation as four (4) bytes separated by a dot, from zero to 255 inclusive. The IP address contains two parts, each uniquely specifying a physical network and host ID. As long as these conventions are followed, within a private network, the IP addresses can have any assignment. However, connection to the public Internet requires use of registered IP addresses. There are five IP address types: (1) Class A (1.0.0.0–126.0.0.0), which supports up to 16 million hosts on each of 126 networks; (2) Class B (127.0.0.0–191.255.0.0), which supports 64,000 hosts on each of 16,000 networks; (3) Class C (192.0.0.0–223.255.255.0), which supports 254 hosts on each of 2 million networks; (4) Class D (224.0.0.0–240.0.0.0), which is a multicast address; and (5) Class E (241.0.0.0–248.0.0.0), which is reserved for future use. Addresses beginning with 127 (in decimal notation) are reserved for local machine testing.

11. IP subnetting provides a means to specify multiple physical networks with one IP network address by using some of the host-identifier bits. All hosts on the same network are designated by the same subnet mask. If there is a need for more hosts and fewer physical networks, then fewer host ID bits are used in the subnet. Because most Class A (recall that a Class A address block can support over 16 million host addresses) and Class B networks do not support as many hosts as they could theoretically, a lot of address space is wasted. Subnetting provides a way to conserve the IP address space provided under IPv4. CIDR conserves the existing address space even further by simplifying the subnet mask notation, essentially reducing it to a convention for separating an arbitrarily specified network ID from the remaining host portion of the IP address. Once the addressing space is exhausted, IPv6, with 128 bit addresses, will be needed.

12. DNS: the Domain Name System is a distributed database system that translates a mnemonic host name into an IP address.

13. DHCP: the Dynamic Host Control Protocol provides a means to assign a pool of IP addresses either dynamically or statically to a controlled group of IP users.

14. Content switch devices perform intelligent load balancing for multi-server Web sites based on URLs, file extensions, and cookies. They are used to maintain optimal server performance and utilization.

15. IPSec is an IETF network layer protocol for encryption and authentication of IP packets sent between peer devices (e.g., routers) on a public IP network infrastructure.

16. PKI (Public Key Infrastructure) provides a mechanism for establishing digital certificates, and registration and certificate authorities, to verify and authenticate each party of a Web-based, Internet transaction. There is no universal standard for PKI.

17. Network delay is introduced by framing, the transmission process, and other factors. By its very nature, synchronous data mirroring has a very low delay tolerance, particularly for write-intensive applications, such as on-line transaction processing (OLTP) applications. When DWDM MAN solutions are used for synchronous data replication, the lower bound on the transmission latency is the physical speed of light over the optical media. This physical constant is the reason behind the distance limitation of data mirroring.

18. SCSI (Small Computer System Interface) is an ANSI-defined standard providing a parallel interface between personal computers and peripheral devices (such as disk drives, tape drives, CD-ROM drives, printers, scanners, etc.). It is a ubiquitous standard that is supported on almost all personal computers and major operating systems.

19. Some NAS devices use the IP protocol over Fibre Channel for network attachment, but this is much less typical.

20. Direct Access Storage (DAS): storage devices directly connected to a server. Server performance can be seriously degraded if it is heavily loaded with DAS access requests. The preceding section discussed DAS in the form of tape drives connected via a SCSI parallel connection.

21. RAID (Redundant Array of Independent Disks) storage creates a seamless, single logical hard disk out of multiple hard disks. There are over nine types of RAID, each of which has been designed to optimize performance and fault tolerance for specific types of data storage requirements. For example, RAID-0 offers very fast performance but no fault tolerance or data redundancy. RAID-1 supports *disk mirroring* with at least two drives storing duplicate copies of data, and is designed to optimize performance and fault tolerance in a multi-user environment. Other types of RAID include RAID-2, RAID-3, RAID-4, RAID-5, RAID-6, RAID-7, RAID-10, and RAID-53.

22. Common Internet File System (CIFS): a protocol built on TCP/IP to allow programs to request files and services on remote computers connected to the Internet. CIFS can be used with Windows, UNIX, Linux, IPX/SPX, and other operating systems.

23. Network File System (NFS): an application that allows an end user to remotely access, store, or update files as if they are on a local hard drive.

24. RADIUS (Remote Authentication Dial-In User Services) is an open protocol developed by the Internet Engineering Task Force (IETF) RADIUS Working Group, and is defined in RFC 2139 and RFC 2865. This protocol is freely distributed as source code for centralized authentication, authorization, configuration, and accounting (to support billing, resource consumption

tracking, and network analysis) of dial-in users. Network access is controlled based on a set of user-specific attributes (e.g., username, password, configuration information, etc.) stored on a RADIUS security server. These attributes are used to establish and exchange communication between a NAS and a RADIUS server. User authentication can be based on a username/password pair, a challenge/response pair, or both. RADIUS is implemented in a variety of third-party and proprietary products (offering value-added attributes/extensions to the RADIUS standard). The IETF is extending RADIUS with the DIAMETER protocol. DIAMETER supports roaming users, exchange of user accounting information between ISPs, and requests for additional information at the time of user log-on.

25. PAP — The Password Authentication Protocol is defined in RFC 1334. It is a two-way handshake authentication scheme used by PPP servers to authenticate the identity of a connection originator. After a link is established, the originator sends an ID-password pair, which is used by the PPP server for authentication. Passwords are sent in cleartext and can be sent repeatedly, so this is not a strong authentication method.

26. CHAP (Challenge Handshake Authentication Protocol) is defined in RFC 1334. It is a three-way handshake authentication scheme used by PPP servers to authenticate the identity of a connection originator. After a link is established, the PPP server sends a challenge to the originator, who must in turn respond with a value computed using a one-way hash function. If the authentication fails, the connection is terminated. CHAP also allows the PPP server to repeat the authentication challenge at any time during the connection.

27. EAP (Extensible Authentication Protocol) extends the authentication methods of PPP, and is defined in RFC 2284. It supports multiple authentication approaches, including smart cards, Kerberos, one-time passwords, digital certificates, and public key authentication schemes.

28. Disk arrays connect several disks to create a high-capacity configuration. Cached disk arrays organize high-speed external memory, and look like a conventional disk to a server.

29. A *stackloader* is a robot-controlled device that loads tape cartridges according to some prescribed order. It is also referred to as an autoloader.

30. RAIT (Redundant Array of Independent Tape) is somewhat similar to RAID, except that removable tape media are used as the storage device, and not disk. It is designed to provide high performance and fault tolerance.

31. This is a fairly straightforward task. If the HBA is installed on a Windows server, the operating system will automatically detect the HBA and install the appropriate driver software.

32. SCSI versions are generally backward compatible. The differences in versions relate to the maximum allowed number of device attachments (from 8 to 16), the maximum cable length (from 1.5 to 12 meters), the maximum transport speed (from 5 to 160 Mbps), and specialized command sets (needed to support different types of device attachments and to provide different networking functions). These versions include (1) SCSI-1; (2) SCSI-2, which replaces the plain SCSI version 1; (3) Fast SCSI-2; (4) Wide SCSI-2;

(5) Fast Wide SCSI-2; (6) Ultra-2 SCSI; (7) Wide Ultra-2 SCSI; and (7) Ultra SCSI-3, which is available in 8- and 16-bit versions. Ultra SCSI-3 is the latest SCSI standard, designed to support speeds up to 160 Mbps.

33. INCITS is an ANSI advisory group.

34. A Fibre Channel device ID is referred to as a World Wide Name (WWN). The WWN is a 64-bit address. Fibre Channel uses the port address to reduce the size of the device ID to a 24-bit address. This is needed to reduce the system overhead associated with tracking and storing the device IDs.

35. Fibre Channel fabric defines the following categories of port types: (1) N-Port — This is used to designate devices such as host or server HBAs and storage targets. These are used to define Initiator and Target nodes. They can only attach to F-Ports or other N-Ports. (2) F-Port: this is used to designate switch ports. They must be connected to N-Ports. (3) E-Port: this is used to designate an expansion port. It must be connected to the E-Port of another switch. It is used to interconnect switched fabrics. (4) FL-Port: this is used to designate a switch port in an AL configuration. FL-Ports can only attach to NL-Ports. (5) G-Port: this is used to designate a generic port that can operate as either an E-Port or an F-Port. The final port assignment is determined at the time of device log-in. (6) L-Port: this is used to designate an AL fabric port or node. NL-Port refers to an AL (Arbitrated Loop) N-Port, while an FL-Port refers to an AL fabric port. NL-Ports only attach to NL-Ports or FL-Ports. (7) U-Port: this is used to designate a universal port that can operate as an E-Port, F-Port, or FL-Port. A port is defined as a U-Port when it is not connected or has not been assigned a specific function in the fabric. (8) Other ports: other types of ports exist and include vendor proprietary port types (i.e., McData's B-Port, which serves as a bridge port).

36. For detailed ANSI specifications on the Fibre Channel layers, the reader is referred to ftp://ftp.t11.org/t11/pub/fc/gs-4/02-026v0.pdf.

37. iSNS was developed by the IETF to support implementation of block storage IP solutions by providing a centralized point for managing and configuring a storage network via an iSNS server. iSNS was designed to accommodate a diverse mix of SAN devices on the same network, and supports a variety of protocols, including Fibre Channel-over-IP (FCIP), Internet Fibre Channel Protocol (iFCP), and Internet SCSI (iSCSI). iSNS is optional for iSCSI and is mandatory for iFCP. According to the IETF, iSNS performs four main functions: (1) name service providing storage resource discovery, (2) discovery domain and log-in control service, (3) state change notification services, and (4) open mapping of Fibre Channel and iSCSI devices. iSNS performs auto-discovery and address keeping functions analogous to Fibre Channel's SNS. As such, it must store and track device-specific information, (including port address, upper layer protocol support [e.g., SCSI-3], and IP address); FC name service requests; iFCP discovery zones; and other security and access control information. For more information on iSNS,

the reader is referred to the IETF Web page: http://www.ietf.org/internet-drafts/draft-ietf-ips-isns-21.txt.

38. The additional TCP/IP processing significantly adds to the workload of the iSCSI servers. Where this is a significant concern, a TCP Offload Engine Network Interface Card (TNIC) can be installed on the iSCSI server to offload this processing.

39. PPTP (Point-to-Point Tunneling Protocol) was developed to allow a private network (i.e., a VPN) to tunnel over a public network, such as the Internet, while providing secure communications. The public network is used as a replacement for a dedicated, private line network. PPTP is an extension of the Internet Point-to-Point Protocol (PPP). Tunneling involves encapsulating both private data and protocol information within TCP/IP PDUs so that they are carried as data over the IP network. Encryption is needed to ensure message privacy. GRE (Generic Routing Encapsulation) is a similar tunneling scheme developed by Cisco.

40. IKE: the IETF Internet Key Exchange (IKE) protocol is specified in RCF 2409 as a means to secure VPN communications. It provides an automated mechanism for remote host and network access that involves automatic negotiation and authentication based on IPSec Security Associations (SAs define security policies and rules defined for inter-node communication). IKE also supports (1) the ability to change encryption keys during an IPSec session, (2) Certificate Authorities, and (3) anti-replay services, as defined by IPSec. Anti-replay uses packet sequencing to prevent message interception and modification (by a hacker) between communicating devices. If a potential error is detected (i.e., it appears a packet may have been re-sent or tampered with), the "replayed" packet is logged and discarded. IPSec's Encapsulating Security Payload (ESP) and Authentication Header (AH) protocols also employ anti-replay protection.

41. For a detailed description of this mapping process, the reader is referred to *SNIA's Internet Fibre Channel Protocol (iFCP) — A Technical Overview.* This can be found at http://www.snia.org/tech_activities/ip_storage/iFCP_tech_overview.pdf.

42. Virtual Interface Architecture (VIA): is a server messaging protocol specification developed by Intel, Microsoft, Compaq, and others. It is designed to reduce CPU usage and delay between communicating processes running on two server nodes within a computing cluster. It does so by capitalizing on high bandwidth and high availability link technologies, thereby eliminating unnecessary processing overhead in transferring data between a CPU, memory subsystem, and high-performance network. This is somewhat analogous to the difference in message handling used by X.25 (which compensates for potentially unreliable network connections with lots of built-in error correction functions), and Frame Relay (which implements much less built-in error handling, introducing correspondingly less processing delay). A copy of the VIA specification can be found at http://www.viarch.org/.

References

[BENN02] Bennett, C., *From storage virtualization to NAS SAN convergence,* Storage Virtualization New, April 2002, http://www.storage-virtualization.com/sannas.htm.

[CISC03] Cisco Systems, *Data Center Networking: Enterprise Distributed data Center, Solutions Reference, Network Design,* March 2003, http://www.cisco.com.

[CISC03B] Cisco Systems, *Data Center Network Solutions,* 2003, http://www.cisco.com/application/pdf/en/us/guest/netsol/ns304/c654/cdccont_0900aecd80096182.pdf.

[CONN03] Connor, D., Start-up offers InfiniBand iSCSI router, Network World Fusion, July 28, 2003, http://www.nwfusion.com/cgi-bin/mailto/x.cgi.

[EQUA03] Equant, *Welcome to Connectonline Business Continuity,* 2003, http://www.equant.com/content/xml/connectonline_june_business_continuity.xml.

[FIBR03] Fibre Channel Industry Association, Business Continuity When Disaster Strikes, http://www.fibrechannel.com/technology/index.master.html.

[FRAU03] Frauenheim, E., *Microsoft gains ground in storage, CNET News.com,* July 7, 2003, http://news.com.com/2100-1015-1023583.html.

[FRAU03B] Frauenheim, E., *Microsoft wants larger piece of storage market, CNET News.com,* August 21, 2003, http://news.zdnet.co.uk/business/0,39020645,39115861,00.htm.

[GARD04] Gardner, W. D., Falling Prices, Regulations Drive SAN Market, *Techweb News,* January 6, 2004, http://www.techweb.com/wir e/story/TWR20040106S0001.

[HORW03] Horwitt, E., *Windows-powered NAS market maturing, ComputerWorld.com,* July 8, 2003, http://www.computerworld.com/hardwaretopics/storage/story/0,10801,82830,00. html.

[INFI04] InfiniBand Trade Association, *About InfiniBand® Trade Association: An InfiniBand™ Technology Overview,* http://www.infinibandta.org.

[INFI04B] InfiniBand Trade Association, *Specifications: FAQ,* 2004, http://www.infinibandta.org/spec/faq.

[INT02] Intel Corporation, *iSCSI: The Future of Network Storage,* 2002, http://www.intel.com/network/connectivity/resources/doc_library/white_papers/iSCSI_network_storage.pdf.

[JACO02] Jacobs, A., Vendors Rev Infiniband Engine, *Network World Fusion,* March 4, 2002, http://www.nwfusion.com/cgi-bin/mailto/x.cgi.

[KING03] King, E., Storage Market Remains Flat, Competition Fierce, December 15, 2003, *Windows &.Net Magazine,* Document #41179, http://www.winnetemag.com/rd.cfm?code=edep273lup

[MEAR04] Mearian, L., IBM to resell InfiniBand switch, *Computer World,* January 13, 2004, http://www.computerworld.com/printthis/2004/0,4814,89037,00.html.

[META02] Meta Group, Disaster and Business Continuity Planning: Key to Corporate Survival, 2002.

[MICR03] Microsoft.com, *Windows Storage Server 2003 is Here!,* 2003, http://www.microsoft.com/windowsserversystem/wpnas/default.mspx.

[OLAV03]   Olavsrud, T., *Gartner: U.S. Firms Unprepared for Disasters*, Gartner Research Group, March 3, 2003, http://www.cioupdate.com/news/article. php/2034101.

[SCHR03]   Schreider, T., US Hot Site Market Analysis & Forecast, *Disaster Recovery Journal*, September 29, 2003, http://www.drj.com/special/stats/tari.htm.

[SEAR03]   searchStorage.com Definitions, Karen Lefkowitz *business continuance*, February 11, 2002, http://searchstorage.techtarget.com/sDefinition/0,,sid5_gci801381,00.html.

[SHAN03]   Shankland, S., Server makers tout InfiniBand sequel, *C.net News.com*, May 5, 2003, http://news.com/com/2102-010_30999617.html?tag= st_util_print.

[SHAN04]   Shankland, S., IBM signs on InfiniBand switchmaker, *C.net News.com*, January 13, 2004, http://news.com/com/2102-1015_35139712.html?tag=st_util_print.

[SNIA04]   SNIA, *SNIA's Internet Fibre Channel Protocol (iFCP) — A Technical Overview*, 2004, Storage Network Industry Association and SNIA IP Storage Forum White Paper, http://www.snia.org/tech_activities/ip_storage/iFCP_tech_overview.pdf.

[SNIA04B]   SNIA, The Benefits of Internet Fibre Protocol(iFCP) for Enterprise Storgage Networks, 2004, *Storage Network Industry Association and SNIA IP Storage Forum White Paper*, http://www.snia.org/tech_activities/ip_storage/iFCP_user_overview.pdf.

[SPAN03]   Spangler, T., McData Sticking with iFCP, *Byte and Switch,* September 19, 2003, http://www.byteandswitch.com/document.asp?doc_id=40494&site=byteandswitch.

[STRO03]   Strohl Systems Group, Survey: BCP Budgets, Employee Involvement on the Rise — Quarterly Survey Investigates Business Continuity Planning Budgeting and Staffing, June 13, 2003, http://www.strohlsystems.com/CompanyInfo/NewsRoom/FullText.asp?ID=61.

[SUNG03]   Sungard Planning Solutions, Planning Outlook for the Financial Industry: Analysis and Recommendations on Trends and International Guidelines for Business Continuity, 2003, http://www.technologyforfinance.com/PDF/SunGardPS.pdf.

[SHAN03]   Shankland, S., Server makers tout InfiniBand Sequel, *News.Com,* Mary 5, 2003, http://news.com.com/2102-1010_3-999617.html?tag+st_util_ print.

[SHAN04]   Shankland, S., InfiniBand company wins further funds, *News.Com,* November 10, 2003, http://news.com.com/2102-7341_3-5105373. html?tag+st_util_print.

[SHAN03]   Shankland, S., Server makers tout InfiniBand sequel, *C.net News.com,* May 5, 2003, http://news.com/com/2102-1010_30999617.html?tag= st_util_ print.

[SHAN04]   Shankland, S., IBM signs on InfiniBand switchmaker, *C.net News.com,* January 13, 2004, http://news.com/com/2102-1015_3-5139712. html?tag=st_util_print.

[WITT01]   Witty, R. and Scott, D., Disaster Recovery Plans and Systems are Essential, Gartner Research Group, September 12, 2001, http://www3.gartner.com/5_about/news/disaster_recovery.html.

Appendix A

Return on Investment (ROI) Calculation

This appendix demonstrates how the costs and benefits of a network or IT initiative are evaluated from a financial perspective using return on investment (ROI) analysis. This analysis helps the decision maker compare and contrast the proposed network/IT project with alternative investment options. It is also helpful in justifying the funding for the endeavor.

A.1 Case Study Overview

For the purposes of illustration, this appendix presents a case study involving the development of a Web-based office suite system that will automate the sales and operations of a small importing and wholesale company.

Currently, all records are kept on paper, and orders and order tracking are done manually. When customers call in orders, customer service representatives (CSRs) write down the orders, check inventory records to determine if orders can be filled, and verify customer files to determine pricing and other information.

The proposed system will automate a large part of this process. Orders will be entered into the system as the customer phones them in. Inventory and pricing information will be stored in the system and automatically checked at order entry time. Part of the Web-based system can be linked

TABLE A.1   Logical Overview of Web-Based Project Network Architecture

Tier 1	Tier 2	Tier 3
GUI	Application Server	Database
The GUI will be developed using HTML, JavaScript, and possibly some Java applets. It will run on a standard Web browser.	App server will be written in PHP4 and run on a Dell Poweredge 4400 server.	MYSQL will be used for the database engine. This will initially run on the same server as the application. The option of having the database run on a dedicated server will be explored in year 3 when a new server is added to the system. We may also look into using NuSphere MYSQL which has better transaction processing support.

to the company's Web site to allow customers to place orders themselves via the Web.

The system will be written using HTML and PHP, with MYSQL as the database. Because these are open source technologies, available free of charge, software costs will be considerably less than for a similar system written using purchased software. The system will have an estimated life of five years. A logical overview of the proposed network architecture is presented in Table A.1. The ROI analysis for the system follows.

A.2 ROI Analysis

Calculating the ROI involves a thorough investigation of the project costs and benefits. The ROI analysis attempts to assess all the costs and benefits of a proposed investment or project over its expected life. The first step in analyzing a project's ROI is to develop an inventory of all the areas and activities that it will affect, and to quantify the associated costs. This is not an easy task but, nonetheless, it is an essential one for the ROI analysis.

A.3 The ROI Equation

ROI is expressed as a ratio of the benefits that an investment will provide (i.e., the potential savings and income expressed in dollars), divided by the total investment needed to derive those benefits (expressed in dollars). The ROI formula is expressed as:

$$\text{ROI} = \frac{\text{Savings} + \text{Income (Revenue} - \text{Costs)}}{\text{Investment Cost}} \quad \text{(A.1)}$$

where:
Savings = how much money will be saved after the system is in place
Income = how much money the investment will generate
   (i.e., revenue − costs)
Investment costs = how much money is needed to build the system

All components of the ROI formula should be computed over the total life expectancy of the project, asset, or investment and should be expressed in the same dollar and time units.

A.4 Depreciation and Capitalization

When the investment will occur over a period of years, depreciation must be taken into account in the ROI analysis. Several methods of depreciation are available for amortizing the investment costs over the expected life of the project.[1] Straight-line depreciation is used here to preserve the simplicity of this presentation. The annual straight-line depreciation expense is computed as:

$$\text{Annual Depreciation Expense} = \frac{\begin{array}{c}\text{Total Capitalization Costs}\\ \text{over Life of Project}\end{array}}{\text{Expected Life of Project in Years}} \quad \text{(A.2)}$$

Throughout the useful life of the project, additional investment costs will occur in the form of hardware upgrades (such as hard disks, memory, etc.) and various software expenses. These expenditures are referred to as capitalization (new investments) costs. They must be added to the denominator of the ROI formula when they are present.

A.5 Expanded ROI Formula

Taking into account the depreciation and expenses incurred to build and maintain the intranet, one can amend the ROI formula accordingly. A more accurate expression of the ROI equation is stated as follows:

$$\text{ROI} = \frac{\text{Savings} + \text{Income (Revenue} - \text{Cost)} - \text{Depreciation}}{\text{Depreciated Investment Cost}} \quad \text{(A.3)}$$

634 ■ *Network Design: Management and Technical Perspectives*

where:
Savings = how much money is saved after the system is in place
Income = how much money the investment will generate
Depreciation = the devaluation of capital expenditures over time
Depreciated investment costs = how much money is invested (capital) to
   build, depreciated over time

All components of the ROI formula should be expressed in the same units
of time and dollars.

A.6 ROI Numerator

A.6.1 Net Benefits: Additions

To calculate the ROI numerator, the benefits of the project must be defined
and calculated. Benefits may take two forms: (1) savings from costs
avoided and (2) returns from efficiencies created. For example, if a cost
is no longer necessary due to the project implementation, this can be
counted as a benefit. Similarly, if a job can be completed faster due to
the project implementation, the company has saved money, which can
also be counted as a benefit.

A.6.1.1 Costs Avoided

Many of the cost savings of the system will result from reduced expenditures due to electronic publishing. The company can place virtually
anything that it prints and distributes on paper onto its internal Web site,
thereby reducing printing costs. Visitors to the company Web site can
complete and return materials in an electronic format.

   In addition, the storage area required for the boxes of paper and toner
cartridges can be reduced. When hard copies are distributed to the
employee offices, they must be filed and stored. In addition, the system
should help reduce distribution expenses (i.e., postage, shipping, and
courier costs), which can be significant.

A.6.1.2 Efficiencies Created

The system will shorten the time between a customer request for information or order and the response and order delivery. It will also reduce
the time spent processing customer and order information. Although major
advantage of a Web-based approach lies in its ability to disseminate
information across many platforms, the ease with which users can access
the information is another crucial advantage. In the example that follows,

the time employees save performing their required assignments as a result of using the automated system will be factored into the ROI analysis. If using the system saves an hour, then that hour of salary should be factored in as a benefit.

A.6.2 Costs Incurred: Minuses

Costs fall into two major categories: (1) "hard" costs and (2) "soft" costs. Hard costs are measurable and concrete. For an IT project, the hard costs include hardware and software purchases, and the costs of installing the hardware or software. Soft costs relate to intangible factors, such as lost productivity or inefficiency during the period the users are learning to use the intranet. Soft costs may be easy to identify, yet difficult to quantify. For example, it is difficult to estimate how long it will take employees to learn the new system and to become as proficient as they are with the existing technology and systems. However, both hard and software expenses must be quantified in concrete numbers for the ROI analysis.

Common expenses associated with the deployment of intranet/Internet projects include:

- *System personnel.* A system administrator is needed to keep the intranet running, install new equipment, upgrade desktop computers, and maintain system security. Although current staff may be able to absorb these responsibilities, a new position may be necessary for larger installations. As the number of employees using the intranet increases, there may also be an increase in the number of customized software applications needed. An application administrator may also be needed to manage existing software programs and to write custom applications.
- *Training.* The system and application administrators will need to attend classes and certification programs to stay on the cutting edge of the technology. Users should also have ongoing instruction on the use of the intranet.
- *Ongoing planning.* The company should put in place a cross-functional team that continuously guides the project's development. While these team members are involved in planning activities, they are not accomplishing their regular assignments. These costs should be calculated based on their salaries.
- *Authoring.* The intranet is only as useful as the information it provides. The company should encourage employees to put information on the internal Web. When employees are authoring such information, they are not performing their regular tasks. The cost of these activities should be based on employee salaries.

- *Miscellaneous hardware and software.* This should include budget allowances for anticipated and unexpected (i.e., this is a form of fudge factor) hardware and software expenses.

In summary, the ROI numerator (net benefits) can be expressed by the following equation:

$$\text{Project benefits} = \text{Cost avoided} + \text{Efficiencies created} \\ - \text{Cost incurred (including depreciation)} \qquad \text{(A.4)}$$

A.7 ROI Denominator

A.7.1 Project Investment

The project investment consists primarily of hardware and software expenditures. Some companies may have enough surplus equipment to implement an intranet or small Web-based project such that they do not need to incur hardware costs. In some cases, appropriate software can be downloaded free of charge from Internet sites to keep start-up costs to a minimum. Even in this situation, the most balanced and accurate ROI analysis will use retail prices for the hardware and software used in project deployment. Investment costs for this case study include:

- *Pre-installation planning.* The company should formulate a cross-functional team to design the system. While these team members are defining the project plans, they are not accomplishing their regular assignments. This lost productivity cost should be calculated based on each employee's salary.
- *Hardware.* The Web-based system will require at least one server computer with adequate hard disk space and speed to support the Web sites. The servers will require replacement or expansion during the expected life of the project. In addition, additional servers will be needed to serve as security firewalls.
- *Software.* The project will require server software, as well as client software (browsers), firewall software, HTML authoring tools, collaborative groupware, a document-management system, a search engine, and possibly more, depending on the company's needs.
- *Installation.* The company must install the proper software on each server. The company may also need to install software (i.e., browser software, etc.) on each user's computer system. One or more skilled technicians may spend many hours completing this work.
- *Depreciation.* As discussed previously, annual depreciation should be subtracted.

To finalize the calculation of the ROI denominator, add all the costs incurred in the original investment and any subsequent years, and then subtract all the annualized depreciation expenses. In summary, the ROI denominator is calculated as follows:

Depreciated investment costs = Investment costs − Depreciation   (A.5)

All components of this equation should be expressed in the same time and dollar units.

A.8 Sample ROI Analysis

The following example provides the details of an ROI analysis for this case study. Recall that this analysis is based on an expected project life of five years.

A.8.1 ROI Numerator

Refer to Table A.2 for the specific items and costs that are relevant for the calculation of the ROI numerator for this sample problem. The discussion that follows provides a detailed explanation of each benefit and cost line item listed in Table A.2. The items listed in Table A.2 represent the revenue opportunities, costs avoided, and efficiencies created that are factored into the ROI numerator.

A.8.1.1 Savings and Cost Avoidance Related to Web-Based Project

The savings and cost avoidance related to a Web-based project include:

- *Salary savings.* This item represents the money saved by making employees more efficient through the use of an automated process. The company has three employee types: partners (two), customer service representatives (four), and warehouse personnel (six). Salaries are based on job category. For this example, an informed assumption is made on the amount of time the system will save each worker during a typical week. A conservative estimate is that the system will save partners ten hours per week, customer service representatives twenty hours per week, and warehouse personnel three hours per week. The associated salary savings are summarized in Table A.3. The formula below shows the calculation of the annual labor savings:

TABLE A.2   Total Benefits Calculated for ROI Numerator

Item	Year 1	Year 2	Year 3	Year 4	Year 5	Cumulativ
Annual Savings and Income						
Salaries	$168,500	$181,980	$196,538	$212,261	$229,242	$988,52
Costs avoided	$1,200	$1,320	$1,452	$1,597	$1,757	$7,32
Increased cash flow	$48,000	$48,000	$48,000	$48,000	$48,000	$240,00
Data mining	$30,000	$31,500	$33,075	$34,729	$36,465	$165,76
Better decisions	$10,000	$10,000	$10,000	$10,000	$10,000	$50,00
Addl. Business		$50,000	$52,500	$55,125	$57,881	$215,50
Total annual savings	$257,700	$322,800	$341,565	$361,712	$383,345	$1,667,12
Annual Expenses						
Training	$6,000	$3,000	$3,000	$0	$0	$12,00
Web hosting	$3,600	$3,600	$3,600	$3,600	$3,600	$18,00
System admin.	$22,500	$23,625	$24,806	$26,047	$27,349	$124,32
Depreciation	$68,075	$44,249	$37,862	$34,410	$31,466	$216,06
Total annual expenses	$100,175	$74,474	$69,268	$64,057	$62,415	$370,38
Total Net Benefits	$157,525	$248,326	$272,297	$297,655	$320,930	$1,296,73

TABLE A.3   Annual Labor Savings due to Web-Based System

Item	# of	Annual Hours Saved	Average Salary/Year	Total Salary	Salary/ Hour	Annual Savings
Partners	2	1000	$150,000	$300,000	$75.00	$75,000
CSRs	4	4000	$40,000	$160,000	$20.00	$80,000
Warehouse personnel	6	900	$30,000	$180,000	$15.00	$13,500
Total	**12**	**5900**		**$640,000**		**$168,500**

Productivity savings by job category =
(Number of employees * Hourly salary) ×
(Number of hours saved per year per employee)

For example, for the Partners category, the first year's productivity savings is:

Year 1 = (2 * \$75) * 500 = \$ 75,000

In each successive year, the system will become increasingly useful, and therefore the company will accrue more salary benefits over time. We estimate that the time savings associated with increased employee productivity due to the system usage will increase 8 percent a year.

■ *Savings from improved flow through better inventory management.* The new system will allow the company to better manage inventory such that only slightly more inventory than that needed to meet customer requirements will be kept on hand. The additional inventory will be used as a small buffer to fill unexpected orders. The use of the data collected by the system to forecast inventory will help the company avoid tying up cash in inventory when it is not necessary. It is estimated that this will save the company approximately \$4000 per month.

■ *Increased sales revenue through data mining and product advertising.* Customer purchasing history can be used to notify customers of new products they may be interested in or to announce that a product they wanted is now in stock. It is estimated that this will increase sales by an additional \$30,000 the first year and increase by 5 percent per year thereafter. This will also allow the company to provide better service to the customer.

■ *Savings due to improved cash management and better business decisions.* Management can use system reports to get an accurate picture of the state of the business at any point in time. These reports will allow management to make better decisions on products and services, and when to rein in costs. A conservative estimate suggests realizing a savings of \$10,000 per year from this benefit.

■ *Additional revenue through increased customer sales.* A conservatively estimate suggests that additional sales received from the Web site will net the company \$50,000 per year, starting the second year, and increasing at a rate of 5 percent per year thereafter.

■ *Costs avoidance savings.* An estimate suggests that the system will allow savings of \$100 per employee per year. These avoided costs are a result of electronic information storage as opposed to the storage of paper that would be necessary otherwise. Although the system will not completely eliminate paper and other supplies from the office, it should greatly reduce the amount currently used. This cost savings is estimated to increase at 10 percent per year.

A.8.2 ROI Denominator

The ROI denominator represents the total net investment made in the project. Table A.4 lists the items relating to the investment costs and the depreciation costs associated with the Web-based system, while the discussion that follows describes these items.

ROI denominator costs related to the intranet project include:

- *Project-related salary costs* include:
  - Pre-installation planning costs are estimated based on a team of four working 40 hours per week at $65 per hour.
  - Server/desktop installation costs are estimated at one person spending 40 hours at $65 per hour.
  - Application planning and development are estimated on the basis of four people working 40 hours per week at $65 per hour for four months.
  - Ongoing application development costs are estimated based on one person working part-time for 500 hours per year at $50 per hour, starting the second year.

TABLE A.4   ROI Denominator Items for Web-Based Project

Item	Year 0	Year 1	Year 2	Year 3	Year 4	Year
Planning	$10,400	$0	$0	$0	$0	
Server	$1,999	$0	$0	$1,999	$0	
Server software	$1,000	$0	$1,000	$1,000	$1,000	$1,0
Server/desktop installation	$2,600	$0	$0	$0	$0	
Document mgmt. software	$2,500	$0	$0	$0	$0	
Application dev.	$166,400	$0	$25,000	$25,000	$25,000	$25,0
Desktops/peripherals	$9,600	$0	$0	$0	$0	
Total investment	**$194,499**	**$0**	**$26,000**	**$27,999**	**$26,000**	**$26,0**
Depreciated inv. cost		**$126,424**	**$108,176**	**$98,313**	**$89,904**	**$84,4**

- System administrator costs are estimated at one person working part time at $45 per hour for ten hours per week, increasing at 5 percent per year thereafter. This person must also be available for emergencies.

■ *Training costs:* employees will have to be trained to use the new system. The employees will be split into two groups and training will be conducted at a rate of $1500 per week per group. Two weeks of initial training will be required, with one week per year thereafter for two years.

■ *Equipment-related costs* include:
  - Server costs for the purchase of one Dell Poweredge 4400 with 256 MB RAM, one Intel Pentium III processor with a 38-GB hard drive.
  - Additional servers: the estimated requirement is for one additional server during the third year.
  - Desktop computers and peripherals expenses: the estimated requirement is for eight desktop systems at $1000 each and two network printers at $800 each.

■ *Software-related costs* include:
  - Apache Web server along with MYSQL and PHP will be used. All are available free of charge.
  - Document management software is estimated at $2500.
  - Incidental software needs are estimated at $1000 per year.

■ *Web hosting expense:* virtual Web hosting using a dedicated server provided by a hosting company will cost $300 per month. This will eliminate normal maintenance costs associated with having the server on site.

■ *Depreciation:* depreciate the total investment by 35 percent per year. The denominator of the ROI formula will be computed by multiplying 0.35 by the previous year's total investment plus the current year's total investment.

A.9 Results

The typical ROI approval guideline for IT investments requires a positive return on investment in the range of 12 to 18 months. However, some tactical initiatives may be required to demonstrate a positive ROI in three to six months. Based on this, the results of the preceding ROI analysis indicate that the project is an excellent investment opportunity. From the first year of the project, to the final year of the project, the ROI is estimated at between 124 percent and 380 percent. Thus, the Web-based system promises to be an important tool to augment company productivity, and

TABLE A.5   Final ROI Calculations for Web-Based Project

	Year 1	Year 2	Year 3	Year 4	Year 5
Annualized ROI	124%	230%	277%	331%	380%

a strong business case can be made for proceeding with the project. The final ROI calculations for years 1 through 5 of the project are summarized in Table A.5. A sample project schedule for realizing the project implementation is presented in Table A.6.

Although the results calculated for this hypothetical case study may be somewhat optimistic, the wise deployment of Internet/intranet technology often provides substantial savings to an organization. One area where Web-based technology has had a significant impact on corporate savings is E-training. In large companies, more dollars are going to technical skills training than any other type of training, and E-learning provides very substantial ROI cost savings over traditional classroom instruction. McDonald's, with 1.5 million employees worldwide and the nation's largest training organization (surpassing the U.S. Army's), estimates that its E-learning program will reduce training time by 40 to 60 percent, and that "just-in-time" (JIT) learning will double employee retention rates for the training material.[2] There are numerous examples of successful IT initiatives with large ROIs, but perhaps more important are the initiatives that are not embarked upon because an ROI analysis revealed that they would not provide a good return to the organization.

As CIOs and IT departments face increasing scrutiny, ROI analysis plays an important role in demonstrating the business value of technology investments. ROI computes returns, risks, and rewards in very tangible terms that can be readily understood by business executives and technologists alike. ROI clearly shows what and where investment will yield real dollar savings, cost avoidance, and incremental revenue opportunities.

TABLE A.6   Sample Project Schedule for Web-Based Project

Date	Task Description*	Duration
1/01/02	Prepare project overview and return on investment analysis.	2 days
1/10/02	**Present overview to partners for approval. Review this project plan with partners and project team members. Modifications to be made as necessary.**	1 day
1/14/02	Meet with partners and key users to develop requirements document.	5 days
1/21/02	Distribute user requirements partners and key users for review. Walk through systems requirements with partners and key users to eliminate ambiguity. Make modifications as necessary.	7 days
1/21/02	Hardware/software installation.	5 days
2/01/02	**Begin systems development.**	4 persons for 4 months
	Schedule weekly status meetings to gauge progress and to discuss issues that may affect schedule.	1 hour per meeting
3/01/02	**Demonstrate Web prototype, with limited functionality. Possibly demonstrate one or two reports if ready.**	2 hours
4/15/02	Major functionality completed. Test plan development to be started.	5 days
4/22/02	**User testing to begin. Track all issues and assign a priority for resolution.**	15 days
5/13/02	**Demonstrate system functionality to partners and key users. Development continues.**	1/2 day
5/13/02	Formal user training to begin.	4 weeks
6/10/02	User training completed.	
6/10/02	Final preproduction testing and issue resolution.	2 weeks
6/28/02	**Project team celebration dinner.**	
7/1/02	**Go live.**	
8/1/02	**Post-production project review.**	2 days

* Task entries in bold indicate milestones.

Notes

1. Generally, Accounting and Finance departments can provide guidance to the analyst preparing an ROI analysis to suggest the most appropriate form of depreciation calculation, given the type of investment under consideration and the business context. For an excellent and comprehensive treatment of various depreciation methods, the interested reader is referred to *Economic Analysis for Engineering and Managerial Decision Making*, by N. Barish and S. Kaplan, McGraw-Hill, Inc., New York, copyright 1978.
2. Gotschall, M., E-Learning a Strategic Imperative for Succeeding in Business, *Fortune Magazine*, April 2004, http://www.fortune.com/fortune/services/sections/fortune/edu/2001_05elearning.html.

Appendix B

Overview of OSI Model and Comparison with TCP/IP

This appendix reviews the OSI Model[1] and compares it with TCP/IP. The OSI Model was developed by the ISO (International Organization for Standardization). ISO is the world's largest standards body, comprised of a Central Secretariat (in Geneva, Switzerland) which performs coordination functions, and members from 146 other countries. Only one member body or institute per country is allowed to participate in the development of and voting on ISO standards. Members are typically part of their respective government sector or have strong affiliations with industry. The intent of the organization is to develop standards for the greater good of society that embody a consensus across a broad spectrum of users, consumers, suppliers, and service providers.

In contrast, the IETF (Internet Engineering Task Force)[2] is responsible for Internet and TCP/IP protocols. The IETF is an international "community" comprised of an enormous cross-section of public, private, research, vendor, and government representatives. However, unlike the ISO, the IETF is open to anyone. Working groups, organized by specialized areas of interest, are responsible for recommendations and technical activities. The IAB (Internet Architecture Board) provides oversight of the IETF.

ARPANET, the precursor to the Internet, was developed in the late 1960s and was based on the TCP/IP packet-switching protocol. From its inception, TCP/IP was designed to promote fault-tolerant resource sharing and connectivity between multiple, autonomous, nonhomogeneous networks. Its development was sponsored by the United States Defense Advanced Research Project Agency (DARPA).

The OSI Model was conceived in the late 1970s as a means to promote device connectivity and interoperability across many types of vendors and platforms. It is based on a seven-layer hierarchy that, in practice, is only partially implemented. The OSI Model has served an essential role in established standards for interoperability; however, it is extremely complicated and difficult to implement fully. Given TCP/IP's history and simplicity, it is not surprising that it has emerged as a predominant networking protocol.

B.1 Overview of the OSI Model

The OSI Model provides a conceptual, layered stratagem for network implementation. The actual implementation of OSI reflects each vendor's unique interpretation of the standard and its own value-added offerings. Thus, the products of different vendors following the same OSI standard are like snowflakes: no two are truly identical.

Each OSI layer performs different functions. The functions of each layer are designed to operate in isolation and as independently as possible, thus facilitating multiple vendor implementations and management of complexity. Thus, by layering the network functions, changes in service at one level are transparent to the rest of the layers. See Figure B.1 for a summary of the OSI model. Note that the bottom two OSI layers — the physical and the data-link layers — are implemented in hardware, while the other layers are implemented in software.

Data passes from the top application layer and successively moves downward through the OSI layers toward the physical layer. As data passes through each successive OSI level — in a process called *encapsulation* — header data is appended according to the protocol rules operating at that level. After the data has been transmitted across a physical media (e.g., fiber, coaxial cable, etc.), it begins an upward transmission back through the protocol stack. This involves a *deencapsulation* process, whereby each layer on the receiving/destination side of the protocol stack removes the header created by its corresponding source layer.

Layer 7	Application	Operating System & Application Software (e.g., E-Mail, Word Processing, Database Software)
Layer 6	Presentation	Network Services: Translation, Compression, Encryption, etc. (e.g., AFP)
Layer 5	Session	Establish, Maintain, Terminate Sessions Between User Apps (e.g., FTP, SNMP)
Layer 4	Transport	Flow Control, Error Recovery, & Reliable End-to-End Network Communication (e.g., TCP, UDP)
Layer 3	Network	Data Transmission & Switching (e.g., IP, IPX)
Layer 2	Data Link	Data Encoding & Framing, Error Detection & Control (MAC, LAN Drivers)
Layer 1	Physical	Physical Interface (e.g., FDDI, Token Ring, Ethernet)

FIGURE B.1   OSI seven-layer model.

B.2 Overview of the TCP/IP Model

In contrast to the OSI Model, the TCP/IP protocols are designed to operate at the network, transport, and applications levels. TCP/IP does not define a layer corresponding to the OSI session or presentation layers, and the physical and data layers are not tied specifically to the TCP/IP protocol suite. See Figure B.2 for a comparison of TCP/IP with the corresponding OSI Model.

The network (also known as the Internet) layer performs a variety of functions, including network addressing to identify communicating nodes; packetization; link management; routing functions; flow control; error detection, control, and correction; and frame synchronization. Four protocols have been defined to perform these functions: (1) Internet Protocol (IP), (2) Address Resolution Protocol (ARP), (3) Reverse Address Resolution Protocol (RARP), and (4) Internet Control Message Protocol (ICMP). These protocols are implemented in routers and customer premise end-devices (CPE).

The transport layer supports communications between two hosts. The protocols defined for this layer are the Transmission Control Protocol (TCP) and the User Datagram Protocol (UDP). These protocols support flow control, as well as reliable, transparent data ordering and delivery.

OSI Layer 7	Application	Includes Network Applications (e.g., FTP, SMTP, & http)
OSI Layer 4	Transport	Host to Host Data Transmission (e.g., TCP, UDP)
OSI Layer 3	Network	Data Transmission & Switching (e.g., IP, IPX)
OSI Layer 2	Data Link	Data Transmission Between Network Node (e.g., ppp and Ethernet)
OSI Layer 1	Physical	Physical Interface (e.g., FDDI, Token Ring, Ethernet)

FIGURE B.2   TCP/IP five-layer model.

The application layer supports communications between user applications. A few examples of protocols operating at this layer include File Transfer Protocol (FTP), which is used to allow the transfer of files between local and remote host computers; Simple Mail Transfer Protocol (SMTP), which is the standard for exchange of e-mail over the Internet; Hypertext Transfer Protocol (HTTP), which is used as a standard for exchanging information over the Internet; and the Domain Name Service (DNS), which is used to translate domain names (e.g., www.tcrinc.com) into IP addresses.

TCP/IP uses a similar encapsulation/deencapsulation process as described for OSI.

Notes

1. For more information on ISO and the OSI model, the reader is encouraged to visit the official ISO Web site at http://www.iso.ch/iso/en/ISOOnline.frontpage
2. For more information on the IETF, the reader is encouraged to visit the official IETF Web site at http://www.ietf.org/overview.html

Appendix C

Subjective Evaluation and Ranking Procedures

This appendix discusses the problem of scoring and ranking options or actions according to the subjective (and objective) features they possess. This type of problem appears in many forms in risk management and decision making. For example, one might wish to identify "high-risk" business objectives. In another case, one might wish to rank network alternatives according to the potential benefits they provide to the organization.

Many treatments for problems of this type are described in the literature. Statistical techniques can be used to produce predictive models based on analysis of historical data. However, these techniques are not effective in predicting behavior that has not been observed in the past. Alternatively, a "fuzzy" approach can be used to evaluate options or assign ranking scores. This approach is practical if the risk factors and corresponding actions are sufficiently well understood and identifiable. It involves defining the following components:

- *Fuzzy membership functions:* which represent a definition of each decision variable. One fuzzy membership function must be developed for each decision criterion. Fuzzy membership functions provide a means to define linguistic measures that may be difficult to quantify precisely but are, nonetheless, important considerations in the design process. For example, one design may be preferable to another because of "convenience," "ease of maintenance," etc. A fuzzy membership function provides a natural way to represent both objective and subjective preferences in a manner that allows numerical analysis.

- *Decision priority weights:* which represent the importance of each decision variable relative to each other.
- *Aggregation operators:* which aggregate the results of the fuzzy ranking function and are used to calculate an overall score for a particular option under consideration.

These components are used to construct a mathematical scoring equation of the following form:

$$F(X) = \text{Operator } (F_1^{w1}, F_2^{w2}, \ldots, F_4^{w4}) = \text{Score}$$

where w_n is proportional to the relative criteria weight for each respective decision parameter 1, ..., n, and is analogous to a linguistic qualifier (e.g., "very"); F_n represents fuzzy membership values for each respective decision parameter 1, ..., n; and "Operator" represents a generic aggregation procedure for computing a final score based on the weighted decision parameters. Many types of aggregation operators are described in the literature. For simple trade-off comparisons of the type suggested in this text, a "Mean" operator generally suffices, although other operators can be used.

After computing the fuzzy score for each option under evaluation using the formula above, the options can be sorted and ranked. It is important to note that although the rankings provided by the formula described above indicate a preference for one option over another, they do not indicate how *much* one option is preferred over another.

The discussion now turns to the derivation of each of the formula components. A two-step method to define fuzzy membership functions is suggested:

- *Interview an expert analyst or key decision maker.* For each decision parameter, ask the analyst to assign fuzzy scores to selected parameter values. The fuzzy score represents the degree to which the expert agrees (or does not agree) that the given parameter value is consistent with the linguistic variable to be defined. A fuzzy score is always between zero (0) and one (1), inclusive. A fuzzy score of one (1) indicates complete agreement with the linguistic measure, while a zero (0) score represents complete disagreement. These results should be collected as (parameter value, fuzzy score) pairs. The first term is the value of the parameter to be evaluated,

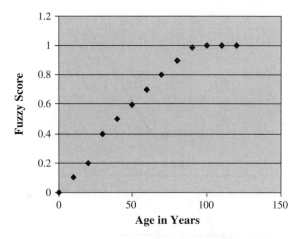

FIGURE C.1   Defining a fuzzy membership function: "Old."

and the second term of the pair in parentheses is the fuzzy score assigned by the analyst. A minimum of three (parameter value, fuzzy score) pairs should be captured, covering the entire range of the solution space. This means that a fuzzy score should be assigned to the minimum possible parameter value and that a fuzzy score should be assigned to the maximum possible parameter value. Thus, at least one pair must have a fuzzy score of 0, and at least one pair must have a fuzzy score of 1 (which implies complete agreement with the term being defined). The fuzzy membership functions can be derived by interviewing resident experts and asking them to assign fuzzy scores to selected values. For example, to define "old," an expert would be given a series of ages and then asked to provide a fuzzy score for each age. The fuzzy score represents the degree to which the expert agrees, or not, that the given age is consistent with the linguistic variable to be defined (i.e., "Old"). This is illustrated in Figure C.1. Figure C.2 and Figure C.3 illustrate this process on a second linguistic decision variable: "High Price."

■ *Use curve fitting techniques to derive an estimation of the fuzzy membership function equation based on the (parameter value, fuzzy score) pairs.* This process can be automated fairly easily, and spreadsheets are particularly well suited to this purpose. For example, linear regression is one of the most common and easily applied curve-fitting techniques. Piece-wise linear approximation techniques are also suitable for this purpose. The resulting equation derived by curve-fitting techniques becomes the definition of the fuzzy membership function. This is illustrated in Figure C.4.

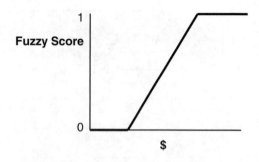

FIGURE C.2   Defining a fuzzy membership function: "High price."

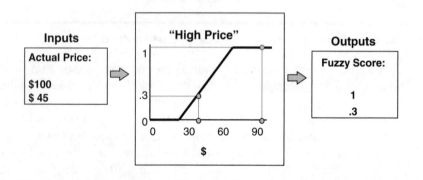

FIGURE C.3   Translating a fuzzy membership function: "High price."

The decision priority weights can be defined in a number of ways. The easiest way is to solicit user preferences or suggestions and to incorporate them directly into the fuzzy model after they have been normalized into the unit interval (i.e., all the weights of all the decision variables should sum to one). For problems that do not require a high degree of precision, this approach may provide adequate results.

Saaty's method of pair-wise comparison is another, more rigorous way to derive the values for the decision priority weights. Saaty's method assumes that the person developing the model can specify the relative importance of each variable relative to every other variable. Saaty [SAAT72] observed that is it conceptually simpler for most people to make pair-wise judgments than to perform a complete prioritized ordering of multiple objects. Based on this insight, Saaty developed a procedure for constructing a vector $W = (W_1, W_2, ..., W_n)$, where each component of W represents the relative importance of the decision criterion under consideration. More formally, the components of W are defined to have the following properties:

$$W = \sum_{i=1}^{n} W_i = 1$$

Where W_i represents the relative importance of decision criteria n, and all W_0, W_2, ..., $W_n > 0$.

To construct the components of W_i, expert analysts are asked to order n decision criteria by building an n×m matrix composed of O_{ij} comparisons (where O_{ij} represents the comparison made by the analyst between criteria i and criteria j), according to the following rules:

1. Assign O_{ij} a value based on Table C.1.
2. Assign O_{ii} and O_{ji} such that $O_{ij} = (1/O_{ji})$, and $O_{ii} = 1$.

This will produce a symmetric matrix. The components of W (representing the weights to be given to the decision criteria) are obtained by computing the largest eigenvalue associated with the eigenvector of the symmetric matrix.

A variety of software packages and routines are available for computing eigenvalues. A few of these include:

- *OCTAVE.* This is a free, public domain software package. It can be found at: http://www.octave.org/
- *MATLAB.* For more information, visit the Mathlab Web site: http://www.mathworks.com/
- *EULER* — This is a free interactive computer for matrix computation on UNIX and for UNIX/Linux systems. For more information, visit the Web site: http://euler.sourceforge.net/
- *JAMA.* This is a basic linear algebra package for Java that supports eigenvalue matrix calculations. Both NIST and Mathworks support a public-domain reference Java implementation. For more information, visit the Web site: http://math.nist.gov/javanumerics/jama/

The largest eigenvalues found are then multiplied by the number of elements in the matrix (i.e., the number of decision criteria). This normalization is done to ensure that the average of the weights will equal one, in the event that all the weights are found to be equal. The normalized eigenvalues are then used as the final components of W. Detailed treatment of this process can be found in [RUBI92].

A third way to determine the decision weights is to use automated discovery techniques, such as genetic algorithms (GAs) or artificial neural networks (ANNs). A GA or ANN can be used to automate selection of

Note:
- Data is first collected to estimate fuzzy membership function. This is usually from interviews with decision makers.
- The data are then fit to a continuous curve in the interval [0,1] using EXCEL, linear regression, or other curve fitting tools.
- The X-axis represents actual values and the y-axis represents the associated fuzzy score. A fuzzy score of 0 means complete lack of agreement, 1 means complete agreement, and .5 means moderate agreement with the linguistic term being defined.

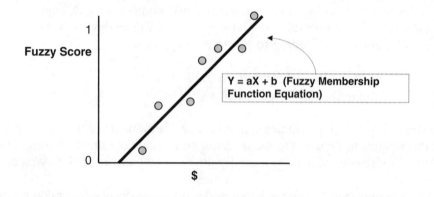

FIGURE C.4   Computing a fuzzy membership function.

the decision parameters, finely tuning them to solve a particular ranking problem. To use a GA or ANN, large numbers of sample cases (showing, for example, high risk, medium risk, and low risk scenarios) must be available for training. The training and model development process using these techniques is beyond the scope of this text; however, more details on this approach can be found in [RUBI99, YAGE97]. When the problem at hand is especially important and when the relative importance of each decision variable is unknown or uncertain, then automated techniques may be needed to derive or tune the weights to the required level of precision. In these cases, the extra complexity of this approach may be warranted.

The final component of the fuzzy ranking equation is the aggregation operator. A simple but robust aggregation operator is the "mean" operator. This operator is simply the arithmetic mean of the weighted decision factors. Although more sophisticated operators may be required for very rigorous evaluations, the authors have found that in many cases, the simplicity and directness of the mean operator make it very appealing for the type of planning and risk management analyses discussed in this book.

To illustrate the use of this operator and others, an example is warranted. Table C.2 lists three network options under consideration. Each

TABLE C.1 Saaty's Pair-wise Comparison Matrix

Assigned Rating (To indicate corresponding weakness of first criteria relative to the second, assign an inverse rating, i.e., 1/Rating)	Meaning
1	Equal importance between decision criteria
3	Weak importance of first criterion over the second decision criterion
5	Strong importance of first criterion over the second decision criterion
7	Demonstrated importance of first criterion over the second decision criterion
9	Absolute importance of first criterion over the second decision criterion
2, 4, 6, 8	Intermediate judgment values between first and the second criteria

TABLE C.2 Evaluating Network Options

Option	Price*	Functionality*	Ease of Use*
Option #1	0.7	0.8	0.9
Option #2	0.9	0.8	0.5
Option #3	0.1	1	1

* Table entries are determined by a fuzzy membership function for each decision variable (i.e., price, functionality, and ease of use). An actual price is translated into a fuzzy score, according to Figure C.3, while functionality and ease of use are assigned based on a subjective assessment.

option is evaluated on the basis of price, functionality, and ease of use. The scores assigned to each of these criteria for each option are listed in the table, and are determined by a fuzzy membership function. Table C.3 lists three possible aggregation operators: (1) the Minimum operator (which assigns the final score based on the lowest valued criteria), (2) the Maximum operator (which assigns the final score based on the highest valued criteria), and (3) the Mean operator (which assigns the final score based on an arithmetic average of all the criteria values). Table C.4 shows

TABLE C.3   Some Simple Aggregation Operators

Operator	Linguistic Interpretation	Features	Scoring Example
Minimum	"AND": goal is to satisfy option 1 and option 2 and option 3, etc.	Noncompensatory Min/max game strategy (minimize the worst risk)	min (0.7, 0.8, 0.9) = 0.7
Maximum	"OR": goal is to satisfy option 1 or option 2 or option 3, etc.	Noncompensatory Maximize one thing, no mater what it is!	max (0.7, 0.8, 0.9) = 0.9
Mean	"ALL": goal is to satisfy all options to some extent	Compensatory Balance objectives in making a choice	mean (0.7,0 .8, 0.9) = 0.8

**TABLE C.4   Ranking Network Options
with MEAN Operator**

Find the fuzzy score for each option, and then sort in descending order.	
F(option #1) =	Mean (0.7, 0.8, 0.9) = 0.800
F(option #2) =	Mean (0.5, 0.8, 0.9) = 0.733
F(option #3) =	Mean (0.1, 0.1, 0 1) = 0.400

the scoring and ranking of the three network options using the MEAN aggregation operator. If decision weights are used, the fuzzy scores for each criterion are augmented accordingly before the final aggregation and ranking. Table C.5 provides an example to demonstrate this process.

In cases where more precision is desired, the Ordered Weighted Averaging (OWA) operator may provide better results than the simple AND, OR, and MEAN operators. The components of the OWA aggregation operator can be derived directly from user-supplied data or using a GA or ANN process of "trial and refinement" described above for the decision criteria weights. The approach suggested by Yager [YAGE88, YAGE97] relies on user-supplied data to set the model parameters (which include A_j (x), w_i, and α_j, as defined below). However, in real life, people are notoriously inaccurate and unreliable in reporting their preferences. This is especially true as the dimensionality and complexity of the problem

TABLE C.5   Ranking Network Options with MEAN Operator and Decision Weights

Find the fuzzy score for each option, and then sort in descending order, given: ■ Criterion 1 has a weight of 0.5 ■ Criterion 2 has a weight of 0.25 ■ Criterion 3 has a weight of 0.25	
F(option 1)	= Mean $(0.7^{.5}, 0.8^{.25}, 0.9^{.25})$ = $(0.8367 + 0.9457 + 0.9740)/3 = 0.9188$
F(option 2)	= Mean $(0.7^{.5}, 0.8^{.25}, 0.9^{.25})$ = $(0.7071 + 0.9457 + 0.9740)/3 = 0.8756$
F(option 3)	= Mean $(0.7^{.5}, 0.8^{.25}, 0.9^{.25})$ = $(0.3162 + 0.5623 + .05623)/3 = 0.4803$

increases. Rubinson and Geotsi [RUBI99] describe genetic algorithm techniques that can be used to automate the derivation of fuzzy multi-criteria function parameters to consistently find good solutions to reconcile what people say they want with how they act.

As presented in [YAGE88], the formal definition of an Ordered Weighted Averaging (OWA) operator is:

A mapping F from $I^n \to I$ in [0,1] is called an OWA operator of dimension n if associated with F is a weighting vector W = $(w_1, w_2, ..., w_n)$ such that W_i in [0,1] and sum(W_i) = 1 and where $F(a_1, a_2, ...a_n)$ = $W_1b_1 + W_2b_2 + ... + W_nb_n$ where b_i is the ith largest element in the collection $a_1, a_2, ...a_n$

The OWA operator aggregates multiple subjective criteria and assigns an overall evaluation score to a particular alternative under consideration. The first step in the aggregation process is to compute the individual criteria weighting a_j, as shown below:

$$a_j = (\alpha_j \vee p) * (A_j(X))(\alpha_j \vee q)$$

where:
a_j = final score
α_j = criteria weight
$A_j(x)$ = individual fuzzy criteria score n
1 = p + q
p = degree of "andness" = 1 − q
n = number of decision criteria
w_i = weighting vector, as defined above
q = degree of "orness" = $(1/(n-1)) * \left[\sum_{i=1}^{n} ((n-i) * w_i) \right]$

The a_j are used, in turn, to compute an overall evaluation score. The overall evaluation score is computed according to the definition of the OWA operator given above. The OWA operator is unique in its ability to represent the relative degree of importance of *each* criterion in the decision process, and the overall importance attached to progressively satisfying "more and more" criteria. Note that in this context, "more and more" is a linguistic qualifier representing the decision maker's desire to satisfy all, or most, or half, or some other subjective number of criteria.

After deriving the components of the fuzzy ranking function, it is a simple matter to substitute them into a mathematical equation. This mathematical equation — when solved using the data values for a particular individual — provides an overall ranking score for that individual. This score can be used, in turn, to sort individuals on a continuum of higher to lower risk.

Fuzzy logic also offers a convenient method to build subjective models of individual and group preferences. When the fuzzy subjective models do not perform as expected in ranking problems, it is usually indicative that the initial preferences indicated are not the true underlying preferences. It is also possible that the ranking problem is too complex to be fully understood without performing sensitivity analysis. Fuzzy logic approaches are very useful in demonstrating (1) when the stated preferences are out of sync with the actual preferences, and (2) the impact of alternative subjective preferences. For example, different individuals may have different preferences, the effects of which can be effectively demonstrated with a fuzzy model. The fuzzy approaches provide an explicit linguistic expression of the problem and the underlying preference model. This, in turn, provides many useful insights for discussion and analysis in the decision making process.

References

[ALL709]Allais, M., "The foundation of a positive theory of choice involving risk and a criticism of the postulates and axioms of the American School," *Expected Utility and the Allais Paradox*, Reidel, Dordrecht, 1979, pp. 27–145.

[BERN05]Bernstein, Peter, *Portable MBA in Investment*, John Wiley & Sons, 1995.

[MICH96]Michalewicz, Zbigniew, *Genetic Algorithms + Data Structures = Evolution Programs*, 3rd edition, Springer-Verlag, 1996.

[RUBI92]Rubinson, Teresa C., A Fuzzy Multiple Attribute Design and Decision Procedure for Long Term Network Planning, Ph.D. dissertation, Polytechnic University, June 1992.

[RUBI99]Rubinson, Teresa C. and Geotsi, Georgette, M., Estimation of Subjective Preferences Using Fuzzy Logic and Genetic Algorithms, Information Processing and the Management of Uncertainty in Knowledge-Based Systems, Granada, Spain, July 1999.

[SAAT72]Saaty, T., Hierarchies, priorities, and eigenvalues, University of Pennsylvania, Philadelphia, 1972.

[YAGE88]Yager, R.R., On Order Weighted Averaging Aggregation Operators in Multicriteria Decisionmaking, IEEE Trans. *on Systems, Man, and Cybernetics*, 18(1):January/February 1988, pp. 183–190.

[YAGE97]Yager, R.R. and Kaprzyk, J., eds., *The Order Weighted Averaging Operators*, Kluwer, 1997, pp. 155–166.

Index

Voice-over IP, market threat to telco business
model, 71
VPN. *See* Virtual private network

W

Wal-Mart, use of technology, 19–20
War, rights of, 130
Wavelength division multiplexing, 244–246
Web-based network management, 479–483
Web-based retailing, 23–24
Web browsers, 330–331
Web server software, 329–330
Web Services Definition Language, 404–406
Wide area network, local area network,
contrasted, 275
Wide area network design, planning, 95–270
backbone networks, 171–192
add algorithm, 179–185
case study, 189–192
center for mass algorithm, 173–175
concentrator/backbone placement,
179–189
design considerations, techniques,
172–189
drop algorithm, 185–189
sample COM problem, 175–179
built-in network management
capabilities, 103
carrier selection, 98
centralized network design, 99–114
legacy system network architecture,
103–108
system network architecture
networks, open networking
standards, 108–114
distributed network design, 114–140
Ethernet-to-MAN connectivity,
119–120
IEEE MAN standards, 117–119
key users, MAN configurations,
127–133
metropolitan area networks, 116–118
private wide area networks, 134–136
public wide area networks, 136–137
Synchronous Optical NETwork,
DWDM, RPR, MPLA
technologies, MANs, 120–127
wide area networks, 133–134
wireless wide area networks, 137–140
end users, 104

Esau-Williams Capacitated Minimal
Spanning Tree Algorithm,
156–162
in-house staff, resources, *versus* outside, 98
Kruskal's Capacitated Minimal Spanning
Tree topology, 154–156
logical units, 104–105
management, 95–98
mesh networks, 192–208
organizational goals, 97
path control network, 105
physical units, 104–105
public *versus* private leased lines, 98
star topology networks, 140–148
case study, 148
design considerations, 141–148
Synchronous Optical NETwork signal
categories, 238
Synchronous Optical NETwork
standards, 238
system services control points, 104–105
technical overview, 98–264
tree networks, 148–170
bin-packing algorithms, 163
design, 152–170
unified algorithm, 162–163
type of service, 98
uptime performance, 103
value-added network services, 208–264
access technologies, 236–264
Asynchronous Transfer Mode
networks, 226–233
frame relay networks, 218–226
Integrated Services Digital Network,
networks, 257–264
packet-switched networks, 210–218
virtual private network networks,
233–236
Wavelength division multiplexing,
244–246
Wi-Fi5, IEEE 802.11a wireless standard, 274
Wi-Fi Protected Access protocol, 274–275
Windows, 328
Distributed InterNet Applications
Architecture, 394
Web Servers, 330
Wireless Application Protocol, 137–138
Wireless local area networks, 78, 79,
272–275
Wireless technologies, 78–80
security risk, 31–32